Astronomy and Cosmology

The 4-meter telescope of the Kitt Peak National Observatory.

Astronomy and Cosmology

A MODERN COURSE

Fred Hoyle
CALIFORNIA INSTITUTE OF TECHNOLOGY

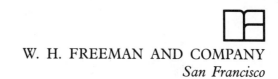

W. H. FREEMAN AND COMPANY
San Francisco

Library of Congress Cataloging in Publication Data

Hoyle, Sir Fred.
    Astronomy and cosmology.

    Includes bibliographical references and index.
    1. Astronomy. 2. Cosmology. I. Title.
QB43.2.H69   520   74-28441
ISBN 0-7167-0351-3

Printed in the United States of America

3 4 5 6 7 8 9

Preface

Since several very good astronomy texts are already available (I refer to them at the ends of the sections into which this book is divided), I felt that a new departure in presentation would be desirable here. For this reason, I have concentrated more on the relationship of astronomy to physics than is usual in an introductory text. Astronomy was the first of the physical sciences to develop—indeed, we owe the rise of the whole of modern science to critical discoveries made in astronomy some three to four centuries ago. However, and somewhat paradoxically, historical factors have tended to keep astronomy a science apart, studied almost in isolation from the rest of science. Today we see this isolation to be no advantage. The time has come for astronomy to take its place as a major branch of physics. Such is the point of view from which I have tried to write this book.

An irritating detail concerned me from the outset. The quantity we call *energy* is widely important both in science and in our daily lives; yet there is no well-known unit of energy. When we buy gasoline, we are buying energy; yet we specify a quantity of gasoline by volume (so many gallons), not by its energy content. Electrical devices are rated according to *power*, not energy; yet we pay our bills to electric companies according to the energy we have consumed. Perhaps the inability of our society to think explicitly in terms of energy has had much to do with the recent development of the energy crisis.

To be sure, scientists always work professionally in terms of a unit of energy, often a unit called the *erg:* an object with a mass of 2 grams moving with a speed of 1 centimeter per second has 1 erg of energy because of its motion. But this definition seems rather remote from everyday life. To find a definition nearer everyday usage, let us start from the concept of the power rating of an electrical device (so many watts, or so many kilowatts). To calculate the energy used by such a device, we multiply the power rating by the time for which the device is employed:

$$\text{Energy} = \text{Power rating} \times \text{Time.}$$

A 10-kilowatt device used for 100 hours requires the same amount of electrical energy as a 1-kilowatt device used for 1,000 hours—the bill from the electric company would be the same. In this book, however, we will consider time to be measured in seconds rather than hours; so the unit of energy is that of a 1-kilowatt device used for 1 second, a unit known as the *kilowatt-second.* The relationship of the kilowatt-second to the erg is expressed by the simple equation,

$$1 \text{ kilowatt-second} = 10^{10} \text{ ergs.}$$

I must also comment on a far more basic topic. Physicists will be surprised to find that, although radiation and the quantum theory are discussed fairly extensively in this book, *quanta* are never mentioned explicitly. The idea of a quantum of radiation appears at first sight to be a simple and useful concept; but later on, serious confusion emerges when one tries to relate the quantum concept of radiation to the wave description of it. This confusion is the inevitable price to be payed for oversimplification. An accurate description of the specific quantum associated with a specific transition in a specific atom is not simple at all, but is instead quite complicated. It is glossing over this complication which causes the later difficulties. The treatment adopted here, although it can be developed in a way entirely equivalent to the usual treatment, avoids this pitfall. The quantum picture of radiation discussed in Chapter 4 connects to the wave picture without confusion.

Certain technical words have been italicized in the preceding paragraphs. This practice will be followed throughout this book wherever the reader is not expected to know the meaning of such a word beforehand. Instead, the meaning is intended to be explained by the discussion at that place in the book.

This book is divided into six main sections. At the end of each section, I have included additional material in the form of Appendixes, partly to give technical support to the main text, partly to extend the range of the discussion where doing so seemed worthwhile. This additional material can be omitted in an introductory course.

A few equations are used in the main text. These will not give trouble so long as the reader remembers that all equations are of the kind $a = b$, where *a* and *b* are simply the same number. Why bother to write *a* and *b* separately if they are the same? In mathematics and in science it is sometimes found that two apparently quite different ways of constructing a number lead to the same value. One way of constructing the number we call *a*, the other *b*. Then we express our surprise and delight at finding them equal by triumphantly writing *a* = *b*. All the unexpected regularities we find in the world are expressed in this way. So if we refuse to write any equations at all, we simply lose all the remarkable equalities which have been discovered between things that, to begin with, seemed different from each other. Our understanding of the world would then have become quite unnecessarily blurred.

It is a pleasure to express my thanks to Robert M. Blanchard for preparing the artwork for this book, and to Aidan A. Kelly for editing a far-from-easy manuscript. I also wish to thank Evaline Gibbs, Jan Rasmussen, and my wife, Barbara, for typing and for helping organize the material of the book.

Pasadena, California   *Fred Hoyle*
January 1975

Contents

Astronomy and Cosmology

Section I:
A First Look at
the Universe

Galileo Galilei, 1564–1642. An engraving after the painting by Allan Ramsay, courtesy of the Ronan Picture Library.

Chapter 1:
The Earth's Orientation to the Sun and the Stars

§1.1. The Year and the Seasons

From our experience of the sequence of day and night, it is clear that the relation of the Earth to the Sun undergoes cyclic change. Nowadays we explain this diurnal progression in terms of a rotation of the Earth. If for the moment we think of the Sun as being always in a fixed direction, and of the Earth as spinning about an axis, as in Figure 1.1, it is easy to see that each complete rotation of the Earth will produce a cycle of day and night.

Astronomical history shows that even this simple concept was not easily arrived at. Figure 1.1 implies that we are all being continuously whirled around by the Earth's motion. Why are we not aware of this motion? This question was a serious puzzle to astronomers in ancient times. It remained a mental obstacle until the time of Copernicus (1473–1543). In modern science we know that it is *irregularities* of motion of which we are aware, not smooth motion itself. A large, expensive car moving at 60 M.P.H. on an excellent highway gives much less awareness of motion than a small car driven at 30 M.P.H. along a rough road. By everyday practical standards, the Earth's motion is exceedingly smooth, not only in its rotation but in its revolution around the Sun.

A plane through the center of the Earth, taken at right angles to the axis of rotation, meets the surface of the Earth at an imaginary circle called the

4

*The Earth's
Orientation to the
Sun and the Stars*

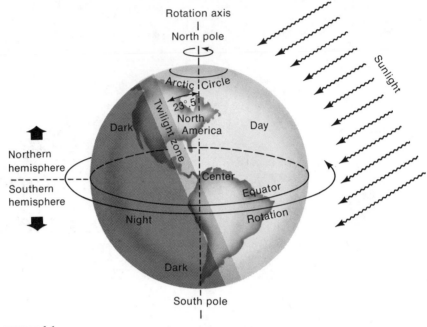

FIGURE 1.1.
The Earth, with a diameter of about 12,750 kilometers, rotates with respect to the Sun once in 24 hours, which causes the sequence of night and day. Here it is dawn across the middle of the North American continent.

equator. The Earth is separated into a northern hemisphere and a southern hemisphere by the equator, as shown in Figure 1.1. Because the direction of the Sun is not perpendicular to the axis of rotation, the situation in Figure 1.1 is not the same for the two hemispheres.

PROBLEM:
For the configuration of Figure 1.1, which hemisphere will be the warmer? Satisfy yourself that the northern hemisphere will have longer days and shorter nights than the southern hemisphere.

By changing the direction of the Sun, as in Figure 1.2, the asymmetry between the two hemispheres becomes inverted. The situation for the southern hemisphere in Figure 1.2 is the same as that for the northern hemisphere in Figure 1.1.

The configurations of Figures 1.1 and 1.2 both occur, but they are separated in time. They occur because the Earth moves around the Sun in a year of about

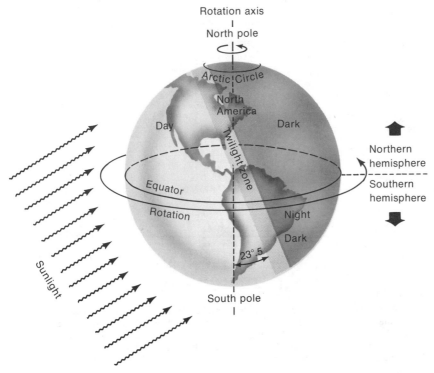

FIGURE 1.2.
With the direction of the Sun reversed, the situation in the southern hemisphere is the same as that for the northern hemisphere in Figure 1.1.

$365\frac{1}{4}$ days. The orbit is nearly a circle, as in Figure 1.3. During the course of a single year, the axis of rotation scarcely changes its direction—in Figure 1.3 it is not considered to change at all. The position of the Earth has been drawn at four particular moments, spaced essentially a quarter of a year apart, known as the *vernal equinox,* the *summer solstice,* the *autumnal equinox,* and the *winter solstice.* These terms were chosen to refer to the seasons of the year as experienced in the northern hemisphere. Spring in the northern hemisphere begins at the vernal equinox, summer at the summer solstice, autumn at the autumnal equinox, and winter at the winter solstice.

The equinoxes are defined as the days on which the lengths of day and night are equal. The summer and winter solstices are the days on which the direction of the Sun makes the least and greatest angles, respectively, with the Earth's axis of rotation. Contrasting with these clear-cut astronomical definitions, the division of the year into months is a matter of convention only. It is easy to imagine other ways of dividing up the year, into ten approximately equal

6

*The Earth's
Orientation to the
Sun and the Stars*

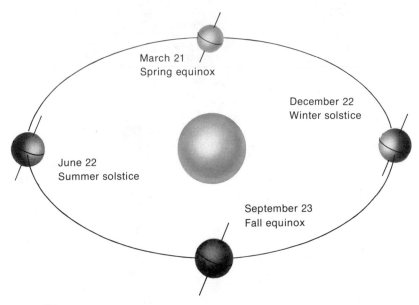

FIGURE 1.3.
The Earth moves around the Sun in an orbit that is nearly a circle, and the axis of rotation of the Earth maintains an effectively fixed direction. The orientation of the Sun to
the two hemispheres of the Earth changes during the year, and this change causes the seasons. The sizes of the Sun and Earth are both grossly exaggerated in this figure. (From
J. C. Brandt and S. P. Maran, *New Horizons in Astronomy.* W. H. Freeman and Company.
Copyright © 1972.)

intervals, for example, thereby fitting our modern passion for counting in units
of ten. With the months defined in the usual way, the equinoxes occur on March
21 and September 23, and the solstices on June 22 and December 22. One
might wonder how a convention leading to such clumsy dates for the important
moments of the astronomical year came to be chosen. The answer lies in a
complex and often confused history. The concept of the month came from
the motion of the Moon around the Earth. The confusion arose from mixing
up this motion with that of the Earth around the Sun. These two motions really
have little to do with each other, although the ancient mind seemed to perceive
a profound relationship between them.

PROBLEM:
At the winter solstice, the northern hemisphere receives less heat from the Sun
than at other times in the year. Yet experience shows that the coldest part of
a northern winter usually occurs in February. Why?

It is interesting to notice that much of the richness of our lives would have been lost if the axis of rotation of the Earth had happened to be perpendicular to the plane of its orbital motion around the Sun. The length of the day would then always have equalled that of the night. Climate would have been steady—steady wind, steady temperature. Individual plants and animals would very likely have become specialized to quite narrow ranges of latitude. But if the angle the rotation axis makes with the orbital plane had been significantly less than its actual value of about 66.5°, the seasons would have been more extreme, with hot tropical summers and cold arctic winters occurring over much of the Earth, and with storms attaining extremes quite outside our experience. The actual angle of 66.5° produces a fortunate balance between undue violence and undue monotony. How did the Earth come to have this angle, we might ask—and in doing so we leap away from simple astronomy to the complex modern problem of the origin of the planets themselves. This is a question to be considered at a later stage.

Figure 1.3 is not drawn to scale, of course. The diameter of the Earth is about 8,000 miles, whereas the radius of the Earth's orbit is some 93 million miles. So the Earth should be thought of as being very tiny compared to the size of its orbit. Indeed, in many problems it is useful to think of the Earth as a mere speck that traces out an annual orbit around the Sun, as in Figure 1.4.

In describing their *relative* motions, it makes no difference whether we consider the Earth to move around the Sun, or the Sun to move around the Earth, as in Figure 1.5. The relative motion of the Earth and Sun is the same in these two figures. From a physical point of view, however, is the situation the same? According to the physical theory developed from the work of Isaac Newton (1643–1727), the two are not the same; Figure 1.4 is correct. Figure 1.5, although giving a correct description of the apparent motion of the Sun

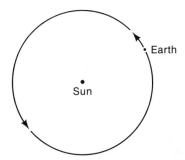

FIGURE 1.4.
In many problems it is useful to think of the Earth as a speck moving in its annual orbit around the Sun. The diameter of the orbit is about 3×10^8 kilometers.

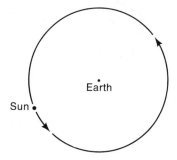

FIGURE 1.5.
The relative motion of the Earth and Sun is the same here as in Figure 1.4.

8

*The Earth's
Orientation to the
Sun and the Stars*

in the sky, does not lead to a correct understanding of the situation. By "understanding" we imply the ability *to predict ahead of time* what the relation of the Earth and Sun is going to be, not only to each other but to other bodies in the universe. Prediction is a far more powerful achievement than description, since description is nothing more than a restatement of what has already been observed to occur.

However, according to the physical theory developed by Albert Einstein (1879–1955), Figures 1.4 and 1.5 are indeed physically equivalent to each other. Since Einstein's theory is technically superior to Newton's, the two descriptions must therefore be regarded as entirely equivalent. In Einstein's theory we could use Figure 1.5 as a basis for prediction if we wished to do so.

§1.2. *The Celestial Sphere*

The concept of *projection* arises naturally from the observation of shadows cast by the Sun. In Figure 1.6 we have the shadow of a disk cast on a flat screen.

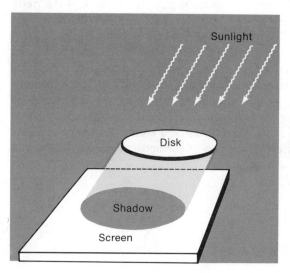

FIGURE 1.6.
Sunlight casts the shadow of a disk onto a flat screen.

The rays of sunlight are usually thought of as being parallel to each other, because the Sun is far away. We have a similar effect for a point source of light not so far away, as in Figure 1.7. The outline of the shadow is here obtained from a cone having the light source as its vertex, the cone being formed by lines from the vertex that pass through each point on the circumference of the disk. The boundary of the shadow is determined by the closed curve in which the cone intersects the screen.

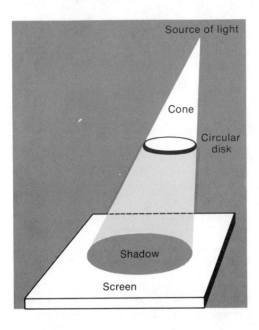

FIGURE 1.7.
A point source of light casts the
shadow of a disk onto a flat screen.

PROBLEM:
What is the shape of the shadow cast by a circular disk?

ANSWER:
This problem is illustrated in Figure 1.8. Although the shape of the shadow

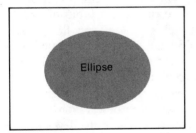

FIGURE 1.8.
When the disk in Figure 1.7 is circular, the
shadow is an ellipse, with a shape that depends
on the orientation of the disk to the source of
light.

depends in detail on where the light source and the screen happen to be located,
the boundary of the shadow is always an *ellipse*.

Figure 1.9 shows a practical method for drawing an ellipse. Indeed, by
varying the distance between the points to which the string is attached and

*The Earth's
Orientation to the
Sun and the Stars*

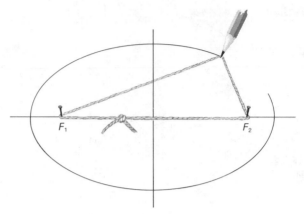

FIGURE 1.9.
How to draw an ellipse. The pins at F_1 and F_2 mark the two
foci of the ellipse. (From A. V. Baez, *New College Physics*.
W. H. Freeman and Company. Copyright © 1967.)

by varying the string's length, one can draw in this way every possible kind
of ellipse. The two points of attachment are called the *foci* of the ellipse.

PROBLEM:
Consider the shape of the ellipse when the string is only slightly longer than
the distance between the points of attachment. How does the ellipse change in
shape as the string becomes longer and longer?

We shall find in a later chapter that the orbits of planets around the Sun
are ellipses, with the Sun at one of the foci. The planetary ellipses are all of
the "long string" kind—i.e., they approximate to circles. The orbits of comets
are also ellipses, but mostly of the "short string" kind—i.e., very flat ellipses.

Instead of using a light source, we could obtain the same projection of a
disk by using straight pieces of wire to construct the cone of Figure 1.7. Or
we could simply imagine constructing the cone in this way, not troubling
ourselves with practical details. Such an imaginary construction gives the
mathematical concept of projection. The progression from observation (shadow)
to construction (pieces of wire) to abstract idea is typical of the development
of a mathematical concept.

It is possible to make similar projections onto a curved screen, for example,
onto the surface of a sphere. Imagine a sphere of large radius concentric with
the Earth. Using the center as vertex, project both the Earth's equator and
the orbit in which the Sun can be considered to move around the Earth (Figure
1.5) onto the sphere, as in Figure 1.10. Both projections are great circles,
intersecting at the points γ and Ω. By "great circles" we simply mean that
a plane through either circle passes through the center of the sphere, as it must
do from the nature of the construction. During the year the projection of the
Sun goes around one of these circles, the *ecliptic* as it is called, being at γ on

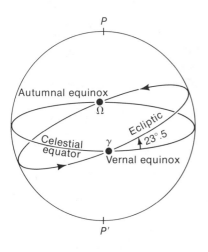

FIGURE 1.10.

The apparent path of the Sun in the sky, known as the ecliptic, is inclined at an angle of about 23.5° to the celestial equator. The ecliptic and the equator intersect at the vernal and autumnal equinoxes. The arrows on the ecliptic show the direction of the Sun's annual motion.

the day of the vernal equinox, and at Ω for the autumnal equinox. The axis of the Earth's rotation intersects the sphere at the points P and P', the *celestial poles* as they are called. The sphere itself is called the *celestial sphere*.

A similar projection can be made for every object we see in the sky: stars, planets, comets, and the galaxies. When an object appears to us as a point, we obtain a point on the celestial sphere. Where an object is extended—i.e., spread out—we obtain a patch on the celestial sphere. The situation is illustrated in Figure 1.11, from which it will be seen that, although the celestial sphere has a large radius compared with the scale of the Earth's orbit around the Sun, the distances of the stars are still larger.

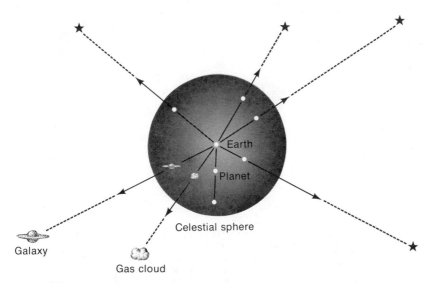

FIGURE 1.11.

All manner of astronomical objects are projected together on the celestial sphere, which is taken to have a radius that is large compared to the scale of the solar system.

12

*The Earth's
Orientation to the
Sun and the Stars*

Using the celestial equator and the poles P and P', one can easily set up a system of latitude and longitude on the celestial sphere. Lines of longitude and latitude are drawn analogously to the usual geographical system. For zero longitude, we take the arc of the circle through P, γ, and P'. This arc is known as the *meridian*. Several minor changes from the geographical system are used in practice. Instead of measuring longitudes both east and west, we measure all longitudes in an easterly sense, going through all angles from 0° to 360° (see Figure 1.12).

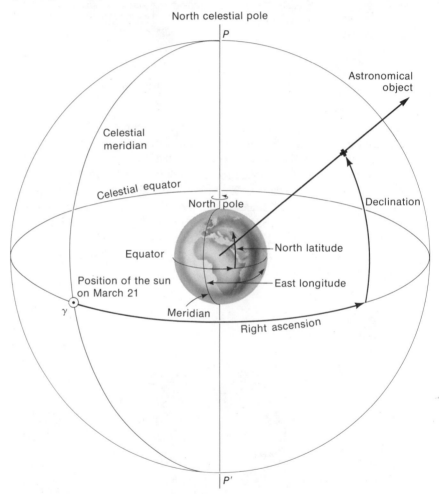

FIGURE 1.12.
The astronomical system of Right Ascension (R.A.) and Declination (Dec.) is similar to the usual geographical system of longitude and latitude. The point from which the R.A. of an astronomical object is measured is the point γ of Figure 1.10. The scale of the Earth here is obviously grossly exaggerated. (From J. C. Brandt and S. P. Maran, *New Horizons in Astronomy*. W. H. Freeman and Company. Copyright © 1972.)

PROBLEM:

The geographical longitude of New York is 73° 56′ W. If geographical longitudes were measured only in an easterly sense, as in astronomy, what then would be the longitude of New York?

Longitudes given as angles can be converted into equivalent times, in the following way. Divide the range 0° to 360° into 24 equal steps of 15° each. To each such step, assign 1 hour. To a portion of a step, assign minutes and seconds of time according to the following rules: 1 minute of time $\equiv$ 15 minutes of angle, 1 second of time $\equiv$ 15 seconds of angle. Using these equivalences, we can express longitudes in hours, minutes, and seconds of time (h m s).

Another minor difference between the geographical and astronomical systems is that, whereas geographical latitudes are given as north or south of the equator, + and − signs are used in the astronomical system. With these small changes, the latitude and longitude of an object on the celestial sphere, are referred to as *right ascension* and *declination:*

right ascension (R.A.) $\equiv$ longitude,   declination $\equiv$ latitude.

Although this situation is straightforward enough in principle, a vast collection of objects, large both in number and in variety, are thus projected together on the celestial sphere. To distinguish different classes of objects, astronomers have constructed a library of catalogues: catalogues of stars (sometimes for separate kinds of star), of galaxies, of patches of glowing gas, of patches of radio emission known as radiosources, and of sources of unusual forms of radiation—of x-rays and of infrared, to mention two recently developed branches of astronomy. Indeed, whenever a new variety of object is discovered, one of the first steps taken by the astronomical community is to construct a catalogue of them. The listing of the objects in each such catalogue is by right ascension and declination. The astronomer wishing to observe a particular object will look it up in a catalogue, where he will read off the appropriate right ascension and declination. This information determines the point on the sky toward which he must direct his telescope in order to make the desired observation.

PROBLEM:

Given the right ascension and declination of an object, a well-constructed telescope can be turned to the appropriate position with an accuracy which ensures that the object lies somewhere within the field of the telescope, although not necessarily at the center of the field. If the object is now visible in an eyepiece or on a television tube, the astronomer can use the controls of the telescope to center it, and so can make his observation. In some cases, however, the object

14

*The Earth's
Orientation to the
Sun and the Stars*

may be too faint to be seen by eye, although not too faint to be photographed, or to be examined with sensitive instruments. How in such a case might the astronomer find the object he is seeking to observe?

From the projection of an object onto the celestial sphere, we know its direction but not its distance. The projections of distant objects and near objects can fall adjacent to each other. The Sun, Moon, and planets are close by compared to the stars, while the stars are close by compared to distant galaxies. Criteria involving the nature of the objects are required to determine distances. We shall consider the more important of these criteria in a later chapter.

§1.3. Fixed Directions

From the beginning we have supposed that an observer on the Earth can determine fixed directions relative to which he is able to observe the changing direction of the Sun during the course of the year. What are these fixed directions?

The usual answer to this question is that the directions of the stars are to be considered as fixed, and this is certainly an adequate answer for observations of the Sun, Moon, and planets. But the directions of the stars are not truly fixed. The stars possess motions which cause their projections on the celestial sphere to change slightly from year to year. Nearer stars with fast motions change their positions by amounts that are quite observable. Such *proper motions,* as they are called, play an important role in methods used to determine the distances of astronomical objects. This again is an issue we shall take up at a later stage.

Although the directions of stars change only slightly in the short term, the changes are cumulative; they add inexorably from year to year. An observer of the sky 100,000 years hence will see well-known constellations, like the Big Dipper, with shapes very different from the present-day configurations, as we can see from Figure 1.13. A million years hence the bright-star constellations will be largely different from their present-day forms.

Directions effectively fixed on the celestial sphere do exist, however. These are determined by galaxies other than our own. The properties of such galaxies are the main topic of Chapter 3. It is interesting that apparently we have no way to orient ourselves in space except by using observations of distant objects—in other words, by relating local experience to the large-scale structure of the universe. There are important reasons for this, which we shall consider in Section VI.

What emphatically cannot be done is to use the points γ and Ω of Figure 1.10 as fixed points. These points *precess* in a manner indicated in Figure 1.14, because of a steady change in the orientation of the celestial equator. This change is caused by a slow precession, like that of a spinning top, in the direction

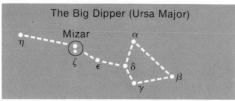

The Big Dipper (Ursa Major)

100,000 years ago

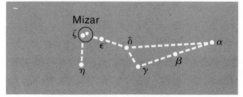

Today

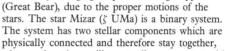

FIGURE 1.13.
The slowly changing shape of the Big Dipper (Great Bear), due to the proper motions of the stars. The star Mizar (ζ UMa) is a binary system. The system has two stellar components which are physically connected and therefore stay together, whereas the others will eventually move apart with the passage of time, although five are moving in nearly the same way. Which five? (The same star is referred to by the same Greek letter in each diagram.)

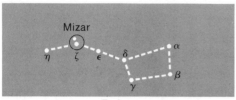

100,000 years hence

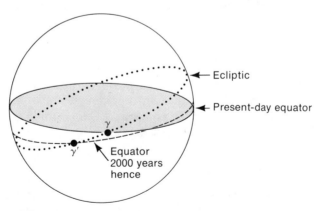

FIGURE 1.14.
In relation to fixed directions on the sky, the celestial equator is subject to a slow change that alters its points of intersection γ and Ω with the ecliptic. This effect is known as the precession of the equinoxes. It is due to a toplike precession of the Earth itself, as shown in Figure 1.15.

16

*The Earth's
Orientation to the
Sun and the Stars*

of the Earth's axis of rotation, an effect illustrated in Figure 1.15. Precession comes from the gravitational influences of the Moon and Sun, which apply a twist to the Earth. The time required for a complete precession around the cone of Figure 1.15 is about 26,000 years. As the Earth's axis moves around this cone, the projection of the geographical equator on the celestial sphere changes correspondingly, and this leads to the motion of the points γ and Ω. Since γ and Ω determine the equinoxes, this motion is often referred to as the *precession of the equinoxes,* a phenomenon known to the Greek astronomer Hipparchus in the second century B.C.

Because of the changing directions of γ and Ω, the catalogue values of the right ascensions and declinations of astronomical objects cannot be regarded as fixed. Thus γ and Ω swing around through $\dfrac{360°}{26,000}$ per year, about four-fifths of an arc minute per year. This means that catalogue values must refer to a particular date, the usual accepted moment of time being 0.0 hours on January 1, 1950.

PROBLEM:
The length of the solar day is measured with respect to the Sun. That is to say, we measure the time interval between two successive moments when the Sun is highest above the horizon (meridian passage), and the result is 24 hours. Because of the annual motion of the Earth in its orbit, the direction of the Sun is not fixed, however. How does this annual motion affect the length of the solar day? What would be the situation if the time interval were measured between two successive meridian passages of a star? How is the difference from the solar day dependent on the sense of the Earth's rotation in relation to that of its annual motion?

---

§1.4. Star Maps

We can think of the stars and other astronomical objects projected on the celestial sphere in the same way that we think of places on the surface of the Earth. Just as we display places on the Earth in the form of maps, so we can construct maps of the sky, as in Figure 1.16.

PROJECT:
Compare the maps of Figure 1.16 with a celestial globe, noticing how the globe has been divided to produce these maps. Notice also that in order to relate the details of the star maps to the globe, you must imagine yourself to be *inside* the globe, at its center. Compare them further with actual observations of the sky, familiarizing yourself with as many constellations as you can identify. The names of the constellations are given in Table 1.1.

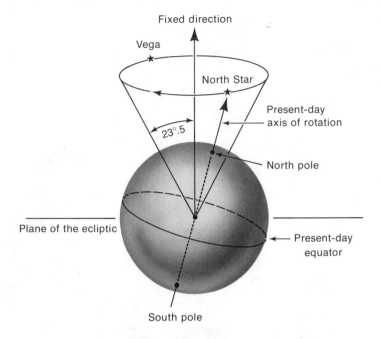

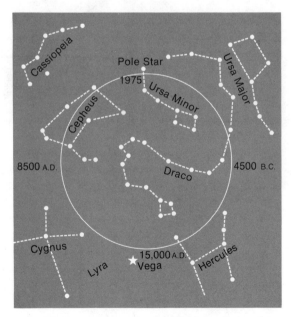

FIGURE 1.15.
The celestial north pole revolves around a point in Draco in a period of about 26,000 years. This revolution causes the precession of the equinoxes.

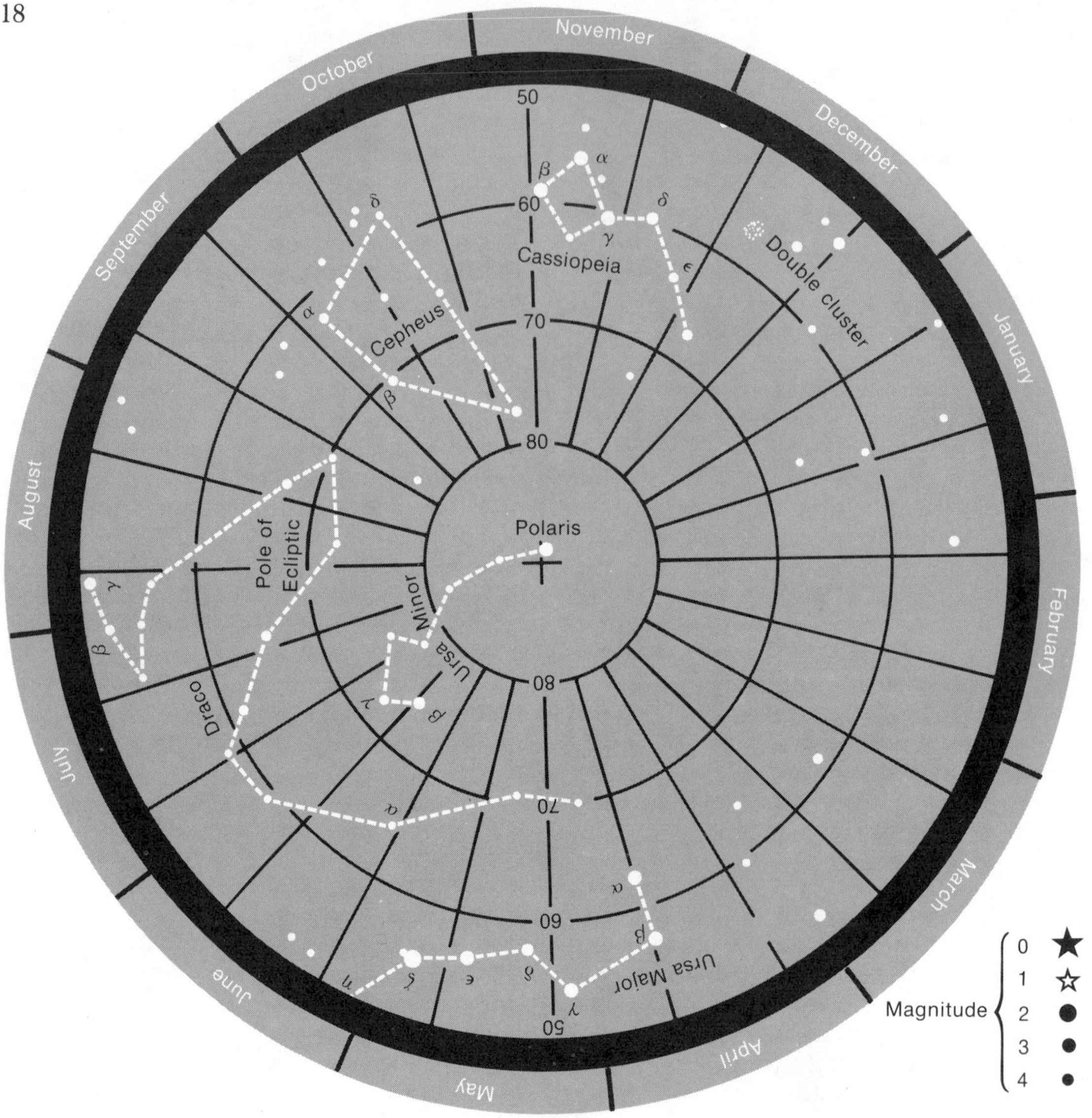

FIGURE 1.16(a).
The constellations of the north circumpolar sky.

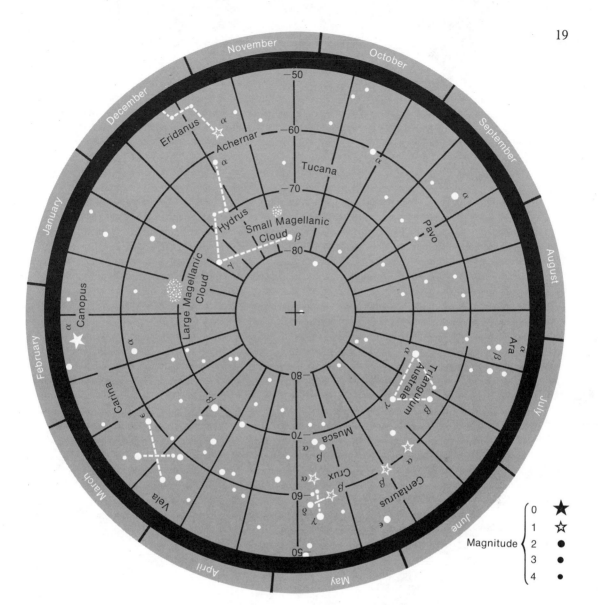

FIGURE 1.16(b).
The constellations of the south circumpolar sky.

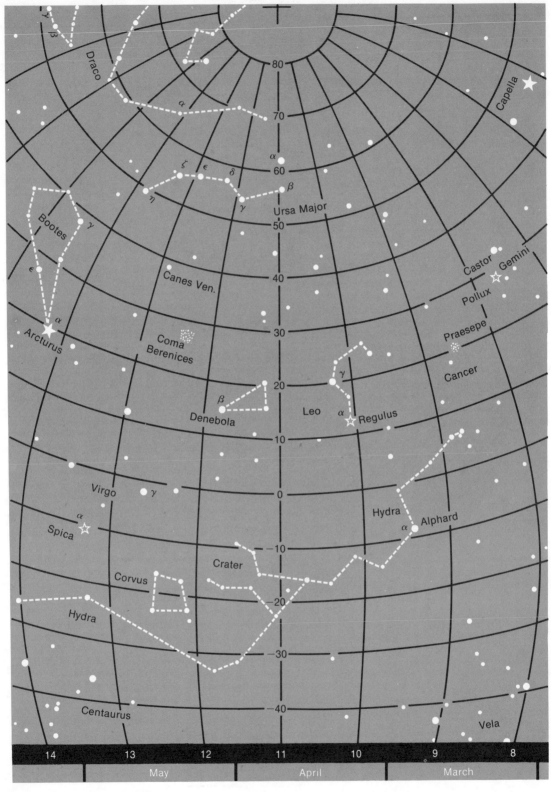

FIGURE 1.16(c). Night sky in Spring (northern hemisphere).

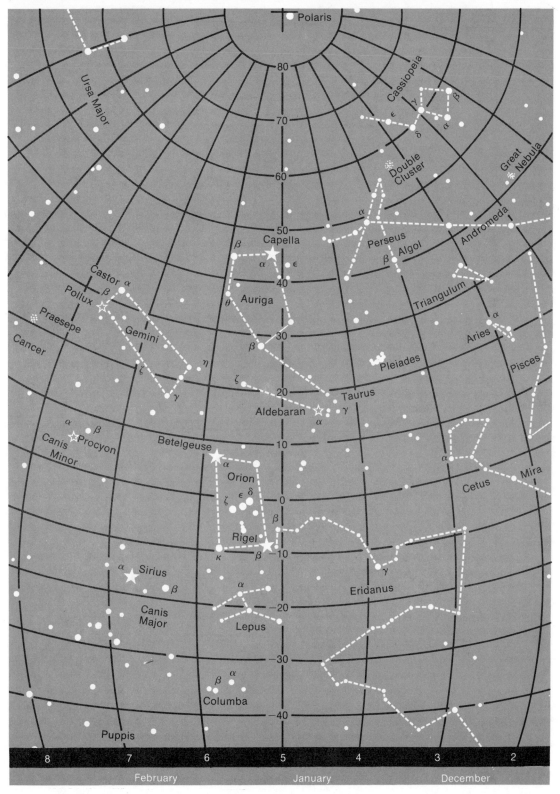

FIGURE 1.16(d). Night sky in Winter (northern hemisphere).

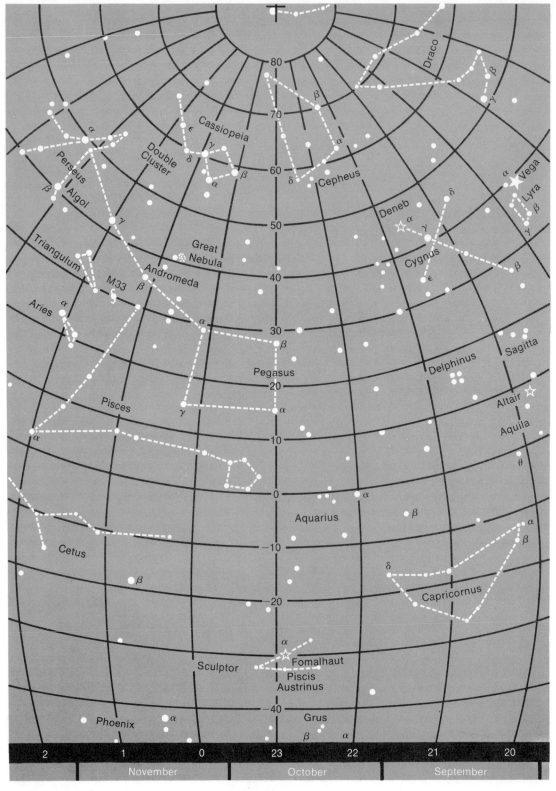

FIGURE 1.16(e). Night sky in Autumn (northern hemisphere).

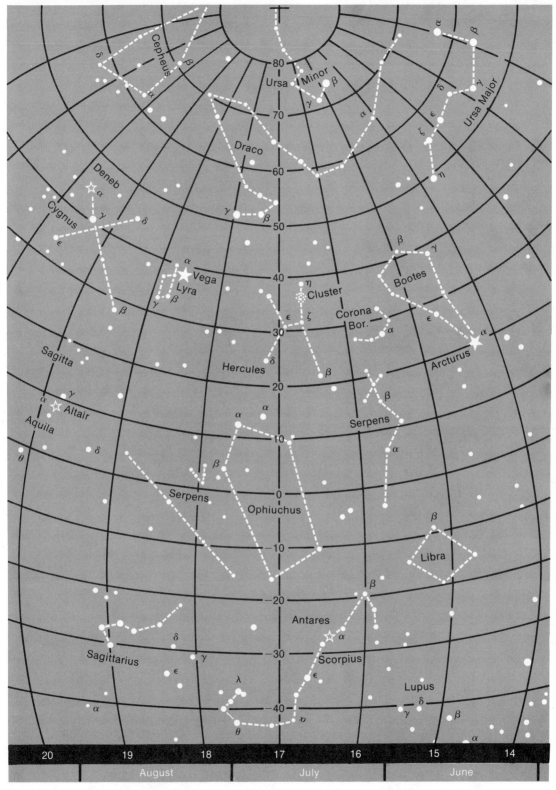

FIGURE 1.16(f). Night sky in Summer (northern hemisphere).

24

*The Earth's
Orientation to the
Sun and the Stars*

TABLE 1.1.
The constellations

Name	Approximate right ascension	Approximate declination	Intended meaning
Andromeda (And)	01	35	Andromeda
* Antlia (Ant)	10	−30	Pump
* Apus (Aps)	17	−75	Bird of Paradise
Aquarius (Aqr)	22	−15	Water Bearer
Aquila (Aql)	20	05	Eagle
Ara (Ara)	17	−55	Altar
Aries (Ari)	02	20	Ram
Auriga (Aur)	05	40	Charioteer
Boötes (Boo)	15	30	Herdsman
* Caelum (Cae)	05	−40	Chisel
* Camelopardus (Cam)	06	70	Giraffe
Cancer (Cnc)	09	20	Crab
* Canes Venatici (CVn)	13	40	Hunting Dogs
Canis Major (CMa)	07	−25	Big Dog
Canis Minor (CMi)	07	05	Small Dog
Capricornus (Cap)	21	−15	Sea Goat
* Carina (Car)	09	−60	Ship's Keel
Cassiopeia (Cas)	01	60	Cassiopeia
Centaurus (Cen)	13	−50	Centaur
Cepheus (Cep)	21	65	Cepheus
Cetus (Cet)	02	−5	Whale
* Chamaeleon (Cha)	11	−80	Chameleon
* Circinus (Cir)	16	−65	Compass
* Columba (Col)	06	−35	Dove
* Coma Berenices (Com)	13	20	Berenice's Hair
Corona Austrina (CrA)	19	−40	Southern Crown
Corona Borealis (CrB)	16	30	Northern Crown
Corvus (Crv)	12	−20	Crow
Crater (Crt)	11	−15	Cup
* Crux (Cru)	12	−60	Southern Cross
Cygnus (Cyg)	21	40	Swan
Delphinus (Del)	21	15	Dolphin
* Dorado (Dor)	05	−60	Swordfish
Draco (Dra)	18	60	Dragon
Equuleus (Equ)	21	10	Small Horse
Eridanus (Eri)	03	−25	River Eridanus
* Fornax (For)	03	−30	Furnace
Gemini (Gem)	07	25	Twins
* Grus (Gru)	22	−45	Crane
Hercules (Her)	17	30	Hercules
* Horologium (Hor)	03	−55	Clock
Hydra (Hya)	10	−15	Water Monster
* Hydrus (Hyi)	01	−70	Water Snake
* Indus (Ind)	20	−50	Indian

TABLE 1.1 (*continued*)

Name	Approximate right ascension	Approximate declination	Intended meaning
*Lacerta (Lac)	22	40	Lizard
Leo (Leo)	10	20	Lion
*Leo Minor (LMi)	10	35	Small Lion
Lepus (Lep)	05	−20	Hare
Libra (Lib)	15	−15	Balance
Lupus (Lup)	15	−45	Wolf
*Lynx (Lyn)	09	40	Lynx
Lyra (Lyr)	19	35	Harp
*Mensa (Men)	06	−75	Table (Mountain)
*Microscopium (Mic)	21	−35	Microscope
*Monoceros (Mon)	07	00	Unicorn
*Musca (Mus)	13	−70	Fly
*Norma (Nor)	16	−55	Square
*Octans (Oct)	22	−85	Octant
Ophiuchus (Oph)	17	0	Snake Bearer
Orion (Ori)	05	00	Orion
*Pavo (Pav)	20	−60	Peacock
Pegasus (Peg)	22	20	Pegasus
Perseus (Per)	03	40	Perseus
*Phoenix (Phe)	01	−45	Phoenix
*Pictor (Pic)	07	−60	Easel
Pisces (Psc)	00	10	Fishes
Piscis Austrinus (PsA)	23	−30	Southern Fish
*Puppis (Pup)	07	−35	Ship's Stern
*Pyxis (Pyx)	09	−35	Ship's Compass
*Reticulum (Ret)	04	−65	Net
Sagitta (Sge)	20	15	Arrow
Sagittarius (Sgr)	18	−30	Archer
Scorpius (Sco)	17	−35	Scorpion
*Sculptor (Scl)	01	−30	Sculptor
*Scutum (Sct)	19	−10	Shield
Serpens (Ser)	16	05	Snake
*Sextans (Sex)	10	00	Sextant
Taurus (Tau)	05	20	Bull
*Telescopium (Tel)	18	−45	Telescope
Triangulum (Tri)	02	35	Triangle
*Triangulum Australe (TrA)	16	−65	Southern Triangle
*Tucana (Tuc)	23	−60	Toucan
Ursa Major (UMa)	11	50	Great Bear
Ursa Minor (UMi)	15	75	Small Bear
*Vela (Vel)	09	−50	Ship's Sails
Virgo (Vir)	13	00	Virgin
*Volans (Vol)	08	−70	Flying Fish
*Vulpecula (Vul)	20	25	Fox

* Of modern origin.

26

*The Earth's
Orientation to the
Sun and the Stars*

The identification of star patterns with quaint terrestrial images has no scientific significance, of course. Yet there is a continuing attraction in many of these fanciful descriptions. Astronomers still use them in referring to different parts of the sky, and in catalogue descriptions of stars. Thus the brightest stars in the constellations are classified sequentially by letters of the Greek alphabet, α being in general the brightest, β the next brightest, γ the next, and so on. However, the very brightest stars have specific names, many of them of Arabic origin. Some of these special cases are marked on the maps of Figure 1.16. The star Vega is also referred to as α Lyrae, for example.

It is a curious thought that modern man is probably less familiar with the night sky than were his remote ancestors of 100,000 years ago. Life in cities has so changed our mode of existence that we are hardly aware of the sky, or indeed of many other aspects of our natural environment. Yet the orientation provided for us by the stars has played a critical part in the emergence of the physical concepts on which our society is based.

It is a deeply rooted problem why the large-scale structure of the universe should be necessary to provide a critical standard of reference, even for local events on the Earth. We are enjoined by the highway authorities to slow our speed in driving around a curve, because of the forces we experience in doing so. These rotary forces come into play whenever we swing around in an arc of a circle. By "swinging round" we mean turning with respect to fixed directions determined by the large-scale structure of the universe. It was speculated almost a hundred years ago, by the Austrian mathematician and philosopher Ernst Mach (1838–1916), that common experiences of this kind require the universe in the large to be capable of producing forces that dominate our daily lives. We shall consider questions relating to the large-scale structure of the universe in more detail in Section VI.

After we observe the stars for an hour or so, it becomes clear that the heavens seem to be turning. The motion during a day of the Big Dipper is shown in Figure 1.17, for example. Stars rise in the east, and others set in the west. This effect is, of course, due to the rotation of the Earth, not to any actual turning of the sky. Although Heraclides in the fourth century B.C. described this phenomenon correctly, later Greek astronomers believed the Earth could not spin around at the necessary rate without experiencing break-up and disintegration because of the rotary forces mentioned above. This argument, urged strongly by Ptolemy in the second century A.D., held sway until the time of Copernicus, who answered Ptolemy, somewhat caustically, by asking how much more surely would the heavens break up, if called on to rotate once around every day. Copernicus understood that, for a given rate of spin, rotary forces become greater the larger the dimensions of the system.

The situation created by the rotation of the Earth is illustrated in Figure 1.18. At a particular moment the observer sees only the stars above his horizon, but as the Earth turns, the horizontal plane determining his horizon also turns.

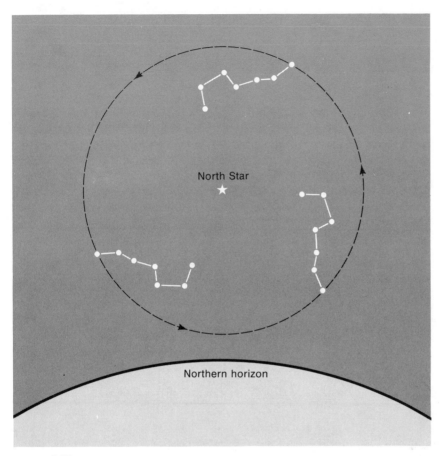

North Star

Northern horizon

FIGURE 1.17.
Three positions of the Big Dipper (Great Bear), 8 hours apart, drawn for latitude 40°N.

PROJECT:
Tie a weight to the end of a piece of string and whirl it around a circle in the manner of a sling, say, once around in a second. Now double the length of the string and again whirl the weight around once a second. The pull in your arm will be found to double itself. This is the point made by Copernicus. The point made by Mach was that the pull in your arm arises from the distant universe.

28

*The Earth's
Orientation to the
Sun and the Stars*

This causes new stars to rise into view from below the horizontal plane, and others to sink below the horizontal plane, becoming then obscured by the bulk of the Earth. Because of this obscuration, there are parts of the sky that can never be brought into view by the rotation of the Earth—unless the observer happens to be located at the geographical equator, in which special case the whole sky is in principle observable. The unobservable part of the sky is determined by the cone shown in Figure 1.19, half the angle at the vertex being determined by the observer's latitude. Most of the world's important observatories are located at sites with latitudes of about 35°, so that the unobservable parts of the sky from these observatories are determined by cones with half angles of about 35°, as in Figure 1.19.

If it were not for the Sun, we would see the whole of the sky accessible from our latitude once every day, but stars become invisible to the naked eye when they are located in the daytime sky. Although some stars might still be picked up with a telescope, observations of them usually cannot be made because

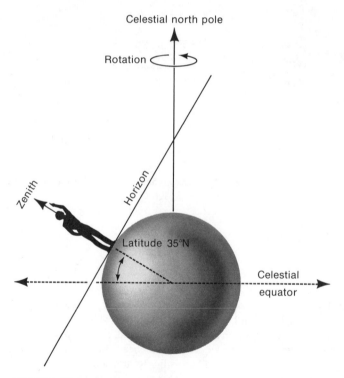

FIGURE 1.18.
During the night an observer sees the stars above his horizon. The figure is drawn for latitude 35°N.

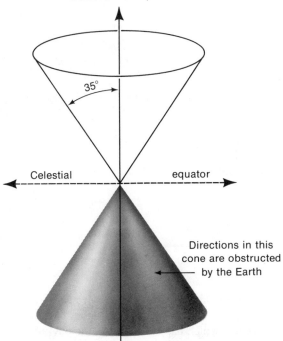

Celestial north pole

35°

Celestial ————————————— equator

Directions in this
cone are obstructed
by the Earth

FIGURE 1.19.
Stars in directions within the shaded cone can never
be seen by an observer in latitude 35°N, because
they are obstructed by the Earth. Stars within the
unshaded cone never set, and are said to be circumpolar.

of the scattering of sunlight by the Earth's atmosphere, illustrated in Figure
1.20. The glare of this scattered light floods the viewing field of the telescope.
This effect would not arise in space, a circumstance that would be an important
advantage for a telescope mounted on a space vehicle.

Fortunately, the position of the Sun on the celestial sphere changes throughout the year, as illustrated in Figures 1.21 and 1.22, so that the daytime and
nighttime parts of the sky change slowly from day to day. Although a star
may be in a direction close to the Sun on a particular day, several months
later the Sun will have moved sufficiently for the star to become observable
in the night sky. By making observations of the night sky *during the whole
year*, one can view all of the sky that is accessible from our particular latitude—i.e., the part falling outside the shaded cone of Figure 1.19.

The Earth's
Orientation to the
Sun and the Stars

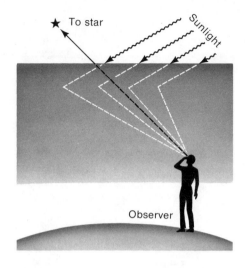

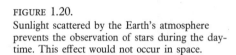

FIGURE 1.20.
Sunlight scattered by the Earth's atmosphere prevents the observation of stars during the daytime. This effect would not occur in space.

Astronomers who are studying particular objects need to plan their observing programs according to the seasons of the year. Should an astronomer's interest focus itself on an object which happens to be in the daytime sky, he may need to wait several months before the object in question passes to the night sky. Such a situation arises from time to time, causing the astronomer to suffer from a good deal of frustration.

Only at the equinoxes are the lengths of night and day equal. The nights are longest near midwinter, giving the greatest proportion of nighttime sky. Conversely, the nights are shortest near midsummer. It is important therefore to choose sites for observatories that have good weather in the winter months, a requirement not easy to meet.

PROBLEMS:

1. Except precisely at the equator, there are some stars which always remain above the observer's horizon. Such stars are said to be *circumpolar*. They are contained in a certain cone. How is this cone determined, and how does the angle at its vertex relate to the latitude of the observer?

2. Suppose you were magically transported in your sleep to an unknown place on the Earth. Consider how by observing circumpolar stars, and by relating the direction of the axis of the cone of Problem 1 to the direction of sunrise, you could determine the hemisphere in which you were located. How would you determine the north-south direction?

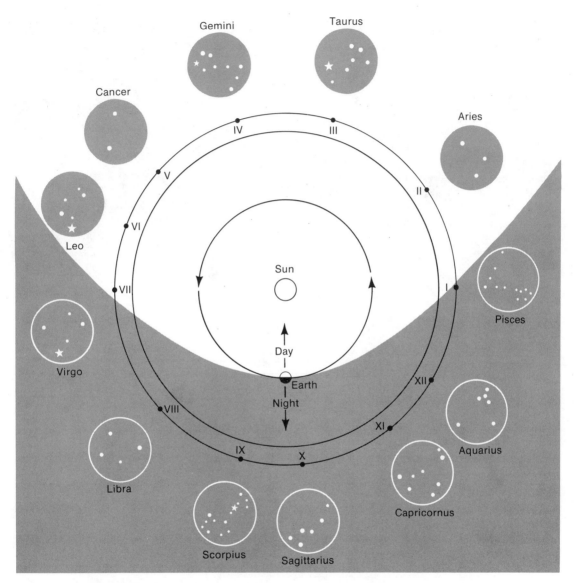

FIGURE 1.21.
The positions of the twelve constellations of the Zodiac marked along the ecliptic. The sizes of the Sun and Earth, and the radius of the Earth's orbit, are grossly exaggerated. This is the position of the Earth at midsummer, and the constellations of Libra, Scorpius, Sagittarius, and Capricornus, are visible in the nighttime sky.

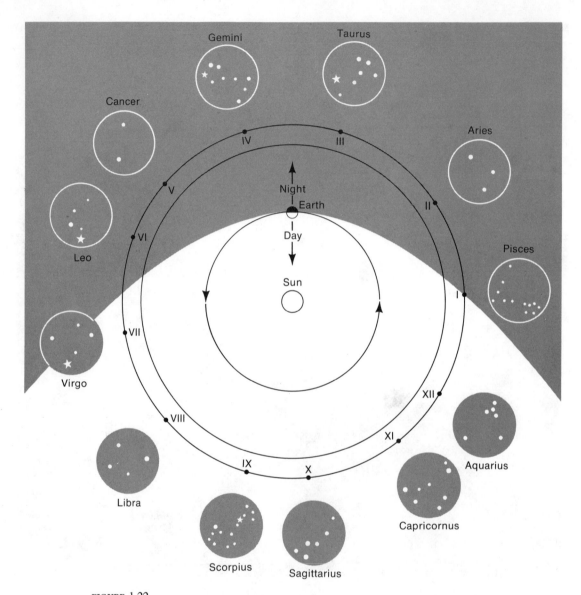

FIGURE 1.22.
The position of the Earth at midwinter. The constellations of Taurus, Gemini, and Cancer, are visible in the nighttime sky.

FIGURE 1.23.
A wide-angle montage of photographs of the Milky Way ranging from Sagittarius to Cassiopeia. The stars and gas clouds seen here are mostly at distances from about 100 light years up to about 5,000 light years. (Courtesy of Hale Observatories.)

§1.5. The Milky Way and the Plane of the Galaxy

Observation of the night sky, well away from the glare of city lights, reveals a remarkable band of diffuse light extending as a vast arch from horizon to horizon. This band of diffuse light consists of a profusion of stars too faint to be distinguished separately by the naked eye. Viewed through binoculars, many stars in this band become visible, although most of them are still too faint to be seen individually. Photographs, given in Figures 1.23 and 1.24, succeed in resolving a veritable multitude of stars. This band of light is a portion of our galaxy, known colloquially as the Milky Way.

FIGURE 1.24.
A detail of the Milky Way, in a direction toward the center of our galaxy. Notice the two star clusters, which are actually about 300 light years apart. (Courtesy of Kitt Peak National Observatory.)

34

*The Earth's
Orientation to the
Sun and the Stars*

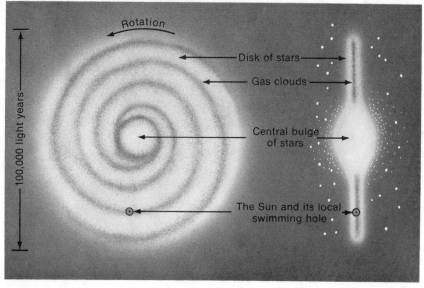

FIGURES 1.25 and 1.26.
A schematic representation of the Milky Way, seen face-on in Figure 1.25 (left) and edge-on in Figure 1.26 (right). Most of the stars we see by naked eye belong to the local "swimming hole" immediately surrounding the Sun and solar system.

PROJECT:
Notice in observing the Milky Way that the two places where it meets the horizon are 180° apart.

Figure 1.25 shows a schematic representation of the whole galaxy seen face-on. The Sun and its planets are far removed from the center. The galaxy has the general form of a disk, so large that light, traveling at 300,000 kilometers per second, takes some 30,000 years to make the journey from the center to the solar system. The stars we see scattered over the sky, the stars that form the constellations, are all comparatively nearby—belonging to the small local "swimming hole" shown in Figure 1.25.

The galaxy is shaped rather like a pancake, but with a developing bulge toward the center, as in Figure 1.26. We see the Milky Way by looking in directions along the plane of the pancake.

PROJECT:
Examine the course of the Milky Way on a celestial globe. Notice that it forms a great circle.

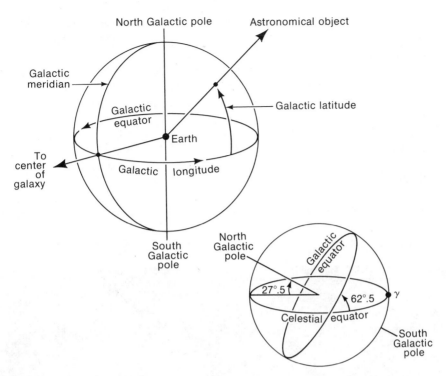

FIGURE 1.27.
With the galactic equator replacing the celestial equator, this is like Figure 1.12. The galactic
equator is defined by the plane of the Milky Way. The galactic meridian is determined by
the direction toward the center of the galaxy, shown in the photograph of Figure 1.24. The
orientation of the galactic equator in relation to the celestial equator is shown in the inset
figure.

It is possible to set up a system of latitude and longitude on the celestial
sphere by using the Milky Way as "equator," and by taking the direction of
the center of the galaxy to define zero longitude. This system is illustrated
in Figure 1.27. The position of an object on the celestial sphere is then said
to be specified in terms of *galactic latitude* and *galactic longitude*.

In Figure 1.28 we have an interesting representation of a galactic system
of reference, showing the form of the Milky Way as it is seen in different
galactic longitudes. These are marked along the long axis of the diagram, with
0° representing the direction toward the center of the galaxy. Galactic longitudes are determined from the curved lines, galactic latitudes from the straight
lines. The grid divides the sky into 10° × 10° regions. Except near the poles,
such regions seen on a celestial globe would approximate to squares. Notice
the two objects below the Milky Way and to the right of center. These are
the Magellanic Clouds, to which we shall refer in Chapter 8.

36

*The Earth's
Orientation to the
Sun and the Stars*

Figure 1.28 contains some distortion, of course. Every attempt to represent the surface of a sphere in terms of a flat map contains distortion, as we know from geographical maps. For example, in the Mercator projection of Figure 1.29, the distance between Pt. Barrow and Spitzbergen appears to be quite large, whereas a glance at a geographical globe shows that these two places are actually quite close together. The distortion of Figure 1.28 in this respect is much less than that of Figure 1.29.

To end this chapter, we can use Figure 1.29 to make a point of some importance. Imagine a thin spherical shell of rubber with the geographical continents, the oceans, and the poles painted on its outer surface. Simply by stretching the rubber there is no way of arriving at Figure 1.29. Yet, if one were permitted to make a cut along longitude 180°, stretching of the remaining rubber would indeed enable one to arrive at Figure 1.29. This example shows that puncturing or tearing is a more drastic form of distortion than mere stretching. Mathematicians express the difference by saying that puncturing

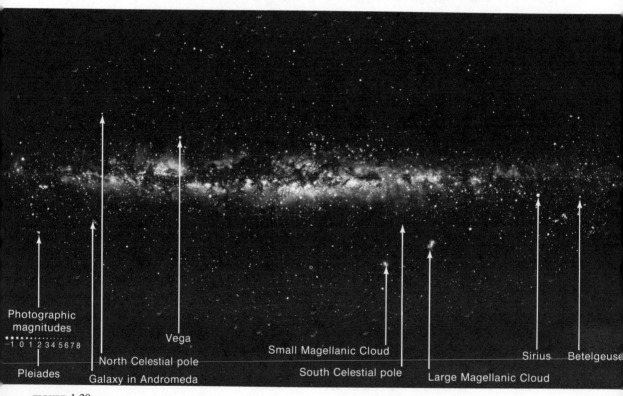

FIGURE 1.28.
Panoramic map of the Milky Way built up from many separate photographs, with certain objects of interest marked. The stars and gas clouds are at the distances mentioned for Figure 1.23. The Magellanic Clouds and the galaxy in Andromeda are much farther away, however (see Table 3.1). (Courtesy of Observatorium, Lund.)

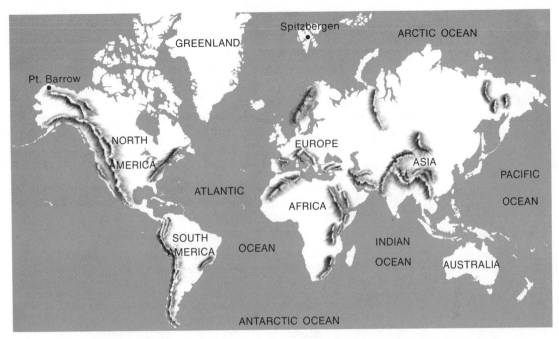

FIGURE 1.29.
The Mercator projection of the Earth's surface topography onto a flat plane. Distances toward the poles are grossly distorted in this projection.

or tearing changes the *topology*. Stretching leaves the topology unchanged. A central problem of present-day astronomy is to discover the topology of the universe, i.e., to discover those properties of the universe that cannot be changed by stretching.

General Problems and Questions

1. If the Earth's orbit around the Sun were exactly a circle, the four time intervals, winter solstice to vernal equinox, vernal equinox to summer solstice, summer solstice to autumnal equinox, and autumnal equinox to winter solstice, would be of the same duration. By actually counting the days in the intervals, find out whether they are the same or not.

2. If the answer to Problem 1 was negative, what explanation do you offer for your result?

38

*The Earth's
Orientation to the
Sun and the Stars*

3. How are the seasons of the year changed by the precessional motion of the Earth? (Refer to Figures 1.14 and 1.15.)

4. Taking the angle which the axis of rotation makes with the plane of the Earth's orbit around the Sun to be 66.5°, for what range of geographical latitude is the Sun not seen at all at the winter solstice?

5. During what parts of the year does the Sun remain hidden below the horizon at (a) the North pole, (b) the South pole?

6. In what ways are the precise answers to Problems 4 and 5 affected by (a) the size of the Sun, (b) the fact that rays of sunlight are slightly bent as they pass through the Earth's atmosphere?

7. The radius of the Sun is 6.96×10^5 kilometers, and on the average the distance from the Earth to the Sun is 1.496×10^8 kilometers. What is the average apparent angular diameter of the Sun?

   (If you have trouble here, try the following preliminary questions:

   a. Taking the Earth's orbit around the Sun to be a circle of radius 1.496×10^8 kilometers, what is the length of the circumference of this circle?

   b. Remembering that *by definition* there are 360° around a circle, what bit of this circumference viewed from the center would have an apparent angular diameter of 1°?

   c. How does the diameter of the Sun, namely $2 \times 6.96 \times 10^5$ kilometers, compare with this 1° bit of the circumference?)

8. The Moon moves around the Earth in a nearly circular orbit of radius 384,400 kilometers, and the radius of the Moon is 1,738 kilometers. What is the apparent angular diameter of the Moon?

9. The object M31 is a galaxy lying outside our own Milky Way system. When photographed it appears as an approximately elliptic form, with a long axis which has been determined to be about 80,000 light years. The distance of M31 is some 2 million light years. What is the apparent angular size of this object? (Note the perhaps unexpected result that an external galaxy turns out to have a larger angular size than the nearby Sun and Moon.)

10. There is an important sense in which the simplicity of the three preceding problems is illusory. You were expected to solve these problems by arguing that a circle of radius *a* has a circumference of length $2\pi a$, with π a *known number* equal to 3.14159. . . . If you did not know the numerical value of π, you could not solve the above problems. Try to find someone who will tell you how π is known to be 3.14159. . . . (This knowledge was a major breakthrough in technology—consider how many practical problems depend on it. Technology is not always connected with machines. When you have tracked this matter down, notice that π was *not* finally determined by practical measurement, but by abstract reasoning. Without powerful abstract reasoning, there could be no technology as we know it.)

11. On a sunny day, the shadow cast by a vertical stick has a length that varies with the time of day. At what moment of the day is the length least?

12. At what moment of the year is the length of the shadow least?

13. Explain how Stone-age man could have determined the solstices. Is there any comparably simple way of determining the equinoxes? (The use of a clock is forbidden!)

14. A shadow is cast on a flat screen by interposing a circular disk between the screen and a point-source of light. Irrespective of the position of the light source and of the disk, is it possible to orient the screen in such a way that the shadow is circular? If the disk were elliptic, would the answer to this question be different?

15. At what time of day does the center of the Sun lie on the meridian?

16. At what moment of the year does the center of the Sun have longitude 90°? What are the right ascension and declination values of the solar center at this moment?

17. Two astronomical objects have the same declination but different right ascension values. Under what relative conditions of the right ascension values does the shortest route between their positions on the celestial sphere go directly over the pole? How does the shortest route go when the right ascension values are nearly equal?

18. Two objects have the same right ascension but different declination values. What is the shortest route between their positions on the celestial sphere? Does the answer to this question, and to the previous one, have any relation to the shortest journey between the actual objects themselves?

19. Are the right ascension and declination values of an object defined in such a way as to be independent of the time of day? Of the day in the year? Of the precession of the axis of rotation of the Earth?

20. How many seconds are there in the year? Taking the Earth's orbit around the Sun to be a circle of radius 1.496×10^8 kilometers, and using the known value of π, show that the average speed of the Earth in its orbit must be about 30 kilometers per second.

21. How many seconds are there in a day? The equatorial radius of the Earth is 6,378 kilometers. What is the speed of the Earth's rotation at the equator? At the pole?

22. Suppose you were to take off immediately after sundown in an airplane which maintained a steady speed of 1,000 kilometers per hour, and suppose the flight to proceed due west. Above what geographical latitude would you expect to see the Sun rise again in the West?

23. Of the seven brightest stars in the constellation of Ursa Major (The Big Dipper), the one of smallest declination is η CMa with a value of 49° 49'. Explain why, on a journey from the Earth's north pole to the equator, it is this member which first breaks the circumpolar condition. Verify that η CMa is just circumpolar for an observer in New York (flat horizon!) but not for an observer in Washington, D.C. It was from a similar observational comparison that the ancient Greeks reasoned that the Earth could not be flat.

*The Earth's
Orientation to the
Sun and the Stars*

24. Name the main constellations distributed along the celestial equator.

25. In order to describe the motions of objects on the celestial sphere in a coherent way, fixed directions are needed. How are such directions determined? To our senses it seems as though the heavens are turning and the Earth is fixed. Yet the concept of fixed directions implied the opposite. What arguments support this opposite point of view?

26. For an observer located at a fixed position on the Earth's surface, any one star rises from day to day at the same point on the horizon. Is this also true for the Sun and Moon? Why not?

27. An astronomer happens to be interested in the observation of an object with right ascension 15h 48m. What time of the year would be best for such observations? If the object were very faint, what other observational requirement would be important?

28. Because of atmospheric effects, it is usually preferable to avoid observing faint objects when their elevations above the horizon are less than 20°. What limitation of declination values does this place on an observatory built in geographical latitude 35° N?

29. Describe the general structure of our galaxy, with particular reference to its scale and content.

30. Instead of using right ascension and declination values to locate objects on the celestial sphere, one can use galactic latitudes and longitudes. How are galactic latitudes and longitudes defined?

Chapter 2:
The Sun and the Stars

§2.1. The Solar Constant

In this chapter we shall regard the Sun as a typical star. Figure 2.1 shows the Sun surrounded by an imaginary sphere of radius r. Write L for the rate at which energy is being emitted from the solar surface, L kilowatts if we elect to use the well-known kilowatt as our unit of power.* The rate at which energy crosses a unit area of the sphere of radius r is then $L/4\pi r^2$, because $4\pi r^2$ is the area of the sphere. With r chosen equal to the radius of the Earth's orbit (assumed here to be circular), this quantity is called the *solar constant*.

The solar constant, having been measured, turns out to be 1.39 kilowatts per square meter. The radius, say, a, of the Earth is 6.378×10^6 meters, and the Earth intercepts sunlight on a surface area πa^2, i.e., an area of 1.278×10^{14} square meters. Allowing for some 40 per cent of the solar radiation incident on the Earth being reflected back into space and so "lost," the power absorbed by the Earth is thus about 10^{14} kilowatts. This amount is vast, greatly exceeding

---

*Astronomers usually work in terms of ergs per second, one kilowatt being the same as 10^{10} ergs per second.

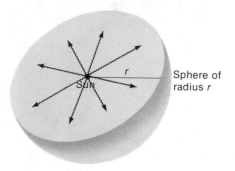

FIGURE 2.1.
The rate at which energy crosses a unit area of a sphere of radius r, concentric with the Sun, is $L/4\pi r^2$, where L is the luminosity of the Sun.

man-made power, which is about 3×10^9 kilowatts. It is this great supply of power from the Sun to the Earth's biosphere, seen from space in Figure 2.2, that maintains life on the Earth.

For FIGURE 2.2, see PLATE I.

PROBLEM:
The observed value of the solar constant is 1.39 kilowatts per square meter, and the measured average radius of the Earth's orbit is 1.50×10^{11} meters. Show that the solar luminosity L is about 3.9×10^{23} kilowatts.

Notice the form taken by this problem. Writing F for the quantity $L/4\pi r^2$, $F = L/4\pi r^2$. It is F and r that are measured directly. Then the luminosity L is *inferred*, $L = (4\pi r^2)$ **F**, where the boldface type denotes the measured quantities. There is nothing in nature that will tell us the value of L directly. Instead, the solar luminosity must be determined by inference in this way.

TOPIC FOR DISCUSSION:
Largely because of cloud, the reflectivity (*albedo*) of the Earth is about 40 per cent; so the energy reaching the Earth's equator at midday averages about 0.8 kilowatts per square meter of ground. A further reduction, to a midday value of about 0.6 kilowatts per square meter of ground, is necessary in the latitude of New York. What is the reason for this further reduction at a higher latitude? Consider if it would be possible to satisfy the power requirements of a normal household by utilizing solar energy, and discuss the practical problems as you see them that would be involved in such a project. Explicitly, what kind of absorptive device might be used, and how could cloudy days be averaged with clear ones?

The observed quantity F is often referred to as a *flux*. Our way of describing F as so many kilowatts per square meter is a sensible procedure. Unfortunately, the system of *magnitudes* still in widespread use is not as sensible. The concept of a magnitude was developed far back in the history of astronomy, and has not yet been abandoned.

Corresponding to F, we could derive another quantity K by means of the equation $10^K = F$. This process is called "taking the logarithm of F," and is carried out by means of a set of logarithmic tables, or a pocket calculator. Having determined K, the apparent magnitude m is obtained by adding a certain fixed specified number to $-2.5\,K$.

The same procedure can be followed for stars. Thus for a star there is a flux f, which is measurable and hence written in boldface type. Once again we take the logarithm of f to obtain k, $10^k = f$. Now we have $m = -2.5\,k + n$, n being the same fixed number. The apparent magnitudes of other kinds of astronomical object, galaxies, for example, are determined similarly.

The "fixed number" appearing above arises in the following way. A glance at the heavens shows some stars to be brighter than others. One can set up a subjective standard of judgment, according to which stars are classified into rank 1, rank 2, and so on sequentially, with rank 1 brighter than rank 2, rank 2 brighter than rank 3, and so on. Owing to the properties of the human eye and brain, it turns out that k in the equation $m = -2.5\,k + n$ must be multiplied by -2.5 in order that m shall differ by unity as we go from rank 1 to rank 2, and from rank 2 to rank 3 sequentially. The fixed number n is chosen to make $m = 1$ for those stars we assign to be exactly of rank 1.

We have bothered to describe this odd procedure in some detail because much of the data in astronomical literature is still expressed in terms of magnitudes. From time to time we shall need to examine such data, and must then perforce remember what magnitudes mean. Magnitudes are essentially flux values, like the solar constant F, but disguised in an obscure way. Using the observed value of the solar constant, together with the "fixed number" employed in the magnitude system, the apparent magnitude of the Sun is found to be -26.8, some 28 magnitudes brighter than a typical first-magnitude star—i.e., a star with $m = +1$. A similar system used for the Moon leads to a magnitude of -12.7, so that the Sun is about 14 magnitudes brighter than the Moon.

Remember that magnitudes are defined so that increasing magnitude means decreasing brightness. Negative values of the magnitude are also permitted. Only two stars have appreciably negative magnitudes, however. Sirius, the brightest star, has -1.4, Canopus has -0.7, and Arcturus, Capella, α Centauri, Vega, and Rigel all have magnitudes close to zero. Planets can have markedly negative values. At their brightest, $m = -4$ for Venus, -2 for Mars, and -2.5 for Jupiter. In the night sky planets can thus appear considerably brighter than any of the stars. In contrast to these bright stars and planets, the faintest objects currently worked on by astronomers have apparent magnitudes of about $+23$.

TABLE 2.1.
Number of stars brighter than a given magnitude

Magnitude limit (visual)	Number of stars
4	500
5	1,600
6	4,800
7	14,000
8	40,000
9	120,000
10	300,000
12	2,000,000
15	30,000,000
18	300,000,000

The numbers of stars visible down to certain magnitude levels are shown in Table 2.1. Since by eye we can only distinguish stars brighter than about magnitude $+6$, it follows that we resolve some 5,000 stars by eye alone. Even a small telescope is capable of resolving to magnitude $+10$, where the number of distinguishable stars rises to more than 100,000. The number becomes very large indeed at the faintest magnitudes, from which we see that the field of a very large telescope can become quite crowded with stars.

§2.3. The Distances of Stars

For a star at distance d, the measurable flux value f is related to the luminosity l by an equation similar to that for the Sun, namely, $l = 4\pi d^2 f$. If we assume the star in question to be exactly like the Sun, then $l = L$, and $4\pi d^2 f = 4\pi r^2 F$, the measurable quantities again being in boldface type. By inserting these measured quantities into the equation, we can calculate the distance d of the star. Since f is much less than F, it follows that d must be much larger than r, i.e., than the distance of the Earth from the Sun. For a star of first magnitude, some 28 magnitudes fainter than the Sun, d turns out to be about $4 \times 10^5 \, r$. With $r = 1.5 \times 10^{13}$ cm, this gives $d = 6 \times 10^{18}$ cm.

PROBLEM:
The speed of light is 3.00×10^{10} cm per second. The length of the year is 3.15×10^7 seconds. On the assumption that other stars are similar to the Sun, $l = L$, show that the light from a first-magnitude star takes some six years to travel to the Earth. Referring to Figure 1.23, which shows a portion of the Milky Way, and taking neighboring resolved stars to be six light years apart, satisfy yourself by counting resolved stars that the scale of this photograph is to be measured in thousands of light years.

It is worth pausing to ask how f values are actually measured. The older established method was to expose a photographic plate to the star, or the galaxy. From the density of the plate image, and from knowing how the plate behaved when exposed to a light source of known intensity, one could calculate the total energy required to produce the observed image. Then a calculation involving the time duration of the exposure, and the aperture of the telescope, gave the required value of f. This rather crude method worked surprisingly well for objects that were not too faint—for objects brighter than about magnitude $+17$. Serious errors occurred for the faintest objects, however; so the need to deal with objects at apparent magnitude $+20$, or fainter, has led recently to the development of more refined techniques. Nowadays the older photographic method is still used for obtaining approximate determinations of flux values, but intricate electronic methods are employed whenever high accuracy is required.

These electronic methods were not available before about 1950. Indeed, before 1850 not even photography was available. Being thus unable to measure f values for the stars, and unable to measure the solar constant F, early astronomers could not use the method discussed above for estimating the distances of the stars.

§2.4. *An Excursion into Astronomical History*

According to the earliest speculations of the Greeks, the sky consisted of a dark shield, pierced in numerous places by small holes. Through the holes an outer fire could be seen. These were the stars. This early notion, probably derived from pre-Greek culture, was later replaced by the rotating spheres of Eudoxus. The universe consisted, according to this view, of eight concentric spheres, all translucent except for the outermost one, which was dark and had points of light, the stars, mounted on it. The spheres moved in various ways, the outermost sphere having the least complex motion, that of a simple daily rotation. The spheres of the Sun, Moon, and planets turned about various axes, rather like a mariner's compass in gimbals. Since every sphere was centered at the Earth, it followed that all the stars were at the same distance from the Earth, a view which may have influenced even the great Kepler (1571–1630), who is said to have believed that all the stars were contained in a shell surrounding the solar system and only a few miles thick.

It is interesting to recall an argument of Hipparchus (circa 150 B.C.) directed against the theory of Aristarchus, who had suggested in the third century B.C. that all the planets, the Earth included, move around the Sun. If this were so, Hipparchus argued, the direction in which we observe a particular star should change during the year, as for example in Figure 2.3. Hipparchus concluded from his observations that no such variations occur, and that consequently the Earth does not move around the Sun. This argument was quoted for more than a thousand years as giving strong support to a geocentric view of the universe.

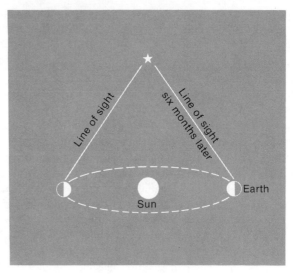

FIGURE 2.3.
The motion of the Earth around the Sun changes the direction in which a star is seen. Hipparchus was unable to detect such changes and was led to argue that the Sun moves around the Earth, not the Earth around the Sun.

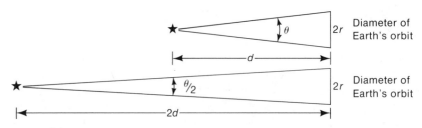

FIGURE 2.4.
When the long sides are much bigger than the short side of the triangle, doubling the distance halves the angle.

Hipparchus might have arrived at a different conclusion, namely, that the stars were extremely distant. Thus the greater the distance, the smaller the expected variation due to the Earth's annual motion around the Sun. From Figure 2.4 we can see that doubling the distance has the effect of halving the expected variation in direction. If θ is the angle of swing and d the distance, the product θd is the same for every star.*

---

*Notice that, whatever the direction of a star, a diameter of the Earth's orbit can always be found that is perpendicular to the direction of the star. This is the diameter shown in Figure 2.4. Whereas Figure 2.3 refers to the direction of a star that is perpendicular to the whole plane of the Earth's orbit, Figure 2.4 is quite general.

PROBLEM:

The distance d of a star being much larger than the radius r of the Earth's orbit, prove that, with θ measured in degrees, $\theta d = 360 \, r/\pi$. [Hint: Consider a circle, centered at the star, with radius d. Since the circumference of the circle is $2\pi d$, then the diameter $2r$ of the Earth's orbit, if thought of as a bit of the circle, represents a fraction $(2r)/(2\pi d)$ of the whole circumference. Hence the angle θ required in the problem must be a fraction $r/\pi d$ of the angle around a complete circle—i.e., of 360°.]

If r is considered known and if, using modern methods, which are more sensitive than those of Hipparchus, θ can be measured, then d is determined. Thus, using the result stated in the problem, we get $d = 360r/\pi\theta$, with $\pi = 3.14159 \ldots$, a known number. Such determinations are indeed made, and are known as *trigonometric distances*. Trigonometric distances tend to become unreliable, however, for stars with values of d greater than about 300 light years, because θ is then less than one fiftieth of an arc second, and such a small annual change of direction is difficult to measure, even with modern equipment.

The values of θ for the nearer stars are about 0.6 arc seconds, corresponding to stellar distances of about 10 light years. The nearest one or two stars, the system of α Centauri, for example, have θ values larger than 1.0 arc second. The value of d that would give θ precisely equal to 2 arc seconds, $d = 3.26$ light years, is often used as a unit of astronomical distance and is known as the *parsec*.* However, throughout this book we shall use the light year as our unit. To sufficient accuracy we have

$$1 \text{ light year} = (\text{speed of light}) \times (\text{length of year})$$
$$= (3 \times 10^{10} \text{ cm per second}) \times (3.156 \times 10^7 \text{ seconds})$$
$$= 9.46 \times 10^{17} \text{ cm}.$$

The technology required to measure values of θ in tenths of an arc second became available to astronomers in the first half of the nineteenth century. The first determination of d, for the star 61 Cygni, was made in 1838 by Friedrich Wilhelm Bessell (1784-1846). The value of d turned out to be about 10 light years.

Going back for the moment to Hipparchus and to the later astronomers of ancient Greece, their equipment could not measure an angle of less than about 100 arc seconds; so they were quite unable to detect the small annual variations of angle required to determine θ values for the stars. Did the possibility that

---

*This apparently peculiar name arises because $\frac{1}{2}\theta$ is called the *parallax* of the star. Thus $\theta = 2$ arc seconds means that the parallax is 1 arc second. The latter can then be referred to as "a second of parallax" or as a "parallax-second," which has become shortened to "parsec."

the distance d is very large, and θ consequently very small, occur to Hipparchus? Presumably it did, and was dismissed as being absurd. It seems a fair inference that the Greeks did not conceive of the stars as being extremely distant.

In mathematical studies we encounter a method of proof by contradiction. Suppose we wish to prove that some statement is true—that a certain line passes through a certain point, for example. We assume the contrary, and then proceed to prove that this contrary assumption is false. This method of proof is logically satisfactory, provided we can be sure that the statement in question and its converse are mutually exclusive to each other. In football, in a finished game between clubs A and B, the knowledge that B did not win is insufficient to establish that A won—the game might have been a tie. In baseball, on the other hand, the knowledge that B did not win establishes that A won.

A similar kind of argument is often used in astronomy. Possibilities P and Q seem mutually exclusive to each other. P is found to imply some "strange" conclusion. "Therefore" Q is correct. In the argument of Hipparchus, we have

$$P \equiv \text{the Earth moves around the Sun,}$$

$$Q \equiv \text{the Sun moves around the Earth.}$$

The consequence of P, illustrated in Figure 2.3, taken with the failure to detect annual changes in the directions of the stars, led to the conclusion that the stars would need to be extremely distant. This conclusion was too "strange" to be true. "Therefore" Q had to be correct.

The weakness of this kind of argument is that a "strange" conclusion is not the same thing as a mathematically incorrect conclusion. The whole history of science that each generation finds the universe to be stranger than the preceding generation ever conceived it to be. A "strange" situation is merely one we have not encountered before—but such situations are only too likely to arise whenever we explore new phenomena. The point is worth some notice, because the same incorrect style of argument as that used by Hipparchus is still quite frequently employed today.

Since the early astronomers could not measure angles with sufficient accuracy to determine θ values even for the nearest stars, and since they had neither photographic nor electronic methods for measuring flux values, the distances of the stars were unknown to them. The first correct statement that the stars lay at enormous distances seems to have been made at the beginning of the seventeenth century, by Giordano Bruno. Bruno's statement gave no proof, however, and was not therefore in the same class as the achievements of Copernicus and Kepler. Kepler determined the orbits of the planets by precise calculations. His results had to be true. Bruno's statement, on the other hand, was a speculation which might or might not have been true. Indeed, Bruno made many pronouncements, some turning out well, others ill. He argued strenuously that the angular size of an object depended on its brightness. Two objects of the same size at the same distance were supposed to subtend different angles if one was brighter than the other—an entirely incorrect idea. Bruno was not, of course, alone in his adherence to propositions we now regard as absurd; we have noticed Kepler's views on the stars. Newton in a later generation was to hold curious ideas about the pyramids of Egypt. To the early pioneers, the world was a remarkable place, full of scientific unicorns, some of which have turned out to be real beasts. Yet among the early pioneers there was one who never took a path which today, with hindsight, we regard as absurd. This was Galileo, whose perception always shone with a clear steady light.

§2.5. The Achievement of Cheseaux

The first general calculation of the absolute distances for the nearer stars was given by Cheseaux, the eighteenth-century Swiss astronomer. Without being able to measure flux values, Cheseaux was nevertheless able to adapt the argument given earlier in this chapter in a most ingenious way.

FIGURE 2.5.
Jean-Philippe Löys de Cheseaux, from a painting by J. P. Heuchoz. (Courtesy of Bibliotheque Cantonale, University of Lausanne.)

The Sun and the Stars

FIGURE 2.6.
The configuration of the Sun and
Mars at opposition—Mars and the
Sun are in opposite directions with
respect to the Earth. The radius of
Mars' orbit is 1.524 that of the
Earth's orbit.

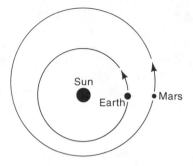

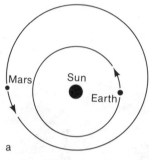 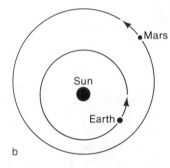

a   b

FIGURE 2.7.
(a) The configuration of the Sun and Mars at "conjunction." Notice
that Mars is about five times farther away from the Earth at conjunction than at opposition. (b) In such a configuration as this, Mars
would be like a first-magnitude star in brightness.

Even though one may not have the equipment necessary to measure individual flux values, it is still possible to judge when two flux values are the same,
or nearly the same. This indeed can be done without serious error even by
the naked eye. What Cheseaux did was to compare the flux value from the
planet Mars with that of a first-magnitude star. As can be seen from Figures
2.6 and 2.7, the distance from the Earth to Mars varies markedly according
to the configuration of the two planets. At its nearest, Mars has a magnitude
of about -2, brighter than any star, but in the configuration of Figure 2.7(b),
Mars appears like a star of about first magnitude. Write **D** for the distance
between Earth and Mars in this configuration, using boldface type because **D**
was known to Cheseaux, as was the distance **R** of Mars from the Sun.

The flux of solar energy at distance **R** is $L/4\pi \mathbf{R}^2$. Denoting the radius of
Mars by **b**, which again was known to Cheseaux, the rate of interception of
solar energy by Mars is $(\pi \mathbf{b}^2) \cdot (L/4\pi \mathbf{R}^2)$.

Cheseaux assumed Mars to be a perfect reflector—meaning that the energy
from the Sun would be reflected uniformly in all directions—so that Mars would

appear as a light source with luminosity $(\pi b^2) \cdot (L/4\pi R^2)$. The flux f_M from Mars, as observed on Earth, would then be given by $f_M = (\pi b^2) \cdot (L/4\pi R^2) \cdot (1/4\pi D^2)$. Next, taking the stars all to have the solar luminosity L, the flux from a star at distance d would be $L/4\pi d^2$. Since for a star of first magnitude this flux was required to be equal to f_M, it followed that d for such a star had to satisfy the equation

$$\frac{L}{4\pi d^2} = (\pi b^2) \cdot \left(\frac{L}{4\pi R^2}\right) \cdot \left(\frac{1}{4\pi D^2}\right).$$

The remarkable point now emerges that the unknown luminosity L cancels from this equation; so that with a little simple algebra we get $d = 2R \cdot D/b$. With R, D, b all known, Cheseaux thus obtained an estimate of a few light years for d, in excellent agreement with the actual distances of the nearest stars.

EXERCISE:
Using the values $R = 2.28 \times 10^{13}$ cm, $D = 3.12 \times 10^{13}$ cm, $b = 3.38 \times 10^8$ cm, obtain d in centimeters, and also in light years.

PROBLEM:
How would Cheseaux's demonstration be affected if Mars is taken to reflect, not the whole of the incident solar radiation, but only a fraction of it? The actual albedo of Mars is 0.15.

PROJECT:
Cheseaux wrote a book called *Traité de la Comète* in which the above derivation formed only a part of a still more remarkable argument. Attempt to gain access to the book in order to discover the nature of Cheseaux's further achievement.

---

§2.6. The Solar Cycle

The very high temperature conditions that exist inside the Sun will form a major topic in Section III; so for now we will confine ourselves to the surface regions and the atmosphere of the Sun, except for noting one interior problem that bears thinking about. The solar luminosity L is 3.9×10^{23} kilowatts. The burning of coal or oil—if the Sun were made of coal or oil—would supply energy at this enormous rate for only a few thousand years. Yet we know from geological evidence that the Sun has been shining, pretty much as it does now, for several billions of years. By what manner of process does the Sun achieve its great energy output? The brief answer is that the Sun is a vast, stable thermonuclear reactor. It does what man would dearly love to do, generate energy by nuclear processes that are essentially "clean."

When viewed in full sunlight, the Sun appears as in Figure 2.8. With even a small telescope, dark, more or less circular spots can often be seen. These

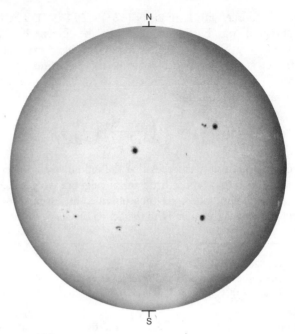

FIGURE 2.8.
Viewed in full sunlight, spots can be seen on the Sun, contradicting the ancient belief that the face of the Sun must be "perfect."
The diameter of the Sun is about 1.4×10^6 kilometers. (Courtesy
of Dr. N. Sheeley, Kitt Peak National Observatory.)

sunspots were discovered by Galileo (1564–1642), whose drawings of them are
shown in Figure 2.9. Their discovery caused a great sensation, since the face
of the Sun was thought to be "perfect," a belief that was probably a remote
survival from an era of prehistoric sun-worship.

Under favorable conditions, large sunspots are easily visible to the naked
eye; so it is surprising that their discovery had to await the dawn of modern
science. On two occasions I have myself been able to observe such naked-eye
spots, even from such an astronomically unfavorable climate as that of the
United Kingdom. Naked-eye spots are best seen near sunset, when absorption
in the atmosphere cuts the normal glare of the Sun down far enough that one
can look directly at the solar disk. Although this condition is realized quite
often, usually the atmosphere destroys clarity of definition for low angles of
elevation, leaving the setting Sun as a "boiling" image. The essential condition
for large sunspots to be seen by the naked eye is that the atmosphere be
absorptive but without distortion, and this condition is encountered only rarely.
Nevertheless, naked-eye spots must have been seen long before Galileo, probably by many thousands of people. The surprise is that their existence does
not seem to have been recorded by monks or by court chroniclers.

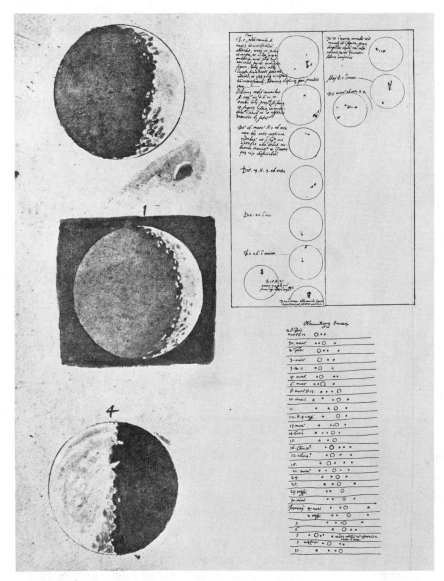

FIGURE 2.9.
Galileo's observations of sunspots. Drawings from a notebook.
(Courtesy of Yerkes Observatory.)

FIGURE 2.10.
Galileo's telescope (courtesy of Prof. ssa. Dott. ssa.
Maria Luisa Righini-Bonelli, Istituto e Museo di
Storia della Scienza, Firenze).

Sunspots are dark only in contrast with the brilliance of the surrounding regions of the solar surface. The temperature within them is actually about 4,000°K, considerably hotter than most industrial furnaces. The surrounding regions have temperatures of about 5,800°K, however.* Since the emission from a hot surface goes as the *fourth power* of the temperature (T^4 if T is temperature), the surrounding regions are appreciably brighter to the eye.

The appearance of the Sun undergoes a dramatic change when it is observed by the light of particular atoms, instead of by the white light of Figure 2.8. In Figure 2.11 the light is that characteristically emitted by hydrogen atoms, and in Figure 2.12 is that characteristic of calcium atoms. The remarkable structures revealed in these photographs are indicative of unusual forces present

---

*The symbol K in the above temperatures denotes what is usually called the "absolute" temperature scale. For the rather high temperatures involved here, the K scale is not much different from the well-known Centigrade scale; so we need not pause here to discuss its meaning further. For a discussion of it, see Appendix I.3 (p. 102).

FIGURE 2.11.
The Sun as in Figure 2.7, but photographed in the light emitted by hydrogen atoms (Hα). (Courtesy of Dr. N. Sheeley, Kitt Peak National Observatory.)

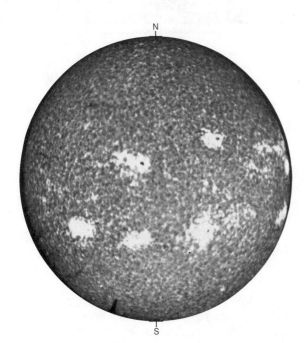

FIGURE 2.12.
The Sun as in Figure 2.7, but photographed in the light emitted by calcium atoms (K line). (Courtesy of Dr. N. Sheeley, Kitt Peak National Observatory.)

FIGURE 2.13.
The solar corona at the total eclipse of June 1973. The form of the corona is indicative of
the presence of magnetic forces. (Courtesy of High Altitude Observatory, Boulder, Colorado.)

at the solar surface. Figure 2.13 is a white-light photograph of the corona,
obtained during a total eclipse, i.e., when the Moon came directly between the
Earth and the Sun, as in Figure 2.14, thereby cutting off the light from the
solar disk. The structure of the corona with its many streamers is indicative
of the presence of magnetic forces. Indeed, Figure 2.13 plainly shows that the
Sun behaves as a great magnet.

Magnetic fields at the solar surface normally have intensities not more than
about ten times the intensity of the Earth's magnetic field, but in and around
sunspots the magnetic intensity can be a hundred to a thousand times greater
than its normal value. Indeed, it is the exceptional magnetic intensity within
sunspots that causes the temperature to be lowered within them, and which
therefore makes them appear dark when projected against the solar disk.

Sunspots wax and wane in numbers in a cycle of about 11 years. Other
phenomena also vary cyclically with the sunspots. Changes in the structure
of the corona, particle streams ejected out through the corona, giant *prominences*
in the solar atmosphere of the kind shown in Figure 2.15, all vary cyclically

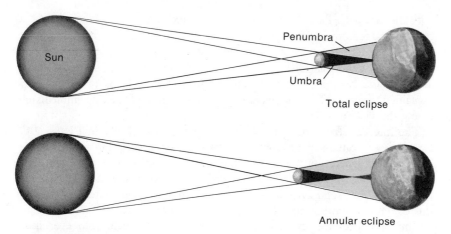

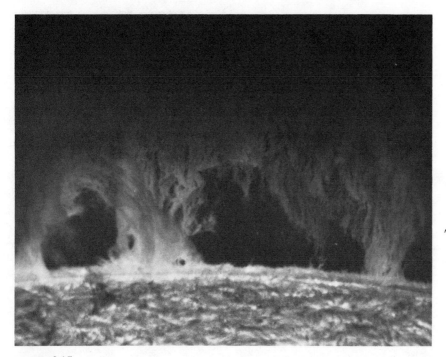

FIGURE 2.14.

An eclipse of the Sun occurs when the Moon comes between the Earth and Sun. For the eclipse to be total, the Moon's direct shadow must reach the Earth. Since the Moon's distance from the Earth varies with time—the Moon's orbit around the Earth not being exactly circular—both the situations shown in this figure occur. Annular eclipses are more frequent than total ones. (From J. C. Brandt and S. P. Maran, *New Horizons in Astronomy.* W. H. Freeman and Company. Copyright © 1972.)

FIGURE 2.15.

A giant prominence, about 200,000 kilometers across, on the Sun. Gases in the corona are cooling and moving under the influence both of gravity and of magnetic forces. (Courtesy of Hale Observatories.)

with the sunspots. Particle streams are associated with *flares,* which appear to be localized regions in the lower atmosphere where intense electrical discharges occur. The discharges are accompanied by heating, causing flares to appear bright against the solar disk, as in Figure 2.16. The origin of sunspots, flares, and prominences, of the corona, and of the solar cycle, is probably to be explained in terms of convective motion in the region of the Sun immediately below the visible surface.

The internal structure of the Sun is shown schematically in Figure 2.17. Energy is generated in a hot core, by processes to be discussed in a later chapter. This energy generation then causes the immediate subsurface regions to take up a circulating convective motion, rather like the contents of a saucepan heated from below. The solar cycle is to be explained in terms of the interplay between this convective motion and the magnetic field of the Sun itself. At first sight we might expect the motions to twist the magnetic field into ever more complex patterns. This would indeed take place if the dissipative effects of sunspots and of flares were not working to simplify the structure of the magnetic field. Thus the existing situation represents a balance between input to the magnetic field from the convective motions and output from the magnetic field into the many phenomena that constitute the solar cycle. Why the cycle has

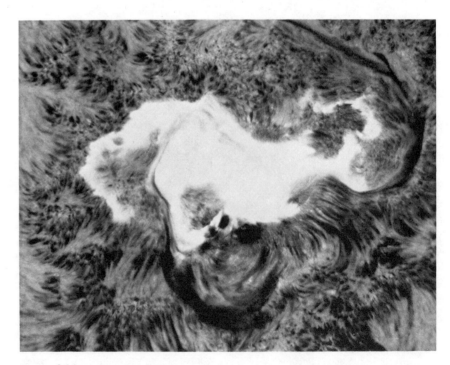

FIGURE 2.16.
A large flare, about 100,000 kilometers across. Flares cause particle streams to be ejected from the Sun. Such jets of particles move rapidly outward and sometimes impinge upon the Earth. (Courtesy of Hale Observatories.)

FIGURE 2.17.
Schematic representation of the structure of the Sun. Energy is generated in a
central core. The energy is then carried by radiation to a subsurface zone, where
it produces convective motions which may well be responsible for the solar cycle.
(From J. C. Brandt and S. P. Maran, *New Horizons in Astronomy.* W. H. Freeman and Company. Copyright © 1972.)

a period of about 11 years remains unclear, however. After much observational
and theoretical work by many astronomers, a comprehensive theory of the solar
cycle is still lacking. The problems involved seem to be among the most difficult
in astronomy.

The average density for the whole of the Sun is about 1.4 grams per cm^3.
The sun's material is mostly hydrogen atoms, each with a tiny mass of
1.67×10^{-24} grams. Thus on the average there are about 10^{24} atoms per cm^3
within the body of the Sun. This density is much like that of terrestrial solids
and liquids—the density of water, for example, being 1 gm per cm^3. In Section
III we will discuss giant stars in which the average density is significantly
lower—rather like the density of the ordinary air we breathe—and we will meet
stars in which the average density is far higher than anything within the range
of terrestrial experience. In the white dwarf stars, the density is such that one
cm^3 contains about a ton of material, and in the neutron stars one cm^3 contains
some 10^8 tons.

In all stars the density is higher than average near the center, lower than
average at the surface. Thus, in the Sun the central density is about 10^2 grams
per cm^3, whereas at the *photosphere*, the surface seen in Figure 2.8, the density
is only 10^{-7} gram per cm^3—i.e., 10^7 cubic centimeters are needed to contain 1

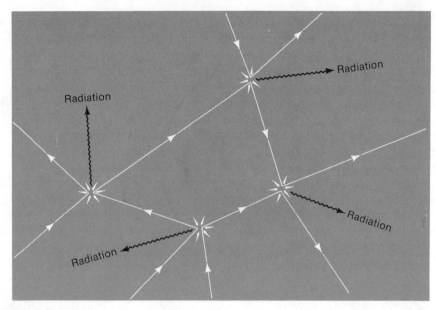

FIGURE 2.18.
Schematic representation of particle motions in a hot gas. When one particle collides with another, the directions of the motions tend to be randomized. In a hot-enough gas, the collisions may be sufficiently violent for x-rays to be emitted.

gram of material. The corresponding number of atoms per cm^3 is about 10^{17}, less than in ordinary air. The density falls still more with increasing height above the photosphere, to about 10^{13} atoms per cm^3 at a height of 1,000 km, and to still lower values, in the range 10^6 to 10^8 atoms per cm^3, for the upper corona.

A hot dense gas soon cools itself by radiation, as we know from practical experience. A hot *diffuse* gas, like that in the corona, also cools itself, but because of the low density it takes much longer to do so. Consequently it is possible for the coronal gases to be very hot without the emission of radiation being intense. By "very hot" we mean that the particles within the gas are moving at higher speeds than particles in the cooler gas at the photosphere (Figure 2.8). Although the particles move fast, they do not go very far in any one direction, because they collide repeatedly with other particles, as shown schematically in Figure 2.18. During such collisions, radiation is emitted, the radiation from the hot gas in the corona being largely different from that of the photosphere. It is interesting that the faster speeds of motion in the corona lead to the emission of ultraviolet light and of x-rays, instead of the ordinary white light from the photosphere. Figure 2.19 is a picture taken in the x-rays emitted by the coronal gases. Once again we emphasize that, although the corona thus emits radiation of an esoteric kind, the radiation is much lower in intensity than the emission from the denser gases at the photosphere.

For FIGURE 2.19, see PLATE II.

This same concept, that a low-density hot gas can emit unusual kinds of radiation, but in low intensity, is one to be kept in mind throughout astronomy. Diffuse gas exists in the regions between the stars. Sometimes it is hot, sometimes cold. Diffuse gas may also exist between the galaxies, and the same principles will apply then on a universal scale. Indeed, the densities on a large scale will be still lower than for the solar corona. Densities of the gas lying between the stars in our galaxy are considered to range from about 1 atom per cm^3 on the low side to about 10^6 atoms per cm^3 on the high side. The latter value approaches the density of the solar corona, but unlike the corona, the interstellar gas is usually cool when its density is as high as 10^6 atoms per cm^3.

Gas lying between the galaxies must have lower densities still, not more than 10^{-5} atoms per cm^3—i.e., only one atom will be found in every 10^5 (or more) cm^3. At such small densities, the particles of the gas can be very fast-moving (hot) and still not hit one another very often; so the gas will not emit radiation intensely. There is a present-day controversy about whether the x-rays which are found to reach the Earth from all over the sky are being emitted by a hot, exceedingly diffuse gas lying between the galaxies.

The gas in the solar corona is hot enough that there is no suitably defined outer boundary to the Sun. A *wind* of particles is constantly streaming outward from the corona, in the manner of Figure 2.20. The density within the wind

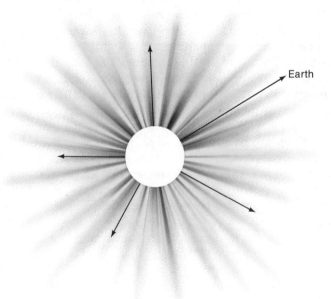

FIGURE 2.20.
The Sun emits a steady wind as well as more intense jets of particles.
A jet incident on the Earth produces a terrestrial magnetic storm.

as it reaches the Earth ranges from about 1 atom per cm³ on the low side to hundreds or even thousands of atoms per cm³ at maximum intensity. The wind at its strongest can appreciably disturb the magnetic field of the Earth, producing what are known as *magnetic storms*. The wind has other observable effects—on planets, comets, and on the Moon.

Why are the diffuse gases of the corona so hot? The answer to this question appears to be rooted in the convective motions we have already noted to be taking place below the photosphere (Figure 2.17). These motions generate waves that travel upward into the solar atmosphere, becoming more and more violent as the density of the gas decreases with increasing height. In other words, the particles are constantly being shaken backward and forward by these waves, the shaking motion becoming more and more intense as the density of the gas decreases. Collisions of one particle with another eventually cause the wave motions to be dissipated into heat, the heating effect being strong where the motions are most violent—i.e., in the corona. The process is very likely a good deal more complicated than it may seem from this simple description. Magnetic fields undoubtedly also play a role, and many problems still lack a precise solution.

I would like to end this chapter on a personal note. I have always found it difficult to believe that the rich aggregation of phenomena associated with the solar cycle could arise from the comparatively slow-speed convective motions of material below the photosphere. From what at first sight seems a very ordinary situation, a vast web of intricate processes comes to be fashioned. My failure of understanding arose, I now realize, from a lack of emphasis on a remarkable circumstance which one learns in the study of the science of *thermodynamics*. Consider a hot gas, say, at a temperature of 10,000°K. The average speed of motion of a hydrogen atom at this temperature is about 15 km per second. Some atoms have speeds above this average, but only one atom in a mass comparable to that of the whole Sun would have a speed as high as 150 km per second, and none would have a speed approaching that of light, 300,000 km per second. Given only such a hot gas at a uniform temperature of 10,000°K, there is no way to obtain particles moving at speeds approaching light, or indeed to obtain remarkable phenomena of the kind found in the solar atmosphere—x-ray emission, for example. But let the gas vary slightly in temperature—say, with one part of it at 10,000°K, another part at 9,900°K—and the situation becomes quite different. It is now possible to do work, to have a "machine" generating mechanical motion: the turning of a wheel or the lifting of a weight. Once a heavy wheel can be turned, electricity can be generated. With electricity available, all manner of complex devices become possible, as is indicated schematically in Figure 2.21. The symphonies of Beethoven can be played on a hi-fi system. Physicists can build accelerators producing particles moving at speeds very close to that of light. A maze of intricate and remarkable phenomena becomes possible, all from the slight difference of temperature in the two parts of our gas. Without this difference,

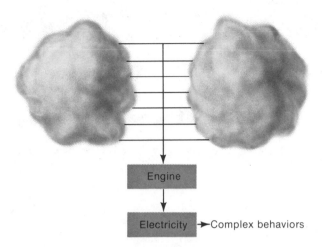

FIGURE 2.21.
From two gas clouds at different temperatures, it is possible to drive an engine, which can be used to generate electricity. No such situation is possible when the two gas clouds have exactly the same temperature.

nothing would be possible. I have known these facts from my student days. I have taught classes in thermodynamics, and yet I have always found it difficult in my thinking to take account of the rich and varied occurrences that may arise when two large masses of gas of slightly different temperatures interact with each other. This is just the situation that arises when convective motions develop within the body of the Sun.

What is the difference here between the motions of the particles of a hot gas and the motion of a wheel? The particles of a gas move randomly, some left, some right, some up, some down, some back, some forward, so that on balance there is no net motion. All the motions of all the particles average to zero, whereas in the case of a turning wheel the particles of the wheel are all going the same way. We have *mass motion*. It is mass motion, whether in the turning of a wheel, or in the convective motions in the subphotospheric regions of the Sun, that permit remarkable phenomena to be generated.

There are also mass motions involving stars. Under special conditions, to be considered later, stars can undergo pulsations, changing their radii cyclically, as indicated in Figure 2.22. This kind of mass motion is capable of generating stellar winds of particles much more intense than that from the Sun, indeed, of such an intensity that the stars even lose an appreciable fraction of their material to interstellar space, from where it can condense later to form new stars.

The Sun is rather unusual in being a star by itself. A considerable fraction of the nearby stars belong to multiple systems, systems with two or more stars.

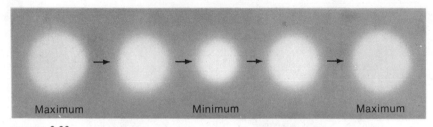

FIGURE 2.22.
An example of mass motion. A star varies its radius cyclically, pulsating between a minimum and a maximum size.

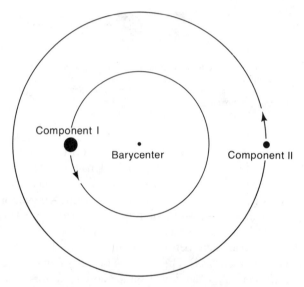

FIGURE 2.23.
Another example of mass motion. Two stars pursue orbits about a common barycenter. In such a situation, component I would be about 1.8 times more massive than component II.

Systems with two stars, usually called *binaries,* revolve in the manner of Figure 2.23. Here we have a mass motion of the stars themselves. It used to be thought that this kind of mass motion produced little effect, but in recent years it has become clear that an array of remarkable events—x-ray emission, explosive outbursts, streams of particles moving at speeds close to light—are generated in binary systems. We will encounter all these processes again at a later stage.

General Problems and Questions

1. What is the area of a square with sides of length *a*?

2. How can the answer to Problem 1 be used to measure any flat area? (Divide the area into many small squares by drawing two sets of parallel lines, one set of lines being perpendicular to the other set. Ignoring incomplete squares, add the areas of all the small squares which lie within the specified area. Except for a small error due to ignoring the incomplete squares, this solves the problem. If you want to make the error smaller still, rule the parallel lines closer together. In this way, at any rate in theory, the error can be made as small as you wish.)

3. What does it mean to say that the surface of a sphere of radius *r* has an area of $4\pi r^2$? (Divide the surface of the sphere into so many small pieces that each of them can be thought of as flat. Use Problem 2 for each such piece, and then add the areas of all the pieces. *Irrespective of how you divide up the sphere, the answer is always $4\pi r^2$.* This way of coping with problems, which at first sight seem quite difficult, belongs to the branch of mathematics called *integral calculus*.)

4. The Sun has a power output 10^{28} times greater than a certain candle. With a large telescope, a star like the Sun could be faintly distinguished at a distance as great as 10^{18} kilometers. How far away could the candle be distinguished? By using ideas and methods to be considered in Section II, it would be possible for an astronomer to know that he was observing a distant candle and not a distant star.

5. What is the volume of a cube with sides of length *a*?

6. Using Problems 2 and 5, work out a plan for determining the volume contained within any specified closed surface. When the specified surface is a sphere of radius *r*, the volume is $\frac{4}{3}\pi r^3$.

7. Suppose a sheet of metal of specified area is to be cut into many pieces which are then welded together to form a closed surface. This can be done in many ways. The resulting surfaces have the common property of all possessing the same area, namely the specified area of the original metal sheet. Which among these different surfaces will contain the greatest volume? (Answer: The case where the surface is a sphere. Many aspects of the world, ranging from spherical water drops to the fact that nonrotating stars and planets take on spherical shapes, are connected with this mathematical property. I was once vividly reminded of it by the following untoward experience. An unpunctured metal can containing pudding was placed in a pan containing water and the whole contraption was heated on a gas burner. Events occurred which caused this set-up to be forgotten, at any rate for awhile. First the water in the pan evaporated. Then water inside the can began to boil. As the temperature rose, the pressure of steam within the can built up inexorably. The can, being well-made, did not explode until after the metal casing had become changed into a spherical shape, thereby

exacting the maximum volume for the steam within it. On explosion, jets of pudding were sprayed to astonishingly great distances, much like the explosion of Krakatoa.)

8. How is the luminosity of the Sun determined?

9. How would you go about measuring the power of the sunlight incident on a specified area?

10. Two stars are similar, but one is 10 times more distant than the other. What would be the ratio of their observed flux values?

11. What is their magnitude difference?

12. Rework Problems 10 and 11 for the situation in which the distance of one star is twice that of the other. Why do you need to use logarithmic tables for this problem, but not for problem 11?

13. Suppose the less-distant star of Problem 12 has 25 times the luminosity of the other. What happens then to the flux ratio and to the magnitude difference?

14. A star has magnitude $+3$. Suppose it to be moved to a distance where its magnitude becomes $+5.5$. How is this second distance related to the initial distance?

15. Repeat Problem 14, but take the initial magnitude to be $+3.8$ and that at the second distance to be $+6.3$. Why is the answer the same as before?

16. If you have access to a pocket calculator, determine the logarithms of 3.14159, 981, 6378.4, 384,400.

17. Assuming you know both the scale of the solar system and the flux value of the Sun, and that you can measure the flux value of the light from a star, how could you obtain a general idea of the distance of the star?

18. Assuming you know the scale of the solar system and that you can measure very small variations of the direction of a star, how could you use the Earth's annual motion around the Sun to obtain the distance of the star?

19. Without being able to measure small variations of angle or flux values, Cheseaux nevertheless was able to obtain a good general estimate for the distances of the nearest stars. How was this done?

20. What is a light year?

21. The temperature of the gases at the surface of a sunspot is typically about $4000\,°K$, whereas the temperature of the normal solar disk is about $5800\,°K$. The power radiated by a unit area of surface goes as (temperature)^4. By what factor is the power radiated by a square kilometer of sunspot less than the power radiated by a square kilometer of the solar disk? Why do sunspots appear dark?

22. Scatter iron filings on a horizontal white surface, and bring a bar magnet to their neighborhood. Complicate the situation by introducing further magnets, and compare the resulting patterns of the filings with the picture of the solar corona shown in Figure 2.13.

23. Neither the Moon's orbit around the Earth, nor the Earth's orbit around the Sun, is quite circular. Given that the Moon's apparent angular size is nearly the same as that of the Sun, what conditions of the Earth and Moon in their orbits will give a total solar eclipse of the longest possible duration?

24. In what ways might the occurrence of sunspots affect the Earth?

25. The radius of the Sun is 6.96×10^{10} cm. The average density of atoms within the Sun is about 10^{24} per cm^3. Show that the whole Sun contains about 10^{57} atoms. Our galaxy has some 10^{11} stars with average properties not much different from the Sun. About how many atoms does our galaxy contain?

26. What are the differences of appearance of the Sun when viewed (a) in white light, (b) in the light of calcium atoms, (c) hydrogen atoms, (d) x-rays?

27. What is the solar wind? How might it be generated?

28. Why does a hot diffuse gas emit much less radiation than a dense gas of the same temperature and volume?

29. Discuss the difference between thermal motions and mass motions. Give everyday examples of mass motions—e.g., a stone falling over a cliff.

30. Give astronomical examples of both mass motions and thermal motions.

Chapter 3:
Galaxies

§3.1. Gravitation

Already we have come across phenomena that raise physical problems. Why do the planets continue to move around the Sun instead of flying off into space? Because of a force of gravitation which holds them to the Sun. In Figure 3.1 we have the elliptic orbit followed by a planet P, the Sun S being at one of the foci of the ellipse. It is important to notice that the gravitational attraction of the Sun on the planet is directed toward S along the line PS. There is no force in the direction transverse to PS. This was a point of great difficulty to astronomers before Newton. It seemed natural, even to Kepler, to suppose that a planet moves around the Sun because it is pushed along in the direction of its motion. Christian Huygens (1629–1695) remarked, after reading Newton's analysis of the problem, that the surprise to him was to find everything working out correctly without any transverse force being required.

The form of the force can be stated quite generally. In Figure 3.2 we have the positions of *any* two particles P and S at a particular moment of time. The gravitational force between them, at the time in question, is along the

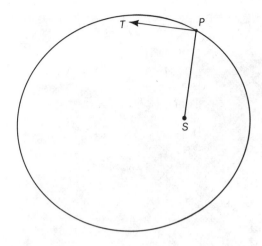

FIGURE 3.1.
The planet follows an ellipse, with the Sun at one of the two foci. The gravitational influence of the Sun on the planet is from P toward S. There is *no* force in the direction PT.

line P to S in such a sense as to pull the particles together—the force on either particle is toward the other one. The magnitude of the force is of the form

$$\frac{GmM}{r^2},$$

where m is the mass of one particle, M of the other, r is their distance apart, and G is a constant to be determined by observing a specific case. With G so obtained, we then use our formula in all other cases. In the planetary problem of Figure 3.1, we simply identify P with the planet and S with the Sun.

It is to be noted that the above formula was arrived at empirically. It was found to work correctly for the planetary motion of Figure 3.1, and so became adopted for every pair of particles in the universe, for example, for every pair of stars in the galaxy. Indeed, just as the planets of our solar system are held to the Sun by gravitational attraction, so the stars of the galaxy are held together

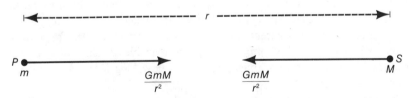

FIGURE 3.2.
For any two particles P and S of masses m and M respectively, there is a force on each, $\frac{GmM}{r^2}$, that draws the particles together.

FIGURE 3.3.
The star cluster M67, about 15 light years in diameter. It is believed that most stars are born in clusters. (Courtesy of Hale Observatories.)

by gravitation. The general problem of the motions of the stars is a much more complicated one than that of the planetary motions, however. There are some 10^{11} stars in the galaxy, and each of them is subject to the gravitational influence of all the others. To calculate the motion of every star in detail would involve a volume of arithmetic, of additions, subtractions, multiplications and divisions, not only outside human capacity, but far outside the capacity of the largest modern digital computer.

With such computers it is actually possible to solve in detail for the motions of a few hundred stars, such as occur in the clusters shown in Figures 3.3 and 3.4. It seems likely that stars are largely born in clusters. Given time, the stars mostly escape from their parent clusters, however, becoming then free to move individually in the galaxy. Computer calculations enable the astronomer to work through this escape process, thus determining how long it takes for a star to achieve "freedom."

By introducing some approximations into the computer calculations, one can give them a still wider range of applicability. Thus if we ask, not for complete detail, but for a general understanding of how the whole galaxy of stars will behave under their mutual gravitational attractions, even this vast problem becomes tractable. What, it may be asked, is the difference between "general

FIGURE 3.4.
The Pleiades, a cluster about 12 light years in diameter, containing several bright, comparatively young stars. (Courtesy of Hale Observatories.)

understanding" and precise calculation? This question is best answered by an example. Every now and then unusual incidents occur; it may happen that two stars move exceptionally close to each other, as if another star were to sweep through our own solar system. A precise calculation would detail every rare occasion of this kind, whereas a more general calculation would tell us how often such incidents might be expected to occur, but would not tell us which particular stars would be involved in them. We would not know whether our system would ever be involved in such an encounter, for example.

It is a quite usual experience in science that, by limiting the depth of inquiry, systems of great complexity can be investigated in lesser depth. But if we insist on knowing all details, then only rather simple systems can be treated.

By restricting ourselves to average properties, we can offer answers to many problems concerning the motions of the stars in our galaxy, the general distribution of which is shaped like a pancake, but develops a considerable bulge toward the center, as indicated schematically in Figure 3.5. The general distribution of stars is said to have *axial symmetry*, because it maintains its shape if the whole system is given a rotation around the axis shown in Figure 3.5. In such an axially symmetric system, the multitude of gravitational interactions of a particular star (for example, the Sun) with all the other stars add together

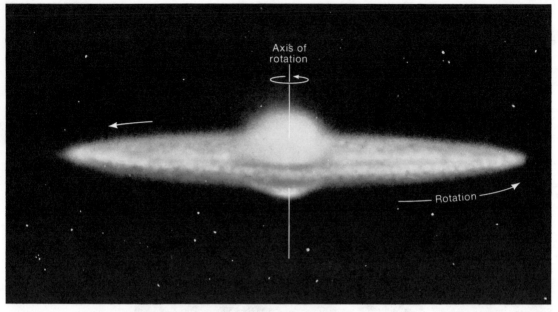

FIGURE 3.5.
The distribution of stars in our galaxy is a disk with a diameter of about 100,000 light years
and with a gradually developing central bulge. A comparatively thinly distributed halo of
stars surrounds this disk. (Adapted from Blaauw and Schmidt, *Galactic Structures*. University
of Chicago Press © 1965.)

to give a combined force that points essentially toward the center of the galaxy.
This centrally directed force permits a star to move in an orbit around the
center rather like a planet moving around the Sun. In particular, the orbit can
be approximately a circle, and this is just the situation for our solar system.
The Sun and planets move together around the center of the galaxy in an
approximately circular orbit, the radius of which is about 30,000 light years.
The time required for us to make a complete circuit of the galaxy is about
200 million years. When the solar system was in much its present position,
the last time around, 200 million years ago, the dinosaurs had still to become
dominant. This time around, the dinosaurs have disappeared and Man has
become dominant. What of the next time around?

Although the total gravitational force exerted by all the stars on a particular
star is directed essentially toward the center of the galaxy, thereby permitting
a star to move around in a circular orbit, there is no necessity for the orbit
to be circular, any more than there is a necessity for a body to move around
the Sun in a circle. It is true that the planets do move nearly in circles, but
the comets, obeying the same law of gravitation, usually move in orbits that
are highly elliptic. Similarly, some stars have orbits in the galaxy that are quite
unlike circles, as in Figure 3.6. It is curious that such stars, when their ages

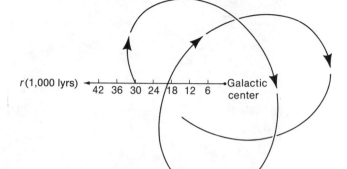

FIGURE 3.6.
The orbit of a high-velocity star in the galaxy. Notice that the orbit is
not closed like that of a planet moving around the Sun. This is be-
cause every star exerts a gravitational influence on every other star.
(From G. Contopoulos and B. Strömgren, *Tables of Plane Galactic
Orbits.* New York: NASA Institute for Space Studies, 1965.)

are judged by criteria to be developed in Section III, turn out to belong to
the oldest class of stars in the galaxy.

All stars possess some degree of motion in a direction perpendicular to the
flat plate which forms the galaxy. These motions become larger as we go from
stars on the outside of the galaxy to stars nearer the center. It is for just this
reason that a bulge develops in the star distribution as we go toward the center.
This property and the age problem mentioned in the previous paragraph are
probably connected with the primaeval process in which the galaxy itself was
formed. This too is a matter for later sections to consider.

§3.2. Spiral Structure and Nearby Galaxies

We have not said much so far about the gas and dust in the galaxy, and we
shall defer our main consideration of it to a later section. Yet it should be noted
that the gas and dust are not distributed with the same axial symmetry as the
majority of the stars. Rather the gas and dust tend to be distributed in lanes
or "arms" having a generally spiral structure. Figure 3.7 shows the way this
structure is thought to go.

Observations of such spiral-like patterns are more readily made for galaxies
other than our own. In Figure 3.8 we have a photograph of a large nearby
galaxy with the catalogue designation M31 (the thirty-first object in a catalogue
compiled by the French astronomer Charles Messier, 1730–1817). This galaxy
is also often referred to as the Andromeda Nebula, since it appears as a nebulous
object in the constellation of Andromeda.

Galaxies

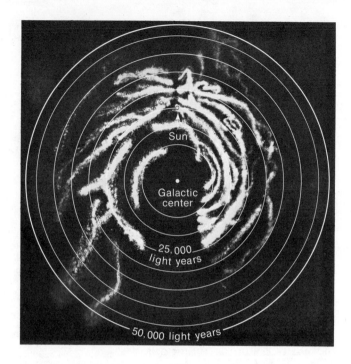

FIGURE 3.7.
The lanes of gas in our galaxy, determined by radioastronomy (Chapter 13). (Courtesy of Leiden Observatory.)

PROJECT:
After locating M31 on a star map, find this galaxy in the night sky. Away from city lights, it can be seen without difficulty by naked eye, although one sees only the inner part of Figure 3.8.

The spiral pattern of M31 shows clearly in Figure 3.8. Light from the spiral arms comes mainly from a comparatively few stars of unusually high intrinsic luminosity which follow the same patterns as the gas and dust, for the good reason that they have only recently formed out of the gas and dust itself and have not yet had time to move away from it.

PROJECT:
A typical point in the star cluster Praesepe has position R.A. 8h 37m, Dec. +20° 10'. Locate this position on a star map and note the configuration of the nearby brighter stars. Then, using binoculars, find Praesepe in the night sky. A typical point in the twin clusters of h and χ Persei has position R.A. 2h 17m, Dec. 57°. Find h and χ Persei also. Given that the latter clusters are at more than ten times the distance of Praesepe, what do you conclude from your observations? The bright stars of h and χ Persei are of the kind that illuminate the spiral structure of M31.

FIGURE 3.8.
The galaxy M31, about 100,000 light years in diameter, situated in the constellation of Andromeda. This galaxy, sometimes referred to as the Andromeda Nebula, is the nearest of the large galaxies. The central regions are visible to the naked eye. (Courtesy of Hale Observatories.)

The galaxy M31 is a companion member to our own galaxy in a small galaxy cluster known as the Local Group. Details concerning these local galaxies are set out in Table 3.1. From the right ascensions, declinations, and distances, it would be possible to construct a three-dimensional model of the Local Group. On paper, however, we can only represent with accuracy two of these three quantities. Thus, in Figure 3.9 we have chosen to represent distance radially with the right ascension as a measure of angle, using the rules 1 hour $\equiv$ 15°, 1 minute $\equiv$ 15′, 1 second $\equiv$ 15″. Information about the declination is lost in this figure.

After M31 and our own galaxy, and apart perhaps from Maffei 1 and 2, the only other major member of the Local Group is the galaxy M33, shown in Figure 3.10. The spiral arms are more regular and more completely formed in M33 than in either our galaxy or M31. The remaining members of the Local Group are so-called dwarf systems, some being small and concentrated (like NGC 205 in Figure 3.11), others being diffuse (like the system of IC 1613 in Figure 3.12). Apart from our galaxy, M31, and M33, the others are minor systems, mere fragments of galaxies, that would not be seen at all if the Local Group were observed from a distance of several hundred million light years.

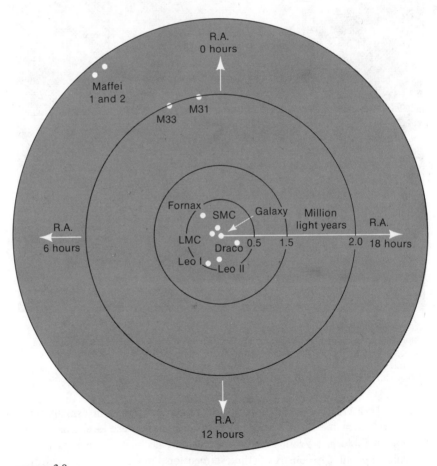

FIGURE 3.9.
A plot of distance versus right ascension for some of the members of the Local Group of galaxies. Declination values are given in Table 3.1.

FIGURE 3.10.
The galaxy M33 (NGC 598), about 30,000 light years in diameter. After our own galaxy and M31 (Figure 3.8), this is the only other major member of the Local Group—except possibly for Maffei 1 and 2. (Courtesy of Hale Observatories.)

TABLE 3.1.
The local group

Description	Approximate distance (light years)	Right ascension		Declination		Class
Our Galaxy						Sb
NGC 224 (M31)	2×10^6	00h	40m	41°	00′	Sb
NGC 221 (M32)	2×10^6	00	40	40	35	Ell (Dwarf)
NGC 205	2×10^6	00	38	41	25	Ell (Dwarf)
NGC 598 (M33)	2×10^6	01	31	30	25	Sc
NGC 147	2×10^6	00	30	48	15	Ell (Dwarf)
NGC 185	2×10^6	00	36	48	05	Ell (Dwarf)
IC 1613	2×10^6	01	03	01	50	Irr (Dwarf)
NGC 6822	1.5×10^6	19	42	−14	55	Irr (Dwarf)
Leo I	9×10^5	10	06	12	35	Ell (Dwarf)
Fornax	8×10^5	02	38	−34	45	Ell (Dwarf)
Leo II	7×10^5	11	11	22	25	Ell (Dwarf)
Draco	3×10^5	17	19	58	00	Ell (Dwarf)
Sculptor	3×10^5	00	58	−34	00	Ell (Dwarf)
Ursa Minor	2×10^5	15	08	67	20	Ell (Dwarf)
Large Magellanic Cloud	1.7×10^5	05		−69°		Irr (Dwarf)
Small Magellanic Cloud	1.7×10^5	01		−73°		Irr (Dwarf)
Possible members						
Maffei 1	3×10^6	02	33	59	25	Ell
Maffei 2		02	38	59	25	Ell

78

FIGURE 3.11.
One of the two small companions to
M31, this is NGC 205, about 10,000
light years in size. (Courtesy of Hale
Observatories.)

FIGURE 3.12.
The galaxy IC 1613, about 10,000 light years in diameter, is a
diffuse dwarf member of the Local Group. (Courtesy of Hale
Observatories.)

FIGURE 3.13.
The symmetrical spiral galaxy M81 is seen near the center of a small group of galaxies in the
constellation of Ursa Major. The group is about 500,000 light years in size. (Courtesy of Hale
Observatories.)

§3.3. Galaxies and the Structure of the Universe

Most galaxies belong to groups or clusters, some containing only a few major
members, like the Local Group. Other examples of rather small clusters are
shown in Figures 3.13, 3.14, 3.15, and 3.16.

Particularly rich clusters, like those in Table 3.2, contain more than a hundred
major galaxies. Three such clusters are shown in Figures 3.17, 3.18, and 3.19.
Such very rich groups are rare, however. Much more typical are clusters with
from about 5 to 20 members, as in Figures 3.13 to 3.16.

PROJECT:
Using the data of Table 3.2 draw a diagram for these clusters like that of Figure
3.9.

With increasing distance, the details of structure within the galaxies become
more and more difficult to observe. Ultimately, at the greatest distances to which
photography can be taken, some 10^{10} light years, nothing remains except a
small "soft" image, as in Figures 3.20 and 3.21, which show fields of galaxies

80

FIGURE 3.14.
Group of four galaxies in the constellation of Leo. The group is about 500,000 light years in size. (Courtesy of Hale Observatories.)

FIGURE 3.15.
The remarkable interacting system known as Stephen's quintet. It is clear that such a situation cannot persist for very long, not for a time-scale comparable to the age of our own galaxy. What happened to lead to this situation? (Courtesy of Hale Observatories.)

TABLE 3.2.
Some rich clusters of galaxies

Description	Approximate right ascension[a]		Approximate declination[a]		Approximate distance (millions of light years)	Radial velocity (km/sec)[b]
Virgo	12 h	25 m	12°	50′	70	1,150
Pegasus I	23	17	07	50	230	3,800
Pisces	01	35	32	00	270	4,500
Cancer	08	16	21	20	295	4,900
Perseus	03	15	41	15	325	5,400
Coma	12	55	28	20	400	6,700
Hercules	16	03	17	52	620	10,300
Pegasus II	23	08	07	20	770	12,800
Ursa Major I	11	45	55	59	930	15,500
Cluster A	01	06	−15	36	950	15,800
Leo	10	24	10	39	1,170	19,500
Corona Bor.	15	20	27	54	1,300	21,700
Gemini	07	05	35	06	1,400	23,400
Boötes	14	31	31	46	2,360	39,400
Ursa Major II	10	55	57	02	2,425	40,400
Hydra II	08	55	03	21	3,630	60,500

[a] The right ascension and declination values are for representative points within the clusters. Note that a cluster actually covers a patch of the sky, which (for Virgo particularly) can have a considerable area.
[b] This column is related to the discussion in Appendix II.8.

FIGURE 3.16.
A cluster of galaxies, about 500,000 light years in diameter, in the southern sky, R.A. 10h30m, declination −27°. (Courtesy of Dr. V. C. Reddish, Science Research Council.)

FIGURE 3.17.
A rich cluster in Hercules, about 600 million light years distant from us, containing many spiral forms. (Courtesy of Hale Observatories.)

FIGURE 3.18.
Part of the rich cluster of elliptical and spiral galaxies in Coma Berenices, about 400 million light years distant from us. Clusters of this kind play an important role in establishing very large distances (Chapter 8). (Courtesy of Kitt Peak National Observatory.)

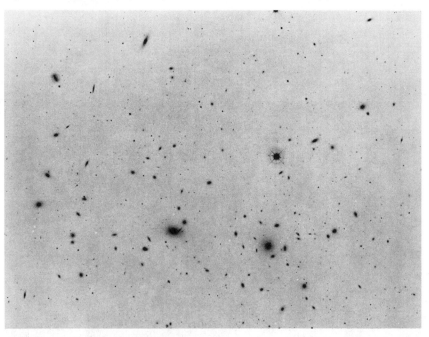

FIGURE 3.19.
Another rich cluster in Corona Borealis. This cluster, similar intrinsically to that of Figure
3.18, is about three times more distant. (Courtesy of Hale Observatories.)

observed with the 200-inch telescope of the Hale Observatories. To avoid the
profusion of stars along the plane of our own galaxy (Figures 1.23 and 1.24),
the field chosen for Figure 3.21 lies in a direction more or less perpendicular
to the galactic plane. In contrast to Figures 1.23 and 1.24, very few of the
images in Figure 3.21 are stars of our own galaxy. Almost all are distant
galaxies, a situation unlike that for the picture of M31 in Figure 3.8, where
the many hard circular images are stars of our own galaxy. At the great distances
of Figure 3.21, faint galaxies have outnumbered the local stars.

It is interesting to compare the hard stellar images of Figure 3.8 with the
softer ones of Figures 3.20 and 3.21. The photographic image of an astronomical object, taken with a large ground-based telescope, has an appearance which
depends on two factors, the angular size and light distribution of the object
itself, and imperfections caused by the terrestrial atmosphere. The situation
in Figure 3.8 is that the whole of the extensions of the star images are due
to the atmosphere, whereas in Figures 3.20 and 3.21 the intrinsic sizes of the
galaxies also make contributions to the images, producing a much "softer"
appearance than do the stellar images.

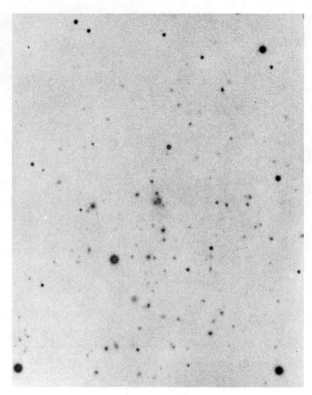

FIGURE 3.20.
In the depths of space, about 3,600 million light years distant
from us, very many galaxies are faintly seen. This cluster is in
the constellation of Hydra. (Courtesy of Hale Observatories.)

We can make an important deduction from these practical considerations.
Figures 3.20 and 3.21 give a roughly valid impression of the relation between
the sizes of galaxies and the distances that separate them, but any impression
we have from Figures 1.23 and 1.24 that stars are close-packed is false. We
must think of Figures 1.23 and 1.24 as if the stars really were abstract points.
Even if the scale of these figures were increased a thousand-fold, the stellar
images should still be points.

The situation is this: if we imagine the average star to be a yard in diameter,
then a star is separated from its nearest neighbor by about 20,000 miles; but
if we imagine a galaxy to be a yard in diameter, the nearest galaxy would not
usually be more than 100 yards away. On this same imaginary scale, the distance
range of Figures 3.20 and 3.21 would be about 100 miles, all the galaxies within
this range being projected together on the sky and thereby producing the
observed moderately close-packed effect.

We are now in a position to discuss an observation, illustrated in Figure

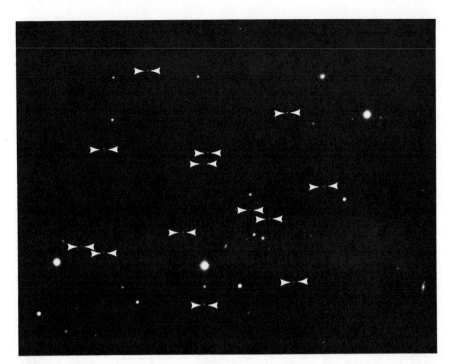

FIGURE 3.21.
Another deep photograph. The marked galaxies are at distances of about 6,000 million light years away from us. (Courtesy of Hale Observatories.)

3.22, which has great importance in the study of the universe as a whole. This figure shows the distribution on the sky of rich clusters of galaxies counted to a distance of about two billion light years, but omitting the region of the sky not visible from Palomar Mt., California, where the survey was carried out, and also omitting a band along the plane of the Milky Way, because the obscuring effects of dust along the Milky Way would otherwise falsify the counting in these directions. Figure 3.22 uses the same kind of galactic latitude and longitude plot as was used in Figure 1.28 to display the Milky Way itself.

PROJECT:
Compare Figures 1.28 and 3.22, noting how the zone of avoidance of Figure 3.22 coincides with the position of the Milky Way in Figure 1.28. Also compare the distribution of clusters towards the south galactic pole with that toward the north galactic pole.

The similarity of the cluster distributions toward the two galactic poles is indicative of *isotropy* in the universe—i.e., that the universe has the same aspect

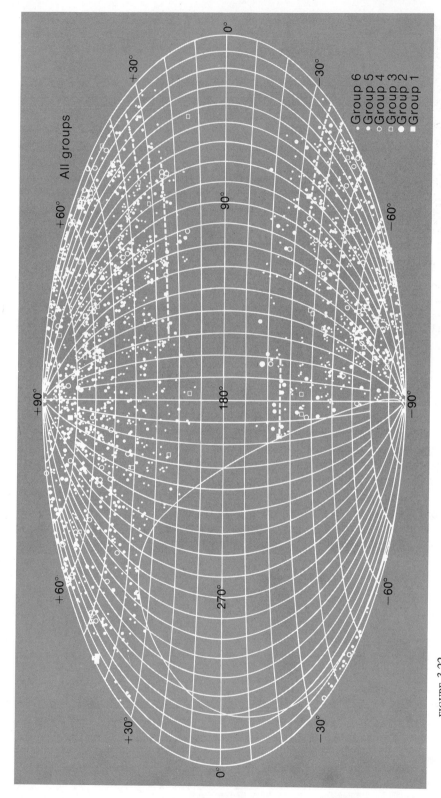

FIGURE 3.22.
Distribution of rich clusters of galaxies, with the larger symbols representing nearer clusters. The oval empty region around galactic longitude 270° is the part of the sky not visible from Palomar Mountain. (From G. O. Abell, *Astrophysical Journal Suppl. Series*, 1957–58, page 211. University of Chicago Press.)

in all directions. The general isotropy of the universe was first discovered by Edwin Hubble (1889–1953), who counted individual galaxies instead of clusters, obtaining a result like that of Figure 3.22.

Would the universe also appear the same on a large scale if we were making our observations from some other spatial position? Without actually making observations from very different spatial positions, which obviously we cannot do, this question cannot be answered. Our observation of isotropy does not settle this further question, since the distant universe would still appear isotropic in the situation of Figure 3.23. However, Figure 3.23 would demand a special relation of our own galaxy to the universe, since in this figure we have taken our galaxy to be located in the center of a nonuniform distribution of galaxies. It hardly seems plausible that our galaxy would be in any such privileged position. So we answer the above question affirmatively on intellectual grounds rather than because such an answer is determined by observation. We thus require the universe to be *homogeneous* (similarity with respect to spatial position) as well as *isotropic* (similarity with respect to direction). These two conditions will turn out to play a central role in our understanding of cosmology—the study of the large-scale structure of the universe—as we shall see in Section VI.

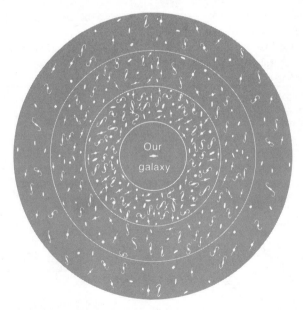

FIGURE 3.23.
Although different concentric shells contain different densities of galaxies, the situation still appears isotropic to an observer at the center, but would not to an observer not at the center. (The density of galaxies is greatly exaggerated in this drawing.)

Examination of photographs of galaxies shows that not all of them are alike. Some galaxies show spiral patterns, and others do not. Indeed, there is a whole class of galaxies, amorphous in appearance, with elliptical outline and hence known as the *elliptical galaxies*, that appear to be quite different from the *spiral galaxies*. The Coma cluster, shown in Figure 3.18, consists in its inner part essentially of ellipticals, whereas the Hercules cluster, shown in Figure 3.17, consists mostly, although by no means wholly, of spirals. Just as on a geographical map contours can be drawn through points having equal height above sealevel, so we can draw contours on the map of a galaxy as it appears on the sky, the contours being through points of equal surface brightness. Such *isophotes,* as they are called, show the two characteristic types of Figure 3.24, one for spirals, the other for ellipticals.

Elliptical galaxies contain very little gas and dust, and this also is a criterion to separate them from spirals, although there is a class, known as the S0 galaxies, which meets this criterion and yet has isophotes of the kind associated with spirals. Such galaxies are believed to result from collisions between spiral galaxies. We saw above that, compared with their own dimensions, galaxies are not very widely spaced, which means that they will collide from time to time, especially galaxies in clusters. Because the stars in a galaxy are very widely spaced compared to stellar dimensions, the stars in one galaxy can pass quite easily between those of another galaxy. But when collisions occur between spirals, the gas and dust clouds in one encounter the clouds in the other. The encounters are at high speed, which causes the gas to become exceedingly hot and to burst explosively out of the parent galaxies, carrying the dust out also. The outcome is galaxies stripped of their gas and dust, but with isophotes still characteristic of spirals.

Hubble classified the galaxies into various categories, illustrated by typical cases in Figures 3.25, 3.26, and 3.27. The sequence E0, E1, E2, . . . , E7, of ellipticals follows a progressive flattening of the profile, with E0 being of

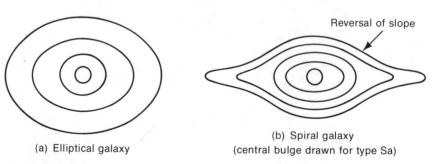

(a) Elliptical galaxy

(b) Spiral galaxy
(central bulge drawn for type Sa)

FIGURE 3.24.
The important difference between the isophotes of elliptical galaxies (a) and spiral galaxies (b) is that the outer contours of equal brightness reverse their slope in (b), but not in (a).

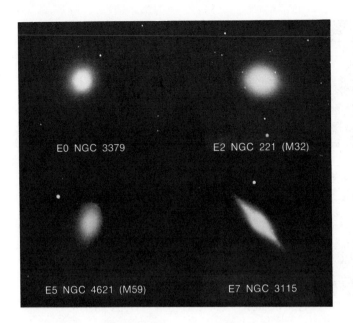

E0 NGC 3379

E2 NGC 221 (M32)

E5 NGC 4621 (M59)

E7 NGC 3115

FIGURE 3.25.
Examples of Hubble's elliptical
sequence E0 to E7. (Courtesy
of Hale Observatories.)

NGC 1201   Type S0

NGC 2841   Type Sb

NGC 2811   Type Sa

NGC 3031 M81   Type Sb

NGC 488   Type Sab

NGC 628 M74   Type Sc

FIGURE 3.26.
Examples of Hubble's Sa, Sb, Sc types, with an intermediate Sab form,
and also with a galaxy of type S0. (Courtesy of Hale Observatories.)

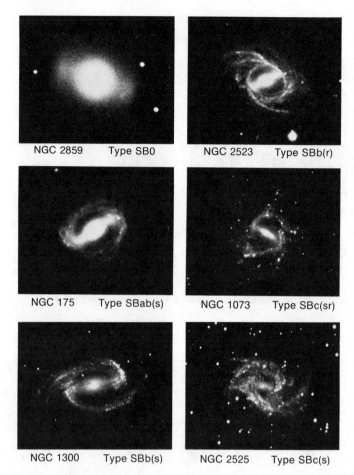

NGC 2859   Type SB0   NGC 2523   Type SBb(r)

NGC 175   Type SBab(s)   NGC 1073   Type SBc(sr)

NGC 1300   Type SBb(s)   NGC 2525   Type SBc(s)

FIGURE 3.27.
Examples of the barred types of galaxy.
(Courtesy of Hale Observatories.)

globular form and with E7 of a markedly flattened lenticular form. The Sa, Sb, and Sc spiral forms are classified according to the importance of the central bulge, or *nucleus*, as it is usually called. Spirals of type Sa have a strongly formed nucleus, those of Sb type have a somewhat smaller nucleus, whereas those of type Sc have only a rather small nucleus. Our galaxy and our companion galaxy M31 are of type Sb; M33 is of type Sc. Because spirals sometimes possess a central bar, as well as a set, or sets, of spiral arms, Hubble also introduced the barred sequence, SBa, SBb, SBc, in which the importance of the nucleus plays a similar role to that of the Sa, Sb, Sc sequence.

Galaxies falling outside the above scheme were described as *irregular*. Such irregular (Irr) galaxies were thought to comprise only a few per cent of all galaxies.

Hubble always insisted that his classification was empirical. He made no claim for any physical or evolutionary connection between one kind of galaxy and another. Today, some forty years after Hubble's scheme was first proposed, we still cannot say with any certainty whether or not such connections exist. We shall find in Section VI that the whole problem of the origin and evolution of galaxies is not at present in a very satisfactory state.

Other, more ambitious, empirical classification schemes for galaxies have been proposed in recent years. Such schemes are helpful to the astronomer, since they cover a wider range of galactic forms than the simple scheme of Hubble. Yet no system of classification covers all galaxies. So many rare, peculiar, and beautiful cases are found that no classification scheme could hope to include them all. To form an idea of the astonishing richness of the system of galaxies, consider the examples shown in Figures 3.28 to 3.32. With the sobering thought that a satisfactory theory of the origin and evolution of galaxies must explain all these curious cases, as well as the more regular forms, we end this chapter.

§3.3. Galaxies and the Structure of the Universe

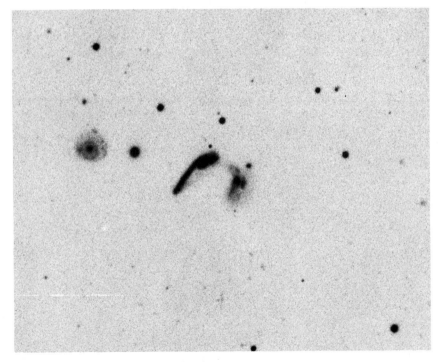

FIGURE 3.28.
A photographic negative of a peculiar chain of galaxies, in which the galaxies appear dark against a light sky. (Courtesy of H. Arp, Hale Observatories.)

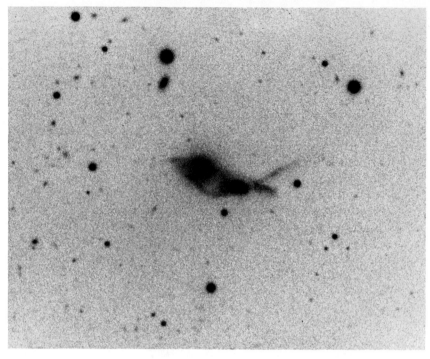

FIGURE 3.29.
Another peculiar system, south of NGC 2841. Again a negative photograph.
(Courtesy of H. Arp, Hale Observatories.)

FIGURE 3.30.
A chain of small galaxies near the corner of NGC 247.
(Courtesy of H. Arp, Hale Observatories.)

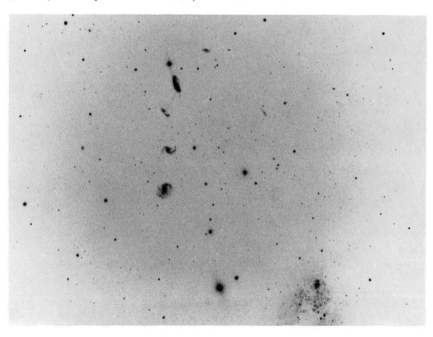

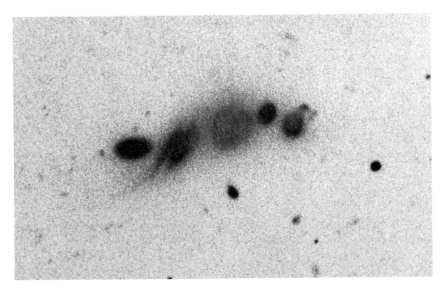

FIGURE 3.31.
The peculiar chain of galaxies known as VV 172. (Courtesy of H. Arp, Hale Observatories.)

FIGURE 3.32.
The remarkable group of interacting galaxies known as Seyfert's Sextet.
(Courtesy of H. Arp, Hale Observatories.)

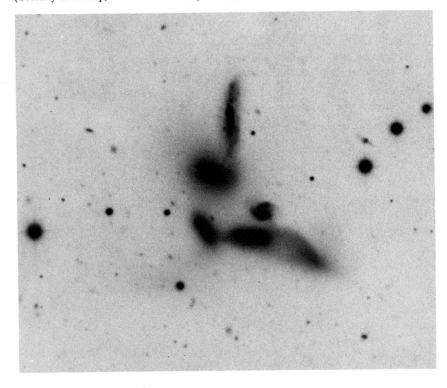

General Problems and Questions

1. What is the form of the gravitational force between two particles, one of mass m, the other of mass M?

2. Two bodies are made of the same material, but one contains 10 times as many atoms as the other. What is the ratio of their masses?

3. In what kind of orbit does a planet move around the Sun? Neglecting the gravitational influence of other planets, what important property does a planetary orbit have that is different from a stellar orbit in the galaxy?

4. Describe the general distribution and motions of the stars in our galaxy. What role does gravitation play in controlling these motions? What would happen to our galaxy if, by some magic, gravitation were to cease to exist?

5. Given that the galaxy M31 possesses approximate axial symmetry, estimate from Figure 3.8 the angle which the line of sight makes with the axis of rotation of this galaxy. (Try drawing a circle on a sheet of paper and then orient the sheet so that the circle appears to have about the same shape as the general outline of M31.)

6. Repeat Problem 5, but for the galaxy M33.

7. Why can a photographic plate not be exposed longer and longer, until faint images, like those of Figures 3.20 and 3.21, become much clearer?

8. What is the approximate ratio of the distances of the faint galaxies of Figures 3.20 and 3.21 to the average distance between neighboring galaxies?

9. Estimate the ratio of the average distance between neighboring galaxies to the scale of the galaxies themselves.

10. Discuss the ideas of isotropy and homogeneity in relation to the large-scale distribution of galaxies and of clusters of galaxies.

11. Describe the structural form of the main classes of galaxies, paying special attention to the differences between spirals and ellipticals.

12. What are S0 galaxies?

Appendixes to Section I

Appendix I.1. The Calendar

The length of the year is not exactly 365 days. And the amount left over, after 365 days are subtracted, is not a simple (rational) fraction of a day. Mathematically, we say that the day and the year are *incommensurable* with each other. Because of this it is awkward to measure time in days, as we must for everyday purposes, and yet prevent the year from getting out of step in an unacceptable way.

There are two possible procedures, of relevance in this problem, for determining the year. The best way, for physics, would be in terms of the fixed directions discussed in Chapter 1. Consider the line from the Earth to the Sun. The time interval required for this line to make a complete circuit with respect to fixed directions is called the *sidereal* year, and has a length of 365.2564 days.

The positions on the celestial sphere of the equinoctial points γ and Ω are changing with respect to fixed directions, due to a slow toplike precession of the Earth's axis or rotation. The time for a complete circuit of γ and Ω around the ecliptic is about 26,000 years. Now, the incidence of spring and autumn are determined by the moments at which the projection of the Sun on the celestial sphere passes through γ and Ω, respectively. Midsummer is determined by the moment when the angle between the direction of the Sun and the Earth's axis takes its minimum value. Precession therefore causes the moment of midsummer, as well as of the equinoxes, to change with respect to fixed directions. So too does the moment of midwinter. It follows that the determination of the year in terms of the interval from midsummer to the next midsummer, or from equinox to equinox, or from midwinter to midwinter, must lead to a result different from the sidereal year by one part in 26,000. The year determined in this way is called the *tropical year,* and is 365.2422 days.

From a civil point of view, there is little interest in keeping the reckoning of days in step with the sidereal year. The interest lies in keeping the reckoning of days—i.e., in keeping a calendar—in step with the tropical year. At the end of the tropical year, we return to the same situation with respect to the seasons, and it is this which is of importance in everyday life. So it comes about that for calendrical purposes we have to cope with a year of 365.2422 days.

According to the calendar introduced by Julius Caesar in 46 B.C., the Julian system, as it is called, a normal year is one of 365 days, made up in a curious way from twelve months with strangely varying numbers of days in them. Every fourth year in the Julian calendar is taken to be an abnormal year of 366 days, a "leap" year, obtained by adding a 29th day to February.

Since the *average* Julian year is thus 365.25 days, it differs on the average from the tropical year by being 0.0078 days too long. Hence the Julian calendar slowly goes out of step with the tropical year, accumulating an extra day about every 128 years. From its inception to the early sixteenth century, the Julian calendar accumulated about 12 extra days. This discrepancy from the tropical year was large enough to cause serious concern, and the Pope of the day, who was regarded as the authority in the matter, began to cast around for suggestions about how the situation could be improved. One of the astronomers canvassed for advice was Copernicus, who replied to the Pope's query by saying that further observations would be required before a satisfactory proposal could be made. Presumably what Copernicus meant was that the tropical year was not at that time known with adequate accuracy, and indeed over the ensuing years astronomers at the Vatican set about determining the length of the tropical year with improved precision. The tower where they did so can be seen to this day.

In the outcome, in 1582, Pope Gregory XIII made a proclamation to the following effects:

1. excess days accumulated in the Julian calendar were to be exorcised;
2. years with a date divisible by 4 were to be leap years of 366 days, as in the Julian calendar, except for the "century years," 1700, 1800, etc.;
3. century years were to be leap years only if divisible by 400.

This Gregorian calendar, as it is called, has an average length of 365.2425 days, differing from the tropical year by only 0.0003 days, and therefore requiring about 3,000 years to go out of step by 1 day.

PROBLEM:

If the human species survives for long enough, extra days will very gradually accumulate in the Gregorian calendar, so that some further calendar reform will eventually become necessary. Consider the effect of applying the following rule:

The calendar year is to be 365 days, unless it is predicted ahead of time that during the course of the year the calendar will go out of step with the tropical year by more than one day, in which case the calendar year must have 366 days.

The value quoted above of 365.2422 days for the tropical year has been rounded off to fourth decimal-place accuracy. By consulting an appropriate reference book, determine the full accuracy to which the length of the tropical year is nowadays known. If the above rule were adopted, how far ahead of time could the incidence of leap years be predicted?

Appendix I.2. Logarithms and Magnitudes

In Chapter 2 we were required to take the logarithm k of the observed energy flux f from a star, and we did so by means of the equation $10^k = f$. The logarithm of any positive number, say, x, can easily be obtained in practice by using either logarithmic tables or a pocket calculator. The resulting number is written as log x; so, if we replace f by x and k by log x, we have

$$x = 10^{\log x}.$$

This equation holds good for any positive value of x. It gives the *meaning* of log x, expressed in algebraic form.

The meaning of log x can also be expressed in a geometrical form. In Figure I.1 we have the curve of the reciprocal of x, $1/x$, plotted against x itself. The curve can be considered in two parts, a part to the right of the ordinate drawn

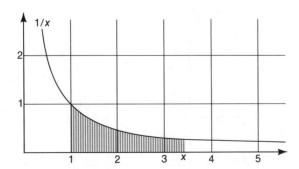

FIGURE I.1.
When x is a number greater than 1, the marked area is known as ln x. The quantity ln 10 is the area for the case 10. The ratio ln x/ln 10 is log x.

at $x = 1$, and a part to the left of that ordinate. We draw also the ordinate at a general point x. The area illustrated in Figure I.1 is called the *natural logarithm* of x, written as ln x. For example, in the case $x = 10$, the area in question gives ln 10. Now, for any general x, take the ratio of areas ln x/ln 10, and define log x to be this ratio,

$$\log x = \frac{\ln x}{\ln 10}.$$

The division here by the area ln 10 is an essentially trivial procedure—it is a normalization process chosen simply to ensure that log 10 shall be equal to unity.

PROBLEM:

Verify that $x = 10$, log $10 = 1$, satisfies $x = 10^{\log x}$. What are the logarithms of 10^2, 10^3, and so on?

The position of the point x in Figure I.1 has been taken to the right of the unit point $x = 1$. We could just as well choose a point x to the left of the unit point, as in Figure I.2. Once again ln x is given by the area shown in Figure I.2, except that a minus sign is added. The convention is that logarithms are positive for x greater than 1, and are negative for x less than 1. At $x = 1$ the logarithm is obviously zero, since there is no area in this particular case.

PROBLEM:

Does the area determining ln x stay finite or does it increase without limit as x decreases more and more closely to zero? For x large and increasing without limit, does the area determining ln x stay finite or does it also increase without limit?

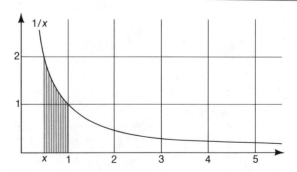

FIGURE I.2.

When x is a number less than 1, the quantity ln x is defined to be minus the area shown here.

Using two well-known properties of numbers,

$$10^a \cdot 10^b = 10^{a+b}, \qquad 10^a \div 10^b = 10^{a-b},$$

we can deduce interesting properties concerning the geometrical interpretation of logarithms. Writing $\log x$ for a and $\log y$ for b, we thus have

$$10^{\log x} \cdot 10^{\log y} = 10^{(\log x + \log y)},$$

$$10^{\log x} \div 10^{\log y} = 10^{(\log x - \log y)}.$$

Since $x = 10^{\log x}$ and $y = 10^{\log y}$, we can write these two equations in the forms

$$xy = 10^{(\log x + \log y)},$$

$$\frac{x}{y} = 10^{(\log x - \log y)}.$$

But like any other positive number xy is equal to $10^{\log xy}$ and x/y is equal to $10^{\log x/y}$, so that

$$10^{\log xy} = 10^{\log x + \log y},$$

$$10^{\log (x/y)} = 10^{(\log x - \log y)},$$

which require

$$\log xy = \log x + \log y,$$

$$\log \frac{x}{y} = \log x - \log y.$$

PROBLEM:
Show that

$$\ln xy = \ln x + \ln y,$$

$$\ln \frac{x}{y} = \ln x - \ln y.$$

In Figure I.3 we have areas corresponding to $\ln x$ and $\ln y$, taking the case where both x and y are numbers greater than 1. The sum of these areas is $\ln x + \ln y$. Suppose we mark off an area equal to this sum, as in Figure I.3. What number do we arrive at along the horizontal axis? The answer must be given by the product xy. If we take the differences of the areas and proceed similarly, we arrive at the ratio x/y. These are curious and by no means obvious geometrical results.

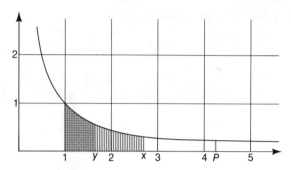

FIGURE I.3.
The area for ln x is denoted by vertical lines, that for ln y by horizontal lines. The area for P is to be the sum of ln x and ln y. What is P?

This diversion on the meaning of logarithmic quantities is of great relevance to astronomy, since data are often given in logarithmic form or in terms of magnitudes, which are really logarithmic quantities but in a disguised form. As an example of the use of logarithms, consider the flux f from a star of luminosity ℓ at distance d. From the discussion of Chapter 2, we have

$$f = \frac{\ell}{4\pi d^2},$$

so that

$$\log f = \log\left(\frac{\ell}{4\pi d^2}\right) = \log\left(\frac{\ell}{4\pi}\right) - \log d^2.$$

Since $\log d^2 = 2 \log d$, we have

$$\log f = -2 \log d + \log\left(\frac{\ell}{4\pi}\right).$$

Recalling the definition of the apparent magnitude m,

$$m = -2.5 \log f + \text{fixed number,}$$

where the fixed number is specified by convention, we get

$$m = 5 \log d - 2.5 \log \frac{\ell}{4\pi} + \text{fixed number.}$$

Suppose we imagine d to change to a different distance D. Then m would change to M, given by

$$M = 5 \log D - 2.5 \log \frac{l}{4\pi} + \text{same fixed number,}$$

and the "fixed number" disappears when we consider the difference $M - \boldsymbol{m}$,

$$M - \boldsymbol{m} = 5 \log D - 5 \log d,$$

i.e.,

$$M = \boldsymbol{m} - 5 \log \frac{d}{D}.$$

Since d is not written here in boldface type, we have not so far considered the distance of the star to be known. But if the actual distance could be determined in some way, we could write

$$M = \boldsymbol{m} - 5 \log \frac{\boldsymbol{d}}{D},$$

which would give the magnitude M that the star would have if it were at distance D instead of at $\boldsymbol{d}$. This procedure is useful for comparing one star with another. We can imagine D to be a standard distance at which we compare one star with another. The choice of D is a matter of convention, the value actually used in practice being $D = 10$ parsecs $= 32.6$ light years. The values of M subject to this convention are called *absolute magnitudes*.

PROBLEMS:

1. Using $\boldsymbol{m} = -26.8$, $\boldsymbol{d} = 1.5 \times 10^{13}$ cm, for the Sun, and remembering that 1 light year $= 9.46 \times 10^{17}$ cm, show that the absolute magnitude of the Sun is $+4.8$.

2. The luminosity L of the Sun is 3.9×10^{23} kilowatts. A certain star in the twin clusters of h and χ Persei has $l = 10^{28}$ kilowatts. What is its absolute magnitude?

Not all radiation is the same, as we are aware from observation of light of different colors. Not all radiation is even visible, as we know from radio waves and from the heat rays we feel but cannot see. It follows that when the magnitude of a star is quoted, we have to be clear what kind of radiation happens to be under consideration. The above discussion, and that of Chapter 2, implied that the radiation in question was all of the visible kind. When this is so, magnitudes and luminosities are prefixed by the word *visual:* visual magnitudes and visual luminosities. However, should radiation of all kinds be involved, invisible as well as visible, the word *bolometric* is used: bolometric luminosities and bolometric magnitudes. For the Sun the distinction is not of overriding importance, because most of the radiant energy emitted by the Sun happens to be visible, so that the visual luminosity of the Sun is nearly the same as its bolometric luminosity. But for red stars emitting a higher proportion of invisible heat radiation than the Sun does, or for blue stars emitting more invisible ultraviolet, the distinction can be very important.

QUESTION:
How far do you consider it to be not accidental that most of the Sun's radiation is visible?

Appendix I.3. Temperatures and the Absolute Scale

When two bodies are placed in contact, energy passes from the hotter one to the cooler one, and energy continues to pass until the two bodies come to the same temperature. A *thermometer* is a measuring device capable of changing its temperature appreciably for a comparatively small energy transfer of this kind. When placed in contact with a larger body whose temperature we wish to measure, a thermometer therefore takes on the temperature of the larger body without disturbing it appreciably. The temperature of the larger body is then inferred from the state of the thermometer—for a mercury thermometer, from the length of the column of mercury within it, which expands and contracts as the temperature rises and falls. However, although practical temperature measurements with a mercury thermometer serve to distinguish different temperatures one from another, the numerical value given to a particular temperature depends not only on the body in question, but also on the expansion properties of mercury, and so contains an extraneous factor in its determination, which it would be preferable to avoid.

The physical meaning of temperature can be better understood by considering a quite different kind of thermometer, one consisting of a moderately diffuse gas composed of suitably simple particles. We place such a gas thermometer in contact with the body whose temperature T we wish to measure, as in Figure I.4. Then we determine T from the equation

$$T = (\text{constant}) \cdot (\text{pressure in gas thermometer}),$$

the pressure here being measured with a gauge. The constant appearing in this equation is fixed by the requirement that two determinations of T, one for melting ice and the other for boiling water, shall differ by exactly 100.

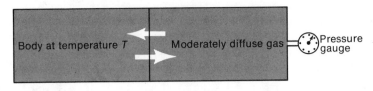

FIGURE I.4.
Measuring the temperature of a body with the aid of a gas thermometer. Heat
can pass freely across the surface of separation of the body and the gas, but the
other surfaces of the body are not considered to be losing, or gaining, heat.

Since the temperature of boiling water depends markedly on the atmospheric
pressure—water boils at lower temperature on a mountain than it does at sea
level—a standard atmospheric pressure must be specified in this determination.
The same standard pressure is also used when measuring T for melting ice—the
temperature for melting also depends on atmospheric pressure, although not
so markedly as for boiling water. A standard atmospheric condition corresponding to a barometric pressure of 76 cm of mercury is used.

Temperatures determined in this way are said to be *absolute,* and are denoted
by the symbol °K. Each integral step of T is called a *degree.* Unlike the mercury
thermometer, which would not give the same result if the mercury were replaced
by another liquid, say, by water, the gas thermometer gives the same temperature determination for all gases composed of suitably simple particles.*

Hot surfaces emit radiation of different kinds according to their temperatures.
Yellow light is emitted at 5,000°K. The light becomes redder at lower temperatures and bluer at higher temperatures. We judge an object of temperature
2,000°K to be very red, and one of 20,000°K to be very blue. Just from the
observed color of a star, the temperature at its surface can be estimated. Very
cool stars with surface temperatures below 2,000°K are known, as are hot blue
stars with surface temperatures above 20,000°K.

PROJECT:
Compare the colors of the two bright stars Betelgeuse and Rigel in the constellation of Orion.

*A gas thermometer cannot be used if changing the temperature changes the chemical nature
or internal structure of the particles. This is the essential limitation on the nature of the particles
forming the gas.

TABLE I.1. *Table of the nearest stars*[a]

Catalogue designation(s)[b]		Right ascension[c]	Declination	Visual magnitude[d]	$B - V$[e]	Distance (parsecs)[f]	Absolute magnitude[g]	Mass ($\odot$)[h]	Radius (solar radius)	Remarks[i]
		1900								
Grm 34 = 43°44	A	00h 13m	43° 27'	8.08	1.55	3.60	10.30			Binary with separation on sky of 38 arc second. A is itself an unresolved binary.
(= CC 19)	B			11.05	1.78		13.27			
β Hyi		00 20	−77 49	2.79	0.61	6.54	3.71			
η Cas	A	00 43	57 17	3.44	0.57	5.52	4.73	0.85	0.84	Binary with period 490 years and separation 12".
	B			7.25			8.54	0.52		
v. Maanen = Wolf 28		00 44	04 55	12.34	0.55	4.26	14.19			White dwarf
UV Cet	A	01 34	−18 28	12.41	1.9	2.67	15.28	0.044		Binary with period about 200 years and separation 5".
	B			12.95	1.9		15.87	0.035		
τ Cet		01 39	−16 28	3.50	0.72	3.64	5.70			
82 Eri = e Eri		03 16	−43 27	4.23	0.71	6.41	5.19			
ε Eri		03 28	−09 48	3.74	0.87	3.30	6.14			
o² Eri = 40 Eri	A	04 11	−07 49	4.44	0.81	5.00	5.96			Triple system B is white dwarf. Period of BC is about 250 years.
	B			9.64	0.03		11.16	0.44	0.018	
	C			11.05			12.57	0.21	0.43	
Kapteyn = −45° 1841		05 08	−44 59	8.9		3.98	10.9			
HD36395 = −3° 1123		05 26	−03 42	7.97	1.48	6.13	9.03			
Ross 47		05 36	12 29	11.58		6.10	12.65			

												Remarks
(= CC 390)	B	06	24	-02	44	14.8			16.8	0.08		16.5 years and separation 1".
Sirius = α CMa	A	06	41	-16	35	-1.46	0.01	2.67	1.41	2.31	1.8	Binary with period 49.7 years and separation 7".6. Component B is a white dwarf.
	B					8.67	0.04		11.54	0.98	0.022	
Wolf 294		06	48	33	24	10.15		5.92	11.29			
Ross 986		07	03	38	43	11.68		5.81	12.86			
Luyten = 5° 1668		07	22	05	32	9.92		3.76	12.02			
Procyon = α CMi	A	07	34	05	29	0.35	0.40	3.48	2.64	1.75	1.7	Binary with period 40.6 years and separation of 4".5. Component B is a white dwarf.
	B					10.8	0.5		13.1	0.64	0.01	
L 745 - 46	A	07	36	-17	10	13.06		6.10	14.14			Binary with separation of 21". Component A is a white dwarf.
	B					17.6			18.7			
L 97 - 12		07	53	-67	30	14.9		5.88	16.1			White dwarf
Ross 619		08	06	09	11	12.88		6.62	13.78			
LFT 571 = L 674 - 15		08	08	-21	15	13.8		6.02	14.9			
53° 1320	A	09	08	53	07	7.68	1.44	6.13	8.74			Binary with about 1,000 years and separation of 19".
53° 1321	B					7.77			8.82			
Grm 1618 = 50° 1725		10	05	49	57	6.60	1.37	4.50	8.33			
AD Leo = 20° 2465		10	14	20	22	9.41	1.55	4.72	11.04			
Wolf 359		10	52	07	36	13.66		2.35	16.80			
Lal 21185 = 36° 2147		10	58	36	38	7.47	1.51	2.51	10.42	0.35		Invisible component of small mass.

TABLE I.1. *Table of the nearest stars*[a] *(continued)*

Catalogue designation(s)[b]		1900 Right ascension[c]	Declination	Visual magnitude[d]	B − V[e]	Distance (parsecs)[f]	Absolute magnitude[g]	Mass (☉)[h]	Radius (solar radius)	Remarks[i]
44° 2051	A	11h 01m	44° 02'	8.76	1.54	5.81	9.94			Binary with separation of 28".
(= WX UMa)	B			14.8			16.0			
CC 658		11 40	−64 17	12.5		4.93	14.0			White dwarf
AC 79° 3888		11 41	79 14	10.92		5.05	12.41			
Ross 128		11 43	01 23	11.13		3.36	13.50			
L 68 − 28	A	12 23	−70 56	15.7		6.58	16.6			Binary with separation of 15".
L 69 − 29	B			17.7			18.6			
Wolf 424	A	12 28	09 34	12.63		4.35	14.44			Binary with separation of 0".5.
	B			12.7			14.5			
15° 2620		13 41	15 26	8.49	1.44	4.98	10.01			
Proxima Cen		14 23	−62 15	10.7		1.31	15.1	0.1		This is the nearest star.
−11° 3759		14 29	−12 06	11.38		6.33	12.37			
α Cen	A	14 23	−60 25	0.00	0.69	1.33	4.39	1.09	1.23	Binary with period of 80.1 years and separation 17".7.
	B			1.38			5.76	0.88	0.87	
−20° 4125	A	14 52	−20 58	5.82	1.12	5.81	7.00			Binary with separation of 20".
−20° 4123	B			8.10			9.28			
−40° 9712		15 26	−40 54	10.1		6.02	11.1			

Name	Comp.	R.A. h	m	Dec. °	′	m	B–V	r (pc)	M	Mass	Notes
	B	16	50	−08	09	9.8			10.7	0.34	C separated from AB by 72″.
(= Wolf 629)	C					11.76		6.25	12.78	0.25	AB a close pair with period 1.7 year.
+45° 2505	A	17	09	45	50	9.95	0.85	6.25	10.97	0.31	Binary with period of 13.1 years and separation of 0″.7.
(= Fu 46)	B					10.31	1.14		11.33	0.25	
36 Oph	A	17	09	−26	27	5.07		5.68	6.31		Separation of AB is 5″. Separation of BC is 12′12″.
(= −26° 12026)	B	17	10	−26	24	5.11		5.81	6.35		
(= −26° 12036)	C					6.34			7.52		
−46° 11540		17	21	−46	47	9.4	1.5	4.69	11.0		
−44° 11909		17	30	−44	14	11.1		4.78	12.7		
68° 946		17	37	68	26	9.13	1.52	4.93	10.67		
UC 48		17	38	−57	14	12.9		5.99	14.0		
Barnard = 4° 3561		17	53	04	25	9.53	1.75	1.83	13.21		
70 Oph	A	18	01	02	31	4.22	0.87	5.21	5.64	0.89	Binary with period of 87.8 years and separation of 4″.5.
(= 2° 3482)	B					5.94			7.36	0.68	
59° 1915	A	18	42	59	29	8.90	1.54	3.53	11.16		Binary with separation of 17″.
(= Σ 2398)	B					9.69	1.58		11.95		
Ross 154		18	44	−23	54	10.6		2.86	13.3		
4° 4048	A	19	12	05	02	9.13	1.49	5.85	10.29		Binary with separation of 1′14″.
	B					18.0			19.2		
L 347-14		19	13	−45	42	13.7		5.92	14.8		
σ Dra		19	33	69	29	4.69	0.80	5.68	5.92		
Altair = α Aql		19	46	08	36	0.75	0.25	5.05	2.23		
δ Pav		19	59	−66	26	3.56	0.75	5.88	4.71		

TABLE I.1. *Table of the nearest stars (continued)*

Catalogue designation(s)[b]		Right ascension[c] (1900)	Declination (1900)	Visual magnitude[d]	B − V[e]	Distance (parsecs)[f]	Absolute magnitude[g]	Mass (☉)[h]	Radius (solar radius)	Remarks[i]
−36° 13940	A	20h 05m	−36° 21′	5.33	0.85	5.81	6.51			Separation 7″.
(= HR 7703)	B			11.5			12.7			
−45° 13677		20 07	−45 28	8.0	1.44	6.29	9.0			
61 Cyg	A	21 02	38 15	5.20	1.21	3.42	7.53	0.59		Separation 24″.6. Period 720 years.
	B			6.03	1.40		8.36	0.50		
−39° 14192		21 11	−39 15	6.69	1.42	3.91	8.73			
−49° 13515		21 27	−49 26	8.9		4.57	10.6			
ε Ind		21 56	−57 12	4.73	1.05	3.50	7.01			
Kruger 60 = DO Cep (= 56° 2783)	A	22 24	57 12	9.83	1.63	4.00	11.82	0.27	0.51	Binary with period 45 years and separation 2″.4.
	B			11.37			13.36	0.16		
L 789-6		22 33	−15 52	12.58		3.38	14.93			
−21° 6267	A	22 33	−21 08	9.3		4.57	11.0			Separation 23″.
	B			11.0			12.7			
43° 4305		22 42	43 49	10.05	1.39	5.05	11.53			
−15° 6290 (= Ross 780)		22 48	−14 47	10.16	1.60	4.85	11.73			

Ross 248	23 37	43 39	12.25	1.8	3.16	14.75
1° 4774	23 44	01 52	8.99	1.49	6.13	10.05
−37° 15492	23 59	−37 51	8.59	1.48	4.57	10.29

$B - V$	Surface temperature (°K)
−0.2	18,800
0.0	10,800
0.2	8,190
0.4	6,820
0.6	5,920
0.8	5,200
1.0	4,530
1.2	3,920
1.4	3,480

[a] Adapted from C. W. Allen, *Astrophysical Quantities*, University of London, 1963. His 1973 (revised) edition gives R.A. and decl. values for the year 1950.

[b] The same star is sometimes listed in more than one catalogue, and these equivalences are given. Notice the peculiar nature of certain of the catalogue descriptions, for example, o^2 Eridani = 40 Eri.

[c] The stars have been arranged not by consecutive catalogue entries, but by increasing right ascension. Why? Which of these stars would you expect to be in the night sky during the month of March? The right ascension and declination values are for 0.00h, Jan. 1, 1900.

[d] This magnitude is for light in the visible range.

[e] The value of $B - V$ determines the surface temperature in accordance with the equivalences

[f] The distances in parsecs may be converted to light years by multiplying them by 3.26, since 1 parsec = 3.26 light years.

[g] This is the visual magnitude which the star in question would have if it were situated at a distance (of 10 parsecs instead of at the actual distance given in column 6. These absolute magnitudes are thus standardized with respect to each other. Their differences represent inherent differences between one star and another.

[h] Mass values are available for only a few stars. They are given as a ratio to the Sun's mass. Notice that most of the nearby stars are less massive than the Sun.

[i] Notice the rather large fraction of the stars that belong to binary and to triple systems. Notice also that there are seven white dwarfs.

TABLE I.2. *Table of the brightest stars*[a]

Name		1900 Right ascension	Declination	Visual magnitude	B − V	Absolute magnitude	Distance (parsecs)
Alpheratz	α And	00h 03m	28° 32'	2.07	−0.07	−0.5	31
Caph	β Cas	00 04	58 36	2.26	0.34	1.5	14
Ankaa	α Phe	00 21	−42 51	2.37	1.07	0.2	27
Schedar	α Cas	00 35	55 59	2.20	1.16	−1.3	50
Diphda	β Cet	00 39	−18 32	2.04	1.01	0.8	18
Cih	γ Cas	00 51	60 11	2.15	−0.2	−0.9	40
Mirach	β And	01 04	35 05	2.07	1.62	0.2	24
Polaris	α UMi	01 23	88 46	2.02	0.6	−4.5	200
Achernar	α Eri	01 34	−57 45	0.49	−0.17	−2.2	35
Almach	γ And	01 58	41 51	2.16	1.3	−2.3	80
Hamal	α Ari	02 02	22 59	2.00	1.17	0.3	22
Mira	o Cet	02 14	−03 26	2.00	1.5	−1.0	40
Menkar	α Cet	02 57	03 42	2.53	1.16	−1.0	50
Algol	β Per	03 02	40 34	2.10	−0.05	−0.5	31
Mirfak	α Per	03 17	49 30	1.80	0.48	−4.1	150
Aldebaran	α Tau	04 30	16 19	0.80	1.55	−0.8	21
Capella	α Aur	05 09	45 54	0.09	0.81	−0.6	14
Rigel	β Ori	05 10	−08 19	0.11	−0.05	−7.1	270
Bellatrix	γ Ori	05 20	06 16	1.63	−0.22	−4.1	140
El Nath	β Tau	05 20	28 32	1.65	−0.13	−2.9	80
Mintaka	δ Ori	05 27	−00 22	2.19	−0.21	−6.0	450
Arneb	α Lep	05 28	−17 54	2.58	0.22	−4.8	300
Alnilam	ε Ori	05 31	−01 16	1.70	−0.18	−6.8	500
Alnitak	ζ Ori	05 36	−02 00	1.79	−0.21	−6.2	400
Saiph	κ Ori	05 43	−09 42	2.06	−0.16	−7.1	700
Betelgeuse	α Ori	05 50	07 23	0.4	1.85	−5.9	180
Menkalinan	β Aur	05 52	44 56	1.89	0.04	−0.2	26
Mirzam	β CMa	06 18	−17 54	1.96	−0.23	−4.5	200
Canopus	α Car	06 22	−52 38	−0.72	0.16	−6.2	400
Alhena	γ Gem	06 32	16 29	1.93	0.00	−0.5	30
Sirius	α CMa	06 41	−16 35	−1.44	−0.01	1.41	2.7

Name	Designation								
Castor	α Gem	07	28	32	06	1.56	0.05	0.8	14
Procyon	α CMi	07	34	05	29	0.36	0.41	2.7	3.5
Pollux	β Gem	07	39	28	16	1.15	1.01	1.0	10.7
Naos	ζ Pup	08	00	−39	43	2.23	−0.27	−7.3	800
	γ Vel	08	06	−47	03	1.85	−0.25	−4.2	160
Avior	ε Car	08	20	−59	11	1.94	1.2	−3.1	100
Suhail	δ Vel	08	42	−54	21	1.93	0.04	0.1	23
	λ Vel	09	04	−43	02	2.23	1.7	−4.3	200
Miaplacidus	β Car	09	12	−69	18	1.68	−0.01	−0.4	26
Scutulum	ι Car	09	14	−58	51	2.24	0.18	−4.2	180
	κ Vel	09	19	−54	35	2.45	−0.16	−3.0	120
Alphard	α Hya	09	23	−08	14	2.05	1.43	−0.7	35
Regulus	α Leo	10	03	12	27	1.34	−0.11	−0.8	26
Algeiba	γ Leo	10	14	20	21	2.02	1.2	−0.5	32
Merak	β UMa	10	56	56	55	2.36	−0.02	0.6	23
Dubhe	α UMa	10	58	62	17	1.81	1.06	−0.6	30
Zosma	δ Leo	11h	09m	21°	04'	2.55	0.12	0.8	23
Denebola	β Leo	11	44	15	08	2.13	0.08	1.6	13
Phecda	γ UMa	11	49	54	15	2.43	0.00	−0.1	32
Gienah	γ Crv	12	11	−16	59	2.58	−0.09	−2.4	100
Acrux	α Cru	12	21	−62	33	0.83	−0.26	−3.7	80
Gacrux	γ Cru	12	26	−56	33	1.68	1.58	−2.5	70
Muhlifain	γ Cen	12	36	−48	25	2.16	−0.01	−1.7	60
Mimosa	β Cru	12	42	−59	09	1.29	−0.25	−4.3	130
Alioth	ε UMa	12	50	56	30	1.78	−0.02	−0.2	25
Mizar	ζ UMa	13	20	55	27	2.12	0.03	0.0	26
Spica	α Vir	13	20	−10	38	0.97	−0.23	−3.1	65
	ε Cen	13	34	−52	57	2.34	−0.23	−3.6	150
Alcaid	η UMa	13	44	49	49	1.86	−0.19	−2.3	70
Hadar	β Cen	13	57	−59	53	0.63	−0.24	−5.0	130
Menkent	θ Cen	14	01	−35	53	2.07	1.02	0.9	17
Arcturus	α Boo	14	11	19	42	−0.05	1.24	−0.2	11
	η Cen	14	29	−41	43	2.39	−0.21	−3.0	120
Rigil Kent	α Cen	14	33	−60	25	−0.27	0.71	4.2	1.3
	α Lup	14	35	−46	58	2.5	−0.22	−2.5	100
Izar	ε Boo	14	41	27	30	2.39	0.93	−0.6	40

TABLE I.2. *Table of the brightest stars (continued)*

Name		Right ascension		Declination		Visual magnitude	$B - V$	Absolute magnitude	Distance (parsecs)
		1900							
Kochab	β UMi	14h	51m	74°	34′	2.04	1.49	−0.6	33
Alphecca	α CrB	15	30	27	03	2.22	−0.02	0.5	22
Dzuba	δ Sco	15	54	−22	20	2.32	−0.14	−4.0	180
Acrab	β Sco	16	00	−19	32	2.52	−0.09	−4.0	200
Antares	α Sco	16	23	−26	13	0.94	1.83	−4.7	130
	ζ Oph	16	32	−10	22	2.56	0.00	−3.4	160
Atria	α TrA	16	38	−68	51	1.93	1.43	−0.4	29
	ε Sco	16	44	−34	07	2.29	1.15	0.6	22
Sabik	η Oph	17	05	−15	36	2.44	0.05	0.8	21
Shanla	λ Sco	17	27	−37	02	1.60	−0.23	−3.2	90
	θ Sco	17	30	−42	56	1.86	0.38	−4.0	150
Ras-Alhagne	α Oph	17	30	12	38	2.07	0.15	0.9	17
	κ Sco	17	36	−38	59	2.39	−0.21	−3.3	140
Eltanin	γ Dra	17	54	51	30	2.21	1.54	−0.8	40
Kans Australis	ε Sgr	18	18	−34	26	1.81	−0.02	−1.7	50
Vega	α Lyr	18	34	38	41	0.03	0.00	0.5	8.1
Nunki	σ Sgr	18	49	−26	25	2.09	−0.20	−2.4	80
	ζ Sgr	18	56	−30	01	2.57	0.09	−0.4	40
Altair	α Aql	19	46	08	36	0.77	0.22	2.3	4.9
Peacock	α Pav	20	17	−57	03	1.94	−0.20	−2.9	90
Sadir	γ Cyg	20	19	39	56	2.22	0.66	−4.8	250
Deneb	α Cyg	20	38	44	55	1.25	0.08	−7.2	500
Gienar	ε Cyg	20	42	33	36	2.46	1.03	0.6	24
Alderamin	α Cep	21	16	62	10	2.43	0.23	1.5	15
Enif	ε Peg	21	39	09	25	2.38	1.56	−4.6	250
Al Na'ir	α Gru	22	02	−47	27	1.75	−0.14	−0.2	25 ?
	β Gru	22	37	−47	24	2.16	1.62	−2.6	90
Formalhaut	α PsA	22	52	−30	09	1.16	0.09	1.9	7.0
Scheat	β Peg	22	59	27	32	2.50	1.7	−1.4	60
Markab	α Peg	23	00	14	40	2.49	−0.04	0.0	32

[a] Adapted from C. W. Allen, *Astrophysical Quantities*, University of London, 1963. Footnotes c, d, e, f, and g of Table I.1 also apply to the equivalent columns in this table. It might at first sight be expected that the brightest stars would be the nearest ones, but a comparison of this table with Table I.1 shows this not to be true. Only four stars appear in both tables. A similar assumption is often made concerning radiosources (see Chapters 13 and 14), and this could also be false. Many of the stars in this table were named by Arabian astronomers, but some were not. Which were not? Which five stars are intrinsically the brightest?

From the fact that our galaxy, an Sb spiral, contains some 10^{11} stars, it evidently possesses a mass of the order of 10^{11} times that of the Sun. Other Sb spirals, such as the Andromeda Nebula, M31, have masses of this same order. The intrinsically brightest of the elliptical galaxies have still larger masses, however, upwards of 10^{12} times the Sun—i.e., 10^{12} times as many atoms as are present in the Sun. Galaxies of type Sc, on the other hand, tend to have lower masses than our galaxy, of the order of 10^{10} times the Sun, whereas galaxies like the dwarf members of the Local Group have masses that are smaller still, of order 10^9 times the Sun. It is not known what the lower limit to the masses of very faint dwarf galaxies might be; there could be dwarfs with masses of 10^8, or even 10^7, times the Sun.

About 10^9 giant galaxies are observable, giants like our own galaxy and like the bright ellipticals. The total quantity of material contained in all these observable galaxies is at least 10^{20} times the Sun. Most of this material is in the form of stars. We do not know what proportion of the stars possess planetary systems, but the considerations of a later chapter suggest that the fraction may well be very appreciable. It seems, then, that the number of planetary systems like our own that exist in the universe may be truly immense.

General Problems and Questions

1. What differences are there between the Julian and Gregorian calendars?

2. Draw the curve of $1/x$ using graph paper. By counting graticule squares, estimate both ln 2 and ln 10. Then by taking the ratio ln 2/ln 10, obtain log 2. Compare your result with the value of log 2 given in tables. (To obtain reasonable accuracy, use graph paper with small squares.)

3. What is an integer? With a and b both integers, prove that $10^a \times 10^b = 10^{a+b}$, $10^a \div 10^b = 10^{a-b}$.

4. Explain the difference between (a) an apparent magnitude and an absolute magnitude, (b) a visual magnitude and a bolometric magnitude.

5. How is temperature measured in the absolute (K) scale?

6. Adding all galaxies, about how many atoms are contained in the galaxies which can be distinguished with a large telescope?

References for Section I

For further reading, see G. O. Abell, *Exploration of the Universe* (New York: Holt, Rinehart, and Winston, 2d ed., 1969); R. H. Baker and L. W. Fredrick, *Astronomy* (Princeton, N.J.: Van Nostrand, 9th ed., 1971); D. H. Menzell, *Astronomy* (New York: Random House, 1970); Ludwig Oster, *Modern Astronomy.* (San Francisco: Holden-Day, 1973).

The classic general texts are: H. N. Russell, R. S. Dugan, and J. Q. Stewart, *Astronomy I and II* (New York: Ginn, 1926–1927); C. A. Young, *Manual of Astronomy* (New York: Ginn, 1904).

The classic book on galaxies is E. P. Hubble, *The Realm of the Nebulae* (New Haven: Yale University Press, 1936).

Section II:
Basic Ideas and
Instruments

Isaac Newton, 1642–1727. From a mezzotint executed in 1740 by
James Macardel from a portrait by Enoch Seeman, courtesy of the
Ronan Picture Library and the Royal Astronomical Society.

Chapter 4:
Particles and Radiation
in General

§4.1. Particle Paths

We can think of matter as made up of discrete units or "particles," with each particle having a trajectory or path. Our first objective will be to learn how to describe such a path in a precise way.

For a particle moving in a plane, we can proceed as in Figure 4.1. At a certain moment, the particle might be at the point P. In terms of the lines OX, OY of Figure 4.1, we could say that the particle lies to the "east" of the line OY by a distance x and to the "north" of OX by the distance y. A specification of x and y would then determine the position of the particle at the moment of time in question. At another time the particle will in general be somewhere else, at Q, say. Again by giving the distance by which Q is "east" of OY and "north" of OX we can describe Q precisely. Proceeding in this way for all moments of time, we can describe completely the path that the particle follows, provided we know how both the "easterly" direction x and the "northerly" direction y change as time goes on.

To allow for the possibility that the path may cross one, or both, of the lines OX, OY, we must consider that x and y can have negative values. For instance, at the point R of Figure 4.1, although x remains positive, y has become negative. The conventions to be adopted for the signs of the distances x and

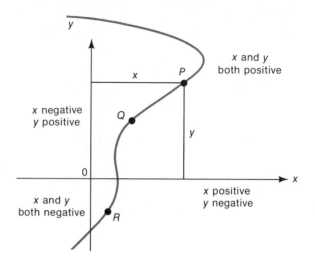

FIGURE 4.1.

P, Q, R, are points on the path of a particle which moves in a plane. The position of *P* is determined by a specification of the distances *x* and *y*, and similarly for all other points on the path. Notice the signs to be attributed to *x* and *y*.

y are shown in Figure 4.1. Without these conventions there would be ambiguities in deciding in which of the four quadrants a point was situated.

As well as having this kind of "north-south," "east-west" motion, a particle can also move "up" or "down." We deal with this more complicated three-dimensional situation by adding a third direction, *OZ*, as in Figure 4.2. The position of a typical point *P* on the path of a particle is now specified by three numbers, *x, y*, and *z*. The first two of these are measured with respect to *OX*, *OY*, in the same way as before, whereas *z* is the distance by which *P* is "above" or "below" the plane formed by *OX* and *OY*, with the convention that *z* is positive if *P* is "above" and negative if *P* is "below." Once again the path of the particle is determined by specifying how *x, y, z* change with time.

It is clear from Figure 4.1 that the lines *OX, OY* are intended to be perpendicular to each other. Because of the perspective problem always involved in drawing a three-dimensional diagram, this is not as clear in Figure 4.2. We note therefore that the lines *OX, OY, OZ* are to be taken as mutually perpendicular to each other.

Only in idealized problems, for example that of a single particle moving around an isolated star, do we have the precise case of motion in a plane. In the actual world, particles always move in all three spatial dimensions, although in some cases—as with an actual planet moving around the Sun—the deviation from motion in a plane may be quite small. Motions in all the spatial dimensions *x, y, z* are commonplace when electrically charged particles move in a magnetic field. For example, the motion of a charged particle in a uniform magnetic

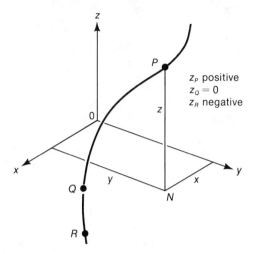

FIGURE 4.2.

P, Q, R, are points on the path of a particle moving in all three of the spatial dimensions. The position of *P* is now determined by a specification of *x, y*, and *z*, and similarly for all other points on the path.

field* has the form of a helix, illustrated in Figure 4.3. This is the kind of motion, possessed by fast-moving electrons, that gives rise to cosmic radio-wave emission. Such emission forms the basis of *radioastronomy*, a branch of astronomy we shall study in detail in a later chapter.

The method used above for drawing the path of a particle, although it tells us the geometrical form of the path, does not display the time-behavior of the particle. We have no idea from Figure 4.2 whether the particle moves uniformly along its path, covering equal distances in equal times, or whether it speeds up over some parts of the path and slows down over other parts. To remedy this defect, we contemplate adding a fourth dimension, of time, to Figure 4.2. Unfortunately most people, myself included, find it already difficult to visualize geometrical relationships in three dimensions, let alone in four dimensions. So such a procedure is not usually very helpful.

When the possibility that life may exist in other places in our galaxy comes under discussion, it is often speculated that creatures more intelligent than ourselves may exist. What form could such an intelligence take? Here is an example. A creature who could think geometrically in four dimensions as easily as we do in two dimensions would have an enormous advantage over us in its understanding of the basic structure of the world.

Rejecting the invitation to add a fourth, time dimension to Figure 4.2, we can adopt a device similar to that used in drawing plans for the construction

---

*A uniform magnetic field is one that is the same everywhere.

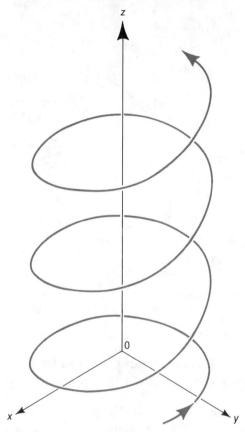

FIGURE 4.3.
The helical path of an electron moving in a uniform
magnetic field which points in the direction *OZ*.

of a building. Such plans consist of a number of drawings, with each drawing in two dimensions only. There are separate drawings for different elevations and for different floors of the building. Still using Figure 4.2 to describe the form of a particle path, we supplement this figure by three more graphs, each in two dimensions only. One supplementary graph gives x plotted against the time t, another gives y against t, and the third plots z against t. Thus for the motion of a charged particle in a uniform magnetic field, in addition to Figure 4.3, which depicts the spatial path, we have the additional (x, t), (y, t), and (z, t) graphs shown in Figure 4.4. These tell us how the particle moves with respect to time.

If we are not called on to represent an explicit case, as in Figure 4.4, it will usually be sufficient to consider only one of the three supplementary graphs, say, the (x, t) graph. This we shall often do for the purpose of general argument. We keep at the back of our minds that a similar argument also applies to the other two graphs. Or, more subtly, we could replace Figure 4.4 by Figure 4.5,

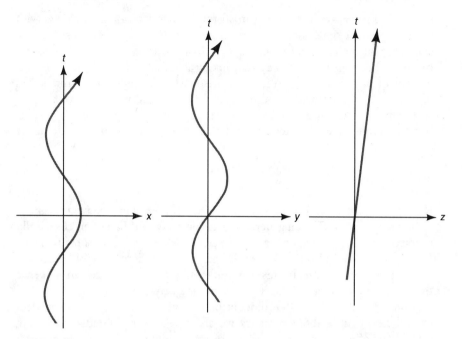

FIGURE 4.4.
To represent the time-dependence of the motion of Figure 4.3, it is necessary either to draw a four-dimensional representation (which is difficult) or to separate the motions with respect to x, y, and z into three graphs. Notice that z increases uniformly with the time t, whereas both x and y oscillate, but not in step with each other.

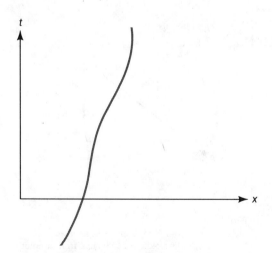

FIGURE 4.5.
Where we are not concerned with the precise specification of a path, it is sufficient to draw only one of the (x, t), (y, t), (z, t) graphs.

imagining Figure 4.5 to be a plot in all four dimensions—that is to say, with the three spatial dimensions, x, y, z, defined in Figure 4.2, collapsed into a single spatial dimension. In spite of the obvious disadvantage of thus losing two spatial dimensions, this way of thinking is frequently useful in the development of physical ideas. With the proviso that the single spatial dimension of Figure 4.5 should really be three spatial dimensions, we now have a representation of both space and time, i.e., of *spacetime*. The path of the particle in this representation is often referred to as the *world line* of the particle.

§4.2. Radiation

Radiation travels from one particle to another in the manner shown schematically in Figure 4.6. From a general point P on the world line of an electrically charged particle a, radiation travels to particle b, reaching it at some point Q. Provided particle b also has an electric charge, the motion of b will be affected by the radiation from a. In this way the motion of one particle can influence the motion of another. We have an *interaction* between particles.

Although radiation is often thought of as having an existence independent of particles, a little thought shows we are never aware of radiation except through its effect on particles. All problems involving radiation can be dealt with in terms of the interaction picture of Figure 4.6—i.e., can be worked out in a clear-cut way. Let us consider an important example.

Suppose the particle a moves only in one spatial dimension, say, x, and suppose the motion is a simple oscillation, as in Figure 4.7. We can count the number of oscillations that occur in some specified time-interval. Then we take

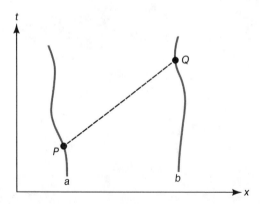

FIGURE 4.6.
The particles a and b are both considered to have electric charge. The radiative influence from a point P on the path of a reaches particle b at some point Q, the time associated with Q being later than that associated with P.

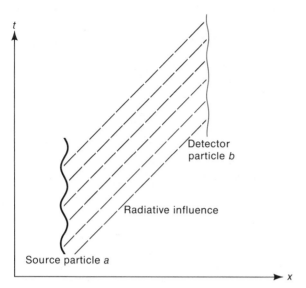

FIGURE 4.7.
Particle a is set in regular oscillation with frequency ν. Provided the speed of motion of particle a is small compared to the speed of light, the radiative effect of the oscillation of a is to cause b also to oscillate with the same frequency ν. Because the oscillations of b occur later than those of a, we refer to a as the source particle and to b as the detector particle.

the ratio, which we denote by ν,

$$\nu = \frac{\text{number of oscillations}}{\text{time-interval}}.$$

This gives the number of oscillations per unit time. The quantity ν is known as the *frequency* of the oscillation.

Next we insert the important condition that the speed of motion of the particle a is to be significantly less than the speed of light. The case in which the speed of motion approaches that of light will be considered separately at a later stage.

PROBLEM:
Writing 2α for the full swing of x in the oscillation of Figure 4.7, show that the average speed of particle a is $4\alpha\nu$. The oscillations of electrons in atoms have values of α that are typically in the range 10^{-10} to 10^{-8} cm. Remembering that the speed of light is 3.0×10^{10} cm per second, and requiring the speeds of electrons in atoms to be not more than one-tenth of that of light, what is the permitted maximum of ν for radiation from atoms?

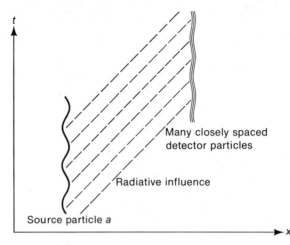

FIGURE 4.8.
A source particle sets many detector particles in oscillation, the swing of the latter being in general very small compared to the swing (amplitude) of the source particle.

With both particles a and b having electric charge, the situation is that the radiative interaction from a causes b to oscillate also with frequency ν, as is indicated in Figure 4.7, where for simplicity it is supposed that particle b has no motion other than that which arises from the influence of a. Depending on the value of ν, we say that the radiation falls into one or another of the categories set out in Table 4.1. These categories are man-made, and have arisen from the differing experimental procedures used in practice for examining the various ranges of ν.

TABLE 4.1.
Forms of radiation according to conventional designations

Name	Frequency (oscillations per second)	Method of detection
Radio	less than 3×10^9	electronic
Short waves (Microwaves)	3×10^9 to 3×10^{11}	electronic
Infrared	3×10^{11} to 3.75×10^{14}	heat
Visible	3.75×10^{14} to 7.5×10^{14}	eye, photography, electronic
Ultraviolet	7.5×10^{14} to 3×10^{16}	photography, electronic
Soft x-rays	3×10^{16} to 2×10^{17}	electronic, photography
Harder x-rays	2×10^{17} to 3×10^{19}	ionization of gases
γ-rays	above 3×10^{19}	ionization of gases

We will refer from here on to particle *a* as the *source* particle and to particle *b* as the *detector* particle. In practice, many particles, not just one, are involved at the detector. This is indicated in Figure 4.8, in which several detector particles have the same frequency of oscillation as the source particle. The problem is now complicated, however, by the fact that radiative interactions arise between the detector particles themselves, as in Figure 4.9. Although these further interactions introduce a conceptual complication, they permit a wide range of practical devices to be constructed. In the example of Figure 4.9, the detector particles are considered to be so positioned in relation to the source that they can themselves influence a further particle *c*, in such a way that the oscillatory swing of particle *c* is considerably larger than the swing of a single detector particle taken by itself would be. The radiation is then said to be *focused*. Indeed, Figure 4.9 illustrates the principle of the *telescope*. Although the practical aspects of telescope design is a topic for the next chapter, we note here that the many detector particles should be so positioned that focusing occurs for a wide range of values of the oscillation frequency *v*. This usually is done by placing the detector particles at the surface of a suitably shaped mirror.

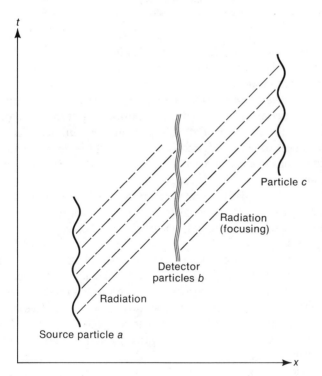

FIGURE 4.9.
The small amplitude of oscillation of many detector particles *b* can be used to make *c* oscillate with a larger amplitude. This is the basic principle of the telescope.

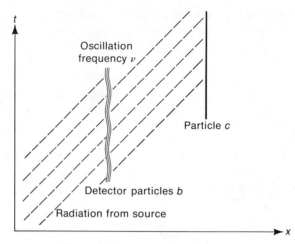

FIGURE 4.10.
When the particle *c* of Figure 4.9 is *prevented* from moving unless the oscillation frequency *ν* of the detector particles *b* is closed to some assigned frequency *ω*, which can be determined by the observer, we have the principle of the spectroscope. In the situation depicted here, *ν* is not the same as *ω*, and so particle *c* cannot oscillate.

It is possible to arrange detector particles in such a way that they produce an effect very different from that of the telescope. They can be so positioned that they do *not* produce an oscillation of particle *c* in Figure 4.10, *unless ν happens to be close to some known value, say, ω*. The value of *ω* in this kind of device is sometimes determined by the positioning of the detector particles, sometimes by the position of *c*, and sometimes in an electronic way. The important aspect of such a device is that *ω* should be known and should be variable at the observer's command. *Spectroscopes, diffraction gratings, electronic receivers,* are all devices of this kind.

This type of device has the important property of determining *ν*. We simply vary *ω* until particle *c* is observed to oscillate. In this way the frequency of radiation from a distant source particle can be determined, say, from a particle in a distant star.

Just as in practice more than one detector particle must be considered, so there are usually many source particles. Remarkable effects occur when the locations of the many source particles relative to each other can be adjusted. Such adjustment is rarely possible—the astronomer is not free to adjust the source particles in a distant star, nor is it normally possible to control the positions of the source particles in a hot gas, even in the laboratory. Exceptional cases do occur, however, in what are called *coherent* phenomena, in which the source particles arrange themselves automatically so as to act in concert with each other. The *laser* is probably the best-known example of a process of this kind. Since the positioning is done by the particles themselves, rather than by the experimenter, coherent processes are actually found to occur occasionally in astronomy, in the emission from gas clouds in our own galaxy. The frequency ν of the radiation in such cases is low. Unlike the laser, where ν falls in the range of visible light, ν for coherent radiation from the galactic gas clouds falls in the radio and microwave ranges. The process, physically similar to the laser, is referred to for these wavelength ranges as a *maser*.

We have been concerned so far only with particles that move slowly compared to light. When a source particle oscillates at a speed close to light, the detector particles b no longer oscillate with the frequency ν of the source particle. The situation, illustrated in Figure 4.11, can be understood by considering the variable-frequency device of Figure 4.10, which had the property that oscillations were *not* set up in the particle c unless the oscillation frequency of the detector particles b was close to an assigned frequency ω. For source particles of high speed, it is found that oscillations occur in the particle c over a wide range of assigned values of ω, even if ω is much larger than ν.

Figure 4.12 shows an object, the Crab Nebula, containing electrons moving at speeds close to light. The electrons follow helical paths in a magnetic field (*cf.* Figure 4.3). In typical cases, the oscillations implied by this helical motion have source frequencies in the range from 10 cycles per second up to perhaps 10^3 cycles per second. Yet oscillations of particle c in a detecting device like that in Figure 4.10 are observed for assigned values of ω from about 10^7 cycles per second in the radio band up to frequencies in the infrared and optical bands, and even up to frequencies for x-rays and γ-rays, i.e., up to more than 10^{18}

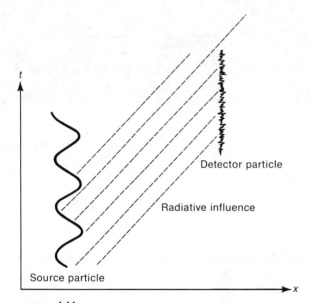

FIGURE 4.11.
The speed of the source particle is here considered to be close to the speed of light. The oscillation of a detector particle b are now much faster and more complex than those of the source particle.

FIGURE 4.12.
The Crab Nebula, about 5 light years in diameter. The light from the inner part of the Crab, shown separately in the lower half of the figure, comes from electrons within the Crab that are moving at speeds close to light. The outer, irregular part of the Nebula shows up because of light emitted by hydrogen atoms, and is different in nature from the inner part (Courtesy of Hale Observatories).

cycles per second. The white light to be seen in the inner part of Figure 4.12 arises from just this process.

How such particle speeds arise is a problem to be considered in a later section. Here we can simply note that fast-moving particles occur not only in our galaxy but also in other galaxies, especially in galaxies that are observed to be strong sources of radio waves (radiogalaxies). Fast-moving particles also occur in a strange kind of object discovered about 10 years ago, the quasi-stellar objects, or *quasars*. These too we shall study in detail at a later stage.

We come now to a deeper question than any we have considered so far. What determines which is the source particle and which the detector? The answer to this question is that radiative interactions always go forward in time. In Figure 4.6 the motion of particle *a* at point *P* influences the motion of particle *b* at *Q*, because *Q* is later than *P*. The motion of *b* at *Q* does not influence the motion of *a* at *P* because this would imply a radiative interaction that traveled backward in time. The answer to our question, therefore, is that we distinguish the source particle from the detector particle through the *time-sense* of the interaction.

Two problems can be solved in exact mathematical terms:

1. Given two points, say, *P* and *Q*, on the path of a particle, and given all the interactions coming from earlier times, the form of the path between *P* and *Q* is completely determined; we can calculate every point on the path, and we know the time when the particle is at every such point.

2. Given the path of a particle, the radiative interaction excited by the particle on other particles is completely calculable.

Notice the separation with respect to time-sense. The interactions we need to know in order to solve the first problem come from the past. The interactions which the particle exerts on other particles, calculated in the second problem, go to the future. This separation, illustrated in Figure 4.13, establishes *causality* in the world.

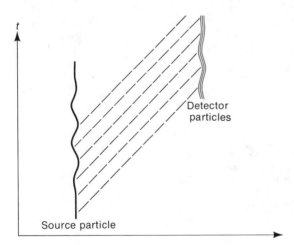

FIGURE 4.13.
The separation in time of the motions of source particle and detector particle establishes causality in the world; the detector oscillates *later* than the source.

We can ask, in relation to the first problem, how can we ever know all the interactions that come from the past? We could seek to answer this question by considering the particles giving rise to the required interactions, by attempting to use the second problem for them. But then we would have to know the paths of these other particles, and so would have to know the interactions to which they themselves were subjected. So the question is simply pushed a stage farther back. Indeed, we have the chicken-and-egg situation of Figure 4.14, in which we have one set of particles being influenced by another set of particles, which in turn were influenced by a further set of particles, and so on. How far back in such a chain must we go before we can reasonably be certain that what happened before was irrelevant to the specific physical system we wish to investigate at the present moment of time?

In principle, there is no limit to how far back we must go, unless it is argued that the universe had a definite beginning, in which case there is nothing before a certain time in the past. In practice, however, it is often possible to study a local aggregation of particles under conditions where the effects of preceding interactions are small. This is the kind of situation the physicist seeks to set up. He designs experiments to be essentially independent of preceding effects, and so succeeds in breaking the causal dilemma of needing to go ever farther and farther backward in the sense of Figure 4.14. The experimental procedure, illustrated in Figure 4.15, is made possible by the local restrictions imposed by the physicist himself. Such a breaking of the causal chain is never strictly

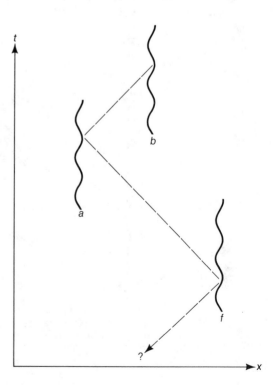

FIGURE 4.14.
The causal chain, in which the motion of *b* depends on *a*, while the motion of *a* depends on *f*, while the motion of *f* depends on still other particles at still earlier moments of time. In principle, this causal chain cannot be broken, at any rate until we have pursued time backward to the origin of the universe—if there was an origin to the universe.

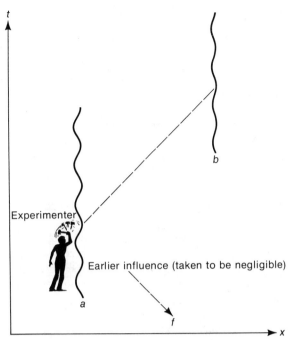

FIGURE 4.15.
The assumption (presumption?) of the experimenter is that the causal chain of Figure 4.14 can be broken by imposing a large and known influence on particle *a*—e.g., by hitting it with a hammer. This is taken to be so large and so drastic that the effect of particle *f* can be ignored in comparison.

accurate. Yet the physicist is satisfied so long as any preceding interactions have only a negligible effect on his results. The situation is less satisfactory, however, when we are concerned, not just with a local set-up, but with the whole universe. The causal chain then cannot be broken, and the logical problem of a "beginning" manifests itself. We will encounter this same problem repeatedly as we go along.

§4.4. Quantum Mechanics and Atomic Structure

Two problems were described above that can be solved exactly, i.e., in precise mathematical terms. The basic theory for doing so was discovered in the late nineteenth century by James Clerk Maxwell (1831–1879). Further important additions to the theory were made in the ensuing thirty years, notably by H. A. Lorentz (1853–1928), and by Albert Einstein.

What, one might ask, is the difference between mathematics and physics? A glance at a few mathematics and physics texts might suggest that the two are really the same. Yet this is not so. The mathematician is concerned entirely with sequential logical statements: If *A* is true, then *B* is true. If *A* is true, then *K* is false. The "pure" mathematician does not seek to establish whether *A* is really true. This is the physicist's job. What the physicist does is to decide which statements about the world are correct. Particularly, he aims to discover statements of maximum content. Instead of proceeding to announce a vast complex of propositions,

A is true,   *B* is true,   *C* is true,   ... ,

the physicist likes to economize as much as he can. He prefers to say,

A is true. Therefore *B, C,* ... , follow immediately from
the sequential connections established by mathematicians.

Statement *A* is to be a basic law, with as short and elegant a form as possible, and with a wide range of consequences, subject of course to the condition that *B, C,* ... , must never contradict the result of experiment. Because physicists thus rely on the logical connections established by mathematicians, physics inevitably appears to contain a lot of mathematics.

QUESTION:
Why is mathematics generally felt to be a difficult subject?

ANSWER:
Because it is difficult to summon up the relentless enthusiasm of the mathematician for following through exceedingly long chains of reasoning. Unless one has a boundless appetite for mathematics, memories of equivalences already established fade quickly from the mind, and consequently are not available for use in later arguments, except through the perpetual and nagging process of "reminding oneself" of what was done a while ago. After "reminding" himself a few times, the student feels himself "lost." This happens inevitably in advanced mathematics unless the subject is pursued with daemonic energy.

From time to time physicists find to their astonishment and agony that among the consequences *B, C,* ... , which follow from the primary *A*, something turns up which contradicts experiment—or the result of a new experiment is found to contradict some old member of *B, C,* .... Now what to do? It follows that our much cherished *A* cannot have been true after all. So a new *A* must be found, preferably having a close affinity with the old *A*, but having an extra twist to it. Unfortunately for the physicist, things do not always turn out this way. Sometimes a new *A* almost disasterously different from the old *A* has to be adopted. This was the case with *quantum mechanics*. Quantum mechanics seemed so much at variance with all the ideas of what is now called *classical physics* that every physicist was deeply shocked by it. Einstein, in particular,

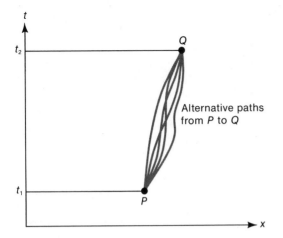

Alternative paths
from P to Q

FIGURE 4.16.
In quantum physics there are alternative paths by which
a particle may go from point P to point Q.

never became reconciled to it, while Schrödinger, who played a big part in
the development of quantum theory, said, "I don't like it, and I'm sorry I ever
had anything to do with it."

Let us return to the first of the two problems stated above, the problems
which are completely solved according to the rules of classical physics:

Given two points say P and Q, on the path of a particle, and given all the
interactions coming from earlier times, the form of the path between P and
Q is completely determined; we can calculate every point on the path, and we
know when the particle is at every such point.

This problem takes a quite different form in quantum theory, in which there
is no *unique* path connecting P and Q. It is possible in principle for the particle
to go from P to Q *by any path*. We can think that the particle must follow
some one, specific path, but we have no means of knowing which one. The
situation is illustrated in Figure 4.16. The system of rules describing quantum
mechanics, our new A, permit a quantity, known as the *amplitude,* to be
calculated for each such path, from which the probability that the particle will
follow the path is easily obtained; the amplitude is a kind of square root
of the probability. This new procedure, startlingly radical in conception,
contains a thankful simplification. Amplitudes can be added. By adding all the
amplitudes for all the paths going from P to Q, we obtain a sum which
represents the total amplitude for paths between P and Q, from which the
probability that the particle will go from P to Q is immediately obtained. Notice
that we do not add probabilities. It is the "square roots" of the probabilities
that are added.

PROBLEM:
Add the square roots of 25 and 36. Square the result, noting that the answer
is not the same as $25 + 36$.

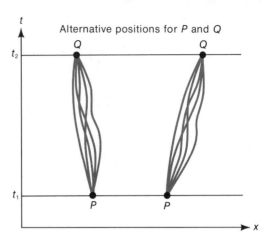

FIGURE 4.17.
In going from time t_1 to time t_2, not only may the particle go from P to Q by any path, but P may be any point at time t_1, and Q may be any point at time t_2. The situation is in principle completely general.

A typical situation in quantum mechanics is illustrated in Figure 4.17. We have two values of the time, denoted by t_1 and t_2, with t_1 earlier than t_2. In the situation of Figure 4.17, we do not even know where the particle is located to begin with. It may be in any spatial position, for example, at P. It may go to any other spatial position, for example, Q, at time t_2, and it may do so by any path. There is certainly a sweeping grandeur in the very generality of this concept. The problem now lies in getting anything precise out of it. In the world of experience, at any rate in the everyday macroscopic world, particles do seem to be precisely located, and they do seem to be able to proceed along definite paths from one place to another. How are we to cope with this?

We begin to answer this question by emphasizing that, although all paths are in principle possible, all paths are not equally probable. To illustrate this point, we can imagine two classes of path, one of comparatively high probability, the other of low probability. If we redraw Figure 4.17, showing only the paths of high probability, the situation can be much more manageable, as in Figure 4.18, where the high-probability paths are confined to a single bundle. And if we imagine the bundle to become more compressed, as in Figure 4.19, we approximate to a situation in which the particle can be thought of as going from one point to another by a unique path. In such a situation, it is found that quantum mechanics leads to exactly the same path as did the older theory of Maxwell, Lorentz, and Einstein. This at least is a port of refuge in the storm. The situation of Figure 4.19 indeed turns out to apply to all large bodies. The Earth's motion around the Sun, or the motion of any macroscopic lump of material, is the same as it was before.

The situation becomes very different from the earlier theory, however, when only a comparatively few particles are involved, as in individual atoms. Let us consider the simplest atom, hydrogen, consisting of a single electron and a single proton. Because the proton has a much larger mass than the electron, the bundle of paths for the proton is less spread out than the bundle for the electron. For simplicity, we can think of the proton as following a definite path,

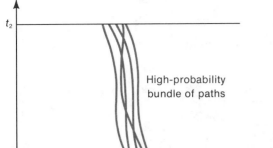

High-probability
bundle of paths

FIGURE 4.18.

The completely general situation of Figure 4.17 is controlled by the circumstance that the particle has a high probability of following a member of only a restricted bundle of paths. The particle *may* follow some other path, but has only a low probability of doing so.

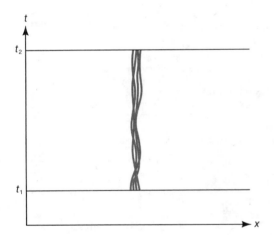

FIGURE 4.19.
As the bundle of paths of high probability becomes more and more constricted, the situation approximates to the classical situation, in which the particle is considered to follow a uniquely specified path.

like a classical particle. But the electron has no unique path; it cannot be thought of as pursuing a definite orbit around the proton. Elementary discussions of atomic physics sometimes state that the electron in the hydrogen atom moves around the proton rather like a planet moving around the Sun, except that, unlike a planet, which can move in any elliptic orbit, the electron can move only in certain specific orbits. This statement is quite wrong, however. The electron can move around the proton *in any orbit*. The orbit does not even have to be an ellipse, as that of a planet does. The most we can say is that some orbits are more probable than others.

This still seems a grotesquely vague picture. Yet the problem turns out to assume a wonderful regularity. If instead of worrying about specific paths, we worry about bundles of paths, the electron in the hydrogen atom will usually have one or another of a certain standard set of bundles, some of which are shown in Figure 4.20. These bundles can be referred to notationally as ψ_1,

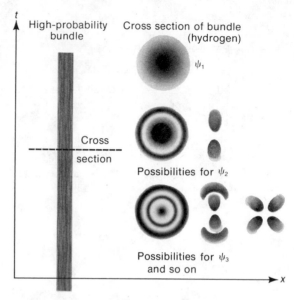

FIGURE 4.20.
The paths of electrons in atoms can be considered in terms of
alternative bundles ψ_1, ψ_2, ψ_3, . . . , in the sense that the electrons will *usually* be following paths in one or another of
these bundles. Sometimes, however, there is a swap from one
bundle to another, in which case the atom is said to undergo
a *transition*. The cross sections of ψ_1, ψ_2, ψ_3, . . . , are taken
for the single electron in the hydrogen atom. As noted in the
text, there are several possibilities for ψ_2, ψ_3, etc. In the righthand diagrams, the paths run through the shaded regions.

ψ_2, . . . , each symbol in this sequence denoting one of the standard bundles.
Expressed in this way, the standard bundles are called *states*.

The electron of the hydrogen atom will usually, but not always, be in one
or another of the states ψ_1, ψ_2, ... Not always, because it is possible for
an electron to jump from one state to another, $\psi_2 \longrightarrow \psi_1$, for example. The
sequence ψ_1, ψ_2, . . . , can be arranged so that we always have a jump leftward
if radiation is emitted by the atom, and a jump rightward if radiation is
absorbed. The states are then said to be arranged in sequence with respect to
energy. So we think of the electron as being mostly in one or another of the
states ψ_1, ψ_2, . . . , but of it occasionally changing its state by jumping to the
left or to the right.

A change of state can be caused not only by the emission or absorption
of radiation, but also by collisions between particles, and such changes can also
go both ways, to the left or to the right in the sequence ψ_1, ψ_2, ... In
a hot gas there are at any moment of time atoms in each of these standard
states, which are called *stationary states* because the atom halts for a while in
any one of them before jumping to another. In a hot gas there is a shuffling

of atoms among the states, some jumping one way in the sequence $\psi_1, \psi_2, \ldots$, others jumping the opposite way, like a vast circus of fleas. If all changes to the right are in exact balance with corresponding changes to the left, $\psi_{37} \longrightarrow \psi_{53}$ being in balance with $\psi_{53} \longrightarrow \psi_{37}$, for example, the gas is said to be in equilibrium, and a temperature can be assigned to it by the procedure described in Appendix I.3.

For a gas, which contains a large number of atoms, suppose that many of the atoms undergo a certain change of state in a given time-interval, say, the transition $\psi_2 \longrightarrow \psi_1$. The radiation emitted by a large number of atoms in such a leftward step in our sequence of states has properties similar to the radiation from the simple particle oscillation of Figure 4.7. However, we must take care to remember that, whereas the frequency ν of oscillation of the classical particle in Figure 4.7 could be assigned to suit ourselves, *we now have a situation in which ν is determined by the transition in question*. A hydrogen atom does not emit radiation at any and all frequencies. It only emits frequencies corresponding to the changes of state to the left in the sequence $\psi_1, \psi_2, \ldots$.

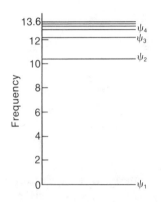

FIGURE 4.21.
The energy diagram of hydrogen, with the states ψ_1, ψ_2, ψ_3, . . . , given levels that are determined by the frequencies of the radiation emitted in the transitions $\psi_2 \longrightarrow \psi_1$, $\psi_3 \longrightarrow \psi_1$, ... The unit of frequency is 2.418×10^{14} oscillations per second. The discrete levels do not go above 13.6.

In Figure 4.21 we have the *energy-level diagram* of the hydrogen atom, each level corresponding to one of the states $\psi_1, \psi_2, \ldots$, with ψ_1 the bottom level, ψ_2 the next level, and so on. The diagram is constructed so that the spacing between adjacent levels is proportional to ν for the transition between the adjacent states in question. For example, the spacing between the bottom level and the level immediately above it is proportional to ν for the transition $\psi_2 \longrightarrow \psi_1$, that between the second and third levels is proportional to ν for the transition $\psi_3 \longrightarrow \psi_2$, and so on.

The levels crowd together as we go to states far to the right in the sequence $\psi_1, \psi_2, \ldots$. This means that a transition like $\psi_{105} \longrightarrow \psi_{104}$ would correspond to a quite low frequency, one which actually falls into the radio band, and which has been observationally detected in the radiation from hot gas clouds in our galaxy.

The bundle of paths corresponding to a member of the sequence ψ_1, ψ_2, . . . , become larger and larger in its spatial dimensions as the state in question

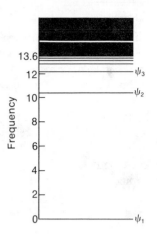

FIGURE 4.22.
There are states such that the electron paths go far from
the proton of the hydrogen atom; these lie above the 13.6
level. Electrons with paths in the bundles corresponding
to these states are said to be *free*. The levels to be given
to these free states, using the same radiation criterion as
in Figure 4.21, form a continuous band in the energy-
level diagram. Same unit of frequency.

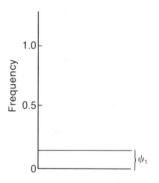

FIGURE 4.23.
The ψ_1 state of the hydrogen atom is actually split into
two very closely spaced levels. Because of this small
spacing, the radiation emitted in transitions from the
slightly higher member of this doublet to the slightly
lower member has an abnormally low frequency, of
1.42×10^9 oscillations per second. This low-frequency
radiation is very important to studies in radioastronomy.
The unit of frequency here is 10^{10} cycles per second.

lies more and more to the right in the sequence. This means that the electron
has an appreciable probability of following paths that take it farther and farther
away from the proton. Ultimately there are paths that take it completely away
from the proton. The electron is then said to be *free*. The level corresponding
to such a free state lies above the regular sequence of "bound" states in Figure
4.21, as in Figure 4.22. Transitions can occur between one free state and
another, and between a free state and a bound state, involving emission of
radiation when the transition is downward in Figure 4.22, and involving
absorption when the transition is upward.

We next cover a point of detail. There are always two or more states at each
of the levels of Figure 4.21. Thus we should think of ψ_1 as representing two
states, ψ_2 as representing several states, and so on, as appeared already in Figure
4.20. To be very precise, the bottom level of Figure 4.21 is actually split
slightly, as in Figure 4.23. Transitions involving a frequency $\nu = 1.42 \times 10^9$
cycles per second occur between the two states of the split bottom level.
Radiation from these transitions is widely observed by radio-astronomers, and
Figure 3.7, which showed the spiral pattern of the distribution of hydrogen
atoms in our galaxy, was obtained using this radiation.

The energy-level diagram for hydrogen, shown in Figure 4.21, can be constructed explicitly in the following way. First, draw a level to represent the limiting situation in which the electron just becomes free. Next, mark the states $\psi_1, \psi_2, \psi_3, \ldots$, below this initial level, at depths that are in the ratios $1 : \frac{1}{4} : \frac{1}{9} : \ldots$, the denominators in the fractions being just the squares of the natural numbers, $2^2 = 4$, $3^2 = 9$, ... The spacings between the bottom level, corresponding to ψ_1, and the levels corresponding to $\psi_2, \psi_3, \ldots$, are then in the ratios $1 - \frac{1}{4} : 1 - \frac{1}{9} : \ldots$. These same ratios appear in the frequencies observed for the transitions $\psi_2 \longrightarrow \psi_1, \psi_3 \longrightarrow \psi_1, \ldots$. Neglecting the small effects mentioned in the preceding paragraph, these frequencies are

$$3.29 \times 10^{15} \left[1 - \tfrac{1}{4}, 1 - \tfrac{1}{9}, \ldots \right] \text{ cycles per second.}$$

Hence the relative spacings of the levels in our constructed diagram correspond to the actual frequency ratios observed for the transitions $\psi_2 \longrightarrow \psi_1, \psi_3 \longrightarrow \psi_1, \ldots$. The radiation emitted in these transitions is known as the *Lyman series*. This series is illustrated in Figure 4.24. Absorption of radiation at the same frequencies occurs for the upward transitions $\psi_1 \longrightarrow \psi_2, \psi_1 \longrightarrow \psi_3, \ldots$, and this is illustrated in Figure 4.25.

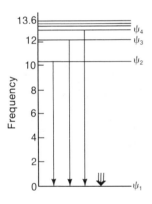

FIGURE 4.24.
The radiation emitted by the transitions $\psi_2 \longrightarrow \psi_1$, $\psi_3 \longrightarrow \psi_1, \ldots$, is called the *Lyman series*. Same unit of frequency as in Figure 4.21.

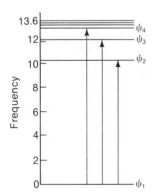

FIGURE 4.25.
Radiation can be absorbed in transitions that are the reverse of those in Figure 4.24. Same unit of frequency.

Similarly we can have downward transitions $\psi_3 \longrightarrow \psi_2, \psi_4 \longrightarrow \psi_2, \dots$. The frequencies associated with these transitions are observed to be

$$3.29 \times 10^{15} \left[\tfrac{1}{4} - \tfrac{1}{9}, \tfrac{1}{4} - \tfrac{1}{16}, \dots\right] \text{ cycles per second.}$$

The observed frequencies again have the same ratios as the spacings of the levels in our constructed diagram.

PROBLEM:

Satisfy yourself that the ratios are the same.

Light emitted at these frequencies by hot hydrogen gas is shown in Figure 4.27, the transitions being illustrated diagramatically in Figure 4.26. The sequence of Figure 4.26 is known as the *Balmer series*. Upward transitions, involving absorption, can also take place at the same frequencies. These are illustrated in Figure 4.28.

It was one of the early triumphs of quantum theory actually to predict the frequency values given above, the number 3.29×10^{15} being obtained by precise calculation and being verifiable by observation.

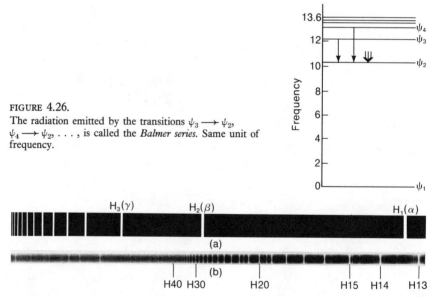

FIGURE 4.26.

The radiation emitted by the transitions $\psi_3 \longrightarrow \psi_2$, $\psi_4 \longrightarrow \psi_2, \dots$, is called the *Balmer series*. Same unit of frequency.

FIGURE 4.27.

In an experimental arrangement that separates radiation according to its frequency, as in (a), with the frequency increasing toward the *left*, the radiation of the Balmer series falls at discrete places with certain regular spacings. Employing such an arrangement for the star HD 193182 gave many of the quite high members of the Balmer series, $\psi_{13} \longrightarrow \psi_2, \psi_{14} \longrightarrow \psi_2, \psi_{15} \longrightarrow \psi_2, \dots$, as can be seen in (b). (Courtesy of Hale Observatories.)

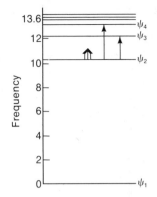

FIGURE 4.28.
Radiation can also be absorbed at the characteristic frequencies of the Balmer series, by $\psi_2 \longrightarrow \psi_3$, $\psi_2 \longrightarrow \psi_4$, ... Same unit of frequency as in Figure 4.21.

PROBLEM:
The energy-level diagram spacings relative to the third level, corresponding to the transitions $\psi_4 \longrightarrow \psi_3$, $\psi_5 \longrightarrow \psi_3$, . . . , are also in proportion to the frequencies of the radiation emitted in these transitions, which give what is known as the *Paschen series*. Write down their values.

We are in a position now to understand how hydrogen can emit radiation both at discrete frequencies and with continuous values. Continuous values come from transitions between states in which the electron is free, whereas discrete frequencies come from transitions between states in which the electron is bound to the proton.

$H_1 (\alpha)$

FIGURE 4.29.
In this photographic negative we again have an observational arrangement that separated the radiation from the quasar 3C 273, with the frequency increasing toward the left. Which are the Balmer lines of hydrogen? (Courtesy of Dr. R. Lynds, Kitt Peak National Observatory.)

PROBLEM:
Figure 4.29 shows light from the quasar 3C 273. Note there is radiation both at discrete frequencies and at continuous frequencies. Three unmarked discrete frequencies belong to the Balmer series of hydrogen. From their pattern, see if you can pick them out.

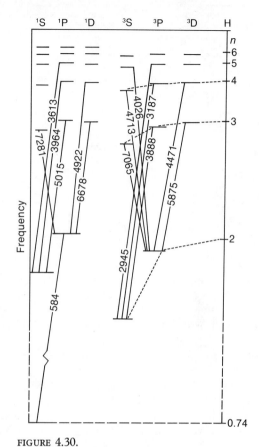

FIGURE 4.30.
The helium energy-level diagram. The unit spacing
in the frequency scale corresponds to 1.49×10^{14}
oscillations per second; the dotted part of the scale
is equal to 31.6 such units. (From H. G. Kuhn,
Atomic Spectra, New York, Academic Press, 1962.
Longman Group Ltd.)

Similar considerations apply to atoms other than hydrogen. Although the
same concepts concerning bundles of paths continue to apply, leading to the
same idea of a sequence of states $\psi_1, \psi_2, \ldots$, among which transitions occur,
the problem becomes much more complex in detail, because more than one
electron has to be considered. Even for the helium atom, which has only two
electrons, the situation is already much more intricate, as can be seen from
the helium energy-level diagram shown in Figure 4.30. Here the levels have
been classified into two branches. The symbols attached to the various levels
are those used in the classification, and need not concern us here.

The study of complex diagrams like that of Figure 4.30 forms the subject
of *atomic physics*. The radiation from each kind of atom has a characteristic

pattern of discrete frequencies that distinguishes it from the patterns of other atoms, a circumstance of great importance in astronomy. When the characteristic radiation pattern of a certain kind of atom is found in the light from an astronomical object, then we know that this kind of atom is present in the object. From Figure 4.29, for example, we know that 3C 273 contains hydrogen. In this way the kinds of atoms present in various objects can be discovered. If atoms only emitted radiation with continuous frequencies, as they would do according to the older, prequantum theory, we could never infer the composition of distant stars, galaxies, or quasars. Without quantum mechanics, most of the power of modern astronomical techniques would be lost.

The radiation emitted in a specific transition, for example, $\psi_2 \longrightarrow \psi_1$ of hydrogen, cannot be absorbed in any transition between the discrete levels of hydrogen except $\psi_1 \longrightarrow \psi_2$, as is illustrated in Figure 4.31, but it could be absorbed from ψ_2 into a state in which the electron was free. Occasionally the discrete radiation emitted by one kind of atom may be absorbed in a transition between discrete levels of another atom, as in Figure 4.32, but such absorption is accidental and therefore rare. The discrete frequencies of one atom do not usually coincide with those of another atom. A more common situation is illustrated in Figure 4.33. Here a downward transition of one atom produces radiation that frees an electron from a second atom. The second atom is then

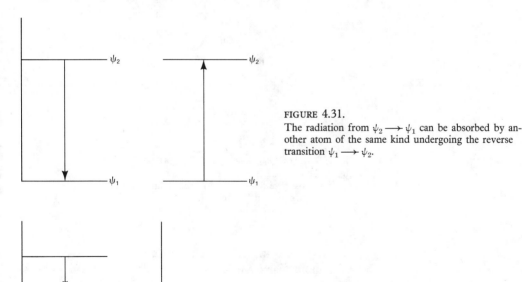

FIGURE 4.31.
The radiation from $\psi_2 \longrightarrow \psi_1$ can be absorbed by another atom of the same kind undergoing the reverse transition $\psi_1 \longrightarrow \psi_2$.

FIGURE 4.32.
Very occasionally, the discrete radiation from one atom can be absorbed in a discrete transition of a different kind of atom. This involves a coincidence in the energy-level diagrams of the two atoms. The same spacing must occur in both.

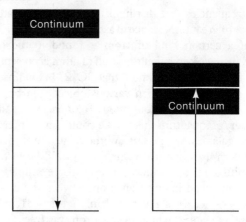

FIGURE 4.33.
The radiation from one kind of atom is more usually absorbed by another kind of atom in a way that causes an electron to become free. This process is called *ionization*. The absorbing atom is said to become *ionized*.

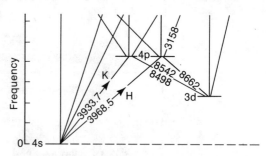

FIGURE 4.34.
The lower part of the energy-level diagram for the calcium atom with one electron removed, showing the transitions that give the so-called H and K lines. Same unit of frequency as in Figure 4.21.

said to become ionized—it loses an electron. This process of ionization is widely employed in practice in all manner of electronic devices. It is known as the *photoelectric effect*.

Calcium atoms, after being ionized, have an energy-level diagram whose lower part is shown in Figure 4.34. The upward transitions marked H and K in the figure involve the absorption of radiation at frequencies $\nu = 7.554 \times 10^{14}$ cycles per second and $\nu = 7.621 \times 10^{14}$ cycles per second. These two absorptions play an important role in astronomy.

PROBLEM:
Show that the frequency of the H line of calcium happens to be very close to that associated with the $\psi_2 \longrightarrow \psi_7$ transition in the Balmer series of hydrogen.

§4.5. *Reflections on the Sophistication of Physical Theories*

Quantum mechanics permits a particle to go between two points P and Q by any path, but with different probabilities for different paths. In the special case where, except for one specific path, the probabilities are all essentially zero, we can assert that the particle would follow that one path. Such a path always turns out to be the same as we would have calculated from the rules of classical physics. Hence classical physics is a special case of quantum mechanics. Classical physics is not "wrong" within its own terms of reference; so long as we stick to appropriate cases and situations, we can continue to use classical methods. However, quantum mechanics applies to a wider range of phenomena than classical theory does. It includes radiation from atoms at discrete frequencies, for example, which cannot be treated at all from a classical point of view.

The development from classical theory to quantum theory illustrates the meaning of "progress" in science. Suppose scientists hold a certain theory, about something or other, which we denote by T. For the theory to be considered "good," it must represent the observed behavior of the world throughout a certain range of phenomena to within a satisfactory tolerance. For example, Newton's theory of gravitation permits the next total eclipse of the Sun by the Moon to be predicted ahead of time to within a tolerance of about one second. Since this tolerance is regarded as satisfactory, Newton's theory is considered to be "good" for the purpose of making eclipse predictions.

If now an improved theory, T', say, should be discovered, we would mean that T' encompasses a wider range of phenomena than T, not that T is to be replaced within its own range of applicability. We can think of the situation territorially: the territory covered by T' includes that covered by T. And if we could find a further improved theory, say, T'', we would mean that the territory of T'' encompasses that of T'.

As an example of this situation, we can identify T with the classical theory of particles and radiation, and T' with the form of quantum mechanics discovered in 1925 by Heisenberg and by Schrödinger. Yet within only a few years, T' became replaced by a T''. According to the 1925 form of the theory, the amplitude for a specific path to go from a point P to a point Q was represented by a complex number. Then Dirac found four complex numbers should be used. The resulting improved theory, T'', is known as *relativistic quantum mechanics,* and the set of four complex numbers denoting the amplitude is known as the *spin* of the particle.

Although the change from the classical theory T to Dirac's T'' represented a major advance in understanding, it did not answer in any sophisticated way the following question:

Not all the particles in the world are of the same kind. What quantities distinguish one particle from another?

Until about ten years ago, this question was answered simply in terms of the mass and the electric charge of the particle. The proton was considered different from the electron because its mass was much larger. The *neutron* was considered different from the proton because, unlike the proton, it had no electric charge. The *neutrino* was different again because it had neither mass nor electric charge. However, as a result of remarkable experiments in *high-energy physics,* the physics of fast-moving particles, it has now become clear that much more is needed to describe a particle. A further major theoretical development, a T''', now in progress, is evidently required.

Is it likely that an ultimate theory, $T''' \cdots$, will ever be discovered that fully and precisely encompasses all phenomena? My own belief is that the answer to this question is no. Whenever we widen the domain of experience, new theoretical ideas always seem to be necessary. For example, the laws which work so perfectly for radiation do not suffice to explain the forces that hold the nuclei of atoms together. The successes of the laws of radiation are achieved by restricting attention to a certain domain of experience. When the domain is widened, the theory becomes inadequate. To achieve a complete and perfect theory, we would need to widen the range of experiment and observation to include everything, the whole universe. If the answer to the above question were affirmative, we would then be able to describe the whole universe in terms of the processes occurring in our own brains—i.e., in terms of a small sub-domain of the universe itself. I doubt the possibility of achieving such a goal. I suspect the truth of the matter to be that no full description of the universe is possible within only a part of the universe. Any complete theory would inevitably encompass the whole universe—our $T''' \cdots$ would be the universe itself. A complete theory, like the Holy Grail, is something we must always seek but will never find.

General Problems and Questions

1. How may the spatial path of a particle be described?

2. Discuss a method of representing the path of a particle in both space and time.

3. Show that the helical path of Figure 4.3 is followed by the particle in a righthanded sense.

4. Would the sense of this helical path be reversed by reversing the arrows of Figure 4.4?

5. Draw a figure corresponding to Figure 4.4 for a particle moving along a helical path in a lefthanded sense.

6. Discuss the way in which the motion of one electrical particle influences the motion of another electrical particle.

7. How is the white light from the inner part of the Crab Nebula generated?

8. Assuming a distant source particle to be in oscillation with a speed small compared to the speed of light, how in principle can the frequency of the oscillation be determined?

9. What is the basic principle of the telescope?

10. Discuss the nature of causality.

11. Under a given external influence, in classical physics a particle always moves from one point to another by a certain uniquely determined path. Is this true in quantum methanics?

12. How does the electron in the hydrogen atom come to be described by a member of a certain set of states? What happens when many atoms undergo transitions from one such state to another?

13. What is the Lyman series?

14. What is the Balmer series?

15. Under what condition does the electron of the hydrogen atom become free?

16. How can an atom become ionized?

Chapter 5:
Particles and Radiation
in Practice

§5.1. Lenses and Refracting Telescopes*

The discussion of the preceding chapter was concerned with the understanding of particles and fields as it has developed during the past hundred years. The classical theory began with the work of Maxwell, published in 1873. The first ideas in quantum theory came at the beginning of this century, and quantum mechanics in the form discussed above followed Heisenberg's work in 1925. From an historic point of view, all this is very modern. Long before Maxwell, scientists had arrived at a picture of the nature of light, which, although crude from our present-day vantage point, still sufficed to enable many practical devices, including the telescope and microscope, to be invented.

According to this older picture, light was "something" that traveled in straight lines, referred to as *rays*. Rays of light of a specific color were found to have certain simple properties which formed the basis for these practical achievements. The properties, described as the "laws of reflection and refrac-

---

*The reader who wishes to press on with astronomical and physical ideas can jump from the end of the third paragraph of §5.1 immediately to §5.4. The details of telescope construction will not be necessary for the understanding of later chapters. Material in this chapter has been adapted from F. Hoyle, *Astronomy* (London: Rathbone Books, 1962).

tion," were first stated early in the seventeenth century by the Dutch scientist Willebrord Snell (1580–1626).

Figure 5.1 shows a ray of light, represented by *AB*, incident on a glass block. There is a reflected ray *BC* that comes off the glass at an angle equal to that made by the incident ray. There is also a refracted ray *BD* continuing on into the glass. The latter ray is bent toward the normal *XY*, the normal being an imaginary line passing through *B* that makes a right angle with the plane surface of the block. The three rays *AB*, *BC*, *BD*, and also the normal *XY*, all lie in the same plane. What Snell discovered is illustrated in Figure 5.2. The two points *A* and *D* are chosen so that the distances *AB* and *BD* are equal. Then the ratio of *DY* to *AX* is always the same. That is to say, if we change the angle which the incident ray makes with the normal, this ratio remains unchanged. Once we know this value, we can work out what the direction of the refracted ray must be for any incident ray. This fact forms the basis of the construction of the refracting telescope.

Light always undergoes some absorption when it passes through matter, becoming progressively weaker the farther it goes. In translucent materials the

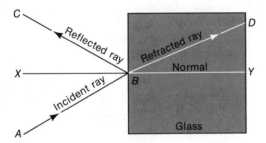

FIGURE 5.1.
At an interface between air and glass, the refracted ray is bent toward the normal, while the reflected ray comes off at an angle equal to that made by the incident ray.

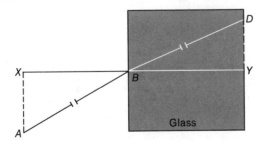

FIGURE 5.2.
If segments of the incident ray *AB* and the refracted ray *BD* are equally long, the ratio of *DY* to *AX* is always the same for a given change of medium.

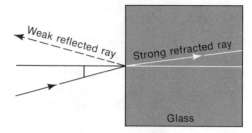

FIGURE 5.3.
At small angles of incidence, glass gives only a
weak reflected ray but a strong refracted ray. It is
thus suitable for refracting telescopes.

rate of loss is comparatively small, however, and so translucent materials—especially glass—are widely used in optical instruments. For astronomical purposes we are almost always concerned with rays at small angles of incidence, and for these a translucent material has the property of giving a weak reflected ray, but a strong refracted ray, as illustrated in Figure 5.3. Thus if we wish to construct a telescope based on the properties of refracted rays, we shall obviously use lenses of glass.

Figure 5.4 shows a cross section through a convex lens with central axis *OC*. (We assume that the lens is made so that the section would be the same for any plane containing the line *OCO'*.) *O* is an object emitting rays of light in all directions. We see that the ray along *OC* travels through the lens without deviation, but all other rays refracted by the lens are deviated. The measure of deviation of any one refracted ray can readily be worked out from the rule

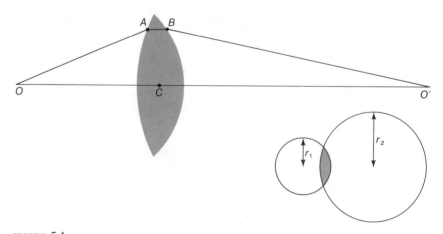

FIGURE 5.4.
Top: Focusing with a convex lens.
Right: Lens surfaces are given curved spherical surfaces of different radii in order to minimize *spherical aberration*.

of refraction which Snell discovered, for if the surfaces of the lens are quite smooth the curvature of the glass at the points A and B can be neglected. The ray is turned toward the normal at A, and away from the normal at B in exactly the way we have discussed. Because of the symmetry of the lens, the ray emerging from B must continue to lie in the plane of OC and OA. It can thus intersect the axis at O', say. The nearer A is to the edge of the lens, the greater is the measure of deviation of BO' from OA.

An important question now arises. Can *all* the rays from O which go through the lens be made to pass through the same point O'? The answer is that if the two surfaces of the lens are shaped correctly the rays can indeed be made to pass through O' to an extremely high degree of accuracy. The accuracy is lost, however, if the distance of the object point O from the lens is changed at all markedly. For this reason, lenses are not usually given the complicated shapes that would be required to produce a well-nigh perfect focus for one particular and precise position of O. Instead, the lens surfaces are made spherical. This leads always to imperfect focusing, the defect known as *spherical aberration,* but for the moment we shall ignore this defect; that is to say, we shall assume a perfect focus at O'. We shall also assume a perfect focus when the object is off-axis, as Figure 5.5, although in actual fact further imperfections are thereby introduced. These are known as *coma* for objects that are slightly off-axis, and as *astigmatism* for objects that are far off-axis.

In a first attempt to understand the broad principle of the refracting telescope, we may indulge in the luxury of ignoring the practical imperfections of lenses, but it is important to recognize that such imperfections do exist and that spherical aberration, coma, and astigmatism are not the only ones. Before we proceed we should look at the others. In Figure 5.6 the plane p is perpendicular to the axis of the lens. Suppose that a number of points of p are emitting light, perhaps in the form of some picture. In accordance with what we have already assumed, each point of emission will be brought to a sharp focus to the right of the lens. Will all the focal points lie in the same plane, p', say? In fact, they will not lie in exactly the same plane but on a curved surface. Will the picture formed on p' be a true representation of the picture on p, or will there

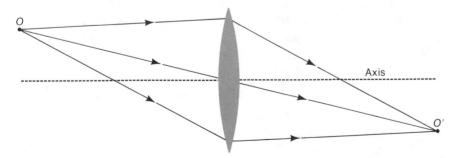

FIGURE 5.5.
When an object O is off-axis, the focus is at O'. In actuality there is always a defect of focus at O': *coma* for objects slightly off-axis, *astigmatism* for objects far off-axis.

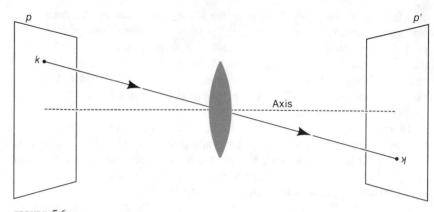

FIGURE 5.6.
Here all points of k (the object) lie in the same plane. Yet there will be some distortion and *field curvature* at the focal plane p'.

be distortion? There will, in fact, be distortion. Lastly, does a lens behave the same way for light of different colors? It does not. Going back to the law of refraction illustrated in Figure 5.2, it is true that the ratio of AX to DY is independent of the angle of incidence, but the value of this ratio changes with the color of the light, because the glass changes its behavior with the color. This means that a given object point, such as O of Figure 5.4, will be brought to a different focal point O' according to the color of the light. This defect is known as *chromatic aberration*.

No practical optical system is entirely free from spherical aberration, coma, astigmatism, curvature of field, distortion, and chromatic aberration. These imperfections can, however, be minimized by a careful attention to the layout of the system and to the conditions under which it is to be operated. The history of the telescope is in large measure the history of attempts to free it as far as possible from these imperfections. For the moment, however, we may ignore all these difficulties, since our immediate aim is to understand the principle of the telescope, rather than the refinements of its design.

With that aim in mind, we might look again at Figure 5.6 and ask how the size of the picture on p' compares with its size on p. Is the picture magnified or is it reduced? The answer depends on the distance of the lens from p. If the lens is far enough away from p, the picture on p' is smaller than the original. But as the lens is moved toward p the size on p' increases, until eventually it is larger than the original. And the size on p' goes on increasing without limit as the lens is brought to a certain critical distance from p known as the *focal length* of the lens. If the lens is brought still closer to p, *no plane p' can be found at all.*

To make this clearer, it must be realized that p' is not a fixed plane. As the lens moves (with p fixed), the plane p' on which the rays come to a focus also moves. As the lens is moved toward p, the plane p' moves farther and farther away to the right. And when the distance of the lens from p becomes equal to the focal length of the lens, the plane p' moves off to infinity. After this, no plane p' can be found.

What is this critical distance, this focal length of the lens, and what does it depend on? Simply on the two surfaces of the lens and on the nature of the glass itself. If these are convex spherical surfaces with radii r_1 and r_2, the reciprocal of the focal length is proportional to the sum of the reciprocals of r_1 and r_2; in other words,

$$\frac{\text{constant}}{\text{focal length}} = \frac{1}{r_1} + \frac{1}{r_2}.$$

The constant in this formula is usually about 2, and is determined by the kind of glass the lens is made of.

We can express all this very simply. The two pictures, the original on p and the *image picture* on p', make the same angles at the center of the lens. If we imagine an observer situated at the center of the lens, he would therefore see the two pictures as having precisely the same size. This means that the image picture is magnified if the lens lies nearer to p than to p', otherwise it is reduced. Figure 5.7 shows that there is an important symmetry between p and p' in the following sense. As the lens moves toward p, the plane p' moves to the right—to infinity when the lens reaches its focal length from p. Similarly,

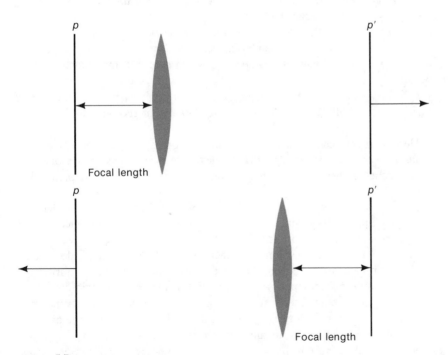

FIGURE 5.7.
When a lens is at its focal length from p, plane p' moves off to infinity. If p moves to infinity, the distance of the lens from p' becomes equal to the focal length.

if p moves off to infinity, the distance of the lens from p' becomes equal to its focal length. It is not difficult to see how this applies to photography. Distant objects we wish to photograph can be thought of as lying on one and the same distant plane, p. Film is placed in the camera on the plane p' and an image is formed with the lens at its focal length from p'. (As Figure 5.6 shows, the image is formed upside down, and left and right are reversed. But because the negative is transparent, we can look through it and turn it in a way that restores the correct orientation of the original picture.) Now it is well-known that nearby objects cannot usually be photographed accurately; some of them will be in focus and some out of focus. This is because we cannot by any stretch of the imagination think of nearby three-dimensional objects as lying in a single plane, whereas distant objects can be thought of as doing so—to an adequate degree of accuracy.

In astronomy, we are concerned with observing objects on the distant celestial sphere, and astronomers are usually content at any one moment to observe only a tiny portion of the celestial sphere. To an extremely high degree of accuracy, this tiny portion can be thought of as belonging to one and the same very distant plane. With p thus very distant, the image plane p' is spaced from the large main lens of the telescope by an amount equal to the focal length of the lens.

Suppose we place a white screen at p'. *The size of the image on the screen will depend only on the focal length of the lens, and not at all on its diameter.* (This is because the image would appear to an imaginary observer at the center of the lens to have the same size as the original object on p.) Hence if we take a series of lenses of increasing diameter but all with the same focal length, the image on p' will have the same size in each case. But the images will not be equally bright. The lens of largest aperture will give the brightest image, simply because it receives the most light from the object plane p; the lens of least diameter will give the faintest image because it receives the least light from p.

Here, then, we can see what is the first important function of a telescope. *It must serve as a gatherer of light.* In fact, a telescope lens with a diameter of 20 inches gathers 10,000 times as much light as the dark-adapted naked eye.

In principle, we could use the large lens of a telescope as a camera lens for photographing the sky, simply by placing a film on p'. In practice, such a procedure fails because the image on p' is too small. Suppose we wish to photograph a fair-sized portion of the Moon, say, the region around the lunar 'sea' known as Mare Imbrium. The size of the image of Mare Imbrium on p' depends on the focal length of the lens. For a small amateur telescope, with a focal length of about three feet, the image has a diameter of less than a tenth of an inch. Even for a big telescope of focal length about 50 feet, the diameter of the image is only about an inch, which may still be inconveniently small. *Hence we must magnify the image on p' before we attempt to make a photograph.*

This is easily done. We simply place a second lens beyond p', as in Figure 5.8. This lens brings the light from p' to a second focus on a second image

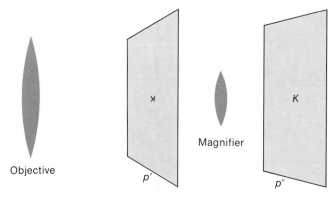

FIGURE 5.8.
The objective of a telescope usually gives only a small image at p'. This image is therefore enlarged for photographing at p''.

plane at p''. And provided the new lens is nearer to p' than to p'', the image on p'' will be larger than that on p'. Indeed, we can ensure that the image on p'' will reach a convenient size simply by placing the second lens at a distance from p' that is sufficiently close to its own focal length. By placing film on p'', we can now photograph a small area of the celestial sphere at the required size. In short, we have a telescope with a camera.

We see, then, that a telescope consists essentially of two parts: a light-gatherer which must have a large diameter, and a magnifier which must be adjusted to give the final image a convenient size. The light-gatherer is usually referred to as the *objective* of the telescope, and the magnifier as the *eyepiece*.

A simpler but less convenient arrangement than that of Figure 5.8 is to place the second lens *in front* of p'. In this arrangement the second lens must be concave, as in Figure 5.9. This arrangement has the effect of increasing the

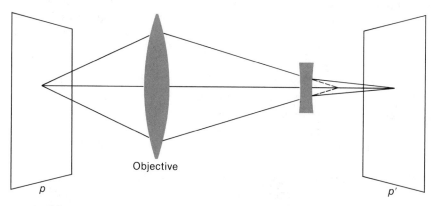

FIGURE 5.9.
The principle of Galileo's telescope.

FIGURE 5.10.
Galileo's telescope.

distance of p' from the objective and of increasing the size of the image. We now have only one image plane, that at p'. The first astronomical telescope, that of Galileo (Figure 2.10, reproduced here as Figure 5.10), was constructed in accordance with Figure 5.9. The better system of Figure 5.8 was invented by Kepler.

Figures 5.8 and 5.9 show only idealized telescopes. To make an actual telescope, the lenses must be mounted in some fashion. The usual arrangement is to place the objective at one end of a tube and the eyepiece at the other, as in Figure 5.11. Because the observer will need to adjust the eyepiece, he must be able to move the eyepiece parallel to the axis of the tube. This is normally achieved with the aid of a ratchet device.

It remains now to mount the telescope. The usual plan is illustrated in Figure 5.12. The polar axis shown in this figure is constructed to be parallel to the Earth's axis of rotation. The actual physical situation is different in different

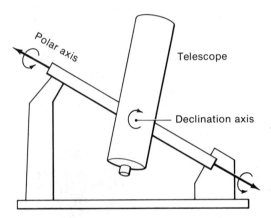

FIGURE 5.11.
The usual placing of objective and adjustable eyepiece.

FIGURE 5.12.
The polar axis of an equatorial mounting is carefully adjusted to be parallel to the Earth's axis of rotation. The telescope itself is mounted on a declination axis which is fastened to the polar axis, the two axes being accurately at right angles to each other.

geographical latitudes. Thus at the equator the polar axis is horizontal, at latitude 30° it is inclined at 30° to the horizontal, and similarly for other latitudes. Because of this structural dependence on latitude, moving a large telescope from one latitude to another is always difficult and sometimes impossible.

The telescope itself is mounted on a second axis, the declination axis, which is arranged to be accurately perpendicular to the polar axis. In older designs, as in Figure 5.13, the telescope itself was mounted toward one end of the declination axis, which necessitated the use of a counterweight at the other end. The direction in which the telescope points is controlled by the degree to which it is turned about the two axes, the polar axis determining right ascension and the horizontal axis determining declination.

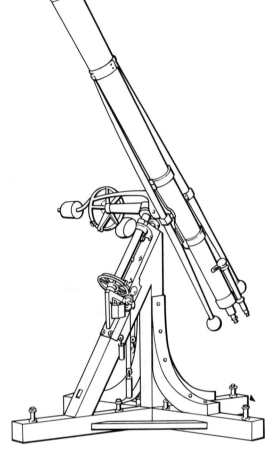

FIGURE 5.13.
A sketch of Fraunhofer's 9½-inch refractor, showing the telescope mounted with a counterweight on the declination axis.

§5.2. Mirrors and Reflecting Telescopes

The telescope has an important advantage over many other optical instruments. Since the astronomer is not usually concerned with light coming into the objective at more than small angles, the imperfections of spherical aberration, coma, curvature, and distortion can be kept within reasonable bounds. But wherever lenses are employed, as they must be in refracting telescopes, there is one imperfection that is more difficult to overcome. An ordinary lens refracts light of different colors in different ways, and does not therefore bring light of all colors to the same focus.

Even in the early days of the telescope this color aberration was considered a serious defect. In 1636, no more than 25 years after Galileo's first telescope, Marin Mersenne, a Minorite friar, proposed the construction of a reflecting telescope. In 1663, James Gregory put forward a design for a reflecting telescope, and in the years 1670–72 Isaac Newton suggested the most practical arrangements for one.

The idea of a reflecting telescope is to replace the objective of the refracting telescope by a mirror. That is to say, a mirror is used as the light-gatherer instead of a lens. This has two important advantages. First, whereas a lens refracts light of different colors in different ways, a mirror reflects all colors in the same way, and thus brings light of all colors from the same object to the same focus. Next, if the mirror is shaped to a special form known as a paraboloid, light from a distant object lying in the direction of the axis of the mirror is brought to a focus without spherical aberration, as is illustrated in Figure 5.14.

But with optical systems, perfection in one respect usually implies a serious imperfection in some other respect. The paraboloidal mirror suffers badly from coma; that is, the focus becomes bad for objects that do not lie in directions very close to that of the axis. In modern instruments this difficulty is overcome by correcting devices located in the magnifier. Instead of using a single lens as magnifier, we now use a complex optical system, and a prime concern in shaping and positioning the various lenses of the system is to correct errors due to the coma produced by the mirror.

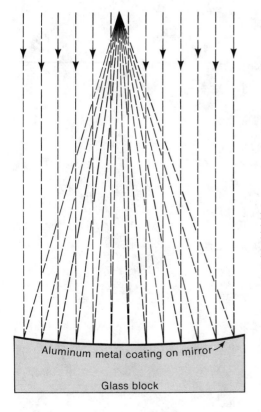

FIGURE 5.14.
Parallel rays of light from a distant source are brought to a focus by a mirror with the shape known as a paraboloid. In practice, the focal point would be about three times further from the mirror than it is shown to be here.

The simplest form of reflecting telescope is shown in Figure 5.15. The mirror brings objects on our distant plane p to a focus on p'. Exactly as before, the picture on p' is then magnified on to p'', where it can either be photographed or viewed by eye. In Newton's time, however, only viewing by eye was possible, and here there was the overriding difficulty that placing the human eye at p''

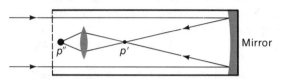

FIGURE 5.15.
The simplest form of reflecting telescope—impractical in Newton's time.

FIGURE 5.16.
Edwin P. Hubble at the prime focus of the 200-inch Hale Telescope. (Photograph by J. R. Eyerman, Time-Life Picture Agency.)

would have blocked light from reaching the mirror, since the mirrors of those times were small compared with human dimensions. So the simple device of Figure 5.15 was not then practicable. Yet this system is now employed in the 200-inch reflecting telescope at Mount Palomar as can be seen in Figure 5.16. The mirror of the Palomar instrument has so great a diameter that a human being can indeed sit inside the telescope without blocking out very much of the light! The observer does not view the image directly with the eye, however. He rides inside the telescope in order to operate the camera (and other instruments) and to ensure that the image plane p'' is kept in the correct position in relation to the camera.

In many respects large modern telescopes are extremely simple in their design. The mounting is more elegant than that shown in Figure 5.13. Instead of a polar axis between the bearings on the fixed pillars, there is a cradle, inside which the telescope itself can ride, as shown in Figure 5.17. It again

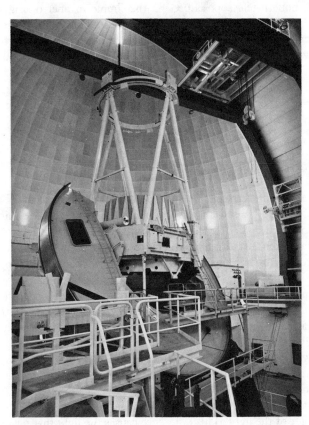

FIGURE 5.17.
The tube of the 150-inch Anglo-Australian telescope is mounted within a massive horseshoe. (Courtesy of the Anglo-Australian Telescope Board.)

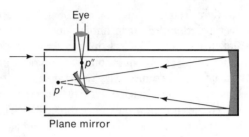

FIGURE 5.18.
Principle of the Newtonian telescope.

turns on a declination axis which passes through bearings fixed to the cradle. This system dispenses with the need for a counterweight on the declination axis.

When a modern telescope is used in the simple manner of Figure 5.15, it is said to be operated *at the prime focus*. But early reflecting telescopes were far too small to be used in this way. The simplest arrangement for a small telescope was suggested by Newton, who placed a flat mirror inclined at 45° in front of p', as in Figure 5.18. All light rays coming to p' are reflected in the flat mirror and are thereby caused to form an image plane p'' at right angles to p'. The observer views the image on p'' with a normal eyepiece mounted in the side of the telescope.

§5.3. Refractors versus Reflectors

The early reflecting telescopes certainly overcame the problem of chromatic aberration, but they raised another problem just as grave. The mirrors used in them were solid disks of metallic alloy, which were subject to gross changes of form due to temperature fluctuations. This caused such mirrors to deviate from the required paraboloidal form and led to very poor focusing. So it is not surprising that when, in the mid-eighteenth century, a method was found of overcoming chromatic aberration in the refracting telescope, the reflector fell into immediate disfavor. Interest in the reflector did not revive until about a century later, when Léon Foucault discovered a method of depositing a thin layer of silver on a glass surface.

Before we can understand how refracting telescopes overcame the problem of chromatic aberration, we shall need to examine more closely just how the problem arises. In Fig. 5.2 we saw that the ratio of the distances AX to DY is always the same for light of a particular color. But this ratio differs slightly with the color of the light. This difference causes the light that passes through the outer part of a lens to be separated into its constituent colors (Figure 5.19). It is said to be *dispersed*. In contrast, light of all colors passes straight through the center of the lens, and is therefore neither refracted nor dispersed. If the

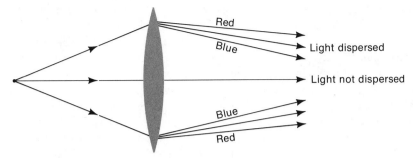

FIGURE 5.19.
Color dispersion produced near the periphery of a convex lens.

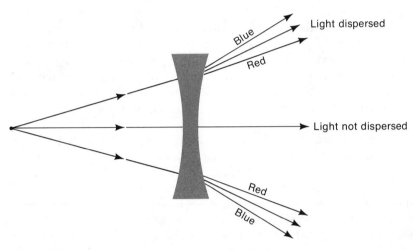

FIGURE 5.20.
A concave lens reverses the dispersion.

lens has concave instead of convex faces, the dispersion is simply reversed, as in Figure 5.20.

Thus one method of correcting for dispersion readily suggests itself. It is shown in Figure 5.21. We simply place a concave lens to the right of the convex one.

This explains a point that might otherwise seem puzzling. How was it that astronomers were so disturbed by the chromatic aberration introduced by the objective of a refracting telescope, and yet were undisturbed by the chromatic effects of the eyepiece? For even reflecting telescopes make use of lenses in the eyepiece!

The answer is that eyepieces were made with two lenses even in the time of Newton. The two lenses produced something of the effect shown in Figure 5.21, so that the chromatic distortions produced in the eyepiece were much less serious than those produced by the objective. The reason why an objective could not readily be corrected by the use of a second lens is that the two lenses, *if made of the same glass,* would need to be very widely spaced—a serious inconvenience. Eyepieces, on the other hand, being small, permit adequate separation without any such inconvenience arising.

But to come back to the objective of a refracting telescope: how is this to be corrected for chromatic aberration, widely spaced lenses being forbidden? Two quite different considerations are involved: the actual value of the ratio of AX to DY (Figure 5.2) for light of a certain color; and the degree to which that ratio alters when the color is altered. These two factors do not change in exactly the same way when the material of a lens is changed, for example, from one type of glass to another. This means that two lenses of different materials can have different ratios of AX to DY in yellow light, but the same degree of dispersion of the ratios with change of color. Then by making a convex lens from the material of larger ratio and a concave lens from the material of smaller ratio, we can produce the desirable situation in which the opposite dispersive effects of the two lenses (Figures 5.19 and 5.20) compensate each other, but in which there is still a net degree of refraction. If, moreover, the lenses are suitably shaped they can be fitted together into a doublet of the form shown in Figure 5.22, which then gives a focal plane p' that is substantially the same for all colors.

There seems little doubt that the man who discovered this method of making achromatic objectives was Chester Moor Hall, a London barrister whose hobby was making optical experiments. Being by nature a somewhat secretive man, Hall, in 1733, approached two different London opticians, one to grind the convex half of the doublet and the other to grind the concave half. Oddly enough, both of them subcontracted the work to the same craftsman, George Bass. Discovering that both lenses were destined for the same customer, Bass fitted them together and recognized their achromatic property. Bass was less reticent than Hall, and within the next few years several London opticians were in possession of the new idea and had begun to make achromatic lenses for themselves. Among them was John Dolland, a man of very high reputation

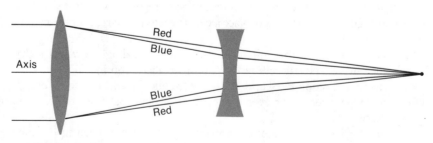

FIGURE 5.21.
Two widely spaced lenses of the same glass can cancel the dispersion effect.

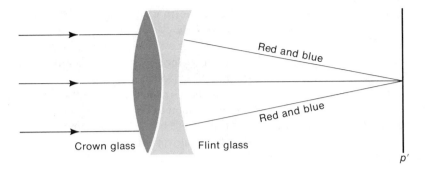

FIGURE 5.22.
By using lenses of different glass, one can avoid the need for spacing.

in the scientific world, who eventually joined his more commercially minded son Peter in a business enterprise at *The Sign of the Golden Spectacles and Sea Quadrant* in the Strand.

Peter Dolland persuaded his father to apply for a patent on the new device, and, although nobody ever claimed that John Dolland was the inventor, the patent was duly granted. Nevertheless, throughout the remainder of John Dolland's life, other British opticians seem to have gone on making achromatic objectives without let or hindrance. But soon after his father's death, Peter Dolland brought an action against one of them and was successful. Thereupon the London Opticians presented a petition to the Privy Council asking for the patent to be revoked. The legal proceedings which followed were long and complicated, but the upshot was that the Dolland patent was upheld. The court, presided over by Lord Camden, held that Chester Moor Hall, "the person who locked his invention in his scritoire," was not the person who ought to benefit by the patent. The right person to benefit was Dolland, "who brought it forth for the benefit of the public."

In fact it is to be doubted whether the granting of such sweeping patent rights is ever an expedient policy, for the interplay of ideas is thereby discouraged, and in the absence of competition the monopolist is apt to become lazy. Certainly, patent rights are hard to justify on moral grounds, for the bigger an idea the less it is patentable. You may make a fortune by patenting a better way of clipping an eraser to a pencil, but you will not make a cent in patent rights through the discovery of a new scientific theory of the scope and power of Einstein's. Society is well aware that only a king's ransom could pay for a really great scientific idea; so it makes no payment whatever.

In any event, the granting of the Dolland patent had an all but disastrous effect on the course of the optical industry in Britain. With the invention of the achromatic objective, the stage was set for the ultimate struggle between the refracting telescope and the reflector. But the British, who had played so large a part in the early development of the reflector and who had produced the first achromatic objective for the refractor, scarcely took any further part in the technological development of either form of telescope. The monopoly accorded to the Dollands allowed them, without any great effort, to produce better refracting telescopes than their immediate rivals could produce. Their rivals, discouraged by being debarred from using the correct technique, tended to wither away. Some fifty years after the Dolland case, the government, becoming alarmed by the rapid rise of the German optical industry, at last attempted through the Royal Society of London to encourage the manufacture of better optical glass in Britain. But the project failed ignominiously, for by then all really high-grade professional optical work in England was well-nigh destroyed.

In Germany things took a very different course. When, in the early years of the nineteenth century, the Dolland telescopes were critically examined by the young Joseph Fraunhofer (1787–1826), he found that none of the really important problems of the refracting telescope had been solved during the fifty years since Lord Camden's decision. The essential problem of the refracting objective was to choose the material of the lenses and to shape their surfaces in such a way as to give not only chromatic correction but also freedom from spherical aberration and coma in a distant object. This was one of the first problems in human history to demand both an accurate mathematical insight and a skilled practical technology. The mathematical insight was available in England, but the practical technology was sadly lacking.

The two requirements were combined in the person of Fraunhofer. It is not too much to say that Fraunhofer carried through, essentially by himself, a research program that would nowadays be assigned to a substantial team of scientists. A poor boy, Fraunhofer was trained as a glass technologist. Later, he acquired mathematical knowledge. In an energetic young man of genius, the combination proved irresistible. Fraunhofer realized that he must *begin* by measuring how much different kinds of glass would refract light of single colors, not light of mixed colors. This led him to the basic technological discovery that one particular kind of glass—flint glass—does not give reproducible results unless the conditions of its manufacture are controlled with extreme care; impurities in it produce variations of behavior. Now Fraunhofer's early training came to his aid. Furnaces were designed and built in which glass disks of stable optical quality could be produced. Fraunhofer's practical skill as a lens grinder and his mathematical knowledge of optics did the rest. The resulting telescopic objectives were sensibly free from chromatic aberration, spherical aberration (distance objects), and coma.

To Fraunhofer it was a simple matter to improve the rigidity and the accuracy of the normal equatorial mounting of the telescope. The final product was of a degree of excellence far surpassing anything that had been seen before. His $9\frac{1}{2}$-inch Dorpat refractor earned him freedom from the taxes of Munich. It did more than that. It shook the complacence of the British government sufficiently for the aforementioned glass-making project to be set under way. But all to no avail. With the failure of the glass-making project, the British government relapsed once more into technological somnolence.

Throughout most of his short life (he died of tuberculosis at the early age of thirty-nine), Fraunhofer was regarded by the scientific savants as a "mere technologist." He was allowed to attend scientific meetings but not to speak. It is therefore pleasant to record that in the course of his work he made a basic discovery which carried him far beyond the science of his own day, right into the science of the twentieth century.

In Figure II.5 (page 201), we have the complex "spectrum" of the Sun. William Wollaston (1766–1828) had observed a few of the dark lines of the solar spectrum, attributing them to natural divisions between the colors. A decade later, Fraunhofer showed that the number of these lines is very large indeed—he observed more than 600 of them. Throughout the nineteenth century, the nature and origin of these "Fraunhofer lines," as they became called, remained a quite unsolved problem. At the height of its success, in the last years of the nineteenth century, the classical theory of particles and of their radiative interactions failed to come to grips with this problem. Only with the development of modern quantum mechanics could we understand the meaning of the dark lines that Wollaston and Fraunhofer discovered more than 150 years ago.

During the era of the Dolland refractors, the reflector was by no means entirely eclipsed. In the last quarter of the eighteenth century, William Herschel (1738–1822), famed for his discovery of the planet Uranus, constructed with consummate skill a series of reflecting telescopes, culminating in one of 48-inch aperture. But although great results were achieved with these instruments, they all suffered from the defects already mentioned.

The magnificent Fraunhofer refractors transformed the situation. Professional astronomers the world over now had no doubt that refractors were much to be preferred to reflectors. Everybody wanted a Fraunhofer refractor. It was true that reflectors could be made with larger apertures, but because of the inefficiency of reflection at metallic mirror surfaces, a reflector of a given aperture was reckoned to have no greater light-gathering power than a refractor of only half the aperture.

Yet, ironically, what Fraunhofer's discoveries had really demonstrated was the ultimate impracticability of the refractor. Fraunhofer's success was based on the superb optical quality of his glass. It had of necessity to be free from bubbles and internal striae. It had to have precisely defined refracting properties. *And these characteristics are extremely hard to achieve in lenses of large aperture.* The Dorpat refractor, Fraunhofer's masterpiece, had an aperture of only $9\frac{1}{2}$ inches. In spite of the poor reflective efficiency of the mirrors of that time, there was no great difficulty in building a reflector with greater practical light-gathering power than that of the Dorpat refractor. If a refractor were to achieve equality with the largest reflectors, the aperture would have to be pushed up to more than 25 inches. Therefore strenuous efforts were made to increase the diameter of refractor objectives. In fact this effort succeeded only during the last decades of the nineteenth century. During the 1870s, two American observatories (Washington and McCormick, Charlottesville) installed

26-inch refractors, while Vienna had one with a 27-inch aperture. In the middle 1880s the Pulkovo Observatory in Russia and the Bischoffstein Observatory in France both had 30-inch instruments. Not until 1888 was a refractor with a still-larger objective installed—the 36-inch telescope at Lick Observatory in the U.S.A.

By that time Foucault had discovered how to silver a glass mirror. From then onward, reflectors were no longer subject to gross losses of light nor to the serious deformations temperature changes had caused in the early metallic mirrors. The reflector now went rapidly ahead, for it made far less exacting demands upon glass technology than did the refractor.

The glass disk out of which a large mirror is to be made must certainly satisfy the requirements of rigidity and of a low temperature coefficient of expansion, but the glass itself need not be of high optical quality. There can even be a plethora of bubbles and striae inside the glass so long as they do not interfere with the grinding of the surface. In contrast, the glass required for a refractor objective must satisfy the most stringent optical requirements. Hence very large mirrors can be made more easily and with less risk of inaccuracy than can very large lenses. For sound technological reasons, therefore, we seem to have reached the ultimate end of the race between refractor and reflector. The world's largest refractor, at present, is the 40-inch instrument at Yerkes, Williams Bay, Wisconsin. By way of contrast, reflectors with apertures in excess of 80 inches are listed in Table 5.1.

TABLE 5.1.
*Observatories owning reflectors with
apertures of more than 80 inches*

Location	Telescope(s)
Mt. Palomar, California	200″ reflector 48″ Schmidt
Mt. Wilson, California	100″ reflector
Mt. Hamilton, California	120″ reflector
Fort Davis, Texas	107″ reflector 82″ reflector
Kitt Peak, Arizona	150″ reflector 90″ reflector 84″ reflector
Cerro Tololo, Chile	150″ reflector
Siding Spring, Australia	150″ reflector 48″ Schmidt
Mauna Kea, Hawaii	84″ reflector
Nauchny, U.S.S.R.	104″ reflector

There is, however, one final issue in the refractor versus reflector struggle. Although the reflector was finally established as the more powerful light-gatherer, the traditional paraboloidal reflector suffers more severely from coma than the refractor does. This means that the reflector cannot be usefully employed when the object rays come into the mirror at more than a small angle to the axis of the mirror. In other words, the reflector necessarily has only a small field of view.

This disadvantage would probably have kept the refractor "in business" had it not been for the invention of a new type of reflector with far less coma than that inherent in the traditional paraboloidal mirror. The optical features of the new system were invented by Kellner in 1910 (American Patent no. 969, 785), but the first telescope embodying Kellner's ideas was not constructed until 1930, by Bernard Schmidt (1879–1935). Such telescopes are now known as Schmidt telescopes.

If rays are admitted through a circular opening onto a spherical mirror, as in Figure 5.23, a change of direction of the object point makes little difference. The rays are brought to a focus without coma, astigmatism, or chromatic aberration, but spherical aberration is now very serious. To overcome this difficulty, a correcting plate of good quality optical glass is placed across the circular opening. The surfaces of the glass are carefully figured to give a weak refraction, just sufficient to compensate for the spherical aberration of the mirror. The glass itself introduces optical defects, of course, but these are not serious unless the aperture becomes very large, in which case chromatic aberration raises new difficulties.

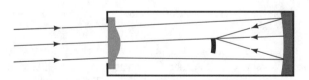

FIGURE 5.23.
Principle of the Kellner-Schmidt telescope, which borrows mirror from reflector and correcting plate from refractor.

A Schmidt telescope is in a sense a cross-breed between the reflector and the refractor. The mirror is borrowed from the reflector, the correcting plate from the refractor. Quite apart from the chromatic difficulty already noted, there is the difficulty at large apertures of obtaining and of shaping a large glass plate of adequate quality. So far, nobody has undertaken the figuring of a plate with a greater diameter than 48 inches. However, it is a less exacting task to shape a plate for only weak refraction than to grind an objective lens of equal aperture.

In recent years, the Schmidt telescope has proved extremely popular. Because of its large field of view, it enables the observer to accumulate astronomical material far more rapidly than with a traditional reflector. It was for precisely

FIGURE 5.24.
The 48-inch Palomar Schmidt. (Courtesy of Hale Observatories.)

this reason that the comprehensive sky survey carried out twenty years ago
by the Mount Wilson and Palomar Observatories was made with the aid of
a Schmidt telescope, shown in Figure 5.24. This form of instrument is
naturally popular with observatories situated in unfavorable climates, for in
the rare periods when astronomical conditions happen to be good, much
material can be obtained. The Schmidt telescope is also well-suited to handling
statistical problems involving large numbers of stars or galaxies. Traditional
reflectors are better suited to examining particular objects, which they can do
in greater detail than the Schmidt.

§5.4. The Wave Theory of Light

It is surprising that the design of such important optical instruments as the
camera, the telescope, and the microscope can be understood in terms of the
simple laws of reflection and refraction as applied to the concept of rays of
light. What are rays? Newton pictured a ray of light incident on a glass block
as a collection of "bullets" which move through the air and then strike the

glass interface, whereupon some bounce back into the air along the direction of the reflected ray while others enter the glass and move along the direction of the refracted ray. But this leads to the puzzling question: what decides whether a certain bullet is going to be reflected or transmitted?

Newton answered this question in a remarkable way. He suggested the bullets worked by "fits," so that a bullet would sometimes bounce back into the air, whereas on other occasions, under identical circumstances, it would continue on into the glass. This is just the concept required by modern quantum mechanics. We saw in the preceding chapter that a particle may go by any path between two points P and Q. Under identical circumstances it can go by quite different routes, which was Newton's idea. Newton's immediate successors, lacking the concepts of modern quantum mechanics, ignored his suggestion; and throughout the eighteenth century and into the nineteenth century, they became more and more impressed by the steadily mounting difficulties that confronted the bullet picture.

One of these difficulties arises when we consider how light travels from a distant source. All the light rays from such a source move essentially parallel to each other, as in Figure 5.25. Some of the light is made to pass through a hole AB in an otherwise opaque sheet and travel on toward a viewing screen. The light that just misses the edge of the hole at A reaches the screen at C, and the light which just misses the edge of the hole at B reaches the screen at D. So we see an area extending from C to D illuminated on the screen. All this can be understood very simply in terms of the Newtonian picture. We can say that the bullets which just miss striking the opaque part of the sheet at A continue to move in a straight line until they hit the screen at C, and those which just miss striking the sheet at B continue in a straight line until they hit the screen at D.

But suppose we decrease the size of the hole AB, as in Figure 5.26; what happens? So long as the hole remains fairly big, the size of the spot of light CD on the screen decreases exactly as we might expect on the basis of the Newtonian picture. But if the diameter of the hole is reduced to a small fraction of a millimeter, something quite different happens. The spot of light on the screen then begins to increase again, so that we have the apparently paradoxical result that as the hole in the opaque sheet becomes still smaller, the area of light on the screen becomes larger. We might attempt to explain this by saying that somehow the light has managed to turn a corner, but this is something

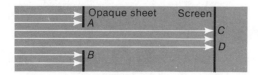

FIGURE 5.25.
Bullet picture of light passing through hole does not explain the observed situation.

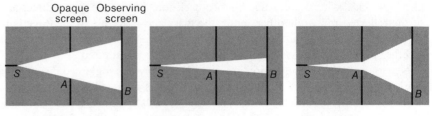

FIGURE 5.26.
Light from a point source passes through a hole in an opaque screen onto a viewing screen.
If light traveled in straight lines, as in (a), the illuminated area should decrease as the size of
the hole decreases. In fact, it does so until the diameter of the hole is reduced to about
0.01 mm. Then it begins to *increase.*

that our bullets are not allowed to do, for the Newtonian picture postulates
that they move *consistently* in straight lines.

Although we cannot concede that bullets can turn corners, we can do so
for waves. Figure 5.27 shows a succession of waves advancing on a breakwater
which has a vertical slit at the point *P*. As the waves reach the breakwater a
disturbance goes through the slit. New waves travel outward from the point
P on the far side of the breakwater, and they travel outward *radially*. That
is, they look just like the ripples that are produced by dropping a stone into
still water. If there is a second obstacle beyond the breakwater—a wall, say—the
disturbance from *P* will spread over a large area on that obstacle, and will not
be simply confined to a central spot at *C* directly opposite to *P*. In other words,
the waves have succeeded in turning a corner.

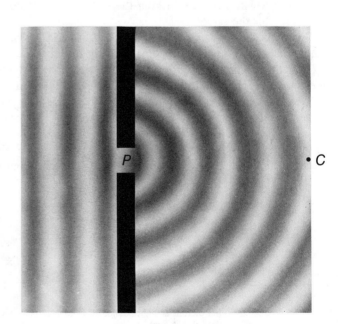

FIGURE 5.27.
Waves turn a corner on passing through a narrow slit in the breakwater.

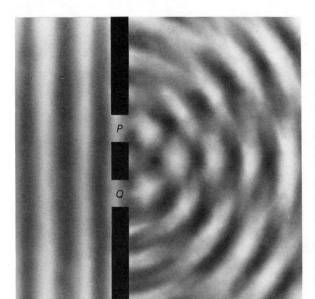

FIGURE 5.28.
If there are two slits, crests will reinforce crests in some directions, troughs will cancel crests in others.

Thus the way in which light travels through very small apertures, coupled with what they knew about ordinary water waves, suggested to many of Newton's successors that some form of wave motion might be needed to explain the nature of light, and that the bullet idea might be completely wrong. The thing to do was to put the matter to further experimental test.

Before looking at the kind of test needed, let us think a bit more about water waves. Suppose we make two vertical slits in the breakwater, at P and Q, as in Figure 5.28. Each point on the obstacle beyond the breakwater will now receive disturbances from both P and Q. What happens at a specific point depends on the timing of the waves. If the crests of the two waves arrive at the same moment, there will be an especially high wave; but if a crest of the waves from P arrives simultaneously with a trough of the waves from Q, then the crest and the trough will tend to cancel each other out and there will be little or no disturbance.

The situation is illustrated in Figure 5.29. It is assumed that P and Q are entirely similar slits in the breakwater, and that the point O is midway between them. From O a number of radiating lines can be drawn, one of them, OC, being along the direction of the original wave motion. At any point along OC the peaks of the waves from P and Q arrive simultaneously. The troughs of the waves also arrive simultaneously. So at all points along OC there are especially high crests and low troughs. Exactly the same is true along the other heavy lines radiating outward from O. But lying between these lines are other, lighter lines. Along these, the peaks of the waves from one slit in the breakwater arrive simultaneously with the troughs of the waves from the other slit, so that there is no disturbance at all. These are lines of still water. To complete

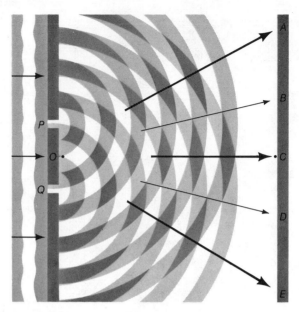

FIGURE 5.29.
Along *OC* and the other heavy lines, reinforcement produces
high crests and low troughs. Along lightly marked lines, cancelation produces the effect of almost still water.

the picture, suppose now that we have a sea wall inside the breakwater, as in
Figure 5.30. Then, at the points *A, C,* and *E,* where the heavy lines meet the
sea wall, the waves will rise high and fall low; but between those points, at
B and *D,* the water will remain still.

Experiment shows that an entirely analogous phenomenon occurs with light.
In fact, we can replace the original water waves to the left of Figure 5.29 by
light incident from a distant source. We can also replace the breakwater by
an opaque sheet in which two parallel slits are cut at the points *P* and *Q,* and
we can replace the sea wall by a viewing screen. On the viewing screen we
then find that we obtain a series of bright bands or fringes, as in Figure 5.31.

But we must be cautious in at least one respect, in applying our water-wave
analogy. The length of the fringes in Figure 5.31 is not produced by the rise
and fall of the waves. It is simply due to the size of the slits at *P* and *Q.* If
these were made longer, then the fringes would be longer. It is the *brightness*
on the screen which is the true analogue of the rise and fall of the water waves.
Points on the viewing screen at which the waves rise and fall by a large amount
appear bright; points where the waves cancel each other out—where a trough
from one slit arrives simultaneously with a crest from the other—appear dark.
So each point of a bright fringe is a place where the waves are rising high
and falling low, whereas each point of a dark region is a place where the waves
are interfering with each other and tending to cancel each other out.

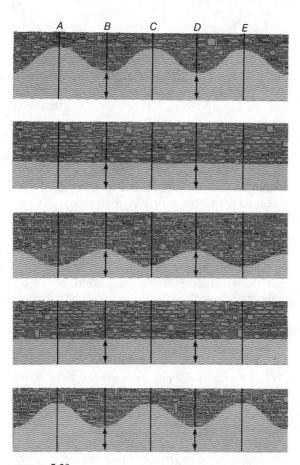

FIGURE 5.30.
This cross section through Figure 5.29 shows that at points *A*, *C*, and *E* on a sea wall inside the breakwater, waves rise high and fall low. At *B* and *D* the water is almost still.

FIGURE 5.31.
If we pass light through fine slits onto a screen, we also get places of reinforcement (the bright bands) and of cancelation (the dark bands).

Let us return for a moment to the behavior of water waves as shown in Figure 5.30. If the distances between the wave crests of the original waves to the left of the breakwater are changed, then the points *A*, *B*, *C*, *D*, and *E* on the sea wall will change also: the wider the spacing of the original waves, the greater the distance from *A* to *B* to *C*, etc. In fact, by carefully measuring the distance between the slits in the breakwater, the distance separating the sea wall from the breakwater, and the distances between the points *A* and *B*, *B* and *C*, etc., we can calculate the spacing of the original waves. In this way an observer on the sea wall can determine the distance between successive crests of the original waves without bothering to look outside the breakwater.

What is the analogue of this for light, and what, especially, is the analogue of the distance between the crests of the original waves to the left of the breakwater? The answer is color. Each pure color consists of a train of waves with the same definite fixed distance from one wave crest to the next. This distance is different for different colors. For blue light it is about 1/3000 part of a millimeter, for yellow light approximately 1/2000 part of a millimeter, and for red light about 1/1600 part of a millimeter. In order to make light turn corners, the width of the slit in the opaque sheet must be not much greater than the distance between the wave crests of the light, in fact, about 1/100 part of a millimeter. By everyday standards this would obviously be a quite extraordinarily thin slit, which explains why we are not used to seeing light turn corners.

For a pure color, the fixed distance between the wave crests is called by the obvious name of *wavelength*. Ordinary white light, as we call it, is a mixture of pure colors. It consists of a whole set of different wavelengths. These different waves can easily be separated, however, by making use of a point we noticed earlier in this chapter. The angle through which a ray of light is bent as it enters a glass plate depends both on the nature of the glass itself and on the color of the light. In particular, blue light is bent more than red light, as we can see from Figure 5.32. If a ray of light containing mixed colors

FIGURE 5.32.
The angle at which a ray of light is bent on entering a given plate of glass depends on the color of the light—that is, on its frequency.

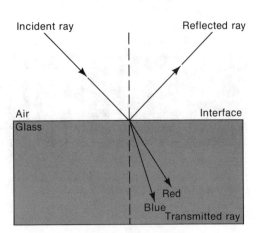

is allowed to hit a glass prism, the various colors as they pass through the prism are refracted differently, and in such a way that they can be separated out as they emerge from the far side of the prism. This is shown in Figure 5.33, where by using a viewing screen the separated colors may be observed. Blue light appears at one extremity and red light at the other, the remaining colors of the spectrum lying between the two extremes. So with this simple device we can separate ordinary white light into the colors of the rainbow. Indeed, in the phenomenon of the rainbow, water drops in the atmosphere act like the prism in our diagram, separating the ordinary white light from the Sun into its constituent colors.

If we are interested only in separating one specific color from all the other constituent colors of white light, then the still simpler method illustrated in Figure 5.34 is available. All we need do is to pass the light through a filter. For example, if we want to obtain yellow light only, we simply pass the original white light through a piece of yellow glass. The yellow glass allows only the yellow light to pass through, and absorbs all the remaining colors.

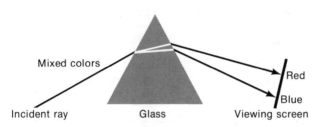

FIGURE 5.33.
We can use a glass prism to separate light of mixed colors, projecting the separate colors onto a viewing screen.

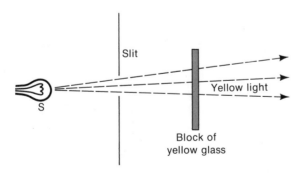

FIGURE 5.34.
Yellow light can be obtained by passing light with many colors through a block of yellow glass.

We may now profitably compare the wave picture of light with the Newtonian bullet picture. We have seen that the wave picture offers a reasonable explanation of how light turns corners—an explanation which the bullet picture does not offer. On the other hand, we have seen above that the bullet picture gives a satisfactory explanation of the construction and operation of the telescope. Can we explain the focusing property of a telescope, or even of a single lens, within the framework of the wave picture? If we can, then all the conclusions drawn earlier will still hold good, and the wave theory will clearly offer a wider range of necessary explanations than the Newtonian bullet theory does.

Let us look first at Figure 5.35, where a train of light waves traveling in the direction of the arrows encounters a convex lens. We make the important assumption that the wave travels more slowly through glass than it does through air. Since some parts of the wave must travel farther through the glass than other parts, and are therefore slowed down longer, the wave crests will be curved when they emerge from the lens, as in Figure 5.36, instead of being ranged in parallel planes as they were before entering it. Provided we make the lens correctly, we can delay the central part of the wave just sufficiently to ensure

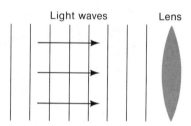

FIGURE 5.35.
Light with parallel wave crests approaches a convex glass lens.

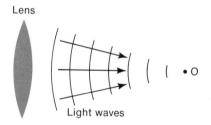

FIGURE 5.36.
Provided light travels more slowly through glass than through air, the wave theory can explain the focusing property of the lens.

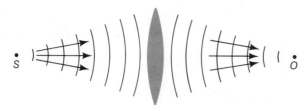

FIGURE 5.37.
The wave theory also can explain how such a lens brings light
from S to a focus at O.

that on emerging to the right of the lens the wave takes on a convergent spherical form. This is very easily understood by recalling how water waves spread out in concentric circles from a stone dropped into water. Here we have exactly the opposite situation: in this case, instead of spreading out, the waves converge. According to the wave picture, it is just this convergence which constitutes the focal property of the lens.

It is noteworthy that the wave picture brings out very clearly the necessity for a correct shaping of the lens. If the lens were made unevenly the waves would emerge in some non-spherical form, in which case they would not converge to a point. The wave picture also explains the necessity for making the lens of perfect optical glass: there must be no uncontrolled variations in the speed at which the wave travels through the glass, such as would happen, for example, if the glass contained bubbles of air.

In an exactly similar way, we can understand the focusing of light from a source S, shown in Figure 5.37. The light wave from S moves radially outward until it encounters the lens. Because of the delay through the central portion of the lens, the shape of the wave is altered as it emerges to the right. So long as the delay at the center of the lens is large enough compared with the delay at the periphery, the emergent wave will be changed into a convergent form, as shown. If, however, the delay were insufficient for this, and a wave were to emerge in a modified but still spherical form, the light would not be brought to a real focus. We should then have the situation shown in Figure 5.38. In this case, the center of the spherical emergent wave lies to the left of S, and is said to be a virtual focus at the point O_1.

Similarly, too, the wave picture enables us to understand the operation of concave lenses. These result in more delay at the periphery of the lens than at the center, and this causes the original spherical form of the wave form S in Figure 5.39 to be accentuated. The center of the diverging spherical wave to the right of the lens must therefore lie nearer to the lens than the point S—at O_1 in the figure. It is clear, therefore, that all the essential features of the operation of lenses can be just as well explained by the wave picture as by the bullet picture, provided that light travels more slowly in glass than in air.

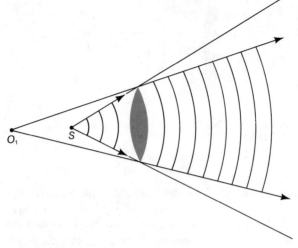

FIGURE 5.38.
The wave theory explanation of how light from S comes to a
virtual focus at O_1.

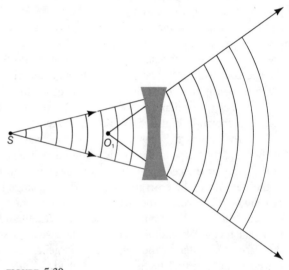

FIGURE 5.39.
A concave lens accentuates the spherical form of the wave
from S. Thus the center of the emergent wave is at O_1.

An interesting point now arises. If we look again at Figure 5.36 we may
reasonably ask whether the wave produces any disturbance at points near O.
We can decide this question by using a simplification first discovered by
Christian Huygens. The effect of the wave at future times can be decided by
first taking the position of the wave at the present moment, and then con-

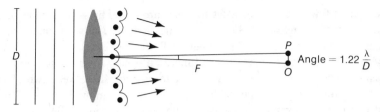

FIGURE 5.40.
At the focal point O, all wavelets augment each other, but they cancel each other at P when the distance $OP = 1.22\,\lambda F/D$, the angle subtended by OP being equal to $1.22\,\lambda/D$.

sidering subsidiary waves to spread out from all points of the present wave front. In Figure 5.40 we have a spherical wave emerging from a lens. This is to be the "present" state of affairs. To calculate the "future," we imagine new wavelets to spread out as shown. If we now wish to find what disturbance occurs at some point P close to O, we must work out how all these subsidiary wavelets affect each other when they reach P. All the subsidiary wavelets add together, by the way, when they reach O, the focal point, so clearly they will all augment each other at O. Calculation shows that the subsidiary wavelets will all cancel each other out at the point P when the distance from O to P is given by the formula $OP = 1.22\,\lambda\,\dfrac{F}{D}$, where λ is the wavelength of the light in question, F is the distance from O to the lens, and D is the diameter of the lens.

This complication is worth grappling with because it has an interesting application to the resolving power of the telescope. Suppose a telescope is pointed toward a distant star. According to the Newtonian picture, the image of the star is formed precisely at a specific point of the focal plane—the point O in Figure 5.40. According to the wave picture, if we place a screen at the focal plane of the telescope, we will obtain not a mere point of light but a circular disk of light. Indeed, not until we reach a distance equal to OP from the center of this circle of light, will the screen appear to be dark. Next suppose that there is a second star lying quite close to the first one. The image of the second star on the focal plane will also be a circle of light. Unless this second circle is well distinguished from the first one, the telescope will not tell us that there is a second star there at all, for the two images will be fused together. For the two circles of light to be well separated from each other, the center of the second must be at least as distant from the center of the first as the distance from P to O. This, in turn, means that the angle between the directions of the two stars must be at least equal to the angle marked in Figure 5.40, namely 1.22 times the wavelength divided by the aperture of the telescope. If two stars are separated by an angle smaller than this, their images will be blurred together, and we shall have no certain indication of their separate existence.

Here are a few instances of what this fact implies. The same applies to the human eye as to the telescope. For the eye, the diameter D is very small—under normal conditions only about 2 millimeters. Remembering that the wavelength of light is only about $\frac{1}{2000}$ part of a millimeter, it is easy to calculate that under normal conditions the human eye is unable to distinguish between two objects that are separated by an angle of less than about one minute of arc. This is about the order of accuracy achieved by the best observers in the days before the telescope. But using a telescope with an aperture of 20 inches, it is theoretically possible to distinguish between two stars separated by an angle of as little as a quarter of a second of arc. And with a telescope as large as the one at Palomar Mountain, with an aperture of 200 inches, the theoretical resolution is about one-fortieth of a second of arc. In point of fact, the slight variations in the positions of stars, caused by the Earth's atmosphere and always present to some degree, even on the clearest and steadiest nights, prevents the theoretical resolving powers of large telescopes from ever being achieved in practice.

These considerations sharply remind us that a telescope is not merely a collector of light. It also overcomes the inherent handicap of the human eye—that it cannot, unaided, distinguish between two objects lying nearly in the same direction.

So great was the prestige of Newton that many people still refused to accept the evidence for the wave nature of light, even after it had been demonstrated by experiments, experiments that followed along lines similar to those we have just considered. Because of this, attempts were made to find a crucial experiment that would finally decide between the relative merits of the Newtonian picture and of the wave picture.

Such an experiment was indeed found. We have seen that the wave theory is tenable only if it is true that waves of light travel more slowly through a medium such as glass than they do through air. The Newtonian picture, on the other hand, can be valid only if light travels faster through glass than through air. It will be recalled that when light which has passed through air is incident on to a glass surface, the transmitted ray in the glass is bent nearer to the normal than was the original incident ray. Newton explained this fact by supposing that his particles, or bullets, gained speed as they entered the glass. The argument was that glass attracted the particles, so for light striking a plane glass surface there was an increase of speed and, moreover, the increase was entirely in the direction normal to the surface. It is easy to see that this would cause the direction of motion of the bullets to become nearer to the normal than it was during their passage through the air. The thing to do, clearly, was to measure the speed of light through a solid or liquid medium, and to compare it with the speed through air. According to the Newtonian picture, the speed should be greater in the denser medium; according to the wave picture, it should be greater in air.

The experiment was actually carried out by Foucault in 1850. The equipment he used is shown schematically in Figure 5.41. Light from a source S passes through a small hole. Part of it then traverses a half-silvered mirror, G. Next

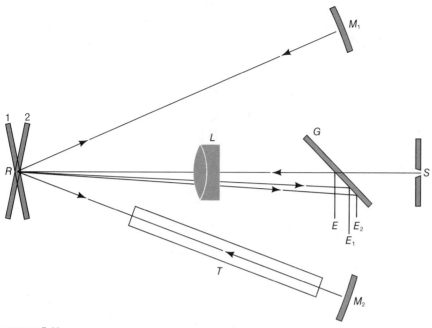

FIGURE 5.41.
The wave theory of light is tenable only if light travels more slowly through a medium such
as glass or water than it does through air. Here is the experiment Foucault used to prove that
it actually does so.

it is focused through a lens L on to a plane mirror at R. When R is in position
1, the light is reflected onto the mirror M_1, which returns the light immediately
to R. The mirror R is rotating rapidly, however, so that although the light
takes very little time to travel from R to M_1 and back again, by the time the
light has made this double journey R is not quite in the same position as it
was before. That is to say, it is not quite in position 1. Hence R returns the
light through the lens L along a slightly different path. Part of this light strikes
the half-silvered mirror G and is reflected now into the eye at the point E_1.
The experiment is repeated, with the rotating mirror R started in position
2. In this case the light is sent toward the mirror M_2 instead of toward M_1.
The distance from R to M_2 is exactly the same as the distance from R to M_1,
but between R and M_2 is a tank (T) filled with water, so that the light has
to traverse the water in order to pass from R to M_2 and also in order to return
from M_2 to R. The mirror R is rotating at exactly the same speed as in the
first experiment, and because of this the light is returned to the lens L again
along a track slightly different from the one it originally traversed on its journey
from the source to R. Again light is reflected from the mirror G into the eye,
but this time at E_2. Here we come to the crux of the matter. If light travels
more rapidly through water than it does through air, as the Newtonian picture
requires, then the point E_2 will lie to the left of E_1. But if light travels more
slowly through water, as the wave picture requires, then the point E_2 will lie

to the right of E_1. Foucault established that E_2 does, in fact, lie to the right of E_1, thus vindicating the wave theory. Thenceforward, for the rest of the nineteenth century, nobody gave any very serious credence to the bullet picture.

§5.6. *The Nature of a Light Wave*

A water wave has three basic properties. First, at each point there is an oscillation; the water moves up and down. This is easily shown by putting a float on top of the water. Second, there is a spatial correlation between the up-and-down motions at different points. This is illustrated in Figure 5.42. A peak

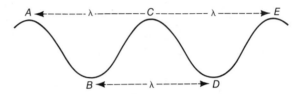

FIGURE 5.42.
At every point of the wave there is an oscillation; λ denotes wavelength.

at A is followed by a trough at B, and that trough is then followed by another peak at C, and so on. Not only is there an oscillation at each point taken by itself, but also different points have an orderly relation with respect to each other. If at one point the wave is up, then at an adjacent point it will be down, and so forth. This spatial ordering is measured by the wavelength (λ), the distance between two adjacent wave crests, or two adjacent wave troughs. As time proceeds, the whole spatial pattern moves along as shown in Figure 5.43. The effect of this motion is to produce the oscillation at each separate point. At one moment, at a given place, the wave is up, and at a later moment it is down. The time required to complete the oscillation at each point is simply the time required by the wave to travel through a distance equal to the wavelength. If the speed of travel of the wave is V, then the time required for the wave to move through the distance λ is simply $\lambda \div V$. This is the time that a float placed on the water takes to move from its highest to its lowest position and back again.

The third feature of water waves is that the motion of the waves can cause the whole train to move through the water, as it does when we drop a stone into a still pond. Waves spread outward—they actually *travel* outward through the water. At one moment waves have not yet reached a post. At a later moment they have traveled outward beyond it. In practice, wave trains are always of finite length. The height of the waves gradually dies away at the edges of the train, as in Figure 5.44.

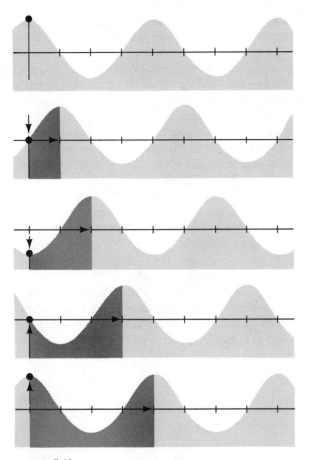

FIGURE 5.43.
Movement of float shows that the oscillation at any one point
is completed in the same time that the whole wave takes to
move through a distance equal to the wavelength.

Finite wave train

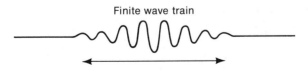

FIGURE 5.44.
Wave trains are always of finite length. Height of waves decreases toward edges of wave train.

So the three basic properties of waves are these: at each point there must be something that oscillates (for water waves, it is the up-and-down motion of the water); then there is a correlation between the states of this oscillation at different points, a regular sequence of peaks and troughs; and as time proceeds, the whole spatial pattern moves along, causing a finite train of waves to propagate itself. Before we decide to accept the wave picture of light, it is reasonable to ask whether light possesses these same three properties.

Let us take them in the reverse order. Light certainly has the ability to propagate itself. A light signal emitted from some source certainly travels outward from that source in such a way that it can be received a moment later by a distant observer. Over small distances we are not very conscious of the time required for a light signal to reach us, simply because light moves so fast; but light emitted by a distant star may take thousands of years to travel across space to us. Light also exhibits the second property of our waves, a spatial correlation expressed in a sequence of regularly arranged peaks and troughs, as depicted in Figure 5.42. But does light exhibit an oscillation at each separate point, and if so, what is it that oscillates?

Figure 4.7, reproduced here as Figure 5.45, gives us the answer to this critical question. Here we have the radiative influence from a charged particle a acting on particle b. With particle a oscillating at frequency ν, and at a speed much less than the speed of light, the radiative influence on a distant charged particle

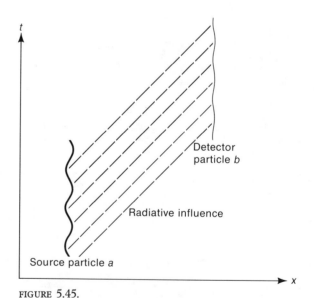

FIGURE 5.45.
Particle a is set in regular oscillation with frequency ν. Provided the speed of motion is small compared to the speed of light, the radiative effect of the oscillation of a is to cause b also to oscillate with the same frequency ν.

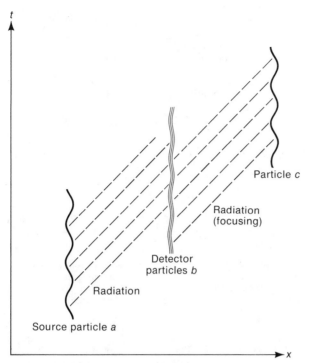

FIGURE 5.46.
The small amplitude of oscillation of many detector particles *b* can be used to make *c* oscillate with a larger amplitude.

b causes it also to oscillate at frequency *ν*. It is this radiative influence, the ability to cause the oscillation of particle *b*, that possesses the wave property. Particle *b* oscillates precisely because it responds to the oscillation of a wave. It is usual to refer to this radiative influence as an *electromagnetic field*, thinking in terms of a wave motion for the field.

Figures 5.46 and 5.47 are also taken from the preceding chapter. In Figure 5.46 we have particle *a* causing a number of particles *b* to oscillate, and these particles in turn causing a further particle *c* to oscillate with increased amplitude. We referred in Chapter 4 to this effect as a focusing process. In this chapter we have had practical examples of this process. The particles of the metallic coating at the surface of the mirror of a reflecting telescope constitute the particles *b*, while the particles of a photographic plate, which we may place at the focal plane of the telescope, constitute *c*. In a refracting telescope, the particles on the focal plane again constitute *c*.

In Figure 5.47 we have the situation in which particles *b* do *not* cause *c* to oscillate unless the frequency *ν* is close to some assigned value *ω*. This effect is achieved by the simple prism device of Figure 5.33. Particles *b* are those

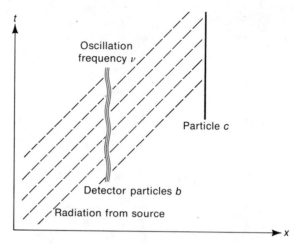

FIGURE 5.47.
When the particle *c* of Figure 5.46 is *prevented* from moving unless the oscillation frequency *ν* of the detector particles *b* is close to some assigned frequency *ω*, which can be determined by the observer, we have the principle of the spectroscope.

constituting the prism itself. The required property with respect to *ω* is achieved by means of the geometrical position of particle *c*. Thus in Figure 5.33, particle *c* lies on the viewing screen, where according to position on this screen we have a separation of "colors," i.e., of frequencies, there being a specific position corresponding to an assigned value of *ω*. A particle at this position will *not* be made to oscillate unless there are incident waves with frequency close to *ω*.

In this way we can relate the general statements of Chapter 4, which already contained the essential logical ideas, to the practical devices discussed in this chapter.

§5.7. Wavelengths and the Speed of Light

From Figures 5.42 and 5.43 we can see that the time required for a complete oscillation in the height of a water wave at a certain point *A* is equal to the time required for the wave structure to move through a distance λ, namely $\lambda \div V$, where *V* is the velocity of propagation of the waves. Now, the time required for such a complete oscillation is just the reciprocal of the frequency, $1 \div \nu$, so that with $\lambda \div V$ equal to $1 \div \nu$, it follows that $V = \lambda\nu$.

A similar relation also holds good for light waves, or indeed for radiation of any frequency, $c = \lambda\nu$, it being usual to denote the speed of radiation by *c* rather than by *V*. Since the value of *c* is accurately known, $c = 2.997929 \times 10^{10}$ cm per second, we evidently can determine λ if we know

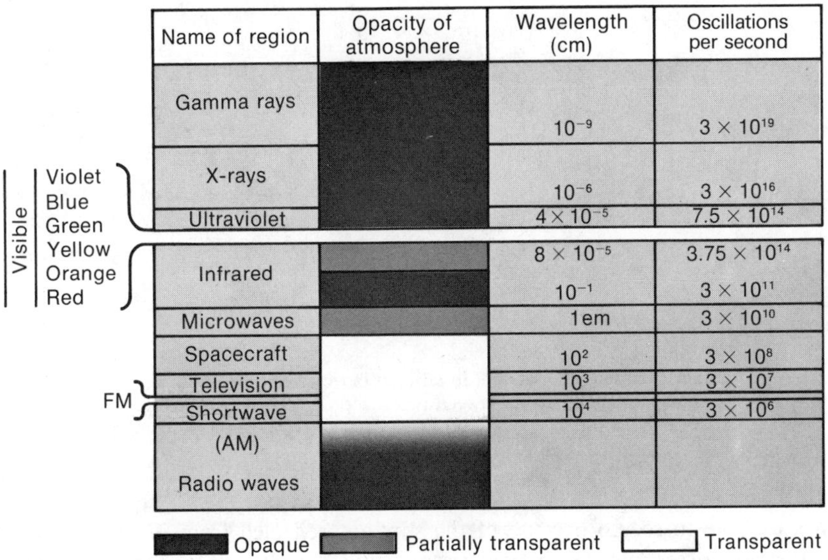

Name of region	Opacity of atmosphere	Wavelength (cm)	Oscillations per second
Gamma rays		10^{-9}	3×10^{19}
X-rays		10^{-6}	3×10^{16}
Ultraviolet		4×10^{-5}	7.5×10^{14}
Infrared		8×10^{-5}	3.75×10^{14}
		10^{-1}	3×10^{11}
Microwaves		1 em	3×10^{10}
Spacecraft		10^2	3×10^8
Television		10^3	3×10^7
Shortwave		10^4	3×10^6
(AM) Radio waves			

Visible: Violet, Blue, Green, Yellow, Orange, Red

FM: Television, Shortwave

■ Opaque   ▨ Partially transparent   ☐ Transparent

FIGURE 5.48.
Wavelength values have been added to the frequencies of Table 4.1. Radiation in the unshaded areas passes through the Earth's atmosphere from outer space.

ν, and *vice versa*. We referred above to yellow light as having a wavelength of about $1/2000$ part of a millimeter, $\lambda = 5 \times 10^{-5}$ cm. The corresponding frequency is about 6×10^{14} oscillations per second. Figure 5.48 repeats the frequency distributions of Table 4.1, but now with wavelengths also included.

We have referred many times to the speed of light as being about 3×10^{10} cm per second. How was this enormous speed measured? We end this chapter by discussing this question.

Observing a man in the distance wielding a heavy hammer, Newton noticed that he could see the blows of the hammer ahead of their sound. From this he concluded that sound traveled more slowly than light, and he proceeded to set up an experiment which measured the speed of sound. But even the ingenious Newton failed to devise an experiment for measuring the speed of light, so fast does it move. Yet the first measurement by Olaus Roemer (1614–1710), a Danish astronomer, resulted directly from an application of Newton's gravitational theory. The innermost of the four large satellites of Jupiter, Io, undergoes frequent eclipses by Jupiter itself whose occurrences can be predicted ahead of time from Newton's theory. A table of such predictions was published in 1668 by G. D. Cassini (1625–1712). The trouble that emerged was that observations deviated from the predictions by several minutes of time. In 1675, Roemer offered the explanation, illustrated in Figures 5.49 and 5.50, for this discrepancy.

Figure 5.49 shows the motion of Io in an orbit around Jupiter. The scale of this orbit is about the same as that of the Moon's orbit around the Earth.

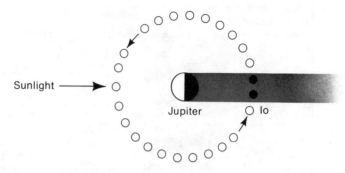

Sunlight

Jupiter

Io

FIGURE 5.49.
To find how long the satellite Io takes to move around its orbit, we measure the time interval between two successive occasions when it moves into eclipse.

We are able to see Io because of the sunlight reflected from its surface, but at times it passes into the shadow cast by Jupiter itself, and we cannot observe it when it lies in this shadow. Suppose now that we wish to determine the length of time that Io takes to move once around its orbit. The obvious method would be to make a note of the moments at which Io moves into eclipse. Then we might say that the time-interval between successive moments of eclipse determines the time taken to complete one circuit. But is this *exactly* right? To examine whether it is or not, we must consider a little more closely just what happens when we make our observations.

At the moment the satellite passes into the shadow, light ceases to be reflected from its surface. The change, from light being reflected to light not being reflected, travels across space from Jupiter to the Earth, so that the cessation of light is recognizable on the Earth only some time after it has actually taken place. But how long after? The answer depends on how far away the Earth is from Jupiter. In fact, the delay is simply the distance between the Earth and Jupiter divided by the speed of light. Provided the delay is precisely the same at two successive moments of eclipse—that is, provided the distance of the Earth from Jupiter is precisely the same—then our method of measuring the time which Io takes to complete one circuit will be correct. But if the distance between the Earth and Jupiter changes between one eclipse and the next, our method will be incorrect, because the amount of delay will be different for the two eclipses.

Does the distance of the Earth from Jupiter change during the time interval between two successive eclipses? It must change if the Earth is moving toward or away from Jupiter. As Figure 5.50 shows, the relationship between the Earth and Jupiter changes throughout the year. At the points *A* and *C* of the Earth's orbit, the Earth is moving transversely to the direction of Jupiter, and the distance of the Earth from Jupiter does not then change appreciably between two successive eclipses of Io. But when the Earth is at *D*, the distance shortens

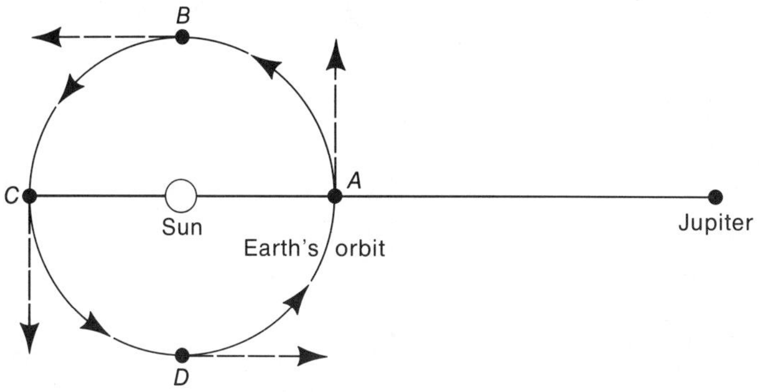

FIGURE 5.50.
When the Earth is at D, the distance between Earth and Jupiter shortens
between successive eclipses of Io. Light, moving at finite speed, takes less
time to reach us, and we thus underestimate the time between two eclipses.
When the Earth is at B, the reverse applies.

steadily, which means that we shall underestimate the time between successive
eclipses. When the Earth is at B, the reverse applies: the distance then lengthens
steadily, which means that we shall overestimate the time between successive
eclipses.

What was found, following the publication of predicted eclipses for Jupiter's
satellites, was that their apparent periods of motion do shorten when the Earth
is moving toward Jupiter and lengthen when it is moving away. And, indeed,
from the amount of the shortening and lengthening, Roemer was able to deduce
the speed with which light moves. In fact, the fractional changing of the orbital
period of a satellite is simply the ratio of the speed of the Earth's motion to
the speed of light. Roemer estimated the fractional shortening of the period
of Io and, since he knew approximately the speed of the Earth's motion, was
able to deduce an approximate value for the speed of light. His result seemed
to his contemporaries to be impossibly large, and it was not widely believed
at the time. Only when Roemer's enormous speed was confirmed by a different
method, about fifty years later, did it come to be generally accepted.

It was more than another century before the speed of light was measured
in the laboratory. In the mid-nineteenth century, Foucault used a method
already illustrated in Figure 5.41. Imagine the tank T, the mirror M_2, and
position 2 of the rotating mirror R removed from this figure, and compare
the point E_1 with the point E, as in Figure 5.51. The position of E is measured
when R is not rotating. Thus with R at right angles to the light rays from
S, the rays are returned toward S, and some of them are reflected to E by
the half-silvered mirror G. On the other hand, E_1 is measured with the mirror
R rotating. As before, the rays travel from S to position 1 of the mirror R.
They are reflected to the mirror M_1, which returns the light to R. Because
of the time required for the light to move from R to M_1, and back again,

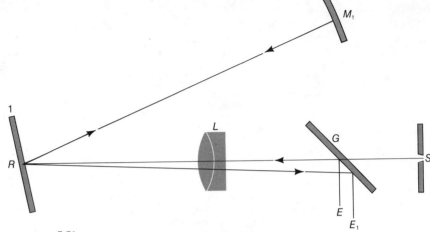

FIGURE 5.51.
When the tank of Figure 5.41 is removed, the resulting arrangement can be used for measuring the speed of light.

R moves slightly away from position 1, causing the rays to be returned to *G* along a slightly different path, which causes E_1 to be slightly displaced from *E*. By measuring this displacement, and relating it to the distance from *R* to M_1, and to the rotation rate of *R*, the speed of the light can be calculated.

PROBLEMS:
1. The distance from *E* to E_1 is measured with the aid of an eyepiece attached to a micrometer, the direction of travel being parallel to the direction from *S* to *R*. Writing *n* for the rotation rate of *R*, show that the distance from *E* to E_1 is $[(2RM)/c] \cdot 2\pi n \cdot RG$, where *RM* and *RG* are the distances from *R* to *M* and to *G*, respectively. How would you design this experiment to obtain the best accuracy in the determination of *c*?

2. For a convex spherical lens, the focal length is given by

$$\frac{\text{constant}}{\text{focal length}} = \frac{1}{r_1} + \frac{1}{r_2},$$

where r_1 and r_2 are the radii of the spheres of which the lens surfaces are portions. If one of these surfaces were made concave, say, the r_2 surface, with r_2 larger than r_1, the formula would be changed to

$$\frac{\text{constant}}{\text{focal length}} = \frac{1}{r_1} - \frac{1}{r_2}.$$

Consider this situation from the point of view of the wave theory of light. If r_2 were less than r_1, this second formula would lead to a negative value for the focal length. What would such a negative value mean?

3. The focal length of the 200-inch telescope is 55 feet. Working at the prime focus, an astronomer wants to photograph a portion of the sky up to 10 arc minutes from the axis of the telescope. Using no magnifier, as one needs to do in smaller telescopes, but only lenses that correct for coma, how large must the plate be? On the 48-inch Schmidt telescope, whose focal length is 11 feet, he wants to photograph the sky as far as 3° off-axis. Under the same condition of zero magnification, how large must the plates be?

4. The moving parts of a large modern telescope weigh about 300 tons. Yet the setting of the polar and declination axes must be accurate to within a thousandth of an inch. Consider the corresponding problems of accuracy of setting for a telescope carried by a satellite into space.

5. A telescope is driven about its polar axis at a rate that is intended to compensate for the rotation of the Earth. If this could be done with complete accuracy, and if the polar axis were exactly parallel to the Earth's axis of rotation, an object once centered in the field of the telescope would remain centered there. In practice, no system of driving gears can be made with perfect accuracy, and in practice there must always be a slight angle between the polar axis and the rotation axis of the Earth. These imperfections, if unchecked, cause the object to become decentered, and indeed to wander slowly out of the field of the telescope. Consider what steps of a manual, and of an electronic, nature might be taken to prevent such an unwanted drift of position from taking place.

6. For mirrors of reflecting telescopes, all of the same focal length, the intensity of illumination of the focal plane is proportional to the square of the mirror diameter. For mirrors all of the same diameter, how does the intensity of illumination depend on the focal length? Is the result the same for refracting telescopes?

7. Consider the reasons why ordinary commercial cameras are built with a lens rather than with a mirror.

8. Many photographic emulsions have the undesirable property that at very low light intensities their ability to register an image depends on about the *square* of the intensity of the illumination. Two telescopes of the same focal length, one of aperture 60″, the other of aperture 120″, are used with such an emulsion to obtain a photograph of a very faint object. How much longer does it take on the 60″ instrument to obtain a picture of the same quality as that given by the 120″?

9. Considering your answer to (8), and the fact that it costs about six times more to build a 120″ telescope than a 60″, would you think it more economic to build several small telescopes rather than a single large telescope?

10. Electronic devices that depend directly on the intensity of illumination, rather than on the square, are currently being developed for the purpose of obtaining pictures of faint objects. In what sense will this development affect the answer to (9)?

1. Describe the laws of reflection and retraction that were discovered by Snell.

2. Using these laws, explain how the image of a distant object can be produced with the aid of (a) a convex lens, (b) a mirror.

3. What is meant by the focal length of a lens?

4. Explain the imperfections of an optical system implied by the following terms: spherical aberration, chromatic aberration, coma, astigmatism, field curvature.

5. Why in small telescopes is a magnifying eyepiece needed?

6. Describe the usual form of mounting of a telescope, and explain the arrangement used to compensate for the apparent rotation of the celestial sphere.

7. Draw a sketch to illustrate the Newtonian method for using a reflecting telescope.

8. How was chromatic aberration corrected in the objectives of refracting telescopes?

9. Compare the practical problems involved in building large reflecting and refracting telescopes.

10. How does a Schmidt telescope differ from the usual designs of both reflecting and refracting telescopes?

11. Why is light better thought of in terms of waves rather than in terms of rays of bullet-like particles?

12. Show how the behavior of a lens can be explained by the wave theory of light.

13. What experiment was devised in the mid-nineteenth century to decide between waves and rays?

14. Discuss the basic properties of a wave, and explain why the product of the oscillation frequency and the wavelength must equal the speed of the wave propagation.

15. Ordinary light contains waves of many frequencies. How may the different frequencies be separated?

16. The range of frequency from red to blue light is very narrow compared to the whole range going from radiowaves to x-rays. Yet visible light is of great importance, both subjectively and in astronomy. What reasons, biological and physical, can you offer for why this should be so?

17. How in the eighteenth century, by observing the eclipses of the satellite Io of Jupiter, was an approximate value derived for the speed of light?

Appendixes to Section II

Appendix II.1. The Emission and Absorption of Radiation

Radiation can be emitted and absorbed by a gas both at discrete frequencies, such as those of the Balmer series of hydrogen, and throughout a continuous range of frequency. Only the changes of state of free electrons can emit and absorb radiation throughout a continuous range of frequency. In contrast, discrete frequencies are emitted and absorbed only by transitions of electrons between bound states, in which the electrons are not free.

There are enough free electrons in the material at the photosphere of the Sun to produce a continuous range of frequencies, as is also true for other stars. It is found that the distribution of frequencies depends in a characteristic way on temperature. For a temperature of 5,800°K, the solar photospheric value, the distribution takes the form given by the curve of Figure II.1. The intensity scale in this figure has been arranged to make the total area under the curve equal to 6.41×10^{10} kilowatts, this being the power radiated by 1

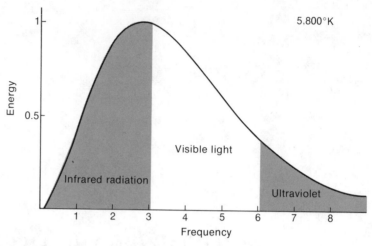

FIGURE II.1.
The frequency distribution of radiation from the Sun, with most energy as visible light, with somewhat less energy in the infrared (heat), and with a comparatively small contribution from the ultraviolet. To obtain the power radiated from 1 square kilometer of the solar surface, the temperature being 5,800°K, the unit of energy should be 1.16×10^{-4} kilowatt seconds. With the unit of frequency equal to 1.21×10^{14} oscillations per second, the area under the curve is then 6.41×10^{10} kilowatts.

square kilometer of the solar photosphere. Similar curves for stars with surface temperatures of 2,500°K, 4,000°K, 10,000°K, 30,000°K are shown in the four graphs of Figure II.2. Rising temperature produces greater relative intensities at the higher frequencies, and also greatly increases the radiated power from each square kilometer of surface. The curves of Figures II.1 and II.2 are often referred to as Planck curves, named for Max Planck (1858–1947), who first found out how to obtain them.

An astronomer could easily determine the photospheric temperature of any star, simply by studying the frequency distribution of its radiation, if the whole of the distribution were available to him. But severe absorption of radiation in the Earth's atmosphere, for frequencies higher than about 9×10^{14} cycles per second, seriously truncates the observable distribution for stars with a high surface temperature. This method works very well for stars whose surface temperature is not too high, however. Temperatures obtained in this way are called *color temperatures*.

In a diffuse gas, there may not be enough free electrons to produce very much continuous radiation, although there can still be enough material to emit significant radiation at discrete frequencies. An example of such a gas has already been seen in Figure 4.27b, which showed the Balmer series of hydrogen accompanied by only a weak distribution of radiation of continuous frequencies. The ratio of such discrete "line" frequencies to the continuous frequencies can vary greatly, depending on the density of the gas, its temperature, and its dimensions. By studying this ratio carefully, astronomers are often able to make important inferences about the density, temperature, and content of the gas clouds lying between the stars, and sometimes even about gas clouds in other galaxies.

Consider now the situation that arises if a diffuse gas should lie in front of a surface which generates a Planck distribution of frequencies corresponding to a certain temperature, say, T. Within the gas we have transitions of electrons between discrete states, for example, the transitions of the Balmer series of hydrogen. If the temperature of the gas is higher than T, downward transitions contribute excess radiation at the discrete frequencies in question, so that the continuous distribution of frequencies from the background surface then has excess radiation at certain frequencies superimposed on it. If, on the other hand, the temperature of the foreground gas is less than T, we have the opposite situation, in which upward transitions in the gas remove radiation from the background surface, again at the discrete frequencies determined by the transitions in question. Both these situations are illustrated in Figure II.3. Given the second situation, an observer sees a radiation distribution with a *deficit* at certain discrete frequencies, as is shown for the Balmer series in Figure II.4.

Now, the gases lying immediately above the photosphere of a star have a lower temperature than the photosphere itself. Consequently we should find radiation to be missing at certain discrete frequencies. That we do can be seen from Figure II.5, which shows the distribution of frequencies emitted by the Sun. These *absorption lines*, as they are called, are produced by many different kinds of atoms, atoms of calcium—notice the H and K lines of calcium—and of iron and other metals playing a particularly prominent role.

After falling for a while as we progress outward toward the top of the photosphere, the temperature rises with increasing height in the chromosphere of the Sun, as we saw in Chapter 2. With the temperature in the chromosphere higher than that at the photosphere, the tendency is for the chromospheric material to add excess radiation at certain discrete frequencies. But because the chromosphere contains comparatively little material, this effect is not very strong. It does occur, however, as can be seen from Figure II.6, which shows the Sun photographed by light at the frequency of the K line of calcium.

PROBLEM:
A star of high temperature, say, 25,000°K, is surrounded by a rotating disk of cooler gas containing hydrogen, as illustrated in Figure II.7. Explain why a distant observer in the plane of the disk sees absorption lines at the frequencies of the Balmer series, whereas a distant observer with a line of sight perpendicular to this plane sees the Balmer series as a set of bright emission lines. A class of stars known as the Be stars possesses such rotating disks.

198

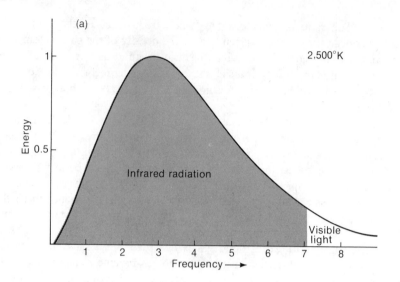

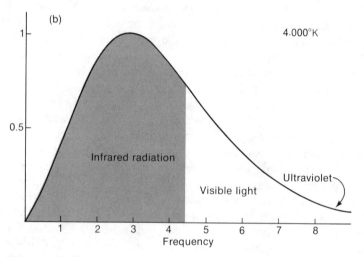

FIGURE II.2.

The basic shape of the radiation distribution is the same, irrespective of the surface temperature of a star. To take account of differing surface temperatures, we must choose different scales for the frequency unit. The scale appropriate for a star is simply proportional to the star's temperature. (For example, the scale for temperature 30,000°K when divided by 12 is the scale for temperature 2,500°K.) The changing frequency scale alters the proportions of infrared radiation, visible light, and ultraviolet, in the manner shown by a comparison of (a), (b), (c), and (d). Most of the radiation for stars hotter than 10,000°K lies in the ultraviolet.

   In (a), the temperature is 2,500°K (a very cool star), and the unit of frequency is 5.21×10^{13} oscillations per second. In (b), temperature is 4,000°K (61 Cygni A), and unit of frequency is 8.33×10^{13} o.p.s. In (c), temperature is 10,000°K (Sirius), and unit of frequency is 2.08×10^{14} o.p.s. In (d), temperature is 30,000°K (blue stars), and unit of frequency is 6.25×10^{14} o.p.s.

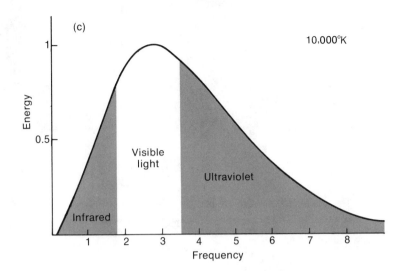

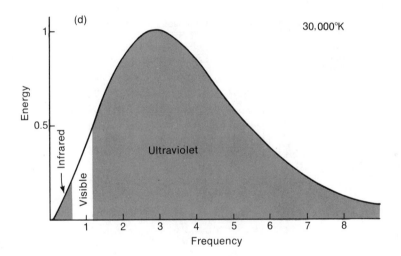

To obtain the power from a given surface area of a star, the energy unit must also be chosen appropriately, being proportional to the *cube* of the temperature. From comparing with the solar case, the power emitted from one square kilometer is given for (a), (b), (c), and (d) by the following choices for the energy unit in kilowatt seconds:

$$1.16 \times 10^{-4} \left[\left(\frac{2,500}{5,800} \right)^3, \left(\frac{4,000}{5,800} \right)^3, \left(\frac{10,000}{5,800} \right)^3, \left(\frac{30,000}{5,800} \right)^3 \right].$$

Taking account of this cubic dependence on the temperature, and also of the temperature dependence of the frequency-scaling factor, how much greater is the power radiated per unit area from a star of surface temperature 30,000°K than from a star of surface temperature 2,500°K?

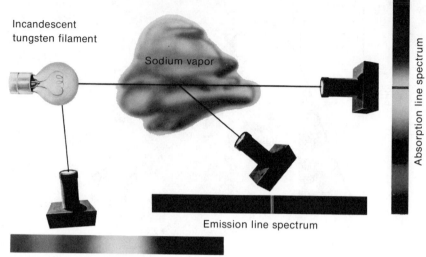

FIGURE II.3.
A gas interposed between the point of observation and a hotter source of continuum light produces *absorption* lines. But when the point of observation is such that there is no background of continuum light, the gas produces *emission* lines. (From J. C. Brandt and S. P. Maran, *New Horizons in Astronomy*. W. H. Freeman and Company. Copyright © 1972.)

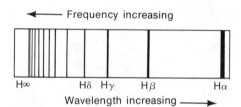

FIGURE II.4.
A schematic representation of the lines of the Balmer series in absorption, with frequency increasing toward the left, and wavelength increasing toward the right.

The spacings of the levels in energy diagrams are different for different atoms. Atoms whose diagrams have wider spacings have transitions that emit and absorb radiation of higher frequencies than is emitted or absorbed by atoms whose diagrams have smaller spacings. The former sort of atoms produce a greater absorption effect on the spectra of stars of high surface temperature than on the spectra of stars of low temperature, because there is more high-frequency radiation from the former stars than from the latter, as we can see

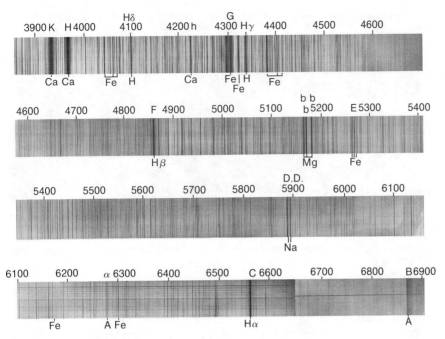

FIGURE II.5.
The many absorption lines (Fraunhofer lines) present in sunlight. This solar spectrum was made with the 13-foot spectroheliograph. The numbers refer to the wavelength of the radiation, measured in a unit known as the Ångstrom (Å), such that one Å = 10^{-8} centimeters. The wavelength increases toward the right, as in Figure II.4. Letters below various absorption lines are the chemical symbols (see Chapter 7) of the atoms that produce the lines. (Courtesy of Hale Observatories.)

FIGURE II.6.
Restricting frequency to that of the K line of calcium, the hotter, more diffuse gas lying above the normal solar surface shows distinctive patches of emission. Figure 2.12 has a similar interpretation. (Courtesy of Kitt Peak National Observatory.)

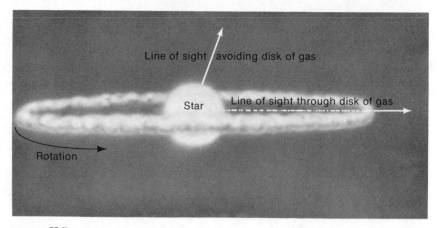

FIGURE II.7.
A star of high surface temperature surrounded by a rotating disk of cooler gas (compare with
Figure II.3).

from Figure II.2. Of the common atoms, the widest spacings occur for helium
atoms in which one electron has become free, *ionized helium.* The next widest
spacings are for helium atoms that retain both their electrons, *neutral helium.*
The next is hydrogen, followed by ionized metal atoms, for example, calcium
atoms in which one electron has become free. It will be recalled that such
ionized calcium atoms give the H and K lines. The narrowest spacings are
for metal atoms with no electrons free, *neutral* metal atoms, for example, iron.

It is possible to classify the stars into categories, *spectral types* as they are
called, according to the main features of their frequency distributions, their
spectra. The categories form a temperature sequence, because the different
spacings of the energy-level diagrams of different atoms cause the atoms'
absorption or emission lines to show up prominently only within certain
temperature ranges, the ranges that happen to match the energy-level diagrams
in a suitable way. We expect the appearance of helium to be associated with
the hottest stars, hydrogen with less hot stars, then ionized metal atoms, and
finally neutral metal atoms. The spectral types are catalogued in Table II.1,
and examples of the types are shown in Figure II.8. The choice of letters to
represent the different categories or *classes* is historic, dating from a time when
the relation of the categories to a temperature sequence was not well-understood. It is possible to divide the categories into subclasses. In category A,
for example, hydrogen absorption lines are stronger in relation to ionized
calcium at the upper end of the temperature range than they are at the lower
end of the range. So, by estimating the relative strength of hydrogen and ionized
calcium for a particular star of category A, it is possible to determine more
closely where the star is in the range 7,500–11,000°K. The study of the various
categories and subclasses, and the assignment of individual stars to them, forms
the subject of *stellar spectroscopy.*

TABLE II.1.
Spectral classifications of stars

Classification symbol	Distinguishing features	Temperature range °K
O	ionized helium stronger than neutral helium	Above 30,000
B	ionized helium weaker than neutral helium	11,000–30,000
A	hydrogen at maximum strength ionized calcium appears	7,200–11,000
F	ionized calcium strong hydrogen weakening neutral metals appear	6,000–7,200
G	ionized calcium strong neutral metals strong	5,200–6,000
K	neutral metals strong ionized calcium weakening	3,500–5,200
M	neutral metals strong absorption bands of molecules	Below 3,500

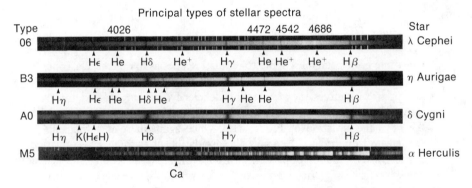

FIGURE II.8.

Examples of the spectra of stars of differing surface temperature. Wavelengths of lines are marked in Ångstroms (one Å $= 10^{-8}$ cm), and with the chemical symbols of the atoms producing the lines also shown. The digit following the class letter is a subclass designation. (Courtesy, Hale Observatories.)

Absorption lines (or *bands,* as they are usually called) caused by molecules appear in cool stars of category M. Beyond remarking that the most prominent molecule in such stars is titanium oxide, we shall not concern ourselves here with these bands. Other cool stars have other prominent molecular bands, cyanogen in R stars, carbon in N stars, and zirconium oxide in the S stars.

The naming of the categories and their ordering can be remembered from the often-quoted mnemonic,

Oh Be A Fine Girl, Kiss Me Right Now.

PROBLEM:

What is wrong with the following argument? Although the gas immediately above the photosphere of a star is cooler than the photospheric gases, and although absorptions at discrete frequencies take place in the cooler gas, with individual atoms experiencing upward transitions of the electrons, such atoms immediately undergo corresponding downward transitions, thereby restoring the absorbed radiation. Consequently absorption lines are not formed by the cooler gas lying above the photosphere of a star.

ANSWER:

It is true that downward transitions in the cooler gas compensate for upward ones, but there is an excess of downward transitions that are *nonradiative*. Such transitions occur when an atom collides with another particle. Energy, instead of being emitted radiatively, goes into the energy of motion of the colliding particles. This heat energy is indeed ultimately radiated, but over a wide frequency range, not at the discrete frequency corresponding to the transitions in question.

Appendix II.2. Radio Telescopes

The same basic principles apply to radiotelescopes as to optical telescopes. The energy fluxes from cosmic radio sources are usually considerably less than those from optical sources, however. The deficiency is compensated for by two factors. First, the detection efficiency is improved by the use of *radio amplifiers*. We have the situation of Figure II.9. Once again there is a focusing from particles b to particle c. But now the oscillation of particle c is converted to an increased oscillation of particle d through the aid of an amplifier. Since radio amplifiers are, of course, familiar from everyday experience, we need not describe details of their mode of operation here. We note, however, that the electronic complexities employed in an ordinary amplifier are contingent on the dimensions of the amplifier being smaller than the wavelength of the radiation in question. With modern "solid-state" techniques, it is possible to meet this condition for all wavelengths λ longer than about 1 cm, i.e., for frequencies ν less than about 3×10^{10} cycles per second. The construction of amplifiers becomes technically difficult for frequencies higher than this, for example, for $\nu = 10^{11}$ cycles per second. Present-day technology is quite unable to construct similar amplifiers for frequencies corresponding to optical radiation, ν about 10^{15} cycles per second, because the corresponding wavelengths are much

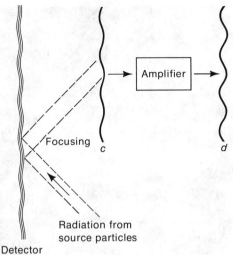

FIGURE II.9.
The oscillation of particle *c*, arising from the focusing
action of the detector particles *b*, produces, by the
aid of an *amplifier*, a still greater oscillation of the
particle *d*.

smaller than any equipment that we are capable of building. Thus the technology of detecting radiation is conditioned by the relation of the wavelength of the radiation to our own human dimensions, since our own dimensions at present control the scale of the equipment we build.

The second compensating factor in a radiotelescope is that the aperture can be much larger than that of an optical telescope. This further advantage is also connected with the larger wavelength of radiowaves, since the accuracy with which a mirror must be constructed is directly proportional to the wavelength. For an optical mirror, the tolerance is only about a millionth of an inch, whereas a tolerance of as much as a centimeter is permissible for a wavelength of 20 cm. The largest mirror at present constructed, or likely to be constructed, at the latter wavelength has a diameter of 1,000 feet, i.e., some sixty times the diameter of the largest optical telescope. However, at the shortest radio wavelengths, about 2 mm, it becomes difficult to construct mirrors larger than about 50 feet. Nevertheless, a mirror of diameter 50 feet still has about 10 times the area of the largest optical mirrors. Examples of radiotelescopes with mirrors analogous to those of optical telescopes are shown in Figures II.10 to II.13. At the shortest radio wavelengths, mirrors are constructed from continuous metal sheeting. At somewhat longer wavelengths, say 1 to 10 cm, the mirror can be made up of sheets or "petals," which can be slightly separated from each other. At still longer wavelengths, a wire mesh gives adequate reflectivity.

FIGURE II.10.
The 1,000-foot fixed reflecting surface of the Arecibo radiotelescope. The structure supported by wire ropes, seen at upper center, is the focusing arrangement (relating particles *b* to *c* in Figure
II.9). To compensate for the fact that, because of its very large
size, the main reflecting surface is fixed, the focusing arrangement
of this radiotelescope is highly sophisticated. (Courtesy of Cornell
University and the National Science Foundation.)

There is a second kind of radiotelescope, known as an *interferometer,* which
operates in a manner analogous to Figure 5.29. Here we had two slits in
a breakwater. Water waves from outside the breakwater were incident in such
a way that the rise and fall of the water was the same at the two slits. That
is to say, the crest of a wave at one slit was simultaneous with a crest at the
other slit, and wave troughs were also simultaneous at the two slits. In radio
terminology, we might say the two slits were *in phase* with each other. Under
this "in phase" condition, waves were transmitted inside the breakwater in the
manner of Figure 5.29. There were strong reenforcements of the wave pattern
for the directions toward points *A, C,* and *E,* but in the directions toward the
points *B* and *D* the water remained still.

FIGURE II.11.
The focusing arrangement of Figure II.10 in more detail. (Courtesy of Cornell University and the National Science Foundation.)

FIGURE II.12.
The 210-foot radiotelescope at Parkes, N.S.W., Australia. The reflecting surface is fully steerable. The mirror is mounted in what is called the alt-azimuth form. Compare this style of mounting with the equatorial form used for optical telescopes. (Courtesy of the Commonwealth Scientific and Industrial Research Organization.)

FIGURE II.13.
The 100-meter radiotelescope of the Max Planck Institut für
Radioastronomie, Bonn, Germany. This is the largest fully steer-
able radiotelescope yet constructed. (Courtesy of Max Planck
Gesellschaft.)

For the radio analogue, in place of the slits of Figure 5.29 we now have
two radiotelescopes, connected together to give an "in phase" condition. The
situation is in reverse, however, because instead of transmitting waves outward
from the slits as in Figure 5.29, the radiotelescopes "compare" incoming waves.
If the waves come from directions that are heavily marked in this figure, there
is a strong response from the two "in phase" telescopes. But if the waves come
from directions that are lightly marked in the figure, there is no
response. This kind of device permits the radioastronomer to distinguish
between waves coming from adjacent directions on the sky, directions that
would be too close to be separated by the more straightforward instruments
shown in Figures. II.10 to II.13. Maps of radio emission from the sky can
thereby be constructed, as in Figure II.14. The contours of Figure II.14
represent points of equal radio brightness, which are analogous to points of
equal height in a geographical map. Figure II.14 is a map of radio emission
from the Crab Nebula, the frequency being 2.7×10^9 oscillations per second.

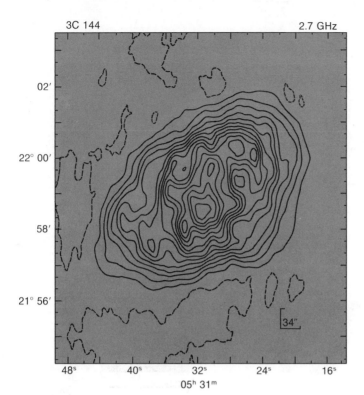

FIGURE II.14.
Contours of equal brightness in the Crab Nebula for a frequency
of 2.7×10^9 oscillations per second. (From A. S. Wilson, *Mon.
Not. Roy. Astr. Soc.,* 157, 1972, 231.)

Appendix II.3. Modern Electronic Devices Used at Optical Wavelengths

Although it is not yet possible to build optical amplifiers that are as electronically sophisticated as radio amplifiers, certain forms of amplification have become possible in recent years even at optical wavelengths. The basis of these optical devices is the photoelectric effect illustrated in Figure 4.33. Light incident on a surface, called a *photocathode,* causes free electrons to be emitted from the material of the surface. Each such electron is then accelerated in an electric field—a simple form of radiation field—so that it strikes a second surface with sufficient energy to knock out several more free electrons from this second surface. The process is repeated, with each one of these further electrons being accelerated toward a third surface, where another multiplication of the number of free electrons takes place. After suitably many such stages, the resulting very large number of electrons is finally collected to yield an electric current, which

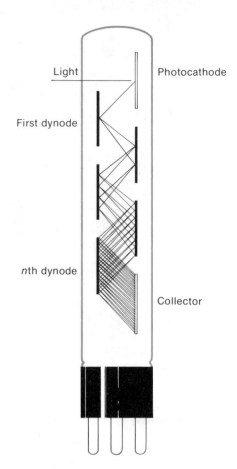

FIGURE II.15.

In the photomultiplier, light incident on a photo-cathode causes electrons to be released by the photoelectric effect (Figure 4.33). These free electrons are accelerated and focused to strike a metallic surface from which many more electrons are then knocked out. The process is repeated in a cascade until the original number of electrons from the photocathode has been increased a millionfold or more. (From J. C. Brandt and S. P. Maran, *New Horizons in Astronomy.* W. H. Freeman and Company. Copyright © 1972.)

can be measured without difficulty. Such a *photomultiplier* is illustrated in Figure II.15.

Compared to the photographic plate, the single photomultiplier tube has major advantages and marked disadvantages. It is much more sensitive, and at low light intensities it responds to the intensity directly, instead of exhibiting the rather complex behavior of photographic emulsions. This makes the photomultiplier particularly suitable for accurate measurements of the magnitudes of stars, galaxies and quasars. On the other hand, so long as the frequency ν of the incident radiation is high enough to remove electrons from the photocathode (see Figure 4.33), the photomultiplier does not discriminate with respect to ν. All frequencies above such a "threshold" contribute to the final current. Thus, to distinguish separate ranges of ν, one must pass the incident light through suitable color filters, or use some equivalent device. But this is a seriously wasteful procedure: while a filter is being employed, light is lost at all values of ν except in the range passed by the filter in question. The photographic plate, on the other hand, used with a *spectrograph* (i.e., replacing the screen of Figure 5.33 by a photographic plate), accepts all values of v over a wide range simultaneously, which tends to compensate for the poorer sensitivity of photographic emulsions.

But why not replace the screen in Figure 5.33 by a row of photomultipliers, as in Figure II.16? This should give the best of both worlds, simultaneous use of light at all v, together with clean-cut response and high sensitivity of each individual photomultiplier. A difficulty with such an arrangement, until recently, was simply to make phototubes that were small enough to permit many of them to be used in this way. This difficulty can now be overcome with the aid of miniaturized equipment. However, a further difficulty remains. A photographic plate continues to add up the effect of light on the emulsion for as long as the observer wishes to expose his plate, but the photomultiplier does not give a larger and larger current output as time goes on. So once again the photographic plate gains, unless some means of cumulative addition can be devised for the scheme of Figure II.16. Such a cumulative addition has indeed very recently been achieved, using a computer to keep track of the addition. With all these improvements, the best modern electronic methods for distinguishing various ranges of v, i.e., for obtaining a spectrum, have become far more sensitive and effective, especially for faint objects, than the old system of photographic plate and conventional spectrograph.

When astronomers wish to obtain a spectrum, they usually want it from a specific object within the field of the telescope, not from all objects that happen to lie in the field. Consequently, the radiation from the unwanted objects must be excluded. In a conventional spectograph, a slit is arranged to admit only the light from the wanted object. A round hole or *diaphragm* is used for a photomultiplier. Such observations, whether by conventional means or by modern devices, are wasteful, in the sense that light from all but the one object is being lost. For many other purposes we may wish not to suffer such loss. We may wish to examine diverse objects over much of the field of the telescope. Can electronic devices also be used in such applications?

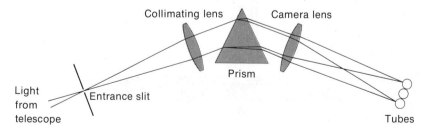

FIGURE II.16.

A spectrograph in which the photographic plate has been replaced by a row of photomultipliers. For simplicity, the spectrograph has been shown here as using a prism; in practice a diffraction grating would be used. (J. C. Brandt and S. P. Maran, *New Horizons in Astronomy*. W. H. Freeman and Company. Copyright © 1972.)

An important problem would be how to obtain a direct picture over the whole field of the telescope, but with better sensitivity than on a photographic plate. Instead of a photographic emulsion being placed at the focal plane of the telescope, imagine a flat surface of the kind employed in phototubes. Free electrons will be emitted from the surface at places corresponding to the positions of objects in the field. To obtain high sensitivity, these free electrons must first be accelerated and then must be detected in a suitable way, and this process *must not seriously distort the final image.* Acceleration presents no problem in itself, nor does detection of the accelerated electrons. Light is emitted when such electrons strike a phosphor screen, and this light can simply be photographed. The difficulty is to obtain acceleration without distortion. This problem is currently being solved. A limited area of the telescope field can be dealt with in the manner we have envisioned—though not yet the whole field—by a device known as an *image tube* (Figure II.17).

Electronic systems already available have produced a major revolution in astronomy. Light from faint objects can now be used at least ten times more efficiently than was possible only a few years ago. Further improvements are

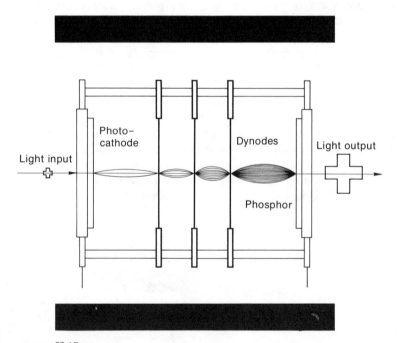

FIGURE II.17.
The image converter tube. Similar in principle to the photomultiplier, but with the additional sophistication that the electron cascade from the photocathode is handled by a focusing device in such a way that a picture is formed on a phosphor screen, the picture having a distribution of electrons on the phosphor similar to the distribution of the original light on the photocathode. (Courtesy of Westinghouse Electric Co.)

well on the way, so that observations in the future will be still more economical in their use of light. Nowadays, the construction of a large telescope involves a dazzling array of techniques, ranging from the mechanical skill required to position several hundred tons of steel to within a precision of a thousandth of an inch, to the intricate details of modern solid-state electronics and of computer technology.

Appendix II.4. "Local" Geometry and Its Relation to the Physical World

Throughout almost everything that has been said in Chapters 1 to 5, we have made free use of the concept of "distance," but without specifying any operational procedure for determining it. Suppose we redraw Figure 4.2, as in Figure II.18, but now think of OX, OY, OZ, as physical posts. The plane formed by OX, OY, can now be thought of as a physical sheet fastened to OX, OY, with sheets also for the planes formed by OZ, OX, and by OX, OY. Then the values of x, y, z, appropriate to a point, say, P, on the path of a particle, are the distances from these planes, which we could think of as being measured with the aid of a ruler.

It might seem at first sight as if we now have a firm grip on the problem of what we mean physically by the numbers x, y, z, but a little consideration soon reveals difficulties. No such set of measurements could be carried out instantaneously; yet x, y, z, are to be measured at a specific moment of time, the moment when the particle is at the point P of its path. If we start by measuring x, the particle will have moved by the time we get around to measuring y. Only if the particle moves very slowly, hardly changing x, y, z, during the time-interval required for all the measurements, will our distances

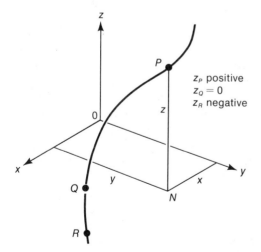

z_P positive
$z_Q = 0$
z_R negative

FIGURE II.18.
To determine the spatial position of a point P on the path of a particle by an actual operational procedure, it is necessary to measure the distances x, y, and z, simultaneously. This is not possible with measurement by ruler, since such a measurement occupies an interval of time.

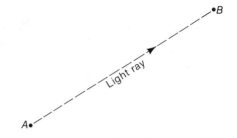

FIGURE II.19.
The straight line joining A to B is determined with the aid of a light ray.

have the meaning we would like them to have. How do we ensure that the posts OX, OY, OZ, are "straight," and how do we know they are at right angles to each other? How, for that matter, do we go about making a "straight" ruler? These questions are not just quibbles. They reveal basic flaws in our proposed operational procedure.

If one were asked to judge whether or not a ruler were "straight," the natural thing to do would be to compare the edge of it with the track of a ray of light, as in Figure II.19. We soon see therefore that we cannot express our geometrical ideas in any practical way except by the use of light, or of some other form of radiation. In Chapter 4 we found that radiation serves as a vehicle of communication between one particle and another. Since geometrical ideas involve the relation of points to each other—points where, in practice, particles are located—it follows naturally that radiation provides the appropriate tool for establishing a practical system of geometry. In what follows in this Appendix, it will be our aim to see how this can be done.

We found in Chapter 5 that the radiative disturbance from a charged particle in continuous regular oscillation, as in Figure II.20, possesses a wave structure that can be understood in terms of the analogy of water waves. We are all familiar with the way ocean waves roll into a beach. If the beach is long and straight, with a smoothly shelving bottom, the waves follow each other in a regular progression. At any moment the crests and the troughs follow lines that are parallel to the beach itself, as in Figure II.21. At any moment we could use the separation from one wave to the next as our unit of distance. Not only this, but by placing a float at some certain distance offshore, we could use the vertical bobbing of the float to determine a unit of time. Light, or some other form of radiation, can be used in just this way. Oscillations at a specific place determine a unit in terms of which we measure the time t, while the regular spacing of waves traveling in a specific direction, for example, the direction of OX in Figure II.18, determine the unit in which we measure spatial distances in that direction. By simultaneously using a wave for the direction OX, another wave for OY, another for OZ, we can determine the lengths x, y, z, in a practical way. The waves serve as both a "ruler" and a "clock."

Suppose we have two points P and Q that happen to lie in the plane formed by the directions OX, OY, i.e., for which $z = 0$, and suppose we measure their distances from O in the direction OX, using the technique outlined in the preceding paragraph. Denote the results by x_P, x_Q, respectively. Then $x_Q - x_P$

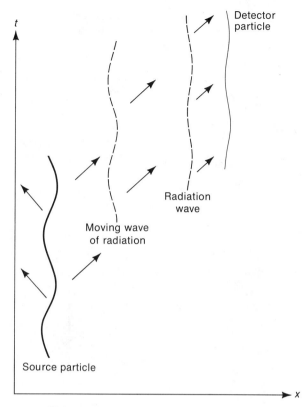

FIGURE II.20.
The radiation from a particle in regular oscillation possesses a
wave structure. Radiation reaching a detector particle causes it to
oscillate.

Waves coming in line abreast

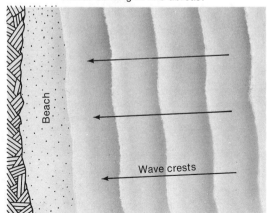

FIGURE II.21.
Waves rolling into an ocean beach are analogous
to the waves of radiation generated by the
regular oscillation of Figure II.20.

is the distance from P to Q, also in the direction OX. We can proceed in a similar way for the distance from P to Q for the direction OY, obtaining a practical answer which we denote by $y_Q - y_P$. Now suppose we measure the distance from P to Q using waves that travel directly from P to Q, say, giving PQ as the answer. The nature of our three measurements giving $x_Q - x_P$, $y_Q - y_P$, PQ, will be clear from Figures II.22, II.23, and II.24. Do we find any relationship between these three measurements? The answer to this question is that, provided P and Q are not too far apart—in practice this means that P and Q must not be more widely separated than several hundred million light years, a very big separation compared to distances in our everyday world—then

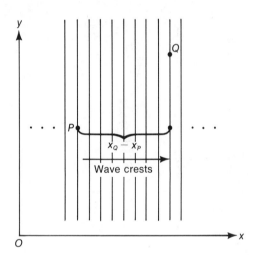

FIGURE II.22.
A wave of radiation traveling in the direction OX is used to measure the distance $x_Q - x_P$; one counts the number of wave crests between P and Q along the OX-direction.

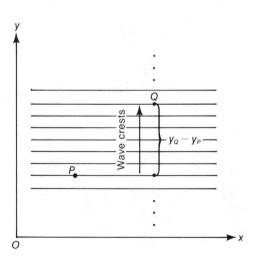

FIGURE II.23.
A wave of radiation traveling in the direction OY is used to measure the distance $y_Q - y_P$; one counts the number of wave crests between P and Q along the OY-direction.

we obtain to within the errors of measurement the result illustrated in Figure II.25. The square of PQ is equal to the sum of the squares of $x_Q - x_p$ and $y_Q - y_P$, namely $PQ^2 = (x_Q - y_P)^2 + (y_Q - y_P)^2$. This result is known as Pythagoras' theorem.

At this point the reader may be inclined to exclaim "But we know Pythagoras' theorem is true! We proved it in a course on geometry. Why all this fuss about radiation and measurement?" Pythagoras' theorem is proved in Euclidean

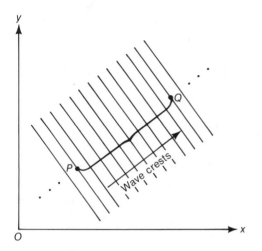

FIGURE II.24.
A wave of radiation traveling in the direction P to Q is used to measure the distance from P to Q; one counts the number of wave crests along the line from P to Q. How does this measurement relate to those of Figures II.22 and II.23?

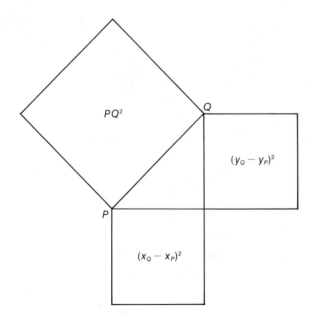

FIGURE II.25.
In Euclidean geometry, the square on the long side of a right-angled triangle is equal to the sum of the squares on the other two sides:
$PO^2 = (x_Q - x_P)^2 + (y_Q - y_P)^2$. This is the famous theorem of Pythagoras.

geometry *subject to certain assumptions,* which may or may not be true in the real world. What we are saying is that observation and measurements show that certain of the tenets of Euclidean geometry are valid in the actual world, provided the distances over which they are considered to hold are not too great.

To give a typical axiom of Euclidean geometry:

> One and only one straight line can be drawn between two
> distinct points, say, *P* and *Q*.

The student may be inclined to accept such a statement, without giving much thought to it, because it appears "obviously true." But by "obviously true" we can only mean that the statement corresponds to *practical experience.* What constitutes "straight"? How do we know when we have actually got the one and only straight line connecting the points *P* and *Q*? What criterion do we use to make such a judgment? Attempts to answer questions like these lead us immediately to the practical problems we considered above. Sometimes the straight line from *P* to *Q* is defined by the statement:

> Of all the lines joining *P* and *Q,* the straight line is
> the shortest.

But the concept of "shortest" involves the measurement of distance, so that any such definition merely takes us into a circular argument.

Euclidean geometry purports to prove Pythagoras' theorem *exactly,* whereas the result of a practical approach to geometry is less far-reaching. We only claim to satisfy Pythagoras' theorem provided the points *P* and *Q* are not too far apart, and then only to within the errors of measurement. How comes it, we may ask, that the geometry of the mathematician can be more assertive than the practical geometry of the physicist? The answer comes from another of the axioms of Euclidean geometry:

> Given a straight line and a point not lying on the line,
> there is one and only one straight line through the
> point, lying in the plane formed by the point and
> given line, which does *not* meet the given line.

The concept of one line *never* meeting another implies unlimited distance, and so goes outside the range of the measurements discussed above. Indeed, if in the actual world measurements are extended to much larger distances than those discussed above, Pythagoras' theorem turns out *not* to be valid, and the mathematical axiom on which it rests turns out *not* to be true. Hence the exactness of the mathematician's geometry is vitiated, at any rate in being applicable to the actual world.

The mathematician defends himself against the criticism that his results are not as strict as he imagines them to be by taking the position that his geometry, and indeed any one of his systems of mathematics, is not intended as a representation of the actual world, but as a "pure" investigation of certain logical relationships. Unfortunately, mathematicians make frequent use of words with emotive significance like "rigor" and "proof," and they sometimes persuade students that their considerations possess a rarified significance not attainable in other fields of intellectual activity. To the physicist, a rigorous system of reasoning would be one that corresponded exactly to what is observed in the world.

We saw above how radiation from an oscillating charged particle can be used to determine intervals of time as well as spatial distances. When we used this method to measure the distance from point P to point Q in Figure II.24, we tacitly assumed that both points were being measured at the same time, indeed, that all the points of Figure II.24 could be measured at the same time. Suppose we consider two points at different times and with different values of x, but with the same values of y and z. This we can represent as usual in an x,t diagram, as in Figure II.26. We can use our radiation technique to measure not only the difference $x_Q - x_P$ (as before) but also $t_Q - t_P$. What can we say now about the direct distance P to Q? Do we say that the square of PQ is equal to the sum of the square of $x_Q - x_P$ and of $t_Q - t_P$,

$$PQ^2 = (x_Q - x_P)^2 + (t_Q - t_P)^2,$$

implying that Pythagoras' theorem remains satisfied, even though the time values we attach to points P and Q are now different? If we were to proceed

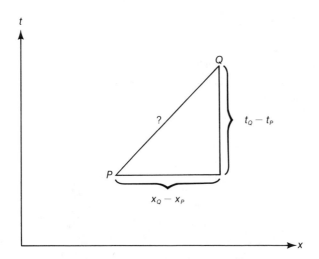

FIGURE II.26.
By referring back to Figure 5.43, notice how a wave can be used both for measuring an interval of time at a particular place, and for measuring the distance between two points at a particular time. The bobbing of a float measures time, while the distance between wave crests defines a measure of spatial length.
    In this figure, suppose $t_Q - t_P$ and $x_Q - x_P$ have been measured in this way. What now is the distance P to Q?

in this way, we would be treating time in the same way as space. All four dimensions of *spacetime* would be on the same footing. From all aspects of experience, we know this to be wrong. Subjectively, we are aware that time is somehow different from space. Yet if we do not accept Pythagoras' theorem as giving a physically significant quantity when time is involved, how else do we proceed?

Einstein's first great contribution to physics, which he made in 1905 while working as a clerk in the Swiss Patent Office, was equivalent to requiring that there be a *sign difference* between time quantities and spatial quantities. Thus writing $S_{PQ}{}^2 = (t_Q - t_P)^2 - (x_Q - x_P)^2$, the physical relationship between points P and Q is to be determined by S_{PQ}, *not* by $PQ^2 = (t_Q - t_P)^2 + (x_Q - x_P)^2$. This may seem a simple enough step in itself, but it turns out to have profound consequences, as we shall see.

In a case where the y and z values of the points P and Q are also different, we keep the same sign for all the spatial dimensions, but give them an opposite sign to the time dimension. The physical relationship between P and Q is now to be determined by S_{PQ}, where

$$S_{PQ}{}^2 = (t_Q - t_P)^2 - [(x_Q - x_P)^2 + (y_Q - y_P)^2 + (z_Q - z_P)^2].$$

The spatial part of $S_{PQ}{}^2$, within the square brackets, has the same structure as it has in ordinary three-dimensional Euclidean geometry.

Appendix II.5. The Speed of Light Again

We come now to an important situation concerning the speed of light, c. In Chapter 5 we saw that for any form of radiation—radio, infrared, light, ultraviolet, x-rays, and γ-rays—c is related to the frequency ν and to the wavelength λ by the simple equation $c = \nu\lambda$. The speed c is the same for all forms of radiation.

Now, the frequency ν of a particular wave *means* the number of oscillations that occur in unit time, analogous to the number of bobbings up and down of a float in water waves [see Figure 5.43]. So if we use the bobbings themselves to determine the unit of time, it follows trivially that the number of oscillations that occur per unit time must be exactly 1. Hence we must have $\nu = 1$ for the particular wave used to determine the time unit. If the *same wave* is also used to determine the space unit, then also trivially we must have $\lambda = 1$. With $\nu = 1$ and $\lambda = 1$, it clearly follows that $c = 1$. The speed of light, using the measuring procedure described above, is inevitably 1. Whatever wave we elect to use—radio, infrared, light, ultraviolet, x-rays or γ-rays—the answer is always 1 exactly.

What, then, do we mean by saying that the speed of light is 2.997929×10^{10} centimeters per second? The answer to this question emerges when we consider in a little more detail how we would actually go about our measurement procedure. Evidently we must choose our standard wave in a readily reproducible way, so that anybody anywhere can obtain it with the aid of a suitable experimental procedure. In Chapter 4 we discussed the discrete quantum transitions of atoms. We saw that, when many atoms of a specific kind undergo the same discrete transition, radiation of a specific frequency is emitted. By choosing a certain atom and a certain transition of it, we specify a certain wave. Notice that we have many possible choices, that each choice will lead to a different ν and λ, and hence that we could vary our units of measurement simply by changing the atom or the transition of it. Suppose next that we make our choice on the basis of practical convenience, choosing a transition that would lead to a time unit of about a second, for example. Then the spatial unit associated with such a wave is found to be much too large; the wavelength λ would be much larger than the whole Earth. Conversely, if we chose an atom and a transition of it to give a wavelength λ of practical size, the associated time unit would be too short, very much shorter than the second.

To overcome this purely practical dilemma, what is done is to choose *two different waves,* one to give a practical time unit and the other to give a practical space unit. The "second" and the "centimeter" are obtained from different atoms, from atoms of *caesium* and *krypton,* respectively. It is this choice of different units that causes the appearance of the number 2.997929×10^{10}, which obviously has no absolute significance. It is simply the ratio of the different units which have emerged, quite artificially, from this procedure. Properly speaking, we should continue to write $c = 1$, so that

$$c = 2.997929 \times 10^{10} \text{ centimeters per second}$$

then becomes

$$1 = 2.997929 \times 10^{10} \text{ centimeters per second,}$$

which tells us that 1 second is equivalent to 2.997929×10^{10} centimeters. It is customary when speaking in this way to include the word "light," 1 light second = 2.997929×10^{10} centimeters, but this addition is not logically necessary, nor is it in accordance with the spirit of Einstein's theory. However, to conform to customary practice, we shall continue to use the terminology of "light second" and "light year."

The understanding of this point is confused in all our minds by certain primitive concepts. From our early years we became accustomed to spatial measurements being made with respect to a hunk of material—a timber merchant measures lengths of wood with respect to a standard rule. In olden times lengths of cloth were measured with respect to a hunk of wood known as the "ell." And we have become accustomed to reckoning time with respect to the Sun: 1 day = 24 hours, 1 hour = 60 minutes, 1 minute = 60 seconds. We tend to think of such procedures as having an absolute quality to them. When we do so, the speed of light becomes a peculiar quantity: so many ells per hour. Yet such empirical procedures have no absolute quality at all. Nor have they satisfactory reproducibility. It is impossible to make two hunks of material precisely equal to each other. And the rotation of the Earth is not entirely smooth, so that one day is not precisely equal to another day. For this reason, time measurement has ceased to be based on the rotation of the Earth, even for practical civil purposes—time is nowadays based on the caesium atom (the change from the old astronomical system of time reckoning to the modern caesium "clock" was made in 1955). There is no way of obtaining space and time units with satisfactory accuracy and reproducibility except by using atoms. It is then logically, if not practically, a weak procedure to use different atoms, one for space, the other for time. The same atom and the same transition should be used, in which case $c = 1$, *independent of which atom and which transition happens to be chosen.* Since no particle moves as fast as light, the speeds of particles then become all simple numbers less than 1, but greater than, or equal to, 0, since speeds are never considered to be negative numbers.

Appendix II.6. The Light Cone and Its Relation to Particle Motions

Next we notice that it is possible for $S_{PQ}{}^2$ to be zero, even though P and Q are distinct points. This is already a quite unfamiliar circumstance, that the "distance" can be zero between separated points. Taking the points P and Q to have the same y and z values, the "distance" will be zero whenever the square of the ordinary spatial distance $(x_Q - x_P)^2$ happens to be equal to the square of the time difference $(t_Q - t_P)^2$. Let us think of P as a fixed point and ask for all the points Q such that the distance P to Q is zero. The result is shown in Figure II.27. The point Q must always lie on one or the other of two lines drawn at an angle of $45°$ to the direction of increasing time.

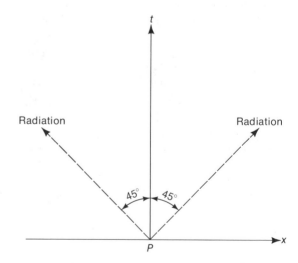

FIGURE II.27.
When the same radiation wave is used to measure both time intervals and spatial distances, radiation from a source at P inevitably travels at $45°$ in the $x, t,$ diagram.

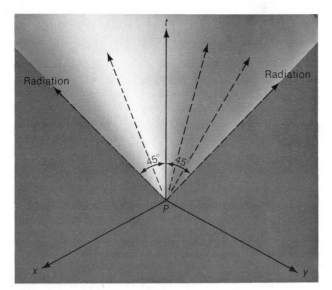

FIGURE II.28.
When a second spatial dimension is added to Figure II.27, radiation travels from P along a cone, with the direction of travel of a light ray always making an angle of 45° to the time axis.

Now let P be a source of radiation. In what directions does the radiation travel? The answer is along the lines of Figure II.27, but only along the segments in the forward sense with respect to time. Radiation does not travel backward in time, as we saw already in Chapter 4. This important property applies to all forms of radiation: radio waves, infrared, light, ultraviolet, x-rays, and γ-rays.

Now let us resurrect one of the missing spatial dimensions, say, y. Here we have $S_{PQ}^2 = (t_Q - t_P)^2 - [(x_Q - x_P)^2 + (y_Q - y_P)^2]$. Once again we can have S_{PQ}^2 zero, namely, when the square of the ordinary Pythagorean spatial distance, $(x_Q - x_P)^2 + (y_Q - y_P)^2$, is equal to the square of the time difference, $(t_Q - t_P)^2$. With P again a source of radiation, the radiation travels in a forward time-sense to points Q such that S_{PQ}^2 is zero, as in Figure II.28. The radiation thus travels along a cone in spacetime, known as the *light cone.*

Nothing is changed in principle if we also resurrect the third spatial dimension, z. We have S_{PQ}^2 zero when the square of the ordinary spatial distance, $(x_Q - x_P)^2 + (y_Q - y_P)^2 + (z_Q - z_P)^2$, is equal to the square of the time difference, $(t_Q - t_P)^2$. Once again, radiation emitted from an isotropic source at P travels to all points Q such that S_{PQ}^2 is zero. The points Q form a "cone" in the forward time-sense, but here we cannot easily draw it, because, as we noted in Chapter 4, the full four dimensions of spacetime are not readily represented on a two-dimensional piece of paper. The "light cone" now has three dimensions, not two, as in Figure II.28.

We have now attached physical meaning to what was previously only a statement. When we wrote

$$S_{PQ}{}^2 = (t_Q - t_P)^2 - [(x_Q - x_P)^2 + (y_Q - y_P)^2 + (z_Q - z_P)^2],$$

one could have responded with a laconic "So what? Why not some other expression for the four-dimensional distance?" For mathematics, this question is an entirely valid one, but not for physics, since we have now used our statement to make an assertion about the world, which may or may not be true. Our assertion is in fact true, provided we do not extend the light cone from the source P to points Q such that $\sqrt{(x_Q - x_P)^2 + (y_Q - y_P)^2 + (z_Q - z_P)^2}$ is too large—i.e., that $t_Q - t_P$ is too large.

Let P be a point on the path of a "classical" particle, as in Figure II.29, a classical particle being one that can be considered to follow a known path. Draw the backward cone from P as well as the light cone. *No point anywhere on the path comes outside this double cone. This statement holds good wherever P is on the path.* We have here the meaning of the often-heard precept that the speed of a particle never exceeds the speed of light.

2. The path of a certain classical particle happens to be a straight line in the (x, t) diagram, as in Figure II.30. P and Q are any two points on the path. What is the meaning of the ratio $(x_Q - x_P) \div (t_Q - t_P)$? Explain why this ratio is never greater than 1.

Another way to state the limitation on the speed of particles is to say that the tangent to the path, illustrated in Figure II.31, can never make an angle of more than 45° with the time direction. But this statement applies only to

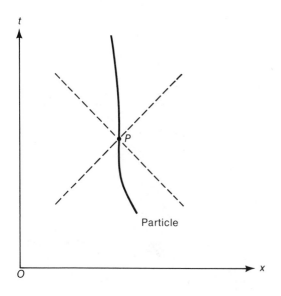

FIGURE II.29.
Here P is *any* point on the path of a particle. All other points on the path lie within the double 45° cone drawn at P. This is the meaning of the often-repeated statement that a material particle "cannot move as fast as light."

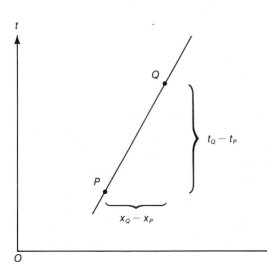

FIGURE II.30.
Path of a particle in uniform motion.

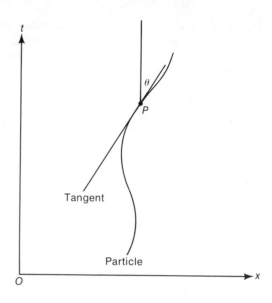

FIGURE II.31.
The tangent at any point P on the path of a particle always makes (in local geometry) an angle θ less than 45° with the time axis.

FIGURE II.32.
The opposite situation to Figure II.29. Here the path lies outside the double cone. This is the situation for a hypothetical particle, invented by theoretical physicists, known as the "tachyon."

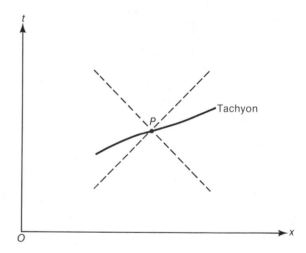

the local geometry here under consideration—it would not apply to the more general geometrical situation discussed later in Section VI. Our first procedure, illustrated in Figure II.29, remains valid, however, in *all* geometrical situations, and is consequently to be preferred.

Recently, physicists have become curious about what a particle would be like if its path lay not within the double cone, as in Figure II.29, but entirely *outside* the double cone, as in Figure II.32. Such particles have been termed *tachyons*. None have yet been found actually to exist, so that consideration of what their properties might be remains an intellectual exercise for now.

In Chapter 4 we saw that macroscopic quantities of matter behave like classical particles; they follow unique paths in an idealized picture in which we think of them as having negligible spatial extension. Thus, in considering the motion of the Earth around the Sun, we often think of the Earth as a mere point that pursues a unique path, as in Figure 1.4. Similarly, we can think of ourselves in terms of a simple point-path which we follow in four-dimensional spacetime. This concept is illustrated in Figure II.33, where B represents birth and D represents death. The points P and Q are neighboring events in our life.

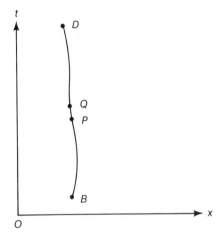

FIGURE II.33.
The life of a person, with point B representing birth and point D representing death.

Suppressing the y, z, dimensions, we have $S_{PQ}{}^2 = (t_Q - t_P)^2 - (x_Q - x_P)^2$. Taking the square root of this equation gives $S_{PQ} = \sqrt{(t_Q - t_P)^2 - (x_Q - x_P)^2}$. Here are some problems involving S_{PQ}:

1. Satisfy yourself that $S_{PQ}{}^2$ is a positive number, and hence that its square root gives a positive number. Provided t_Q is later than t_P, show that S_{PQ} can be written in the form

$$S_{PQ} = (t_Q - t_P) \cdot \sqrt{1 - \frac{(x_Q - x_P)^2}{(t_Q - t_P)^2}}.$$

2. Explain why, when Q is close to P, the square of our speed when at P is given to a good approximation by the ratio $(x_Q - x_P)^2 \div (t_Q - t_P)^2$.

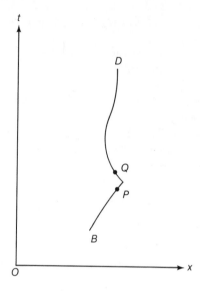

FIGURE II.34.
The path of a person does *not* have
sharp kinks.

3. Problem (2) would be wrong if the particle path had a sharp kink at P, as
in Figure II.34. Why do physical bodies not have kinks like this? We sometimes think of a golf ball experiencing a discontinuity when it is struck, but
when considered on a fine enough time scale, is this really true?

4. The points P and Q being close together, use problems (1) and (2) to write
S_{PQ} in the form

$$S_{PQ} = (t_Q - t_P) \sqrt{1 - v_P^2},$$

where v_P is the speed when at P.

5. It is implicit in the above exercises that we are using the same radiation wave
to determine both x and t values. If different waves are used, leading to a
velocity of light $c = 2.997929 \times 10^{10}$ cm per second, as in practice, there
is ambiguity about whether we elect to measure S_{PQ} in centimeters or in
seconds. Show that if we choose to measure S_{PQ} in seconds, then
$S_{PQ} = (t_Q - t_P) \sqrt{1 - v_P^2/c^2}$, but that if we choose to measure S_{PQ} in
centimeters, then $S_{PQ} = (t_Q - t_P) \sqrt{c^2 - v_P^2}$. Notice how much simpler
things are when the same radiation wave is used.

6. Show that the formula obtained in Problem (4) is correct even when the y
and z dimensions are not suppressed.

Suppose now that we start at birth and proceed to divide our life into many
small bits, as in Figure II.35. As in Problem (4) above, we can work out a
value of S to be associated with each separate small piece. Having done so,
we can easily add together all the resulting values, starting at birth B and going
as far as the point P of Figure II.35, which we take to represent the present
moment. The essential physical result toward which we have been working
is that this addition determines our present *age*.

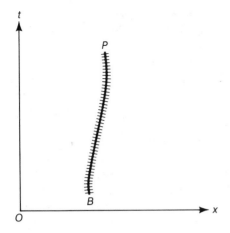

FIGURE II.35.
Our true physical lifetime is determined by dividing our path into many small pieces, and by adding together the four-dimensional distance values calculated for each piece separately. This method is known mathematically as "determining an integral."

PROBLEM:
If in working out the S values associated with the separate pieces of our life, the mistake were made of omitting the square-root factor that appeared in Problem (4) above, show that the result for our age would then be simply $t_P - t_B$.

Usually, when we specify our present age, we state it to be $t_P - t_B$, the time difference between the present and the moment of our birth. The problem just considered shows that this conventional procedure is wrong. Our present age is actually *less* than $t_P - t_B$, because the square-root factor of Problem (4) reduces the S values calculated for each small segment of our lives—unless, of course, we go through the whole of our existence without ever moving, i.e., with v always zero. It follows, then, that we are not really quite as old as we usually think we are. The difference is in practice exceedingly small, however. The difference between our true age and the erroneous $t_P - t_B$ is actually much less than 1 second.

PROBLEM:
A traveler in a jet plane has a speed of about 0.25 km per second. Show that this is less than one millionth of the speed of light, and that the square-root factor $\sqrt{1 - v^2}$ is less than unity for such a traveler by only one part in about 3×10^{12}. Why must v in this square-root factor be reckoned in relation to the speed of light, not in kilometers per second?

It takes light several thousand million years to travel to us from the most distant observed galaxies. So one might think it would be quite impossible to visit distant galaxies within the compass of a human lifetime. Yet suppose an astronaut were equipped with a spaceship capable of accelerating continuously at the equivalent of "1.5 g"—i.e., one-and-a-half times the acceleration of the Earth's gravitational field—and suppose the acceleration to be maintained in a certain direction. The space ship would pursue a path of the form shown in Figure II.36. The acceleration would cause the angle between the time direction and the direction of travel to increase. It would approach more and more closely to the limiting value of 45°, and the value of v to be used in the square-root factor $\sqrt{1 - v^2}$ would approach more and more nearly to 1, with the result that the square-root factor itself would become small. One could then no longer ignore this factor, even for making an approximate calculation. Indeed, as the path of the spaceship approached the critical 45° angle of the light cone, the astronaut would hardly age at all, because of the multiplication by this small square-root factor. From a detailed calculation, it turns out that in an aging period equivalent to what we normally experience in a little over 15 years, an astronaut equipped with such a spaceship would be able to reach any one of the galaxies pictured in this book.

There is no snag anywhere in this conceptual argument. The snag is technological. We cannot make space vehicles with boosters that can deliver a 1.5-g acceleration for more than a few minutes. No booster of which we can conceive could deliver such an acceleration for as long as 15 years.

But let us stay with our idealized spaceship, and imagine a space voyage made up of three parts, an outward acceleration in some chosen direction for

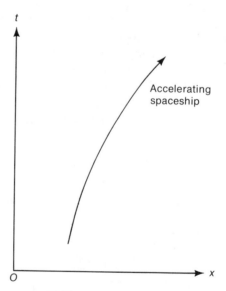

FIGURE II.36.
The tangent to the path of a continuously accelerating spaceship approaches the limiting angle of 45° more and more closely as the time t increases.

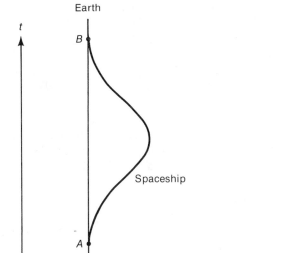

FIGURE II.37.
The paths of the Earth and of a spaceship. The distances along the paths from *A* to *B*, measured by the
integral method of Figure II.35 are not the same.
Which is the shorter path?

a specified aging of our astronaut, say, equivalent to the aging we normally
associate with a span of 15 years, then secondly a reverse acceleration in exactly
the opposite direction for twice this aging span, and finally an acceleration just
like the first one. This flight plan results in a path of the form shown in Figure
II.37. Because of its low speed, the Earth hardly moves, and so its path has
been considered in Fig. II.37 to be simply a line parallel to the time direction.
The situation now is that the astronaut, after visiting a distant galaxy, say,
100 million light years away, returns to Earth. While he himself has aged by
the equivalent of only 60 years, the Earth itself has aged by rather more than
200 million years. (A long-lived dweller on the Earth would observe the
space-ship to make a round-trip journey of 200 million light years. Since for
the most part the ship would be seen to be moving at a speed close to that
of light, the time for the journey, as reckoned by the terrestrial dweller, would
therefore be about 200 million years.)

It is clear from this example that we have two kinds of time. The time *t*,
used in constructing our diagrams, is measured with the aid of radiation in
the manner discussed above. This kind of time is often referred to as *clock
time,* with the radiation supplying the "ticks" of the clock, i.e., supplying the
unit of time measurement. On the other hand, the aging time, obtained by
adding the quantity *S*, calculated for each element of the path of a particle,

is called *proper time. Each particle has its own proper time,* although for slowly moving particles the proper times are all nearly the same, and they are all nearly equal to clock time. But the proper times of fast-moving particles have little similarity to clock time or, in general, to each other. It is important to remember that the basic physical properties of a particle have to be referred to its own proper time, not to clock time or to the proper times of other, differently moving particles. Thus the aging of our astronaut was determined by his own proper time, not by the proper time for creatures remaining on the Earth.

To some people the results we have just described seem so strange that they feel there must be a mistake somewhere. The situation illustrated by Figure II.37, in particular, seems to some to be too odd to be true, and it has been referred to for this reason as the *clock paradox.* But this is really a misnomer, for there is no paradox. A paradox arises when two different but seemingly correct arguments lead to contradictory conclusions. Here we have no plausible rival argument to the one described above. What we do have is a strange and perhaps unexpected result.

Sceptics argue that the circumstances of Figure II.37 can never be realized in practice. So how do we know the claimed result to be true? The answer is that other experiments, equivalent in their significance to Figure II.37, can indeed be carried out. Let us consider one of them.

Fast-moving particles are constantly hitting the Earth's atmosphere from outside. These particles, known as *cosmic rays,* hit the atoms of the outer atmosphere at speeds very close to that of light, generating other fast-moving particles in their collisions. One of these other particles is the muon, written as μ. The muon can also be studied in the laboratory. It is found to decay spontaneously into several particles (of which the electron is one). *When a muon is traveling at comparatively slow speeds,* the characteristic time for this decay is about 2 millionths of a second, 2×10^{-6} seconds. Muons generated by the cosmic rays are mostly produced at a height above the Earth's surface of about 30 kilometers. Now it takes light 10^{-4} seconds to travel 30 kilometers, and the muons cannot move downward toward ground level faster than this. On a naive view—the sort of view held by the people who worry about the clock paradox—one would argue that, since the travel time of 10^{-4} seconds is much longer than the measured characteristic decay time of 2×10^{-6} seconds, the cosmic-ray muons must decay before reaching the ground, and therefore they cannot be observed in an ordinary laboratory. Experiment shows this naive deduction to be wrong. Many muons do in fact reach ground level. Why?

When muons are studied at low speeds in the laboratory, their proper times are not much different from our own proper time, or from clock time. A clock measurement of 2×10^{-6} seconds is then a fair estimate of the muon's own proper decay time. But the clock measurement of 10^{-4} second for the travel time downward from a height of 30 kilometers has no relevance at all to fast-moving muons, *which decay with respect to their own proper times.* To obtain proper time for a fast-moving muon we must multiply the clock time of 10^{-4} seconds by a square-root factor, $\sqrt{1 - v^2}$, where v is the speed of travel.

Evidently for v close to the speed of light, i.e., close to 1, this square-root factor is small, and the result of multiplying 10^{-4} seconds by $\sqrt{1-v^2}$ becomes less than 2×10^{-6} seconds. Thus in the muon's own proper system, there is insufficient time for decay (for most of them, at any rate). Consequently, fast-moving muons should reach ground level, as in fact they are found to do. This experiment shows decisively that our present way of thinking is correct.

Appendix II.8. The Doppler Shift

We are now in a position to emphasize a very important condition, namely, that the atoms we choose to give a particular transition, leading to the radiation whereby we measure both clock time t and the spatial dimensions x, y, z, must be at rest. If motion were permitted, the atoms would oscillate with respect to their own proper system, which would be different from that of atoms at rest. Although this difference would not in itself prevent us from using moving atoms to generate radiation whereby time and space measurements could be made, a problem of reproducibility would arise. Since atoms of the same kind, but moving in different ways, would produce different measuring systems, because they would have different proper systems, reproducibility of a specific measuring system would require us always to move our atoms in a specific way, and this would certainly be awkward to achieve. For convenience, therefore, we take the caesium and krypton atoms that are used in practice for determining the second and the centimeter to be at rest.

Suppose we take two sets of atoms, all of the same kind, with one set at rest and the other set moving. How will the frequency, say, ν', of the radiation emitted in a direction toward us by a transition of the moving atoms compare with the frequency, say, ν, of the radiation generated by the same transition of the stationary atoms? Let us consider this question for two simplified situations: one in which the two sets of atoms move directly toward each other, as in Figure II.38; the other in which the two sets of atoms move directly away from each other, as in Figure II.39. In neither situation is there to be any transverse relative motion, i.e., in the y and z spatial directions.

The atoms can be considered to mark off steps of proper time, i.e., of the quantity S, along their respective paths. Now, because the atoms are of the same kind, and also because their transitions are the same, the steps expressed in terms of S-values are the same for both the stationary and the moving sets of atoms. At each step of the moving atoms, consider the radiation traveling toward the path of the stationary atoms, as in Figure II.40, where the radiation sets up a system of time marks that can be compared with the time marks already established by the radiation of the stationary atoms themselves. The spacings of these two sets of marks determines the ratio we are seeking, namely, that of the frequency ν' of the radiation from the moving atoms to the frequency

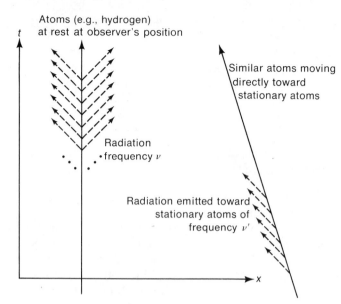

FIGURE II.38.
Two sets of atoms of the same kind (e.g., hydrogen) approaching each other, one of the sets being at rest. How does the frequency ν of the radiation emitted by the stationary atoms compare with the frequency ν' of the radiation emitted by the moving atoms?

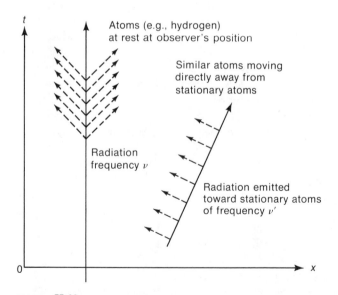

FIGURE II.39.
A similar problem to that of Figure II.38, but with the two sets of atoms receding from each other. The ν' here is not the same as the ν' in Figure II.38.

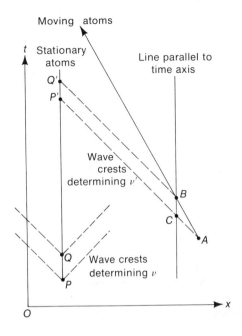

FIGURE II.40.
Typical wavecrests come from
points A and B for the moving
atoms, and from P and Q for the
stationary atoms. Because all the
atoms are of the same kind,
$S_{AB} = S_{PQ}$. It is this basic physical
condition which permits the relation
of the frequency ν' of the radiation
from the moving atoms to the frequency ν from the stationary atoms
to be worked out, in the manner indicated in the text.

ν of the radiation from the stationary atoms. Calculation shows that, for a speed
of approach v, a stationary observer finds the ratio

$$\frac{\nu'}{\nu} = \sqrt{\frac{1 + v}{1 - v}}.$$

Since v is positive, speeds being always positive, the frequency ν' of the
radiation from the moving atoms is higher than the frequency ν of radiation
from the stationary atoms. The radiation from the moving atoms is said in
this situation to be *blue-shifted*, because if ν happens to be a frequency associated
with red light, ν' is shifted in the sense toward blue light. In an obvious notation,
the corresponding wavelength ratio is given by

$$\frac{\lambda'}{\lambda} = \sqrt{\frac{1 - v}{1 + v}},$$

from which we see that the product $\lambda'\nu'$ is the same as $\lambda\nu$, as it must be, since
both products are equal to the speed of light, which remains equal to c, even
for the radiation from the moving atoms.

When I was a student, I grew to detest authors who used phrases like
"calculation shows that," and then proceeded to write down something whose
derivation I couldn't understand. I came to feel that explanations, however
complex, were preferable to nonexplanations. For the sake of any reader who
may feel the same way here, I give the following problems. The ultimate result
to emerge from them explains the formula given above for the ratio ν'/ν. This

derivation is usually supposed to be too hard for a general text like this, but I think the student who "hangs on to his hat," working carefully with the lettering of points given in Figure II.40, may find the various steps not too difficult to make.

Before beginning, note that $x_P = x_Q$, and satisfy yourself that the speed v of the moving atoms is given by the ratio $(x_A - x_B) \div (t_B - t_A)$.

PROBLEMS:

1. Show that the basic physical condition for similar atoms, namely, $S_{PQ} = S_{AB}$, requires

$$t_Q - t_P = \sqrt{(t_B - t_A)^2 - (x_B - x_A)^2}.$$

2. Combining (1) with $v = (x_A - x_B) \div (t_B - t_A)$, show that

$$t_Q - t_P = (t_B - t_A) \sqrt{1 - v^2}.$$

3. From the fact that the light rays of Figure II.40 all make an angle of $45°$ with the time axis, it is easy to see that $t_{Q'} - t_{P'} = t_B - t_C$. Explain why

$$t_C - t_A = x_A - x_B = v(t_B - t_A).$$

Now show that

$$t_{Q'} - t_{P'} = t_B - t_C = t_B - t_A + (t_A - t_C) = (t_B - t_A)(1 - v).$$

4. From problems 2 and 3 show that

$$\frac{t_Q - t_P}{t_{Q'} - t_{P'}} = \frac{\sqrt{1 - v^2}}{1 - v} = \sqrt{\frac{1 + v}{1 - v}}.$$

But $(t_Q - t_P) \div (t_{Q'} - t_{P'})$ is just the ratio ν'/ν, so that

$$\frac{\nu'}{\nu} = \sqrt{\frac{1 + v}{1 - v}},$$

which is what we set out to establish.

In the situation where the two sets of atoms move apart with speed v, as in Figure II.39, the corresponding results are given simply by changing the sign in front of v,

$$\frac{\nu'}{\nu} = \sqrt{\frac{1 - v}{1 + v}}, \quad \frac{\lambda'}{\lambda} = \sqrt{\frac{1 + v}{1 - v}}.$$

PROBLEM:
Show that for v small enough, it is sufficiently accurate to write

$$\frac{v'}{v} = 1 - v, \quad \frac{\lambda'}{\lambda} = 1 + v.$$

In the above work, we have regarded the t and x measurements as being made with the same radiation wave, so that the speed of light is unity. If we revert to the practical system of using one radiation frequency to measure t in seconds and another radiation frequency to measure x in centimeters (caesium atoms for t, krypton atoms for x), we need to write

$$\frac{v'}{v} = \sqrt{\frac{1 - v/c}{1 + v/c}}, \quad \frac{\lambda'}{\lambda} = \sqrt{\frac{1 + v/c}{1 - v/c}},$$

for the situation of Figure II.39, with $c = 2.997929 \times 10^{10}$ cm per second.

PROBLEM:
Write down the corresponding forms of v'/v, λ'/λ, for the situation of Figure II.38, and the approximate forms when v/c is small. What happens where v/c is very close to 1?

Although we might seem to have become absorbed in very abstract problems, we have in fact worked ourselves into a position from which we can understand one of the most powerful observational tools available to the astronomer. The situation we have reached can be summed up in a rather different way with the aid of Figures II.41 and II.42. In Figure II.41 we have atoms emitting radiation of a certain wavelength, say, λ, and frequency, say, v. It will be realized that this figure is not a spacetime diagram of the kind we have just been considering, for the reason that we now wish to display more than one spatial dimension. Figure II.41 refers to a specific moment of time—it is what we referred to earlier as a *time section*. A true time section of spacetime has of course three spatial dimensions, x, y, z, whereas in Figure II.41 we have shown only two such dimensions. To obtain a correct three-dimensional picture, simply take any straight line through the center of the circular pattern and spin the whole figure about this line.

If similar diagrams were drawn at subsequent moments of time, the wave crests of Figure II.41 would be seen to spread out like ripples from a stone dropped into still water—doing so, of course, at the speed of light. The rate at which the wave crests pass a stationary observer gives the frequency v, and the spacing of the crests in Figure II.41 gives the wavelength λ.

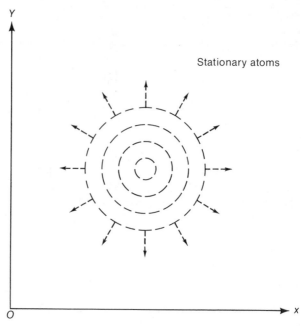

FIGURE II.41.
Whereas Figure II.40 had but one spatial dimension, here we
have two spatial dimensions, showing the wave crests from the
stationary atoms *at a single moment of time.* With the passage of
time, these wave crests move outward at the speed of light in
the manner of ripples from a stone dropped into water.

In Figure II.42 we have radiation from the same atoms with the same
transitions as in Figure II.41, but now with the atoms in motion toward the
direction O_1. To obtain a correct spatial picture here, imagine Figure II.42
to be spun about the line O_1 to O_2. Once again, similar diagrams drawn for
subsequent moments of time would show the wave crests to move outward
everywhere at the speed of light. To a stationary observer in the direction O_1,
the wave crests are more closely spaced than in Figure II.41; so the crests move
past such an observer at a faster rate, and consequently are judged to have
a higher frequency (ν') than before (ν). Just the opposite situation occurs for
a stationary observer in the direction O_2; the wave crests are more widely
spaced, and the frequency with which they pass an observer in this direction
is less than before. Thus the spacing λ' of the wave crests and the frequency
ν' are different in Figure II.42 for stationary observers in different directions.
There are certain directions, shown as O_3 and O_4, in which λ' is equal to λ,
and hence ν' equal to ν. The motion has no effect on the wavelength and
frequency for stationary observers in these particular directions. The angle θ
which these directions make with the line $O_1 O_2$ depends on the speed of the

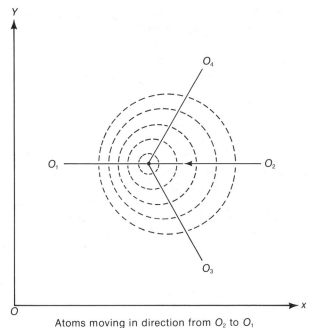

FIGURE II.42.

The corresponding form of diagram for the wave crests (at a single moment of time) of radiation from atoms moving in the direction O_2 to O_1. The wave crests also spread outward with the passage of time, with a speed that is the same as in Figure II.41, namely, the speed of light.

emitting atoms. For low speeds θ is close to $90°$, so that the special directions are then essentially at right angles to the direction of the motion. But as the speed increases, θ changes, with the points O_3, O_4 converging toward O_2. As the speed increases the waves become more closely spaced toward O_1 and more widely spaced toward O_2. The waves are said to be "bluer" toward O_1 and "redder" toward O_2.

Astronomical instruments are especially well-suited to the accurate measurement of frequencies, as we saw in Chapter 5. Objects in the universe—planets, stars, galaxies—are all in motion with respect to the Earth. We would like to use the pattern of behavior shown in Figure II.42, together with suitable frequency measurements, to infer the actual motions of as wide a range of objects as we possibly can. To do this we have still two further problems to solve, however.

When radiation is observed at a certain frequency, ν', say, how are we to know what kind of atom emitted the radiation? If we can identify the atom, how are we to know which transition of the atom caused the observed radiation? These questions cannot be answered from observation at the single frequency

ν'. But in Chapter 4 we saw that atoms have characteristic patterns of transitions. So if we observe radiation from the several frequencies required to identify a well-known characteristic pattern, both the nature of the emitting atom and the transitions themselves will be identified. Examples might be the transitions of the Balmer series of hydrogen, as in Figure 4.26, or the H and K transitions of ionized calcium atoms. The first problem can thus be solved without undue difficulty.

The second problem is more troublesome. We can illustrate it by considering a star in uniform motion with respect to the Earth. If we knew the direction of the motion $O_2 O_1$ in relation to our own direction, say, O, in the representation of Figure II.42, then the measurement of the frequencies of identified transitions from identified atoms would yield the speed of the star. We would need to know the precise form of Figure II.42 for all directions, not just for the directions O_1 and O_2 that were discussed above, but this is a quite soluble problem. However, we do not in general know our direction in relation to the line $O_2 O_1$, although we can determine whether our direction O lies somewhere in the arc from O_3 through O_1 to O_4, or in the arc from O_3 through O_2 to O_4.

PROBLEM:
How can our direction in terms of these two arcs be determined?

Because we do not know the angle between the line of sight from the Earth to an object and the direction of its motion, it might seem as if the Doppler shift could not be of much practical use. Yet it turns out that practical applications are indeed possible in many situations where some unusual circumstance gives us a way around this difficulty. To end this Appendix, we will consider three such situations. In each of these examples, the practical application of the Doppler shift is of critical importance to the whole structure of astronomy.

1. MOVING GROUPS OF STARS

If a star is nearby, as in Figure II.43, its motion relative to the solar system will cause it to change position on the sky by a measurable amount (see §1.3). If such a "proper motion" of the star were accurately known, the direction $O_2 O_1$, in relation to the direction from the solar system to the star, could be found by an easy calculation. Unfortunately, proper motions are technically difficult to measure accurately. Sometimes, however, it is known from independent evidence that a group of stars, perhaps several hundred of them, all have the same direction $O_2 O_1$. Then, by averaging the proper motions and Doppler shift measurements for all the stars of such a group, a much more accurate result is obtained. An adaptation of this idea for a comoving group of stars known

← ★

• Sun

FIGURE II.43.
Accurate measurement of the proper motion
of a nearby star enables the direction O_2O_1
of Figure II.42 to be found.

as the Hyades forms a crucial link in the method of determining large astronomical distances.

In Chapter 1 we saw how the distances of nearby stars can be found by
using the effect of the Earth's motion. In Section III we will discuss the problem
of measuring very much greater distances than those of nearby stars, and will
refer back to this issue of moving groups of stars.

2. ECLIPSING BINARY STARS

We have noted already that stars are often found in pairs which move around
a common center, the *barycenter,* in the manner of Figure II.44. The line joining
the two stars always passes through the barycenter, so the stars move around
their orbits in the same period, say, *P.* Figure II.44 has been drawn for the
simple case in which the orbits are circles—this is actually a good approximation
for most binary systems in which the stars are reasonably close to each other.

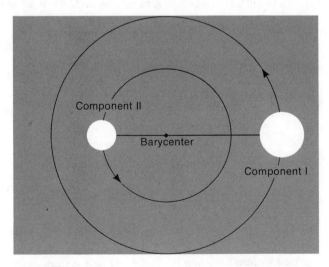

Component II

Barycenter

Component I

FIGURE II.44.
Two stars, forming a binary system, moving in orbits in the same period about a common barycenter.

There is a simple relation between the masses of the stars and their distances from the barycenter. Writing the masses as m_1, m_2, and the distances as a_1, a_2, the relation is $m_1a_1 = m_2a_2$. We do not have to explain this relation, because the barycenter is *defined* so that it is true.

Writing v_1, v_2, for the speeds of motion of star 1 and star 2, respectively, we also have $v_1/v_2 = a_1/a_2$, so that $m_1/m_2 = v_2/v_1$, from which it follows that an observational determination of v_2/v_1 would immediately give the mass ratio of the two stars.

PROBLEM:
Show that

$$P = \frac{2\pi a_1}{v_1} = \frac{2\pi a_2}{v_2},$$

and hence that $v_1/v_2 = a_1/a_2$. Using $m_1a_1 = m_2a_2$, show also that $m_1/m_2 = v_2/v_1$.

Let us see next what can be determined from observation. Because we are considering the binary pair to be not too far apart, we may not be able to distinguish them visually. So how do we separate v_1 and v_2? When the light from such a system is analyzed with the aid of a spectrograph, we hope that spectrum lines from both stars will be present. We will soon know whether or not this is the case, even without bothering to identify the atoms and the transitions that give rise to the observed spectrum lines. How? When one star moves toward us, the other moves away, as in Figure II.45. And as the stars

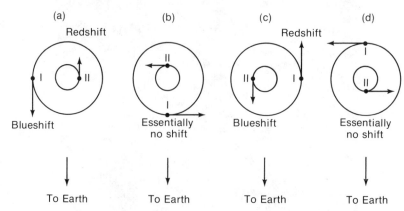

FIGURE II.45.
In position (a) the light from component I is blue-shifted for a terrestrial observer, while the light from component II is red-shifted. The opposite situation occurs in position (c). In positions (b) and (d), the frequency of radiation from both stellar components is essentially unaffected by the binary motion.

λ 4415.1   λ 4528.6

FIGURE II.46.
Two spectra of the double star Mizar in the constellation of the Big Dipper (Ursa Major). The upper spectrum, in which the lines of the two components are superimposed, corresponds to one or the other of positions (b) and (d) of Figure II.45. In the lower spectrum, however, the radiation from one component is blue-shifted, and that from the other component is red-shifted, the difference between the orbital velocities being 140 km/sec. (Courtesy of Hale Observatories.)

move around their orbits, their roles in this respect alternate, as can also be seen from Figure II.45. The orbital motion shown in this figure causes the spectrum lines from each star to undergo a frequency oscillation, and the oscillations for the two stars are exactly out of step with each other. This oscillatory, out-of-step effect on the spectrum lines is illustrated in Figure II.46 for the star Mizar in the constellation of the Big Dipper. Indeed, the oscillatory, out-of-step effect obviously permits the spectrum lines from one star to be separated from the lines of the other star.

Simply by observing the frequency oscillation of one set of spectrum lines relative to the other, assuming both sets to be present, we can determine the ratio v_1/v_2, and therefore the mass ratio of the two stars. Not only this, but the time required for a complete oscillation of the lines determines the period P.

It might be thought that v_1 could be determined separately by measuring the maximum frequency displacement of the spectrum lines from star 1, and that v_2 could also be separately determined in the same way. We could indeed do so, if we knew that the line of sight from the Earth to the star happened to lie in the plane of the orbital motion, as it has been drawn to do in Figure II.45, for then there would be moments when our line of sight was in exactly the same direction as the motions of the stars. However, in general the line of sight will be inclined to the orbital plane, as in Figure II.47. No such deduction is then possible unless we know the angle of inclination, which we usually do not.

PROBLEMS:

1. Kepler's third law, which we will study in Section IV, states that $P^2(m_1 + m_2)/(a_1 + a_2)^3$ is the same in every binary system, including that of a planet moving around the Sun. Satisfy yourself, by comparing a binary star system with the Earth's motion around the Sun, that where the period P of the binary is measured in years, m_1, m_2, are measured in terms of the Sun's mass

as unit, and a_1, a_2, are measured in terms of the Earth's distance from the Sun as unit,

$$m_1 + m_2 = \frac{(a_1 + a_2)^3}{P^2}.$$

The Earth's mass is here to be considered negligible compared to that of the Sun.

2. If v_1, v_2, and P were separately known, we could determine a_1 and a_2 from $P = 2\pi a_1/v_1$, $P = 2\pi a_2/v_2$, and hence obtain $m_1 + m_2$ from the previous problem. Using $m_1 a_1 = m_2 a_2$, show that m_1, m_2, would then be given by

$$m_1 = Pv_2(v_1 + v_2)^2, \qquad m_2 = Pv_1(v_1 + v_2)^2,$$

where P is again measured in years, and v_1, v_2, are measured in terms of the Earth's speed in its orbit around the Sun.

3. Notice that the Earth's annual motion around the Sun leads to an oscillatory Doppler effect. For a star in uniform motion with respect to the Sun, and lying exactly in the Earth's orbital plane, the annual frequency variation $\Delta\nu$ (caused by the Earth's motion) in a spectrum line of average frequency ν satisfies the approximate equation

$$2\frac{\Delta\nu}{\nu} = \frac{\text{Earth's orbital speed}}{\text{Speed of light}}.$$

Why does the factor 2 appear, and why is the relation slightly approximate?

4. The radius a of the Earth's orbit is known from what are called "solar parallax" measurements, to be discussed later. Here we use the result, $a = 1.496 \times 10^{13}$ cm. Knowing also the length of the sidereal year, $T = 3.156 \times 10^7$ seconds, show from the equation $T = 2\pi a/v$ that the Earth's speed v is 29.8 km per second. What fraction of the speed of light is this? Consider how, if the solar-parallax determination of a were not available, the previous problem could be used to find a.

From the second of these problems, it is seen that the masses of the stars of a binary system can be determined separately, provided v_1 and v_2 can be separately determined. For this, we require the direction from the Earth to make only a negligible angle with the plane of the binary motion as in Figure II.48. Are there systems for which we know this requirement is met? Because the stars have a finite size, and because we are considering them to be not too far apart, the binary motion in such a system must cause one star to pass across our line of sight in front of the other. Star 1 will eclipse star 2 during some part of its orbit, and when it does so the light from star 2 will be cut off, either completely or partially. The total light we observe from the binary system must therefore be reduced. There will also be a stage in the motion when star 2 will similarly eclipse star 1, and there must again be a reduction in the light

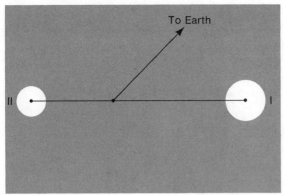

FIGURE II.47.
In general, the line of sight from a binary system to the Earth
is inclined at an appreciable angle to the plane of the binary
motion. In such a system, the Doppler-shift method does not
yield the actual speeds of the component stars in their orbits.

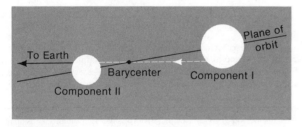

FIGURE II.48.
When the line of sight from a binary system to the Earth happens to make a small angle with the orbital plane of such a
system, the two stars eclipse each other.

we receive from the system. So in every complete period there will be two
reductions of light, two eclipses, which in general will not be equal, because
in general the two stars will not be equally bright, nor will they have just the
same radius. A faint star passing in front of a bright one cuts off more of the
light than the bright star passing in front of the faint one. An example of such
a double light reduction is shown in Figure II.49 for the star WW Aurigae.
Systems whose light fluctuates in this regular and characteristic way are known
as *eclipsing binaries*. The details of the eclipses can vary quite widely from
system to system, as can be seen from the examples illustrated in Figures II.50
and II.51.

The conditions required for the determination of both v_1 and v_2 can thus
be taken to exist in this particular class of binary system, and hence separate
masses m_1 and m_2 can be obtained for their component stars. A great deal of
what we know concerning the masses of stars comes from this method of

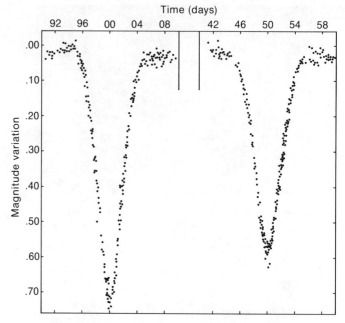

FIGURE II.49.
In exceptional cases, where the line of sight happens to make a small
angle with the plane of the orbital motion, the Doppler method does indeed determine the speeds, and such exceptional cases can be detected
from the periodic eclipses which occur in the manner of Figure II.48.
Here we have the example of the binary WW Aurigae, with variations
of magnitude plotted upward and with time in days plotted horizontally.
What is the period P in this case? (From C. M. Huffer and Z. Kopal,
Astrophysics Journal, 114, 1951, 297. University of Chicago Press.)

investigation. These particular data have played a critical role in the development of astronomy, since stars are found to behave in markedly different ways
according to their masses. It would have been difficult to separate the various
categories of behavior, which we shall study in the next Section, if mass values
had not been available for certain individual stars.

PROJECT:
A table of the brighter stars was given at the end of Section I. From a catalogue
of binary systems, discover which of the stars in the table of Section I are
binaries. If a small telescope is available, make up an observing list. Notice from
observations the differences of luminosity and color between the components
of these systems. Can you think of any reasons why there should be these
differences? Down to what angular separation are you able to distinguish the
components, and how does your answer to this question depend on their relative
brightness?

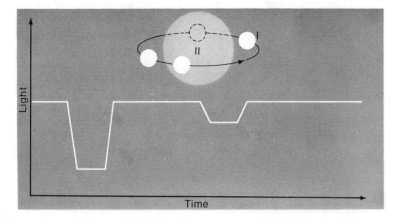

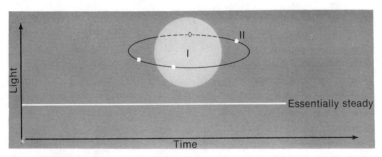

FIGURE II.50.
A schematic illustration of the fact that when the two components are of unequal luminosities, the eclipse of the brighter component by the fainter one produces a deeper eclipse than does the reverse situation. Often the component of greater luminosity is a hot blue star of smaller size than its giant companion. This is the situation for the star Algol.

FIGURE II.51.
Sometimes one of the components is a faint white dwarf, too small in size to produce an effective eclipse of the brighter component. In such systems the eclipses have no significant effect on the light, but the binary motion still gives a Doppler effect.

3. THE EXPANDING UNIVERSE

In thinking about the large-scale structure of the universe, we can consider galaxies to be "particles," representing each of them by a world line in the manner already discussed in Chapter 4.* The situation is illustrated in Figure II.52, where we imagine the single spatial dimension x to represent the three spatial dimensions x, y, z. Proceeding in the same way for every galaxy, we can represent all their paths together in a single diagram, as in Figure II.53.

Suppose we consider a certain moment of time, i.e., a time section of space-time. The world lines of the galaxies intersect such a time section at points.

---

*If several galaxies are permanently associated in a cluster, the whole cluster is to be treated as a "particle." Such clusters do not expand in the sense of the following discussion.

Appendixes to
Section II

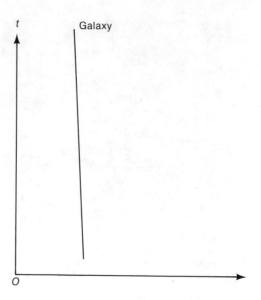

FIGURE II.52.
A galaxy may be represented by a world line in space-time, as if it were a particle.

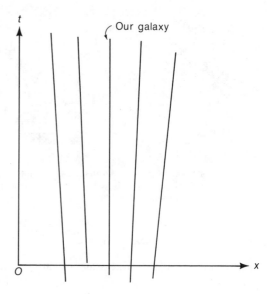

FIGURE II.53.
The world lines of all the galaxies can be represented in a single diagram. Subject to the assumption that the universe on a large scale is homogeneous and isotropic, the world lines of the galaxies must satisfy one of three possible conditions. They could form a uniformly divergent fan, as in this figure.

According to the discussion of Chapter 3, the large-scale distribution of these points must be uniform, in order to satisfy the requirements of *homogeneity and isotropy*. And this condition must be satisfied for *all* such time sections. Subject to this restriction, we ask:

What forms are possible for the distribution of the
world lines of the galaxies?

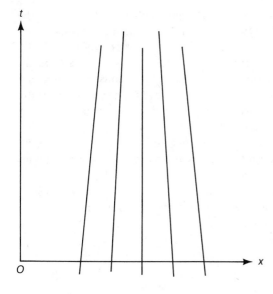

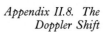

FIGURE II.54.
World lines of galaxies could form a uniformly convergent fan.

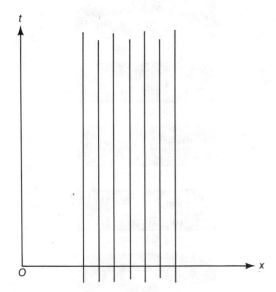

FIGURE II.55.
World lines of galaxies could form a constantly spaced
distribution.

There are three possible answers to this question. The world lines must form
either a uniformly divergent fan as in Figure II.53, a uniformly convergent
fan as in Figure II.54, or a constantly spaced distribution as in Figure II.55.
In the first case we have an *expanding universe,* in the second case a *contracting
universe,* and in the third case a *static universe.* To conclude this section, we
will see how the Doppler shift can be used to establish the first of these
alternatives.

For a static universe there would be no Doppler shift in the light from the galaxies—for any particular galaxy, we would have the situation of Figure II.41. In the other two cases, however, we would have the picture of Figure II.42, with observation always from O_1 for a convergent fan, and always from O_2 for a divergent fan. The data permit us immediately to dismiss the static and the convergent possibilities. Doppler shifts *are* observed, contrary to the static case, and the shifts are of the red-shift kind, requiring observation from O_2, not O_1. Therefore the universe expands.

The red shifts of the galaxies were first discovered by V. M. Slipher (1875–1969), who succeeded in measuring the shifts for more than twenty galaxies during the period 1912 to 1925. To Slipher's surprise they all showed red shifts, a consistent pattern later greatly extended by Hubble and Humason (1891–1972). Data obtained by Hubble and Humason are shown in Figure II.56.

In Figure II.53, one galaxy, which we take to be our own, has a world line

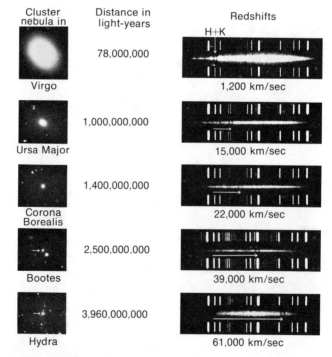

Cluster nebula in	Distance in light-years	Redshifts
Virgo	78,000,000	1,200 km/sec
Ursa Major	1,000,000,000	15,000 km/sec
Corona Borealis	1,400,000,000	22,000 km/sec
Bootes	2,500,000,000	39,000 km/sec
Hydra	3,960,000,000	61,000 km/sec

FIGURE II.56.
The relationship between redshift and distance for extragalactic nebulae. The redshift of the spectrum lines from specific atoms, for example, the H and K lines of calcium (indicated by the arrows), when interpreted in terms of the Doppler effect, gave the speeds shown in this figure. Redshifts are expressed as velocities, $c\Delta\lambda/\lambda$. A light-year equals about 9.5 trillion (9.5×10^{12}) kilometers. Distances are based on an expansion rate of 50 km/sec per million parsecs, a parsec being 3.26 light-years. (Courtesy of Hale Observatories.)

parallel to the direction of increasing time. We take $x = 0$ for this galaxy, and if we resurrect the y,z, dimensions, we also take $y = 0$, $z = 0$. Our galaxy is therefore considered to be stationary, with all the other galaxies moving away from us. Does this mean that our galaxy is a kind of center of the universe? The answer to this question is that we have made our galaxy a center by measuring the four dimensions of spacetime, t, x, y, z, with the aid of radiation from atoms that are stationary with respect to ourselves. An observer in another galaxy, using atoms at rest with respect to himself, would obtain a diagram like Figure II.53, but with his own galaxy represented by a world line parallel to the direction of increasing time. The central placing of our galaxy in Figure II.53 is therefore not unique.

PROJECT:

Consider two time sections of Figure II.53, corresponding to two slightly different moments of time as we measure it. The world lines of the galaxies intersect each of these sections in "points." Take three galaxies, our own, G, and any two others, say, G_1 and G_2. The two triangles formed on the two time sections by the world lines of G, G_1, G_2, have the same shape and orientation. This is exactly what is meant by saying that the fan of world lines of the galaxies is *uniformly* divergent. The triangle on the slightly later time section is a little larger than the other—this is what is meant by saying that the fan is *divergent*. Two such triangles are shown in Figure II.57.

Draw the triangles of Figure II.57 on separate pieces of tracing paper. Keeping the same orientation, superimpose the points G of the two triangles. This shows how galaxies G_1 and G_2 appear to us to have moved in the time interval between the two sections. Next, superimpose the points G_1, always maintaining the orientations of both triangles. This shows how an observer in the galaxy G_1 would see our galaxy and G_2 move, if he were to set up his own system of time and space measurement. Similarly for an observer in G_2. Consider how the same idea can be applied to further galaxies, G_3, G_4, ...

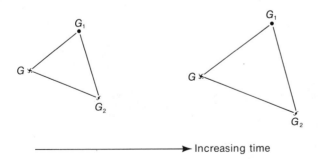

→ Increasing time

FIGURE II.57.

The triangles formed by our galaxy G and by two other galaxies G_1, G_2, drawn for two different moments of time, have different scales. For the divergent fan of Figure II.53, the triangle corresponding to the later moment has the larger scale.

In considering the fan of world lines of the galaxies to be *uniformly* divergent, we have moved ahead of the actual observations of Figure II.56. By uniform divergence, we mean that the greater the distance of a galaxy, the greater the speed of its motion away from us—the speed of recession, say, v, must be proportional to the distance, say, d. Thus for uniform divergence we require $v = Hd$, where H is some constant, with a value to be determined by observation.

PROJECT:
Satisfy yourself, by reconsidering the previous project, that $v = Hd$ would hold good in such a practical construction.

We end the present discussion by considering the evidence for the relation $v = Hd$, which therefore supports the concept of *uniform* divergence. We concentrate on a specific class of galaxies which, from their characteristics, we may reasonably suppose to be similar to each other. As an example of such a class, we might take the brightest member of the rich clusters of elliptical galaxies that we studied in Chapter 3.

Now the flux from a galaxy at distance d is $L/4\pi d^2$, where L is the intrinsic luminosity, i.e., the energy emitted in unit time. If all the galaxies in our class are taken to have the same value of L, then the fluxes f we receive from them, and which we regard as measurable, will vary only with distance, according to the relation $f = L/4\pi d^2$. It will be recalled from Chapter 1 that the "magnitude" m is obtained from f by a certain curious procedure. We take the logarithm, defined by $10^{\log f} = f$. Then we multiply $\log f$ by -2.5 and add a certain constant, chosen conventionally in relation to "first"-magnitude stars, giving

$$m = -2.5 \log f + \text{certain constant.}$$

PROBLEM:
Satisfy yourself that

$$\log f = -2 \log d + \log\left(\frac{L}{4\pi}\right),$$

and that if $v = Hd$, then

$$\log f = -2 \log v + 2 \log H + \log\left(\frac{L}{4\pi}\right).$$

Using the expression for $\log f$ obtained in this last problem, we see that the effect of $v = Hd$ would be to make the magnitude m depend on v:

$$\boldsymbol{m} = 5 \log v - 5 \log H - 2.5 \log \left(\frac{L}{4\pi}\right) + \text{a certain constant.}$$

We can lump together the last three terms appearing on the righthand side of this equation to form a new constant, writing more simply

$$\boldsymbol{m} = 5 \log v + \text{new constant.}$$

Next, since we are dealing with red shifts, the observed frequency $\boldsymbol{\nu}'$ of a line in the spectrum of a galaxy is related to the speed v by

$$\frac{\boldsymbol{\nu}'}{\boldsymbol{\nu}} = \sqrt{\frac{1 - v}{1 + v}},$$

where $\boldsymbol{\nu}$ is the frequency of the same line emitted by stationary atoms. Hence v is known from the observed red-shift ratio $\boldsymbol{\nu}'/\boldsymbol{\nu}$, and we can use our convention of writing observationally determined quantities in bold type,

$$\boldsymbol{m} = 5 \log \boldsymbol{v} + \text{constant.}$$

What we have thus discovered is that *if* $v = Hd$ is true, and *if* we observe a set of galaxies all of the same L, there must be a straight line relationship of the form shown in Figure II.58 between $\boldsymbol{m}$ and $\log \boldsymbol{v}$. Notice that the scale of $\boldsymbol{m}$ has been chosen to be five times smaller than that of $\log \boldsymbol{v}$ in order to make the straight line have a slope in the diagram of $45°$.

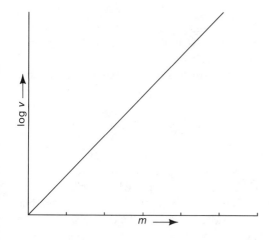

FIGURE II.58.
The theoretical linear relationship between m and $\log v$ for not too distant galaxies.

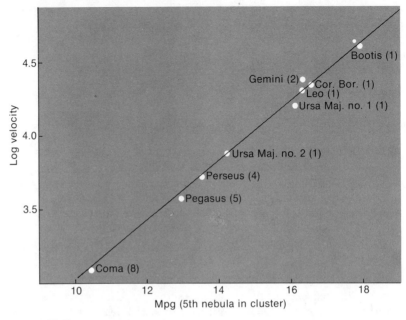

FIGURE II.59.

The observed results of Hubble and Humason support the linear relationship between *m* and log *v*.

Data obtained by Hubble and Humason for the fifth brightest elliptical galaxies in clusters are shown by the marked points of Figure II.59. Instead of specifying *v* by numbers between 0 and 1, i.e., as a fraction of the speed of light, Hubble and Humason expressed the speeds of the galaxies in kilometers per second, and it is the logarithms of these that are plotted in Figure II.59. Thus a speed *v* = 10,000 km per second has a logarithm of 4.0, 1,000 km per second has 3.0, and so on.

The observations fall close to the straight line required by the relation *v* = *Hd*, and hence provides strong evidence that the world lines of the galaxies do form a uniformly divergent fan. In Section VI we will be much concerned with the physical meaning of this result, which is one of the most remarkable achievements of astronomy.

Appendix II.9. Diffraction Gratings and Spectrographs

In Figure 5.33, reproduced here as Figure II.60, we saw how by using a glass prism one can separate the light of a star or galaxy into its constituent frequencies. When the viewing screen of Figure II.60 is replaced by a photographic plate, the resulting arrangement is known as a spectrograph. The spectrographs actually used by astronomers are considerably more efficient, however, than the simple scheme of Figure II.60. In practice, an instrument known as a

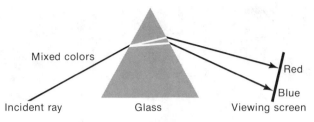

FIGURE II.60.
We can use a glass prism to separate light of mixed colors, projecting the separate colors onto a viewing screen.

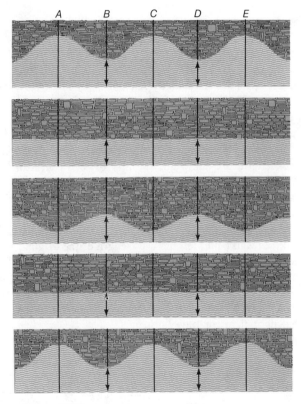

FIGURE II.61.
This cross section through Figure 5.29 shows that at points *A*, *C*, and *E* on a sea wall inside the breakwater, waves rise high and fall low. At *B* and *D* the water is almost still.

diffraction grating is used in preference to a prism. Because the separation of light into its constituent frequencies is manifestly very important—we met some examples of this in Appendix II.8—it is worth our while to consider in some detail how a diffraction grating works. We begin by going back to Figure 5.30, reproduced here as Figure II.61.

FIGURE II.62.
If we pass light through fine slits onto a screen, we also get places of reinforcement (the bright bands) and of cancelation (the dark bands).

Just as the distance between the points *A* and *B*, *B* and *C*, etc., in Figure II.61 depends on the wavelength of the original water wave outside the breakwater, so the distance between the fringes in Figure 5.31, reproduced here as Figure II.62, depends on the wavelength of the light. The longer that wavelength, the wider apart are the fringes. All this can be easily demonstrated by the simple experiment shown in Figure II.63. The lamp *L* has a cylindrical source marked *S*. Because the lamp does not emit light of a pure color, a filter *F* must be used. (In practice, no filter gives completely pure color. A small dispersion of wavelengths still remains after passage through the filter, but the remaining wavelengths are sufficiently similar to each other for the purposes of our experiment.) The "breakwater," marked *D*, consists of an ordinary photographic plate on which slits, spaced about half a millimeter apart, have been ruled with a knife-edge. The interference fringes can be viewed directly by placing the eye immediately behind this photographic plate. Just as our observer on the sea wall could tell the wavelength of the original waves outside the breakwater from the positions of the points *A*, *B*, *C*, etc., in Figure II.61, so we can here calculate the wavelength of the light by measuring the distance between adjacent fringes. By changing from a blue filter to a red one, the distance between the fringes can be changed. The fringes are more widely spaced for red light than for blue light.

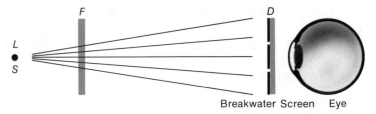

Breakwater Screen   Eye

FIGURE II.63.
In Figure II.62 the distance between the bright and dark bands (fringes) depends on the frequency of the light. Here a filter allows only light of a certain frequency to pass. A breakwater type of screen enables the distance between fringes to be measured and hence permits the frequency to be calculated.

What would happen if we repeated this experiment without using a filter? We should then have fringes formed for the whole range of colors emitted by the lamp; and because fringes for different colors fall in different places, the bright fringes from one color could fall in the dark gaps belonging to another color. So instead of obtaining a series of clearly defined bright bands as in Figure II.62, we should tend to get a continuous strip of light. But the strip would evidently not be uniformly colored. The places where the blue fringes fell would tend to appear blue and the places where the red fringes fell would tend to appear red. This means that the two slits in our breakwater would have served to separate the colors present in the original light emitted by the lamp.

This result offers us a challenge. Is it possible, by a suitable arrangement of slits in the breakwater, to separate the different colors emitted by the lamp in a systematic way, so that the fringes from the various colors fall into an orderly sequence instead of overlapping with each other in a confused jumble? If we can do so, we shall have succeeded in separating the light into its constituent colors, just as with the prism shown in Figure II.60. In fact, we shall have succeeded in producing a diffraction grating.

Think for the moment of the fringes produced by a specific color. If we can make the gaps between successive fringes become large compared to the widths of the fringes themselves, then clearly it will be much easier to lay sets of fringes from different colors side by side without running the risk of their overlapping. Both experiment and calculation show that there is a simple prescription for increasing the distances between successive fringes. To do so we need only rule the two slits in our breakwater closer together than they were before. Unfortunately, however, this also has the effect of increasing the width of the fringes themselves, so that there is still a risk of overlapping.

The solution to the problem turns out to be that we must not only cut the slits very close together, but also have a very large number of slits in our breakwater, as shown in Figure II.64. Although more complicated, the situation is exactly the same in principle as it was before. Now, however, we have waves spreading out from a whole multitude of slits. In some directions the waves from all the slits augment each other, just as they did with two slits, and where these directions impinge on our viewing screen we again have bright bands.

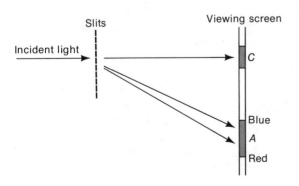

FIGURE II.64.
Fringes for light of various colors fall in different places. By ruling many fine slits close together in the breakwater screen, narrow bright bands are produced for each color, adjacent to each other, without entirely filling up the dark gaps.

In other directions the waves interfere with each other, because crests of some arrive simultaneously with troughs of others, and where these directions impinge on our viewing screen we have dark bands. But it happens that the bright bands are much narrower than the dark spaces between them. This is just the condition we set out to achieve. If we now take light made up of a range of colors, instead of light of one specific wavelength, the bright bands for the different wavelengths can be made to fall adjacent to each other without entirely filling up the dark gaps. This result is more clearly observable by taking the bright bands which fall on the outskirts of the screen, say near A in Figure II.64, rather than those near the center, at C.

Color dispersion by the arrangement shown in Figure II.64 is more complicated both to understand and to produce experimentally than the simple prism dispersion shown in Figure II.60. It may be wondered, therefore, why the astronomer prefers to use a diffraction grating rather than a simple prism for obtaining a spectrum. The reason is that the diffraction grating separates colors far more efficiently than the prism.

Suppose we wish to separate light of two different colors. Provided they are of widely different wavelengths the job of separation is easy, but as the wavelengths become more and more similar the problem becomes increasingly difficult. Indeed, every known method of separating light fails sooner or later as the wavelengths become too close. The prism is a comparatively crude method of separation, and it fails long before the diffraction grating does. With a prism it is possible to separate two wavelengths differing from each other by about one part in 10,000; with a diffraction grating wavelengths differing by as little as one part in 100,000 can be separated. To separate wavelengths with even smaller differences—as little as one part in a million—it is necessary to use highly specialized equipment which need not here concern us.

Before we leave the subject of diffraction gratings it is worth noticing that a similar phenomenon arises if instead of a breakwater with many slits in it we utilize a series of posts, as shown in Figure II.65. Waves are scattered by

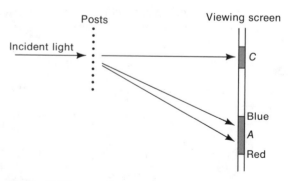

FIGURE II.65.
Waves scattered from posts set close together behave like waves passing through closely spaced slits.

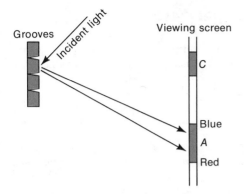

Grooves

Incident light

Viewing screen

C

Blue

A

Red

FIGURE II.66.
A diffraction grating is based on the arrangement of Figure II.65, evenly spaced grooves on glass serving as "posts." Instead of shining light *through* the posts, we shine it from the side onto the glass plate, thus efficiently separating the colors.

the posts and interfere with or reinforce each other in exactly the same way as we have already considered. This fact greatly assists the practical construction of a diffraction grating. The method of making one is to rule on a plane glass surface a very large number of equally spaced lines, the rulings being cut in the glass with a diamond or some other hard point. Great care must be taken to ensure that the rulings are spaced at precisely equal distances apart. They then act like the posts of Figure II.65, but now, instead of shining the light *through* the posts, we can shine it from the side on to the glass plate, as shown in Figure II.66. The grooves in the glass plate now scatter the light waves just as they did in Figure II.65, and the scattered waves reinforce each other in certain directions just as in Figure II.64. By viewing the light scattered in these particular directions we obtain a spectrum; we have at our disposal the essential feature of an instrument which plays a vital part in modern astronomy.

The accurate ruling of diffraction gratings is a technical problem of very considerable difficulty, and the outstanding pioneer work in this field was done by H. A. Rowland, who in 1882 successfully constructed a ruling engine capable of making almost 15,000 lines per inch on the surface of speculum metal, a hard alloy of copper and tin. As we have seen, the chief requirement of a good grating is that the lines should, as nearly as possible, be equally spaced. To get such a result it is necessary that after each groove has been ruled the machine should lift the diamond point and move it forward by a fixed distance determined by a small rotation of a screw. The screw must clearly be of almost perfect construction, and it was Rowland who first achieved such near-perfection.

In modern times gratings are ruled on aluminized glass surfaces instead of on speculum metal. The aluminum causes the grooves to produce a much stronger scattering, so that far less light is lost in the process than would be lost by a grating ruled on an untreated glass surface. This added efficiency is of great importance when very faint astronomical objects are under observation.

1. Draw the Planck curve for a star of surface temperature 20,000°K, stating the unit you use for the frequency scale, and adjusting the intensity scale so that the area under the curve gives the power radiated by 1 square kilometer of surface of the star.

2. After drawing the curve of Problem 1, mark the frequency range of visible light.

3. An experimenter analyses the radiation of a hot gas with respect to its constituent frequencies. Would you expect emission lines or absorption lines to be found?

4. Radiation from a hot solid surface passes through a cool gas. When analysed with respect to frequency, would you expect the radiation to contain emission lines or absorption lines?

5. In which spectral class are absorption lines of the Balmer series most prominent?

6. For which spectral classes are the H and K lines of singly ionized calcium atoms (one electron made free) very prominent?

7. In what spectral class are the lines of singly ionized helium most clearly seen?

8. At first sight, it might seem from Figure II.9 as if infinitely weak radiation from a source particle might be detected. Why is this not so in practice?

9. What is a photomultiplier? What are its advantages and disadvantages compared to a photographic plate?

10. What is an image tube?

11. Discuss the practical measurement of (a) time intervals, (b) spatial distances.

12. State Pythagoras' theorem.

13. How is Pythagoras' theorem changed when a time interval is involved?

14. Discuss the determination of the speed of light.

15. Describe the light cone and explain its physical meaning. How can the light cone be used to give significance to the statement that "particles cannot move faster than light"?

16. What is meant by the age of a person? By writing down appropriate formulas, show how a person's age can be measured precisely.

17. On two occasions in their lives, two persons meet together. Using Problem 16, explain why it is possible for them to age by quite different amounts between these occasions.

18. What observational result involving the muon supports the ideas of Problem 17?

19. Describe the Doppler shift. Give formulas relating the frequency of radiation from moving atoms to similar radiation from stationary atoms, restricting your discussion to the cases where the motion is directly toward and directly away from the observer.

20. Radiation from the ground-level transitions of stationary hydrogen atoms has a frequency of 1.42×10^9 cycles per second. With what fraction of the speed of light must hydrogen atoms move directly toward an observer in order that radiation from these transitions shall appear as ultraviolet light of a frequency of 1.42×10^{15} cycles per second?

21. How can the Doppler shift be used to derive important astronomical information? Illustrate your answer by discussing the case of an eclipsing binary star.

22. Show how the Doppler interpretation of the red shifts observed in the light from distant galaxies implies that the universe expands.

23. Do these red shifts also imply that our galaxy is at the center of the universe?

24. Discuss the evidence for the relation $v = Hd$, relating the recessional speed v of a galaxy to its distance d. What is the largest value of v for which this relation has been established?

25. Satisfy yourself that the straight line of Figure II.59 would have a slope of $45°$ if it were redrawn with the magnitude scale made five times smaller than the scale of log v.

26. Explain the general principles involved in the operation of a diffraction grating.

References for Section II

The most accessible presentation of quantum mechanics, along the lines of the concepts of Chapter 4, is R. P. Feynman and A. R. Hibbs, *Quantum Mechanics and Path Integrals* (New York: McGraw-Hill, 1965).

For a semipopular book on telescopes, see A. G. Ingalls, *Amateur Telescope Making* (New York: Munn, 1947). For a much more technical exposition, see *The Construction of Large Telescopes,* edited by D. L. Crawford (New York: Academic Press, 1966).

For Einstein's theory of 1905 ($E = mc^2$), see "Zur Elektrodynamik bewegter Körper," *Annalen der Physik,* 17 (1905), 891.

Section III:
Astrophysics

Albert Einstein, 1879–1955, at his desk in the Patent Office, Bern, Switzerland, where he was working as a clerk during the period when he developed the Special Theory of Relativity (see Appendix II.4).

Chapter 6:
The Formation of Stars and
the Origin of the Galaxy

§6.1. Gas Clouds

I doubt whether even the most hard-bitten soul can look up at the sky on a clear, dark night without wondering where it all came from, and what it all means. What it *means* is a difficult and intractable problem, but at least we can say something about where the stars came from. Stars are forming even now within clouds of gas, like the Orion Nebula (Figure 6.1). This cloud can easily be detected in the "sword" of Orion by the naked eye; a wide-angle photograph of the constellation of Orion is shown in Figure 6.2. The diameter of the Orion Nebula is about 15 light years, and its distance from us is about 1,500 light years. Since this is much less than the diameter of our galaxy, 100,000 light years, it is clear that the Orion Nebula is a comparatively nearby object. Together with the whole constellation of Orion, it belongs to what in Chapter 3 we called our "local swimming hole."

We can expect that there will be many objects like the Orion Nebula within the galaxy, and that stars will also be forming within them. Indeed, if we extend the distance range to 5,000 light years, quite a number of remarkable gas clouds are found, the Rosette Nebula (Figure 6.3), for example. Observation of the

*The Formation of
Stars and the Origin
of the Galaxy*

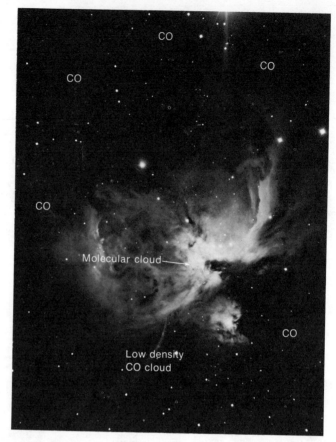

FIGURE 6.1.
The Orion Nebula, a cloud of gas about 15 light years in diameter, in which stars are now forming (courtesy of Hale Observatories).

whole of our galaxy is restricted, however, by small particles, "grains," of solid material that exist everywhere throughout the Milky Way. Both for optical wavelengths and for ultraviolet light, these grains produce an effect like a general fog. Although each individual grain is small, typically about 10^{-5} cm in radius, there are so many of them that their cumulative effect becomes serious when we attempt with optical techniques to observe stars or gas clouds that are more than a few thousand light years away. Indeed, the raggedness of the Milky Way, as it appears to the naked eye in directions like that of Sagittarius, comes from a large-scale obscuration by myriads of interstellar grains. When astronomers wish to observe very distant objects lying far outside our own galaxy, they take care to choose objects in directions more or less perpendicular to the plane of the Milky Way. This involves the least "fog," but still a few hundred light years of it.

FIGURE 6.2.
A wide-angle photograph of the constellation of Orion (courtesy of
Hale Observatories).

FIGURE 6.3.
The Rosette Nebula, a more distant cloud, about 75 light years in diameter, is still close to us
within our galaxy taken as a whole (courtesy of Hale Observatories).

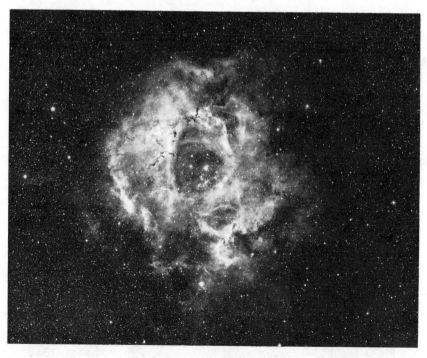

268

*The Formation of
Stars and the Origin
of the Galaxy*

In Chapter 3 we discussed a class of galaxy known as type S0. These are spiral galaxies that have lost their gas clouds and dust grains through collision with one or more other galaxies. An example of such a galaxy was shown in Figure 3.26. If our galaxy were to experience such a collision and were to become an S0, how would the general appearance of the night sky be changed?

Sometimes the obscuring grains are highly concentrated into what are called *dark nebulae*. The Horsehead Nebula (Figure 6.4) is a striking example. Here the "fog" is so thick that we see nothing on the far side of the nebula; it blots out everything lying behind it. In contrast to the dark nebulae, the *bright nebulae*, like Orion, are made to emit light by the stars which have formed within them. Radiation emitted by the stars is first absorbed and then reradiated by atoms of gas within the nebula. The general order of the density of atoms in nebulae like the Orion is about 1,000 per cm^3, but much higher gas densities

FIGURE 6.4.
The Horsehead Nebula, a *dark* nebula about 5 light years in size, is produced by myriads of fine dust grains (courtesy of Hale Observatories).

FIGURE 6.5.
Schematic drawing of the Orion Nebula, showing a dense molecular
cloud, clouds emitting intense infrared radiation, and a much less dense,
extensive region surrounding the optically visible bright nebula of Figure
6.1.

than this occur in localized regions with dimensions of about a light year, i.e.,
about one-tenth of the linear size of the main nebula itself. The presence of
such a very high-density region within the Orion Nebula has recently been
detected observationally using the new techniques of infrared astronomy and
of molecular astronomy; these will be discussed in a later chapter. The situation,
as it has now been revealed, is illustrated in Figure 6.5. From this figure it
will be seen that a larger, more extensive cloud, surrounding the optically visible
part of the Orion Nebula, has also been recently detected, by means of a
characteristic radiation at a wavelength of 2.6 millimeters of the carbon mon-
oxide molecule.

Neither this lower-density outer cloud, nor the much smaller, very high-
density region can be detected by ordinary optical means, i.e., using ordinary
light. Too little light is radiated by the outer cloud, while the inner, very

*The Formation of
Stars and the Origin
of the Galaxy*

high-density region also contains a high density of dust. The inner region, if it were by itself, would be a dark nebula, like the Horsehead. This is an unfortunate circumstance, *because it is inside such very high-density regions that stars are being formed.* Thus the obscuration caused by a multitude of grains prevents us from using the techniques of optical astronomy to study the formation of stars at first hand. Although stars are constantly being born, we are not able to observe the process with the most powerful observational tools now available to us. Once stars have formed, once they emit sufficient radiation to heat the surrounding gas and dust, and to blow it away, then at last we can see them, but by then it is too late—the stars have already formed. This is the infuriating aspect of the whole star-formation problem. We can see the outcome of the star-formation process, but not the process itself.

Of course, even if we could observe star formation directly within the Orion Nebula, we could still wonder about the origin of the nebula itself. The natural argument would be to say that both dark and bright nebulae form by condensation from the general distribution of gas and dust which exists everywhere along the plane of the galaxy. Even so, we could still press our question still further: How did the gas and dust along the plane of the galaxy come to be formed? Indeed, how did the galaxy itself come to be formed?

So we can distinguish a whole series of stages and questions:

1. How did our galaxy form? In particular, how did it develop the disklike structure which we recognize as the Milky Way?
2. How do gas clouds like the Orion Nebula form?
3. How do dense smaller clouds form within objects like the Orion Nebula?
4. How do such dense smaller clouds fragment into protostars?
5. How do the protostars develop into stars of the kind we see in the sky?

If we can answer all these questions, we will have gone a long way toward understanding "where it all came from."

The order in which our questions are set out here suggests an over-all process of increasing compression. Since gravitation is known to pull particles together, it is natural to think of gravitation causing condensation on both a galactic and a stellar scale. Notice, however, the great range of compression required in order to understand the origin of both galaxies and stars. In stage 1, the pre-galactic stage, the density of matter must on the average be less than 1 atom per cm^3. In stages 2 and 3 we are concerned with densities from about 10^3 to 10^6 atoms per cm^3. This still leaves a vast gap to be bridged by stages 4 and 5, since stars have a much higher average density of about 10^{24} atoms per cm^3, the density being about 10^{18} atoms per cm^3 at the lower end of the stellar range and 10^{30} atoms per cm^3, or more, at the upper end. Enough is known about stage 4 for it to be clear that this particular stage plays a crucial role in bridging this enormous density gap, but unfortunately not enough is known, because of the observational difficulties discussed above, for details to be thoroughly understood. Since we lack a complete theory of stage 4, we will not say very much about it. In what follows we begin, chronologically, with

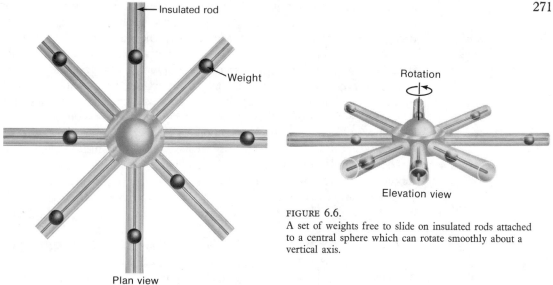

Insulated rod

Weight

Rotation

Elevation view

FIGURE 6.6.
A set of weights free to slide on insulated rods attached to a central sphere which can rotate smoothly about a vertical axis.

Plan view

stages 1, 2, 3, and then we pass to stage 5, where once again our knowledge is reasonably complete. At the end, however, we will consider an issue which shows clearly how very important a satisfactory theory of stage 4 would be.

§6.2. Early Galactic History

It is usually supposed that at the time of its formation our galaxy was a vast condensing cloud, containing sufficient gas to make about 10^{11} stars. The cloud had a simpler chemical composition than the smaller gas clouds of the present day, containing atoms only of hydrogen and perhaps helium. The reason why the composition of the galaxy has changed since its formation, complex atoms like iron now being present, is the main subject of the Appendixes to this Section. It is a consequence of such a change that the earlier stars to form possessed fewer complex atoms than the later stars.*

As the initial gas cloud fell together, it experienced the phenomenon of "spin-up." Imagine a set of weights free to slide on insulated rods attached to a central sphere which can rotate smoothly about a vertical bearing, as in Figure 6.6. Set the arrangement spinning slowly with the weights initially at maximum extension. Now let the weights be retracted without any external influence being exerted by means of an electrical attraction between the weights and the central sphere. The whole system will now spin up: it will rotate faster

---

*The earlier stars are called Type II, the later stars Type I. Chronologically, these type designations are clearly the wrong way round. They were named this way because almost all the stars we see with the naked eye are later stars, of Type I—and these were the ones known in ancient times.

272

*The Formation of
Stars and the Origin
of the Galaxy*

as the weights move inward. This same process is taken to have operated on the large gas cloud which fell together to form our galaxy, except, of course, that gravitation, not electrical attraction pulled the galaxy together.

There is a limit to which spin-up can take place in any practical system. If the device of Figure 6.6 were initially set spinning very fast, the electrical attraction might not be strong enough to retract the weights at all. In fact, by carefully adjusting the initial spin, one could set up a self-balancing situation. The system would then rotate in an equilibrium condition, illustrated in Figure 6.7, with the electrical attraction just compensating the outward rotary forces. Except that in the galactic case we have gravitation instead of electrical forces, the condensing gas cloud eventually experiences a similar equilibrium condition. The primaeval galaxy spins up until the forces of rotation, directed perpendicularly outward from the axis of spin, just compensate for the inward pull of gravity. The spin-up then stops, the situation at this stage being illustrated in Figure 6.8.

Now it is important to notice that the discussion of the preceding paragraph applied only to the situation perpendicular to the axis of rotation. Whereas the rotary forces are directed perpendicularly away from this axis, the gravitational forces act to pull the gas cloud together in all directions, parallel to the axis as well as at right angles to it. Consequently gravitation still continues

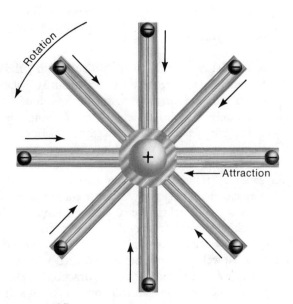

FIGURE 6.7.
The outward rotary forces on the weights are compensated for by the electrical attractions between the similar negative charges on the weights and the positive charge on the central sphere.

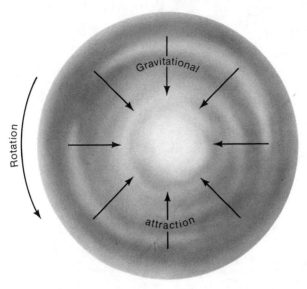

FIGURE 6.8.
For a condensing gas cloud, the rotary forces are compensated for
by gravitation, in analogy to Figure 6.7.

to produce condensation and compression of the gas in a direction parallel to
the axis of rotation, with the result that the gas settles more and more into
a flat rotating disk, as in Figure 6.9. This is clearly a very satisfactory result,
since galaxies like our own do have a general disklike form, as can be seen
simply from observation of the Milky Way. Figure 6.10 gives a plot of the
galactic distribution of bright nebulae that show this disklike property.

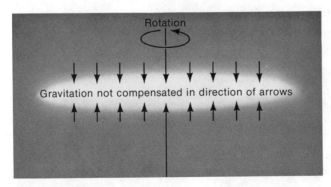

FIGURE 6.9.
Although gravitation compensates for the radially acting rotary forces,
it causes a gas cloud to form itself into a thinner disk.

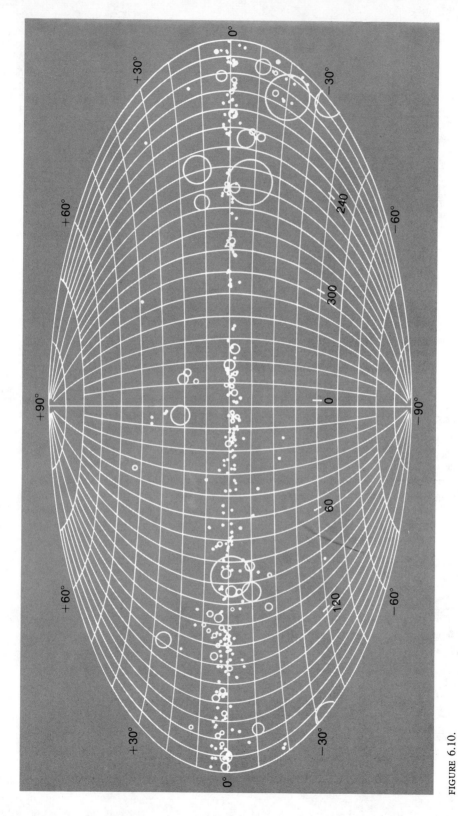

FIGURE 6.10.

This figure gives the directions in the Milky Way (see Figure 1.28) in which bright nebulae are found. (From S. Sharpless, *Galactic Structure*, University of Chicago Press, 1965.)

PROBLEM:

In Figure 6.10, a minority of the bright nebulae appear to lie far outside the plane of the galaxy. Given that this is not so, what explanation of these exceptional cases do you offer?

It is important to notice an effect illustrated in Figure 6.11. Compression of a gas cloud, any cloud, develops heat. In pumping up a bicycle tire, one finds that the base of the pump gets hot—the metal of the pump picks up heat from the compressed air. The energy required for this heating of a pump comes from the action of the forces exerted on the pump handle. For a condensing galaxy, the heat comes from the action of the gravitational forces.

A bicycle pump can be used to illustrate another important point. Should the exit valve become blocked, attempts to continue the pumping become exceedingly frustrating. The downstroke can no longer be completed, and as soon as the force on the handle is released, the compressed gas simply bounces back. The whole operation becomes unpleasantly spongy. What happens is that compression increases the pressure to the point where it is sufficient to balance the downward force which you exert on the handle. Then when the downward force is released, the pressure simply causes the air to bounce back. In normal operation, however, the exit valve opens as soon as a certain pressure is reached, and air then leaves the barrel of the pump, permitting the downstroke to be completed.

Since from the galaxy there can be no escape of gas, as in a normally operating pump, how was the bounce-back effect of a blocked pump avoided? Through radiation. As compression heated the gas, and the more rapidly moving gas molecules began to collide with one another, their kinetic energy became converted into radiation (as discussed in §2.6). Because the density of the gas

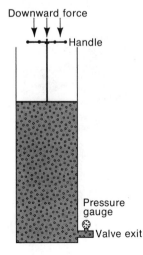

FIGURE 6.11.
Heat is generated by the compression of a gas.

276

*The Formation of
Stars and the Origin
of the Galaxy*

was very low, much of the radiation escaped entirely into outer space. So the energy of compression was largely lost, and the pressure was thereby prevented from rising to the point of stopping compression. Hence the sequence of events described above, leading to the formation of a rotating disk of gas of comparatively small width, was not inhibited by pressure effects.

The situation was otherwise, however, for stars that formed within the condensing gas cloud at various stages of its contraction. Stars, once formed, have no effective way to change their spatial distribution. Stars formed when the disk was wide did not settle into a thin disk, as the gas continued to do, because a star distribution has a form of pressure of its own that acts like that in a blocked bicycle pump. Thus the first stars to form, when the galactic disk of gas was still very wide, continued to occupy a thicker disk than the stars that formed later. For this reason, we expect Type II stars—the early ones—to be much less disklike in their distribution than the later Type I stars, as is indeed found to be the case. Astronomers often refer to Type II stars as "halo"

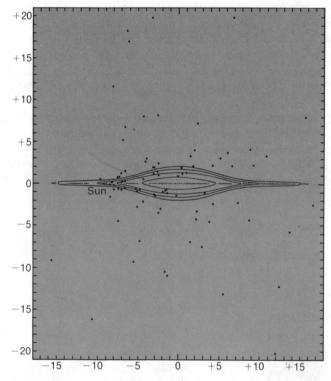

FIGURE 6.12.

A schematic representation of our galaxy, showing halo stars surrounding the thinner disk which forms the main part of the galaxy. The unit of distance here is the kiloparsee, which is about 3260 light years.

objects, because they are observed to occupy a general halo surrounding the
thinner disk of the main part of our galaxy.

In Figures 6.12 and 6.13, we have a schematic diagram of our galaxy and
a photograph of the galaxy M 104, colloquially known as the "Sombrero Hat."
In Figure 6.12, Type II stars occurring in groups known as *globular clusters*
form a halo surrounding the disk of the galaxy. These globular clusters contain
large numbers of stars, which until recently, were thought to be their only
content. Several of the ones in our own galaxy have been detected to be sources
of x-rays, however, implying that an extremely hot gas exists within them (see
page 60). Since this gas is not simply exploding out of the clusters, it is thought
that it is being held by gravitation to within a few light years of a massive central
object, one whose mass may be as much as 1,000 times the mass of the Sun. It
has been suggested that this central mass is very different from a star, may
perhaps be approaching the ultimate condition of a *black hole*. We shall study
black holes in more detail at the end of Section VI.

FIGURE 6.13.
The galaxy M104 (NGC 4694), about 100,000 light years in diameter, is a member of the
Virgo cluster of galaxies, and is referred to colloquially as the "Sombrero Hat" (courtesy of
Hale Observatories).

*The Formation of
Stars and the Origin
of the Galaxy*

Although there can be no doubt that stars actually formed during the condensation of the galaxy—so that the above discussion of their spatial distribution
had a sound practical basis—a serious question remained unresolved, namely,
that a spin-up sufficient to prevent further radial contraction of the galaxy would
have produced rotary forces so large that individual protostars would have been
torn apart by these rotary forces. This dilemma, which seemed an insurmount-
able logical obstacle during my own student days, is now accessible to resolution. By considering this problem next, we will find ourselves arriving at a
further interesting conclusion, namely, that stars must form in clusters of many
hundreds, thousands, or even hundreds of thousands of members.

Let us return to our spin-up model consisting of a set of weights free to
slide on insulated rods, but now with an outer, still more massive ring surrounding the previous structure, as in Figure 6.14. Suppose the system of

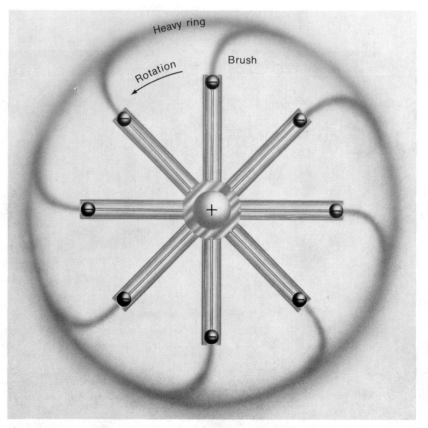

FIGURE 6.14.
An outer massive ring has been added to the arrangement of Figure 6.6, with brushes
attached to the inner structure and able to rub against the ring.

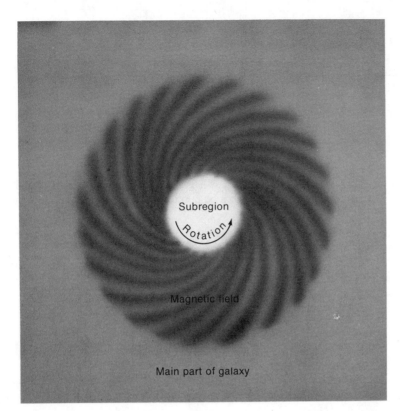

FIGURE 6.15.
The relation of a subregion to the main part of the galaxy is similar to the relation
of the weights and spokes of Figure 6.14 to the outer massive ring. A magnetic
field connecting the subregion to the main part of the galaxy plays a role similar to
that of the brushes of Figure 6.14.

weights to be rotating initially at so fast a rate that the outward rotary forces
on them are too strong to permit any retraction by the inward electrical forces.
Now imagine brushes, fastened to the inner structure, that sweep against the
outer massive ring. What happens? Clearly, the outer ring develops a slow
rotation. As it does so, the rotation of the inner structure gradually slows until
electrical attraction between the weights and the central sphere is at last able
to begin pulling in the weights. We then have a situation with the weights
being steadily pulled in, the outer massive ring slowly speeding up.

The principles of this example are clear, that there must be a massive outer
region to which rotation can be transferred, and that there be some means for
transmitting the rotation, a means equivalent to the brushes in Figure 6.14.
In analogy to the inner part of this figure, we can think of a subregion of the
galaxy, and in analogy to the outer massive ring, we can think of the main
body of the galaxy itself, as in Figure 6.15. But what then is to transmit the
rotation? Herein lies the basic difficulty that caused so much difficulty to

280

*The Formation of
Stars and the Origin
of the Galaxy*

astronomers for so many years. Nowadays an answer can be given: that the rotation is transmitted by a magnetic field. As the subregion separates from the main galaxy, a magnetic field continues to join the two, in the manner of Figure 6.15. The magnetic field acts to produce a braking effect very similar to that caused by the brushes of Figure 6.14.

PROJECT:
Consider a well-known form of toy in which a piece of metal suspended by an arm moves in the magnetic field generated by several magnets. Here, as in a galaxy, geometrically separated aggregations of particles—the magnets and the suspended metal—influence each other through the intermediary of a magnetic field.

It is known from observation, especially observations in radioastronomy, that magnetic fields exist everywhere throughout the galaxy.

QUESTION:
How does the presence of a magnetic field enable electrons, moving at speeds nearly equal to that of light, to emit radio waves?

The ability of a magnetic field to influence a cloud of gas depends on its intensity. For the situation existing in the galaxy, the intensity is not sufficient for the field to have a serious influence on the whole galaxy itself, but the intensity is indeed sufficient to have an important influence on local clouds of gas. Sub-regions of the kind shown in Figure 6.15 are prevented from condensing effectively if their masses are too small. If the mass were less than a few thousand times the Sun's, the magnetic forces would prevent a subregion from forming at all. And at the lower gas densities existing in the early stages of the formation of the galaxy—the stage of the halo stars—magnetic forces seem to have prevented subregions from forming unless their masses exceeded the Sun's by as much as a hundred thousand times.

If we suppose that the subregion of Figure 6.15, after extensive compression, is eventually able to fragment itself into stars, it follows that a very large number of stars must form simultaneously, several hundred in the galaxy under present-day conditions, and very likely a much larger number in the early history of the galaxy when the halo stars were being formed. The fact that the magnetic field, necessary to enable subregions to overcome the spin-up effect, itself imposes a condition of condensation—namely, that the masses of condensing clouds must be hundreds, thousands, or even hundreds of thousands, of times that of the Sun—forces star formation to occur in clusters.

FIGURE 6.16.
Halo stars sometimes occur in globular-shaped clusters with diameters of about 30 light years, each containing perhaps 100,000 stars (courtesy of Hale Observatories).

This deduction appears to be correct. A very large cluster, containing perhaps 100,000 stars, is shown in Figure 6.16. This is a globular cluster, the kind that belongs to the halos of galaxies, which we considered above in connection with Figures 6.12 and 6.13. On the other hand, Figure 6.17 shows the kind of cluster which forms in the plane of the galaxy at the present time. It contains far fewer stars than a globular cluster, in accordance with our expectations. Clusters like that in Figure 6.17 are called *open clusters,* since there is not the same high concentration toward their centers that can be seen in Figure 6.16. We already encountered two other open clusters in Chapter 3; the comparatively nearby clusters of M67 and the Pleiades were shown in Figures 3.3 and 3.4.

How are we to reconcile the idea that all stars are born in clusters with the fact that most stars are not today to be found in clusters? The answer to this question is that the gravitational field of the main body of the galaxy subjects clusters to slow disintegration, so that today we see large numbers of stars which have become free from their parent clusters, especially from open clusters. Globular clusters, being much larger, have a better chance of retaining their identity, but open clusters tend, given enough time, to disappear. Most of the open clusters we see today are therefore comparatively young systems. The well-known cluster Praesepe may have an age of about 10 per cent of that

*The Formation of
Stars and the Origin
of the Galaxy*

FIGURE 6.17.
The open cluster NGC 5897, about 20 light years in diameter, contains several hundred stars. Such clusters are found along the plane of the galaxy (courtesy of Hale Observatories).

of the galaxy, the Pleiades only 1 per cent, and h and χ Persei less than 0.1 per cent.

Yet stars, even after they have become free, retain a kind of "memory" of their cluster origin. This shows itself in the high proportion of them which belong to binary systems, and indeed to multiple systems containing more than two components. One of the most remarkable multiple systems is the star Castor (α Geminorum), which appears visually in a telescope, when seen under good conditions, as a triple system. The two brighter components move around each other in a period of rather more than 300 years. The third, fainter component moves around the other two in a period that exceeds 10,000 years. Furthermore, each of the three components turns out to be itself a closely spaced binary, detectable only by instrumental means—spectroscopically as described in Appendix II.8. Castor, a system thus containing six stars, would seem to be a remnant of an old cluster.

In Figures 6.18 and 6.19 we can see the galactic distributions of open clusters and globular clusters. It is clear that the open clusters are much more concentrated toward the plane of the galaxy, just as we expect them to be. The average distance of the globular clusters in Figure 6.19 is very large, being comparable with the distance of the solar system from the center of the galaxy, about 30,000

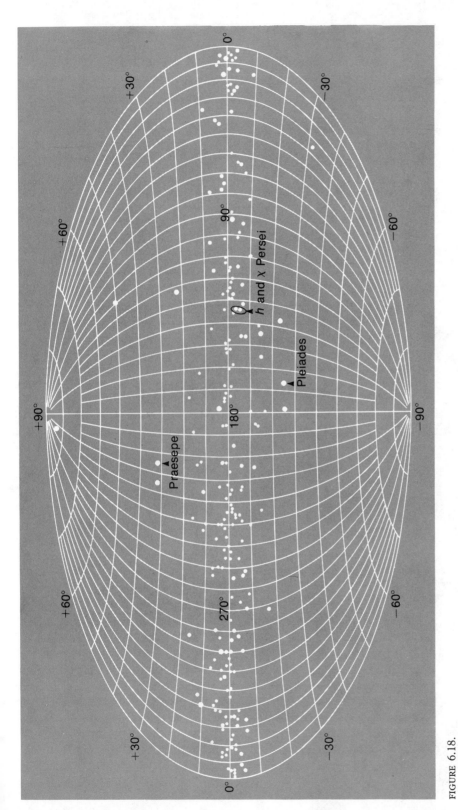

FIGURE 6.18.
This figure shows the directions in which open clusters are found (compare with Figure 1.28).

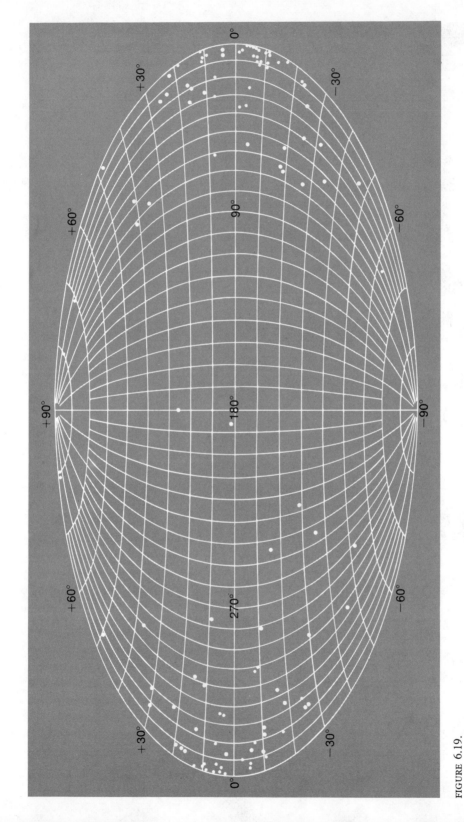

FIGURE 6.19.
The directions in which globular clusters are found are shown in this figure. The direction toward the center of the galaxy has longitude 0°, latitude 0°. The fact that the globular clusters evidently tend to be concentrated in directions toward the center shows them to be distributed on the scale of the whole galaxy.

light years. The average distance of the open clusters of Figure 6.18 is only about a tenth of this, which means that the open clusters mostly belong to our local swimming hole. The detection of more distant open clusters runs into the observational difficulty, discussed in §6.1, of the "fog" formed by the myriads of interstellar grains, the situation to be seen in the Sombrero Hat galaxy, (Figure 6.13). Observations of the globular clusters, which are halo objects, are not as hampered by this difficulty, and hence they can be observed to greater distances.

§6.4. Fragmentation

By now we have considered the first three of the five stages of star formation listed in §6.1. We have discussed how astronomers think our galaxy was formed, and how separate clouds, "subregions" like the Orion Nebula, can develop within the galaxy. A process analogous to that illustrated in Figure 6.14 would also permit subregions to form within the Orion Nebula; so we have covered stage (3) also. What we have not done is to consider stage (4). We simply assumed above that our subregions would ultimately fragment themselves into clusters of stars.

What we can say about stage (4) is that, so long as the magnetic field remains unimpaired within the cloud, the cloud can only shrink slowly as a whole; it cannot break into pieces, as it must if stars are to be formed. The magnetic field remains unimpaired in this sense so long as electric currents can flow freely within the material of the cloud, as they certainly can to begin with. But after a considerable shrinkage of a cloud, when the gas density within it rises to the order of 10^8 atoms per cm^3, it seems that the flow of electric current can become seriously impeded. It is at this stage that the cloud at last fragments itself into a swarm of bodies of stellar mass. Our next objective will be to trace the history of one such fragment which, for definiteness, we will take to be precisely equal in mass to the Sun. Details for fragments of other masses would differ somewhat from the following discussion, but the principles would be the same. The situation concerning stage (4) fragmentation into protostars is illustrated in Figure 6.20.

§6.5. The Condensation of a Protostar

We take up the description of the evolution of our solar-mass protostar, with it having a more or less spherically symmetric form, and with the radius being somewhat greater than the present scale of our planetary system, say 10^{15} cm, more than 10,000 times the radius of the present-day Sun, which is 6.96×10^{10} cm. To avoid problems concerning the origin of planets, we shall also simplify the situation by supposing the protostar to have no rotation (we will consider rotation in Section IV).

The Formation of
Stars and the Origin
of the Galaxy

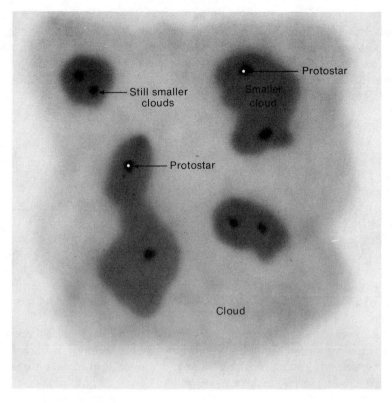

FIGURE 6.20.
The fragmentation of a gas cloud into protostars.

The self-gravitation of the protostar causes it to collapse, the initial time-scale for it to do so being about 100 years. The average density at a radius of 10^{15} cm is 3×10^{11} atoms per cm³, but as the radius decreases the density increases.

PROBLEM:
If the radius of the protostar is written as r and the average density of atoms as N, explain why N must increase like $1/r^3$.

Heat is generated in the interior of the protostar by compression, exactly as in our bicycle pump analogy. Heating causes atoms inside the protostar to emit radiation, but the radiation no longer escapes freely as it did at the much lower density which occurred when the galaxy itself was being formed. The radiation bounces from one particle to another, in the manner of Figure 6.21. With the radiation now essentially trapped, the protostar runs into the blocked bicycle-pump situation—the pressure in its interior rises to the stage where it prevents the inwardly directed gravitational forces from causing further rapid collapse.

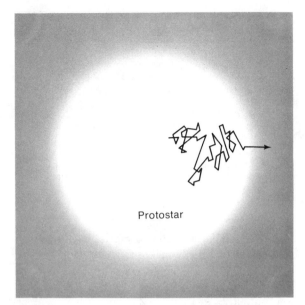

Protostar

FIGURE 6.21.
Because of absorptions and re-emission, and because of scattering inside a star, radiation leaks out of the interior only very slowly.

At this stage the system assumes a state of quasi-equilibrium, with the pressure forces and the gravitational forces nearly in balance with each other. The reason why these forces do not come into exact balance can be understood by again considering what happens with a blocked bicycle pump. It is true that the downstroke of the pump handle is suddenly impeded by the rise of pressure within the air inside the barrel, but if we maintain a steady downward force on the handle, some further compression does occur. This is due to an escape of heat: the air inside the barrel loses heat to the metal of the pump, and the metal then passes the heat to the outside air. Radiation is also lost gradually from a protostar, since the process of Figure 6.21 eventually permits radiation to leak out to the surface, from whence it is at last able to escape into space. This loss slowly weakens the pressure forces, allowing a further slow shrinkage of the protostar to occur.

The gases at the surface have the curious property illustrated in Figure 6.22. Above a temperature of 4,000°K, the atoms can largely deflect the radiation, and so prevent its free escape, but for temperatures below 4,000°K they cannot, and radiation can stream out into space quite freely. This property turns out to require that the surface temperature of the protostar must be maintained at about 4,000°K throughout further condensation, until a radius comparable to that of the present-day Sun is reached.

An object with a surface temperature of 4,000°K would emit light mainly of an orange to red color. The surface temperatures of stars can indeed be

288

*The Formation of
Stars and the Origin
of the Galaxy*

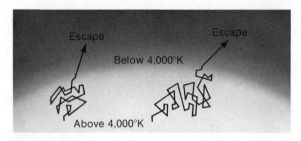

FIGURE 6.22.
When the temperature near the surface of a newly forming
star falls below about 4,000°K, the gases are no longer able to
block the escape of radiation in an effective way.

judged from their colors, quite accurately if precise instrumental measurements
are made, and approximately if one makes the rough color assignments shown
in Table 6.1.

It is useful to represent the behavior of stars in a diagram that relates color
to total luminous output, as in Figure 6.23. The unit of luminosity used here
is the present-day power output of the Sun, 3.9×10^{23} kilowatts. The surface
temperatures are in units of 1,000°K, and are arranged *to increase toward the
left*. Both the luminosity and the temperature are displayed *logarithmically*. A
diagram of this kind, but drawn up in a slightly different form, was first used,
independently, by Henry Norris Russell (1877–1957) and by E. Hertzsprung
(1873–1967), and is referred to nowadays as the *Hertzsprung-Russell diagram*.

Also displayed in Figure 6.23 are contours of equal radius, using the present-day solar radius, $6.96 = 10^{10}$ cm, as unit. It is clear from Figure 6.23 that
if we wish to follow the contraction of our protostar on such a diagram, we
must take a line essentially parallel to the luminosity axis, drawn for a temperature of 4,000°K, as in Figure 6.24. The latter figure also gives further information, namely, the time required for contraction to proceed at various stages.
The contraction is seen to slow down from 100 years, to 1,000 years, to 10,000
years, to 100,000 years, and, for a final, comparatively short, reversed part of
the track, to 10 million years. This reversed part of the track from *A* to *B*
has a special significance which we must now examine.

TABLE 6.1.
Color-temperature relationships

Color	Temperature (°K)
Orange-red	4,000
Green-yellow	6,000
Blue	8,000
Violet	10,000
Ultraviolet	20,000

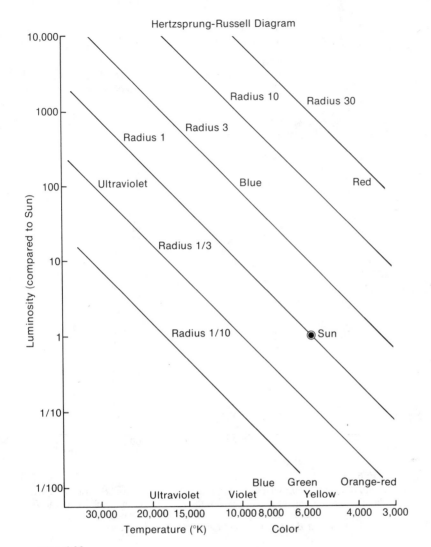

FIGURE 6.23.
The Hertzsprung-Russell diagram, with lines of equal radius marked (solar radius as unit). Note that the lefthand scale refers to the total emission of radiation of all frequencies by the star. This is the so-called *bolometric* luminosity.

PROBLEM:
The total radiant output, the *bolometric* luminosity as it is called, is proportional to the fourth power of the surface temperature and to the square of the radius. Estimate what the slope of the contours of equal radius in Figure 6.23 should be, noting that the temperature scale is just four times the luminosity scale, and verify that the contours have been correctly spaced.

*The Formation of
Stars and the Origin
of the Galaxy*

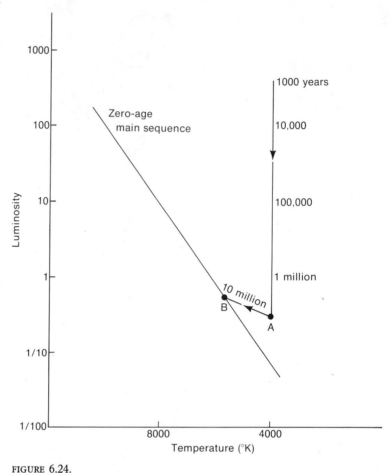

FIGURE 6.24.
The path of a condensing solar-type star in the Hertzsprung-Russell diagram. Note the slowdown in the rate of shrinkage, with the final part of the track from *A* to *B* taking longer than all the rest of the formation process. At *A* nuclear reactions become important; at *B* the star reaches the zero-age main sequence.

Material near the center of the protostar is hotter than material near the surface. It is just this temperature difference which causes a flow of heat to take place from the central regions to the surface, and which supplies the energy that the surface radiates outward into space. As contraction proceeds, the central temperature rises, until eventually a situation is reached where energy in appreciable quantity is delivered by *nuclear reactions,* that is, by reactions that change atoms from one kind to another. Since in the next chapter we will study these reactions in detail, we will not concern ourselves here with their nature, but only with the fact that they deliver energy. Point *A* of Figure 6.24 corresponds to the stage where significant quantities of energy are first released by nuclear reactions, and point *B* corresponds to the stage where this release has proceeded to establish an important new form of equilibrium.

The nature of this equilibrium is such that, if at any radius, say, r, within the star we draw the sphere of radius r, as in Figure 6.25, the outflow of energy across the surface of this sphere is equal to the nuclear-energy production within its interior. Putting r equal to the total radius, say, R, $r = R$, we see that the loss of energy from the outer surface is then just compensated by the total nuclear-energy production within the whole interior of the star. We thus have a complete energy balance, achieved at the expense of converting one kind of atom into another. If we think of atoms as nuclear fuel, the equilibrium is achieved through the *burning of nuclear fuel*. When this stage is reached, at point B in Figure 6.24, the star is said to have reached the *main sequence*. The star is now considered to have "formed"; it is no longer a protostar.

A similar discussion could be given for protostars with masses different from that of the Sun. The endpoints attained, corresponding to point B for the solar case, all lie on the line shown in Figure 6.24. This line is the main sequence itself, and because it gives the positions of stars when the burning of nuclear fuel first establishes an energy balance, it is sometimes called the *zero-age main sequence*.

It will be convenient from here on to use the symbol $\odot$ to denote a mass equal to that of the Sun. We may ask; What is the full range of stellar masses? This question is hard to answer, however, because stars with masses less than $\frac{1}{3}\odot$ are intrinsically so faint that they are difficult to observe, whereas stars with masses greater than $50 \odot$ tend to be confined within the dense, opaque

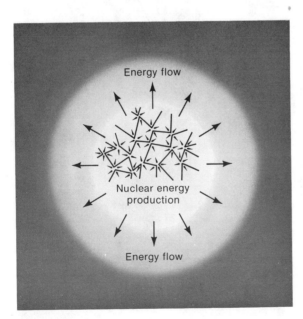

FIGURE 6.25.
A condensing star eventually reaches an equilibrium configuration in which the flow of energy across any sphere is equal to the rate of nuclear-energy generation within that sphere.

clouds in which they are born, essentially because their lives are very short and they have inadequate time to move away from their parent clouds. Observations outside the range from $\frac{1}{3}$ ⊙ to 50 ⊙ are consequently restricted, so that the main sequence is usually taken to extend from about $\frac{1}{3}$ ⊙ on the lower side to about 50 ⊙ on the upper side. We must recognize, however, that stars outside this range exist, probably in large numbers on the low side, but only rather infrequently on the high side.

The simple proportionality $L \propto M^4$, relating the luminosity L to the mass M, can be used as a reasonable approximation over the mass range from $\frac{1}{3}$ ⊙ to 20 ⊙. Thus a star of mass 20 ⊙ is about $(60)^4$ times brighter than a star of mass $\frac{1}{3}$ ⊙.

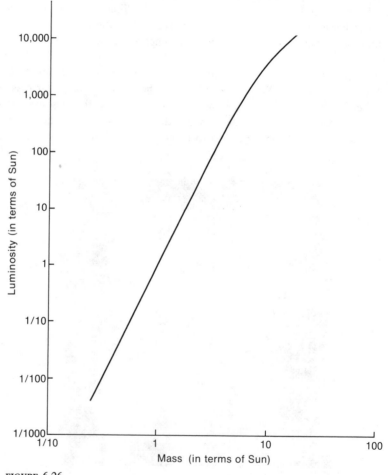

FIGURE 6.26.
The general form of the relation between the luminosity of a newly formed star and its mass. The curve tends to lean over at high mass and high luminosity.

PROBLEM:
Taking the absolute magnitude of a main-sequence star of mass $\frac{1}{3}$ ⊙ to be $+12$, estimate the absolute magnitude of a main-sequence star of mass 20 ⊙.

Since L must be close to the luminosity of the Sun when $M = $ ⊙, it follows that the proportionality $L \propto M^4$ gives $L = (M/⊙)^4$ when the solar luminosity is used as the unit of L (as in Figures 6.23 and 6.24). The relation $L = (M/⊙)^4$ is represented graphically in Figure 6.26, except that as the mass increases above $M = 20$ ⊙ the luminosity increases more slowly than the proportionality $L \propto M^4$. This is also illustrated in Figure 6.26.

PROBLEM:
The amount of nuclear fuel in a star is proportional to M, whereas the rate at which the fuel is burnt is proportional to L. Given $L \propto M^4$ for stars in a certain mass range, explain why the lifetime of such a star, determined by the availability of nuclear fuel, must be proportional to M^{-3}. Given that the lifetime of the Sun is 10^{10} years, estimate the lifetime of a star of mass 20 ⊙.

§6.6   The Salpeter Function

Values of the mass M are plotted in Figure 6.27 along a line that is divided into equal small steps. Notice that Figure 6.27 is *not* a logarithmic plot in M, as in Figure 6.26, but a direct (linear) plot. What we have thus done is to divide the mass range into many equal "bins." Suppose now we consider the outcome of a star-formation process in which a very large population of stars is produced. What fraction of the population will fall into each of the bins of Figure 6.27?

Individual stars in bins have increasing mass.

FIGURE 6.27.
Stars vary in mass, over a range which can be divided into a number of bins. What fraction of the stars fall into each bin?

If we had been able to deal with stage (4) of the star-formation process in a thoroughgoing way, this is just the sort of question we would be able to answer theoretically. But since we only dealt with stage (4) in terms of generalities, we are not in a position to determine an answer *except by an appeal to observation*. The result of such an appeal is shown in Figure 6.28. Here we

*The Formation of
Stars and the Origin
of the Galaxy*

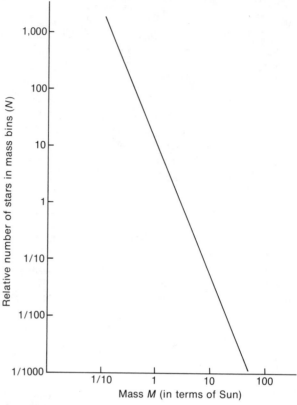

FIGURE 6.28.
This figure shows the way in which the masses of newly formed
stars fall into the bins of Figure 6.27. This distribution is known as
the Salpeter function.

have a quantity N plotted logarithmically against M, also plotted logarithmically. If one reads off from the line the value of N corresponding to any specified M, the result is proportional to the number of stars that fall into the particular bin situated at the value of M in question. Thus the ratio of the values of N for any two values of M gives the ratio of the number of stars in the two corresponding bins. The quantity N is known as the *Salpeter function*.

§6.7. The Colors of Galaxies

For FIGURE 6.29,
see PLATE VI.

Figure 6.29 is a color photograph of the galaxy M31, already seen in black and white in Figure 3.8. Let us see if we can use the ideas developed above to understand why the outside of M31 is blue and the inside yellow to red.

Suppose that, instead of asking for the relative numbers of stars falling into the bins of Figure 6.27, we ask for the relative amounts of light radiated by a population of newly formed stars. The answer to this question is given by a suitable multiplication of Figures 6.26 and 6.28, with the result shown in Figure 6.30, the quantity $\mathcal{L}$ in this figure denoting the required amount of light. What Figure 6.30 shows is that the most light—i.e., the most energy—comes from the bins containing the stars of largest mass. Although there are comparatively few such stars, they dominate the light emitted by a newly formed population, and since stars of large mass mainly emit blue, violet, and ultraviolet light, as can be seen from the main sequence shown in Figure 6.24, it follows that these colors will dominate the radiation from an aggregation of recently formed stars. This is the situation in the outer regions of M31. It is also the situation throughout the small galaxy NGC 253, shown in color in Figure 6.31.

To understand why the inner regions of M31 are yellow to red in color, we next ask what happens to a star population as it ages. Lifetimes are shortest for the stars of largest mass (see the last Problem). Whereas the supply of nuclear fuel in stars of about solar mass is adequate to maintain them for as long as 10^{10} years, the supply in stars of largest mass is enough for only about a million years. With their fuel supply exhausted, such stars no longer contribute effectively to the general light of the population. In effect, as a star system ages, the main sequence becomes progressively truncated, with the stars of highest mass ceasing first to contribute effectively, then with stars of intermediate mass ceasing to contribute, and so leaving only stars of small mass in populations of advanced age. This process is illustrated in Figure 6.32. With the main sequence truncated as far as stars with masses of solar order, the light emitted by the population becomes yellow to red in color. Hence we

For FIGURE 6.31, see PLATE VII.

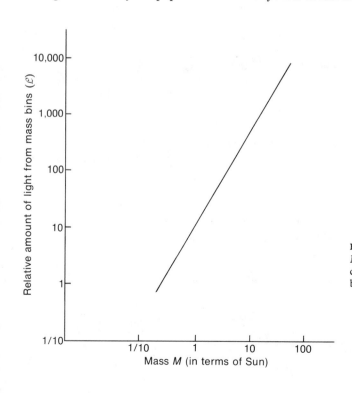

FIGURE 6.30.
Most of the light emitted by a newly condensed population of stars is contributed by the stars of largest mass.

296

*The Formation of
Stars and the Origin
of the Galaxy*

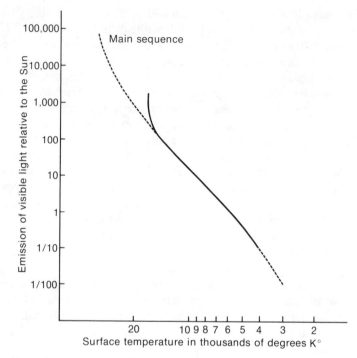

FIGURE 6.32.
As a population of stars ages, the main sequence becomes progressively truncated, as is shown here for the intermediate case of the Pleiades.

conclude that the stars of the inner regions of M31 belong to an old population, whereas those of the outer regions belong to a much younger population.

Why then is the blue outer part of M31 less bright than the yellow inner region? Because the younger population is much less numerous than the old population.

We noticed in Chapter 3 that elliptical galaxies contain very little gas. Hence there can be very few recently condensed young stars in the elliptical galaxies; the light from ellipticals must therefore be dominated by old populations. Consequently the elliptical galaxies are yellow to red in color, very much like the central regions of M31.

We remarked above that the Pleiades are a comparatively young cluster of stars. Hence we expect the Pleiades to be still a blue cluster, an expectation amply confirmed by Figure 6.33.

For FIGURE 6.33,
see PLATE VIII.

PROBLEM:
The straight line of Figure 6.28 corresponds to $N \propto M^{-7/3}$. Taking $L \propto M^4$ for the luminosities of individual stars, show that the function $\mathcal{L}$ plotted in Figure 6.30 satisfies the proportionality $\mathcal{L} \propto M^{5/3}$.

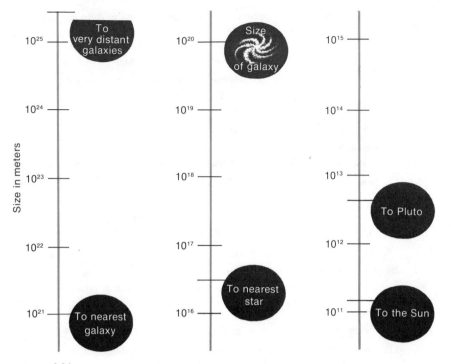

FIGURE 6.34.
The sizes of objects, and the length scales, ranging from stars to the most remote galaxies.
(From L. Pauling, *College Chemistry*, 3d ed., W. H. Freeman and Company. Copyright ©
1964.)

Now that we have encountered the use of a logarithmic plot, it will be useful
to set out a scale ranging from 10^{10} cm, comparable to the radii of the faintest
main-sequence stars, to 10^{28} cm, comparable to the distances of the most remote
galaxies. The various objects we have discussed in this chapter fall between
these dimensions and are shown in Figure 6.34.

General Problems and Questions

1. The Sun contains about 10^{57} atoms. A certain interstellar gas cloud has a
   density of 10^3 atoms per cm^3. Assuming the cloud to be uniform and
   spherical, and taking the atoms within it to be of the same kind as the atoms
   in the Sun, what must the radius be in order that the mass of the cloud
   shall be 10^5 times the mass of the Sun?

2. What would be the apparent angular size of such a cloud if its distance
   from the observer were 1,000 light years?

3. The density of stars appears to be roughly the same in different directions along the plane of the Milky Way. This circumstance was at one time interpreted as demonstrating that the solar system must be located near the center of our galaxy. Why was the argument wrong?

4. Explain why a diffuse condensing gas cloud with a mass of galactic order, and with some measure of rotation, tends to condense into a disk rather than into a spherical ball.

5. What are Type I and Type II stars?

6. Describe the process of star formation, explaining why magnetic fields seem to be important. Why does it appear likely that stars must form in clusters?

7. What are globular clusters? In what respects do they differ from open clusters?

8. If most stars were born in open clusters, what explanation is to be given for the fact that most stars are not now in clusters?

9. Are stars forming at the present time? If so, in what kind of object?

10. A condensing spherical cloud fragments into 1,000 stars, the mass of each being, on the average, equal to that of the Sun. It does so at the stage where the average density within it is 10^{12} atoms per cm^3. What is the radius at this stage? By about how much must each of the resulting 1,000 protostars condense further before normal stars are formed?

11. Once radiation can no longer escape readily from its interior, a condensing gas cloud becomes hotter as it shrinks. Give an example to illustrate this effect.

12. Why do condensing protostars tend to have surface temperatures of about 4,000°K?

13. What is the physical condition which determines that a star has formed?

14. Describe the Hertzsprung-Russell diagram.

15. In a sketch of the Hertzsprung-Russell diagram, indicate the position of the sequence on which newly-formed stars are found to lie. What criterion determines where on this sequence a particular star lies?

16. What is the approximate mass range of stars on the zero-age main sequence? Give the approximate relation between luminosity and mass for such stars.

17. Explain how the lifetimes of stars at various places on the main sequence can be estimated.

18. In a large population of newly-formed stars, what is the distribution with respect to (a) mass, (b) luminosity?

19. Why is an old population of stars redder in color than a population of newly formed stars?

20. Estimate (a) the ratio of the average spacing between stars to their individual radii, (b) the ratio of the dimension of a galaxy to the average spacing between stars, (c) the ratio of the average spacing between galaxies to their individual radii, (d) the ratio of the distances of the most remote galaxies to the average spacing between galaxies.

Chapter 7:
Atoms, Nuclei, and
the Evolution of Stars

§7.1. The Need for a Stellar Energy Source

In the preceding chapter we saw that stars form by condensation in gas clouds like the Orion Nebula. Gravitation pulls both gas clouds and the protostars within them together. However, internal pressure within a protostar resists the gravitational forces, to such a degree that at any moment during the condensation the pressure forces and gravitational forces are nearly in balance. The balance cannot be exact because radiation into space from the outer surface slowly weakens the ability of the pressure forces to resist further condensation. Yet we saw that a stage is eventually reached when the leakage of heat from the surface is at last compensated for by a process of energy production by nuclear reactions, and at this stage we take the star to be formed. What are these reactions? This question will occupy us in this chapter.

Historically, it was a hard question to answer, because a point of view developed during the nineteenth century which held that matter consists of atoms that are *indestructible*. There are many different kinds of atom, which were thought to be all separate from each other, each with its own distinctive properties. If this had been true, there could be no conversion of one form of atom into another atom, which is the essential characteristic of a *nuclear* reaction. Many reactions were studied, but these were all ones that today we call *chemical*, reactions in which certain elementary substances never changed.

THE PERIODIC SYSTEM

1 H																	2 He
3 Li	4 Be											5 B	6 C	7 N	8 O	9 F	10 Ne
11 Na	12 Mg											13 Al	14 Si	15 P	16 S	17 Cl	18 A
19 K	20 Ca	21 Sc	22 Ti	23 V	24 Cr	25 Mn	26 Fe	27 Co	28 Ni	29 Cu	30 Zn	31 Ga	32 Ge	33 As	34 Se	35 Br	36 Kr
37 Rb	38 Sr	39 Y	40 Zr	41 Nb	42 Mo	43 Tc	44 Ru	45 Rh	46 Pd	47 Ag	48 Cd	49 In	50 Sn	51 Sb	52 Te	53 I	54 Xe
55 Cs	56 Ba	57* La	72 Hf	73 Ta	74 W	75 Re	76 Os	77 Ir	78 Pt	79 Au	80 Hg	81 Tl	82 Pb	83 Bi	84 Po	85 At	86 Rn
87 Fr	88 Ra	89† Ac	104	105	106												

*Lanthanides (rare earths)

58 Ce	59 Pr	60 Nd	61 Pm	62 Sm	63 Eu	64 Gd	65 Tb	66 Dy	67 Ho	68 Er	69 Tm	70 Yb	71 Lu
90 Th	91 Pa	92 U	93 Np	94 Pu	95 Am	96 Cm	97 Bk	98 Cf	99 Es	100 Fm	101 Md	102 No	103 Lr

†Actinides

FIGURE 7.1.
The periodic system of the elements. This is a scheme of classification first adopted in the nineteenth century by D. I. Mendeleev.

These substances could be combined into associations called *molecules,* but it was always possible to recover the "elements," as they became called, from such molecules by a process of dissociation, i.e., of breakup. Each element was taken to consist of a very large number of identical units, the *atoms.* The fact that the elements were never found to change in any chemical reaction was then explained by the indestructibility of the basic atoms themselves.

Techniques became available that enabled chemists to measure standard amounts of the elements, i.e., samples containing the same number of atoms, and so it became possible, simply by weighing the samples, to arrange the elements in an order 1, 2, 3, . . ., with element 1 the lightest sample (hydrogen), element 2 the next lightest sample (helium), and so on. Adding the elements that have been discovered in this century, this method of ordering* gives a present-day list of 103 elements, set out in Table 7.1. Also given in this table are the relative abundances of the elements as they are believed to have existed in the gas cloud from which the Sun was formed. These relative abundances are scaled to give the element silicon (Si) a standard value of 10^6. That is to say, relative to a million silicon atoms, the other atoms existed in the numbers given in Table 7.1.

Early in the second half of the nineteenth century, the English chemist John Newlands pointed out that when the elements are arranged in the order of Table 7.1 every eighth element seemed to have similar properties. This idea was developed by the Russian D. I. Mendeleev (1834–1907), by setting out the elements in an array like that shown in Figure 7.1. The distinctive feature of this array is that the elements in each column have similar chemical properties.

TABLE 7.1.
The elements

Z	Name	Chemical symbol	Date of discovery	Abundance in cosmic material[a]
1	Hydrogen	H	1766	3.18×10^{10}
2	Helium	He	1895	2.21×10^9
3	Lithium	Li	1817	49.5
4	Beryllium	Be	1798	0.81
5	Boron	B	1808	350
6	Carbon	C	Old	1.18×10^7
7	Nitrogen	N	1772	3.64×10^6
8	Oxygen	O	1774	2.14×10^7
9	Fluorine	F	1771	2,450
10	Neon	Ne	1898	3.44×10^6
11	Sodium	Na	1807	6.0×10^4
12	Magnesium	Mg	1755	1.06×10^6
13	Aluminum	Al	1827	8.5×10^5
14	Silicon	Si	1823	10^6

*For a reason that will appear later, this ordering by weight has been inverted for the pairs Co, Ni, and Te, I.

TABLE 7.1. (*continued*)

Z	Name	Chemical symbol	Date of discovery	Abundance in cosmic material[a]
15	Phosphorus	P	1669	9,600
16	Sulfur	S	Old	5.0×10^5
17	Chlorine	Cl	1774	5,700
18	Argon	A	1894	1.17×10^5
19	Potassium	K	1807	4,205
20	Calcium	Ca	1808	7.2×10^4
21	Scandium	Sc	1879	35
22	Titanium	Ti	1791	2,770
23	Vanadium	V	1830	262
24	Chromium	Cr	1797	1.27×10^4
25	Manganese	Mn	1774	9,300
26	Iron	Fe	Old	8.3×10^5
27	Cobalt	Co	1735	2,210
28	Nickel	Ni	1751	4.8×10^4
29	Copper	Cu	Old	540
30	Zinc	Zn	1746	1,245
31	Gallium	Ga	1875	48
32	Germanium	Ge	1886	115
33	Arsenic	As	Old	6.6
34	Selenium	Se	1817	67
35	Bromine	Br	1826	13.5
36	Krypton	Kr	1898	47
37	Rubidium	Rb	1861	5.88
38	Strontium	Sr	1790	26.8
39	Yttrium	Y	1794	4.8
40	Zirconium	Zr	1789	28
41	Niobium	Nb	1801	1.4
42	Molybdenum	Mo	1778	4
43	Technetium	Tc	1937	unstable
44	Ruthenium	Ru	1844	1.9
45	Rhodium	Rh	1803	0.4
46	Palladium	Pd	1803	1.3
47	Silver	Ag	Old	0.45
48	Cadmium	Cd	1817	1.42
49	Indium	In	1863	0.189
50	Tin	Sn	Old	3.59
51	Antimony	Sb	Old	0.316
52	Tellurium	Te	1782	6.41
53	Iodine	I	1811	1.09
54	Xenon	Xe	1898	5.39
55	Cesium	Cs	1860	0.387
56	Barium	Ba	1808	4.80
57	Lanthanum	La	1839	0.445
58	Cerium	Ce	1803	1.18
59	Praseodymium	Pr	1879	0.149

TABLE 7.1. (*continued*)

303

Z	Name	Chemical symbol	Date of discovery	Abundance in cosmic material[a]
60	Neodymium	Nd	1885	0.779
61	Promethium	Pm	1947	unstable
62	Samarium	Sm	1879	0.227
63	Europium	Eu	1896	0.085
64	Gadolinium	Gd	1880	0.297
65	Terbium	Tb	1843	0.055
66	Dysprosium	Dy	1886	0.351
67	Holmium	Ho	1879	0.079
68	Erbium	Er	1843	0.225
69	Thulium	Tm	1879	0.034
70	Ytterbium	Yb	1878	0.216
71	Lutetium	Lu	1907	0.0362
72	Hafnium	Hf	1923	0.210
73	Tantalum	Ta	1802	0.0210
74	Tungsten	W	1781	0.160
75	Rhenium	Re	1925	0.0526
76	Osmium	Os	1803	0.745
77	Iridium	Ir	1803	0.717
78	Platinum	Pt	1735	1.40
79	Gold	Au	Old	0.202
80	Mercury	Hg	Old	0.40
81	Thallium	Tl	1861	0.192
82	Lead	Pb	Old	4.0
83	Bismuth	Bi	1753	0.143
84	Polonium	Po	1898	unstable
85	Astatine	At	1940	unstable
86	Radon	Rn	1900	unstable
87	Francium	Fr	1939	unstable
88	Radium	Ra	1898	unstable
89	Actinium	Ac	1899	unstable
90	Thorium	Th	1828	0.058
91	Protoactinium	Pa	1917	unstable
92	Uranium	U	1789	0.0262
93	Neptunium	Np	1940	unstable
94	Plutonium	Pu	1940	unstable
95	Americium	Am	1945	unstable
96	Curium	Cm	1944	unstable
97	Berkelium	Bk	1950	unstable
98	Californium	Cf	1950	unstable
99	Einsteinium	Es	1955	unstable
100	Fermium	Fm	1955	unstable
101	Mendelevium	Md	1955	unstable
102	Nobelium	No	1958	unstable
103	Lawrencium	Lw	1961	unstable

[a] Abundances from a recent compilation by A. G. W. Cameron (*Space Science Reviews*, **15** (1970), 121-146). Notice that the abundances are *relative* to each other, with 10^6 for Si taken as the standard of reference.

Not all the elements set out in Figure 7.1 were known in Mendeleev's time, so that his table had gaps in it. From the dates of discovery of the various elements, given in Table 7.1, the reader will notice that, whereas most of the elements discovered in the twentieth century are heavier than lead, five elements lighter than lead have been discovered in our century. They are: technetium at number 43, found in 1937; promethium at number 61, found in 1947; lutetium at number 71, found in 1907; hafnium at number 72, found in 1923; and rhenium at number 75, found in 1925. By noting the position of gaps, Mendeleev was able to predict ahead of time what elements would eventually be discovered. Nowadays, no gaps are left. New elements will only be found in the future by extending the list of Table 7.1 to still heavier atoms.

PROBLEMS:

1. Mendeleev's *periodic table*, as it is generally called, was first published in 1871. What gaps did it then contain?

2. Consider the abundances given in Table 7.1, particularly for the metals. Notice the general correspondence between low abundance and economic value. For example, tin is a more valuable metal than iron. Yet economic value and low abundance, although generally correlated, are not related precisely. For example, gold has about half the abundance of silver. Yet gold is much more than twice as valuable as silver. Lead is less abundant than copper, yet copper is more valuable than lead. What explanations can you think of for this lack of a precise relationship?

Although the multiplicity of reactions studied by chemists seemed to support the indestructible nature of atoms, even before the end of the nineteenth century there was cause for disquiet. Some form of energy production had to be making good the energy which the Sun was constantly radiating away into space. From fossils, found in rocks, it was known that the Sun must have been radiating pretty much as it does now for a long period of time. Geologists estimated ages of the order of a hundred million years for these rocks. The problem was to understand how the Sun could have maintained itself this long. Already in 1854, H. von Helmholtz (1821–1894) had considered the possibility that the Sun might still be condensing—i.e., might still be a protostar—in which case the radiated energy could be coming from the compression caused by the gravitational forces, along the lines considered in the preceding chapter. But when Lord Kelvin (1824–1907) later examined this idea in detail, he found that, if it were true, the Sun could not be more than about 20 million years old, less than the geologists were claiming for the rocks containing the fossils. So puzzled was Kelvin about this discrepancy that he suggested the geologists must be in error. Indeed they were, but not in the sense argued by Kelvin. The oldest rocks containing fossils of living creatures are very much older than 100 million years—3,000 million years is the present-day estimate! Thus the discrepancy is real, much worse than Kelvin thought it to be, the clear-cut implication being that Helmholtz's idea cannot be correct.

The first step toward a resolution of this problem of the solar energy supply came with the quite unexpected discovery of radioactivity, by Henri Becquerel (1852–1908) in 1896. The discovery arose from an investigation into the properties of x-rays, which had been produced for the first time by Wilhelm Röntgen (1845–1923) only a few months earlier. As we have already noted, x-rays are a form of radiation, like ordinary light or like ultraviolet light. But in 1896 the nature of x-rays was quite unclear—as the designation by the letter x implies! An erroneous idea came to be generally held that x-rays were somehow produced when ordinary light was shone onto atoms, and scientists began to try out various kinds of atom from this point of view. Becquerel was lucky in trying the element uranium. Light was shone onto a sample of material containing uranium, which was then placed below a photographic plate wrapped in black paper, with a thin sheet of silver between, the idea being that only x-rays would be able to penetrate the silver. After storing this simple arrangement for a while, the photographic plate was found to be fogged. Triumph! The idea worked; x-rays really were generated by shining light onto uranium. But notice now the enormous power of experiment to correct wrong ideas. Becquerel repeated the experiment without shining light on his uranium material, and still the photographic plate became fogged. So it followed that the uranium was doing "something," all by itself. But what?

QUESTIONS:
Are there fields of human activity, other than science, in which you consider it possible to discover and correct mistakes with the decisive clarity of Becquerel's second experiment? In matters connected with the large-scale organization of society, has any self-corrective process ever been operative? Discuss war and famine from this point of view.

The next step was to see if other kinds of atom also produced the effect which had been discovered by Becquerel. Mme. Curie (1867–1934) found that, of the elements then known, only uranium and thorium had this property of spontaneously fogging a photographic plate, a property which became known as *radioactivity*. Together with her husband, she went on to concentrate, from large quantities of uranium ore, a small quantity of a substance which became known as *radium*, a name suggested by the intense radioactivity which it was capable of producing. The suspicion that atoms of uranium were changing into different atoms—atoms of radium—soon became a certainty, and indeed radium is atom number 88 in Table 7.1, whereas uranium is atom number 92.

Where did this process of spontaneous transformation of one form of atom into another end? From the abnormal quantities of lead found in association with uranium ores, the answer appeared to be that radioactivity ended at lead: lead was the ultimate product. This answer turned out to be correct. The scheme

TABLE 7.2.
Radioactive decay of uranium

Parent element	Characteristic decay time	Ray emitted	Daughter element
Uranium I[a]	4.5×10^9 years	α	Thorium I
Thorium I	24.1 days	β	Protoactinium
Protoactinium	1.2 minutes	β	Uranium II
Uranium II	2.5×10^5 years	α	Thorium II
Thorium II	8×10^4 years	α	Radium
Radium	1620 years	α	Radon
Radon	3.8 days	α	Polonium I
Polonium I	3 minutes	α	Lead I
Lead I	27 minutes	β	Bismuth I
Bismuth I	20 minutes	β	Polonium II
Polonium II	1.64×10^{-4} seconds	α	Lead II
Lead II	22 years	β	Bismuth II
Bismuth II	5 days	β	Polonium III
Polonium III	138 days	α	Lead III (stable)

[a] The Roman numerals refer to different forms of the same element. Thus Uranium I and Uranium II are different forms of uranium.

of transformations eventually worked out by Ernest Rutherford (1871–1937) and Frederick Soddy (1877–1956) has the principal form shown in Table 7.2. Rutherford had discovered that at each transformation "something" was emitted, and that there were two kinds of "something," which he called α-rays and β-rays. It was just these "somethings," particularly the β-rays, that produced the fogging of a photographic plate in the manner found by Becquerel. In 1909, Rutherford and Royds obtained the result that the α-rays consisted of the nuclei of helium atoms, whereas the β-rays were shown to be fast moving electrons; the "rays" were particles. The surprise in Table 7.2 is to find that starting from uranium, element 92, after it emits an α-particle and two β-particles, we nevertheless arrive back again at element 92. A similar situation occurs also for element 84, polonium, named for Poland, the country of Mme. Curie's birth. These cases provide a strong indication that it is possible to have *more than one form of the same chemical atom.*

At first it was thought that radioactivity was only associated with atoms heavier than lead. It therefore came as a further surprise when N. R. Campbell found that samples of both potassium and rubidium also emitted β-rays, but very weakly, so that they had not been noticed in Mme. Curie's experiment. As a consequence of these emissions, potassium atoms were able to change into calcium atoms, and rubidium atoms into strontium. The nineteenth-century concept of indestructible atoms was dead. Atoms *could* change into one another. When they did so, either an α-ray (helium) or a β-ray (electron) was emitted, *and energy was released in the motion of these emitted particles.* Consequently, here was now a new form of energy production which might be important in the Sun and stars.

The concept of energy derived from the radioactivity of naturally occurring elements was important, not because it turned out to be the right idea, but because it freed physicists and astronomers from the straightjacket of nineteenth-century chemistry. It became possible to think in terms of transformations of all manner of atoms, one into another. The trouble with the idea that naturally occurring radioactivity is the source of stellar energy will be clear from the relative abundance values given in Table 7.1. There is far too little uranium, thorium, or any of the elements heavier than lead, present in stellar material for appreciable quantities of energy to be generated by it.

PROBLEM:

The conversion of 1 gram per second of uranium into lead yields about 2×10^7 kilowatts. The solar mass is about 2×10^{33} grams, and of this about 1 part in 5×10^9 consists of uranium. If all the Sun's uranium were completely transformed into lead, how long could this process maintain the solar power output of 3.9×10^{23} kilowatts?

It was worry over this situation that caused A. S. Eddington (1882–1944) in 1920 to turn in a quite different direction. Instead of being concerned with the heaviest of the elements, Eddington's suggestion involved the two lightest, hydrogen and helium. His proposal was that if four atoms of hydrogen could be converted into one atom of helium, energy would be released in adequate quantity, since, unlike uranium, hydrogen is present in the Sun in great quantity, as can be seen from the abundance values of Table 7.1. From 1905, the time of Einstein's special theory of relativity, scientists had been familiar with the concept that mass and energy are quantities of a similar type. Hence, because four atoms of hydrogen have a mass greater than the mass of one atom of helium, by about one part in 125, it followed that Eddington's proposed transformation would indeed yield energy,

$$4\,H \longrightarrow He + energy.$$

PROBLEMS:

1. The conversion of 1 gram per second of hydrogen into helium yields a power output of about 6×10^8 kilowatts. Taking $\frac{2}{3}$ of the solar mass to be hydrogen, for how long would a total conversion of hydrogen to helium be capable of maintaining the solar energy output?

2. The mass of hydrogen in the terrestrial oceans is about 1.5×10^{23} grams. Industrial power consumption is about 5×10^9 kilowatts. For how long would a total conversion to helium of all oceanic hydrogen be capable of maintaining the present-day human power requirement?

3. Present-day technology is not by any means able to convert hydrogen to helium in the sense of Problem (2). Suppose it will take at least 50 years to develop the necessary technology. If it were your decision to make, what scale of effort would you consider desirable in this problem at the present time?

Eddington's proposal has turned out to be correct. Indeed, it *had* to be correct, for there is no other way, consistent with the composition given in Table 7.1, to explain the energy supply that the Sun clearly has. Yet in 1920 the details of how four atoms of hydrogen could be converted to one atom of helium were quite obscure. It will next be our business to examine these details, and for this we must concern ourselves more closely with the structure of atoms, specifically, with the properties of atomic nuclei. However, to lay some groundwork for Section IV, we will first consider certain properties of the outer electronic structure, especially those related to the basic principles of chemistry.

§7.4. Atomic Structure

In Chapter 3 we saw that the behavior of electrons in atoms is highly organized, even for atoms containing many electrons. Here we will consider some details of the way this organization is achieved. In Figure 7.1, the numbers attached to the elements ordered them with respect to increasing weight. We could also have ordered the elements according to the number of electrons present in each kind of atom, but if we had done so, exactly the same arrangement would have been obtained. We can therefore also think of the periodic table of Mendeleev as a classification of atoms with respect to electron number.

Working from the top of the periodic table, and counting the number of elements in each row, we have 2, 8, 8, 18, 18, 32 (i.e., 18, plus the 14 in rare earths), with the final row incomplete. The elements in the rightmost column are helium (order number 2), neon (order number $10 = 2 + 8$), argon (order number $18 = 2 + 8 + 8$), krypton (order number $36 = 2 + 8 + 8 + 18$), xenon $(54 = 2 + 8 + 8 + 18 + 18)$, and radon $(86 = 2 + 8 + 8 + 18 + 18 + 32)$. Except in certain very special circumstances, these six elements are chemically inert, remaining in gaseous form as individual atoms except at very low temperatures, whence they are known as the "noble" gases. This property of chemical inertness can be explained in terms of the above electron numbers, which represent "closed shells" that rarely interact with other atoms. Helium, with 2 electrons, has the first closed shell. Neon has two closed shells, one of 2 electrons, the other of 8 electrons—and similarly for the others, radon having six closed shells, of 2, 8, 8, 18, 18, and 32 electrons. Other atoms complete the closed shells in the same order, but since they do not have the same number of electrons as the noble gases (i.e., 2, 10, 18, 36, 54, or 86), some electrons are inevitably left over after all the possible closed shells have been filled. Thus sodium (Na), with 11 electrons, fills up two shells of 2 and 8 electrons, and then 1 electron is left over. Similarly, potassium (K), with 19 electrons, fills up three shells of 2, 8, and 8 electrons, and then 1 electron is left over. Omitting, for simplicity, the two subsidiary series of 14 elements given at the foot of Figure 7.1, it is thus easy to see that all the elements in each column of this figure have the *same number of electrons left over after all the possible complete shells have been filled*. In this connection, the shells must always be taken in the same order—2, 8, 8, 18, 18, 32—whatever the element. It is just this property of the electron structures of the elements that determines which of them will be similar in their chemical behavior.

Although there is no way in which the electrons of the sodium atom, taken by itself, can all arrange themselves into closed shells, a sodium atom, by interacting with another, suitable atom can use its extra electron to form a hybrid closed shell. The chlorine atom (Cl) is one electron short of being able to form a complete set of closed shells. Thus if the sodium atom "lends" its extra electron to a chlorine atom, a hybrid closed shell is formed. In the terminology of Chapter 4, the excess electron of the sodium then has paths of high probability that remain close to the chlorine atom, causing the two atoms to form a composite particle, written as NaCl. Such composite particles are called *molecules*, and NaCl is the molecule of sodium chloride, which is common salt.

It is possible for more than two atoms to be involved in this process of electron sharing. In the molecule of ammonia, for example, the three electrons of three hydrogen atoms join the five extra electrons of a nitrogen atom to form a hybrid shell of eight electrons, the molecule in this case being written in the form NH_3.

PROBLEM:
Satisfy yourself that a hybrid shell of 8 electrons is formed for both the molecule MgO and the molecule CaO. Magnesium metal burns brilliantly to produce MgO. The lime used by farmers to sweeten sour land is CaO.

Although atoms with one or two spare electrons, like Na, Mg, K, Ca, can be thought of as "lending" atoms, and atoms like O or Cl, which require only one or two electrons to fill a shell can be considered "receiving" atoms, this concept is not entirely clear-cut. Carbon has four spare electrons, and also requires four electrons to complete a shell of eight electrons. Thus carbon can be a lending atom, as in the molecule of carbon tetrachloride, CCl_4, or a receiving atom, as in the molecule of methane, CH_4. The ability of carbon to operate in both these ways is a reason why carbon is able to join with other atoms in a most prolific way, and thence why carbon forms the basis for the complex chemical processes that occur in biological material.

PROBLEMS:

1. Sulfur is another atom that can operate in both ways. Verify that H_2S has a hybrid closed shell of eight electrons. Verify that the six spare electrons of sulfur are just sufficient, when lent to oxygen in the molecule of sulfur trioxide, SO_3, to give three hybrid closed shells.

2. Explain why H_2S is a chemically similar molecule to that of water, H_2O. Consider the extent to which the inner, comparatively inert shell structures of the O and S atoms are responsible for differences in the macroscopic properties of bulk samples of water and hydrogen sulfide. A world based on H_2S would be very different from our water-dependent world. Do you consider the unpleasant smell of H_2S to be an absolute quality, or could living creatures in an H_2S world have developed in such a way as to give H_2O an equally unpleasant smell?

Hybrid closed shells are not as mutually standoffish as the closed shells of the noble gases. Two molecules, both entirely closed in the present hybrid sense, may well be capable of joining into a composite molecule. For example, CO_2 and CaO are both completely closed in a hybrid sense, and yet they can combine to give the molecule of calcium carbonate, $CaCO_3$. Water, H_2O, and sulfur trioxide, SO_3, combine similarly to give sulfuric acid, H_2SO_4. Such associations are usually not very strong, however, and can be dissociated without undue difficulty, often simply by heating. Thus calcium carbonate is dissociated into CaO and CO_2 in a lime kiln. Much of industrial chemistry is concerned with the associations and dissociations of such "closed" molecules.

It is possible for molecules to be formed in such a way that, although one or more hybrid shells are produced, some spare electrons nevertheless remain. In the simple molecule CO, four electrons from the carbon, together with six electrons from the oxygen, give a hybrid shell of eight plus two spare electrons. These spare electrons are able to join with a second oxygen atom to give a second hybrid shell in the molecule of carbon dioxide, CO_2. Thus CO will always tend to acquire O, a property which produces a kind of suffocation if you breathe too much CO gas. Almost every molecule with electrons "left over" is similarly poisonous, some very markedly so.

Usually, in any given mixture of atoms and molecules, all the hybrid shells that are possible are already formed; spare electrons occur only when there is no way for them to form further shells. Occasionally, however, a solid or liquid can be prepared in such a way that possible hybrid shells still have not been formed. Such a situation occurs for a *chemical explosive*. "Detonation" of an explosive consists of starting up hybrid shell formation. Once started, shell formation proceeds rapidly and spontaneously to completion, accompanied by a sudden energy release. It is this energy release that constitutes the explosion itself. Any sudden energy release produces the phenomenon of explosion, whether or not the energy is generated chemically; a chemically inert stone dropping at very high speed from the sky, that is, a meteorite, produces an explosion when it hits the ground.

Molecules can form even without a single hybrid shell being formed. A hydrogen atom will "lend" its electron to an oxygen atom, for example, giving the molecule OH. Or a hydrogen atom will "lend" its electron to a carbon atom, to give the molecule CH. These very simple molecules are exceedingly active, so much so that they can hardly be studied at all in the laboratory, where they immediately join with some other atom or molecule to produce a closed shell. Such exceedingly active molecules are given the special name of *free radicals*. They are found widespread in astronomy, in gas clouds like the Orion Nebula, for example.

PROBLEMS:

1. The energy released by the combination of free radicals with other atoms and molecules gives the most powerful energy source obtainable by chemical means. The combination of 1 gram per second of OH with H would yield a power output of about 30 kilowatts. Compare such a chemical source of energy with the nuclear conversion of hydrogen to helium.

2. What is the reason why free radicals can exist in the interstellar gas but not readily in the laboratory?

§7.5. Nuclei and Particles

Electrons all have the same electric charge, and consequently repel each other electrically: like charges repel; unlike charges attract. Why then do the electrons in an atom not simply fly apart? Why do they stay in the atom? The answer is that the electrons are bound to the atom by a charge of opposite sign carried in a small but massive nucleus. Nuclei contain protons, and the proton has been found to carry an electric charge equal in magnitude but opposite in sign to that of the electron.

Sometimes atoms contain fewer electrons than protons, in which case the atom is said to be *ionized*. The ionized atom then has a strong tendency to acquire electrons, until the number of electrons comes into balance with the number of protons, when the atom becomes *neutral*. There are a few atoms, hydrogen and oxygen being two of them, that occasionally carry their electron-acquiring tendency too far, so that they end up by having one electron too many—i.e., one more than the number of protons. When in this state, hydrogen and oxygen are said to have formed *negative ions*, an odd property which turns out to have quite important effects in the atmospheres of stars, and in our own terrestrial atmosphere. This oddity apart, atoms have a marked tendency under cool conditions to become neutral, and it is with neutral atoms that the science of chemistry is mainly concerned. What now becomes clear is that, whereas the chemical properties of an atom arise directly from the electron shell structure, the number of electrons and hence the electron shell structure is itself controlled by the nucleus. The chemistry of an atom is determined therefore by the number of protons in the nucleus, *and this is the deeper meaning of the number by which we ordered the elements in Table 7.1.* It is usual to refer to this number by *Z*, which now explains why this letter was used in the heading of Table 7.1.

The electrons of an atom form an extensive lightweight cloud surrounding a small massive nucleus. The mass of the nucleus exceeds that of the electron cloud several thousandfold, and if we think of the nucleus of a typical atom, say, iron, as being a centimeter in diameter, then the electron cloud is about the size of a baseball park. In actual size, however, both the nucleus and the electron cloud are small compared to practical dimensions. In Figure 6.34, we showed astronomical distances ranging from the radius of the Earth's orbit around the Sun to the distances of the most remote galaxies, and we did so with the help of a logarithmic plot. It will be useful here to show a similar plot, but ranging now from the nuclei of atoms at the small end to the main-sequence stars at the upper end, as in Figure 7.2, where some familiar objects and quantities, and some not so familiar, have been marked. By combining Figure 7.2 with Figure 6.34, we can go from the nuclei of atoms to the greatest observed distances in the universe.

We saw above why the electrons of an atom do not fly apart, because they are held to the atom by the electrical attraction of the protons in the nucleus. But since charges of the same type repel each other, and the protons all have the same charge, why do the protons not immediately burst apart? To answer this question, we must postulate the existence of an entirely new force, quite different from electrical and gravitational forces, which we call the *nuclear force*. It must work to bind the protons together. Since no such force is observed when protons are far apart from each other, we have to suppose that, unlike electrical and gravitational forces, which do operate for particles that are far apart, *the nuclear force is active only for particles that are close together.*

The picture is not yet complete, however, for an important reason that emerges as soon as we consider in detail the actual masses of nuclei. For example, the nucleus of the helium atom, which contains two protons, has a mass almost four times greater than the normal hydrogen atom, which contains one proton. Plainly something else besides protons must be present in the nucleus of helium, and a similar argument applies to all elements heavier than

---

FIGURE 7.2.

The range of scales and of lengths, going from the nuclei of atoms up to stars. The relative sizes of planets, and of the smallest stars—white dwarfs and neutron stars—compared to the Sun, are covered here. This figure and Figure 6.34, taken together, go from the smallest lengths to the largest lengths. (From L. Pauling, *College Chemistry,* 3d ed. W. H. Freeman and Company. Copyright © 1964.)

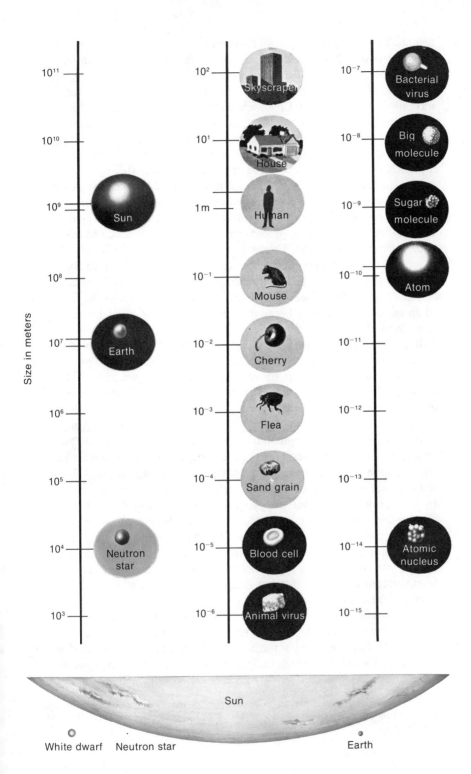

Size in meters

10^{11}
10^{10}
10^{9} — Sun
10^{8}
10^{7} — Earth
10^{6}
10^{5}
10^{4} — Neutron star
10^{3}

10^{2} — Skyscraper
10^{1} — House
1 m — Human
10^{-1} — Mouse
10^{-2} — Cherry
10^{-3} — Flea
10^{-4} — Sand grain
10^{-5} — Blood cell
10^{-6} — Animal virus

10^{-7} — Bacterial virus
10^{-8} — Big molecule
10^{-9} — Sugar molecule
10^{-10} — Atom
10^{-11}
10^{-12}
10^{-13}
10^{-14} — Atomic nucleus
10^{-15}

Sun

White dwarf   Neutron star

Earth

helium. This "something" cannot have any electric charge, otherwise the whole of our chemical scheme would be destroyed. We need a particle without electric charge, but otherwise similar to the proton, that is, similar in mass and in being subject to the nuclear force. Such a particle was in fact discovered experimentally in 1932 by James Chadwick, and was named the neutron. The nuclear force operates equally to hold protons to protons, neutrons to neutrons, and protons to neutrons, so that all the particles of the nucleus have the property of being able to attract each other. Notice, however, that if for some reason a particle were to manage to move appreciably outside the nucleus, it would not be attracted back again, because the nuclear force between the particle and the nucleus must cease to operate when the two become separate. And if such a particle contained one or more protons, it would actually be repelled by the electrical effect of the other protons remaining in the nucleus, and so it would gain speed as it separated. This is just the situation for the α-rays observed to emerge from the nuclei of the uranium atoms, the α-rays which Rutherford and his colleagues identified with nuclei of helium atoms. So the phenomenon of radioactivity, first observed by Becquerel, is now subject to the beginnings of an explanation: when a heavy nucleus of an atom of large Z emits α-rays, it is simply losing bits of itself.

But what, then, were the transformations observed in 1907, by N. R. Campbell, in which the emission of β-rays (electrons) was found to change potassium to calcium and rubidium to strontium? For rubidium we have $Z = 37$, for strontium $Z = 38$, i.e., 37 and 38 protons respectively. So the emission of an electron from rubidium is accompanied by the appearance of an extra proton in the nucleus of the atom, the implication being that a neutron n changes into a proton p with the emission of an electron e, n $\longrightarrow$ p + e. This transformation is balanced in terms of electric charge. The neutron has zero charge, and although the proton has a charge, the proton charge is just balanced by the opposite charge of the electron.* Electric charge is said to be *conserved* by the transformation. But energy was not conserved in this way of writing the transition from neutron to proton. This lack of balance of energy eventually led Wolfgang Pauli (1900–1958) to suggest that a fourth particle, say, $\bar{\nu}$, had to be involved, n $\longrightarrow$ p + e + $\bar{\nu}$. This further particle could not have electric charge, nor could it have appreciable mass, otherwise it would already have been detected experimentally, but it could have energy, and it could serve to maintain the conservation of energy.

The astronomer Walter Baade told me that, when he was dining with Pauli one day, Pauli exclaimed, "Today I have done the worst thing for a theoretical physicist. I have invented something which can never be detected experimentally." Baade immediately offered to bet a crate of champagne that the elusive $\bar{\nu}$ would one day prove amenable to experimental discovery. Pauli

---

*Historically, the proton charge was taken to be positive and the electron charge to be negative. A reversed choice, with the electron taken as positive and the proton as negative, would be equally possible.

accepted, unwisely failing to specify any time limit, which made it impossible for him ever to win the bet. Baade collected his crate of champagne (as I can testify, having helped Baade consume a bottle of it) when, just over twenty years later, in 1953, Cowan and Reines did indeed succeed in detecting Pauli's particle.

Shortly after the dinner with Baade, Pauli's arguments for the existence of $\bar{\nu}$ were being discussed at a seminar in Rome. Confusion arose between $\bar{\nu}$ and the newly discovered neutron. In some exasperation, Enrico Fermi explained to the assembled company that Pauli's particle was not the massive neutron at all. It was only a "little neutron," and the Italian diminutive, neutrino, has been used for Pauli's particle ever since.

If $n \longrightarrow p + e + \bar{\nu}$, why not $p + e + \bar{\nu} \longrightarrow n$, since all other detailed transformations in physics are reversible? It would be hard to observe this reverse form, however, because three particles—one of them the elusive $\bar{\nu}$—would have to come together in order to produce the neutron. But then why always associate two particles with the proton and none with the neutron? Why not symmetrize the situation, writing $p + e \longrightarrow n + \nu$, or $n + \nu \longrightarrow p + e$? Written in this form, ν is called the *neutrino*. Our previous $\bar{\nu}$, the particle detected by Cowan and Reines, is nowadays called the *anti-neutrino*.

Is there actual data to support such a way of writing an inverse transformation from proton to neutron? A close examination of the decay of potassium shows that, although most decays do result in the production of calcium, in about 10 per cent of the decaying atoms something different happens. Instead of calcium, argon is produced. This involves a change from $Z = 19$ to $Z = 18$, the reverse of potassium going to calcium, which requires Z to change from 19 to 20. Evidently then, potassium can go "both ways." A neutron can go to a proton, or a proton can go to a neutron. The proton goes to a neutron by joining with an electron from the innermost shell of the surrounding electron cloud—in short, we have $p + e \longrightarrow n + \nu$.

With clear-cut data thus available for both $n \longrightarrow p + e + \bar{\nu}$ and $p + e \longrightarrow n + \nu$, why not a transition like $p \longrightarrow n + \nu + ?$, where ? is some new particle. Since neither $n \longrightarrow p + e + \bar{\nu}$ nor $p + e \longrightarrow n + \nu$ involves any change of the total electric charge, there should also be no change for $p \longrightarrow n + \nu + ?$, which means that ? cannot be an ordinary electron—otherwise we should have a positively charged proton changing into three particles having a total *negative* charge. So we write $p \longrightarrow n + \nu + e^+$, with e^+ a particle like an electron but having positive charge. Such e^+ particles were in fact detected in 1932, by C. D. Anderson, and by P. M. S. Blackett and G. P. S. Occhialini, at just about the time Pauli was postulating the existence of the neutrino. These positively charged electron-like particles are called *positrons*. The various ways in which these "decays" can occur are shown in Figure 7.3.

By now we have encountered quite a number of different particles, and to keep order it will be useful to classify them into families. In Appendix II.7

$$\begin{array}{ccccccc}
\text{n} & \rightarrow & \text{p} & + & \text{e} & + & \bar{\nu} \\
\text{(neutron)} & & \text{(proton)} & & \text{(electron)} & & \text{(antineutrino)}
\end{array}$$

$$\begin{array}{ccccccc}
\text{n} & + & \nu & \rightarrow & \text{p} & + & \text{e} \\
\text{(neutron)} & & \text{(neutrino)} & & \text{(proton)} & & \text{(electron)}
\end{array}$$

$$\begin{array}{ccccccc}
\text{p} & + & \text{e} & \rightarrow & \text{n} & + & \nu \\
\text{(proton)} & & \text{(electron)} & & \text{(neutron)} & & \text{(neutrino)}
\end{array}$$

$$\begin{array}{ccccccc}
\text{p} & \rightarrow & \text{n} & + & \nu & + & \text{e}^+ \\
\text{(proton)} & & \text{(neutron)} & & \text{(neutrino)} & & \text{(positron)}
\end{array}$$

FIGURE 7.3.
The various ways in which the process of "β-decay" can occur.

we discussed briefly a particle known as the muon (μ). Like the electron, the muon has negative charge. Indeed apart from having a mass rather more than 200 times that of the electron, the muon seems to be exactly like the electron. Recent experiments have shown there to be two kinds of neutrino, written as ν_e and ν_μ, with ν_e the same as ν, p + e $\longrightarrow$ n + ν_e, and ν_μ a neutrino similarly associated with muons, p + μ $\longrightarrow$ n + ν_μ. The four particles e, μ, ν_e, ν_μ, form a family known as the *leptons,* the lightweights. Some physicists think this lepton family should have eight members, but there is as yet no direct evidence of the four other members. What is certain, however, is that the proton and neutron belong to a family of eight particles known as *baryons,* the heavyweights. The two families of leptons and baryons appear to be quite distinct, in the sense that a member of one family never changes into a member of the other family, although connections between them can be established through the agency of a third family of particles called *mesons.* A table of the members of these three families is given in Appendix III.10.

Corresponding to the four leptons e, μ, ν_e, ν_μ, there is a family of antileptons, written as e$^+$, μ^+, $\bar{\nu}_e$, $\bar{\nu}_\mu$, of which we have already encountered e$^+$ and $\bar{\nu}_e$, and corresponding to the family of baryons there is a family of eight antibaryons. There is no separate family of antimesons, however. Certain further families have been experimentally detected in recent years, but they will not concern us in this book. In most astronomical problems only two baryons, p and n, are involved, and only two leptons, e and ν_e, so that the situation in astronomy is simpler than it is in physics generally.

§7.6. *Isotopes and Their Stability*

Write N for the number of neutrons contained in a nucleus, and Z for the number of protons. The sum $A = N + Z$ is called the atomic number, or sometimes the *mass number*. If we imagine Z to stay fixed, but N to vary, we have a situation in which the basic chemical properties of the neutral atom do not change, since the chemical properties are determined by Z alone. Atoms with the same Z, but different N, are called *isotopes* of each other. *Isotopes all belong to the same element.*

PROBLEM:
In Table 7.2 we had the *radioactive series* starting with ^{238}U, the isotope of uranium ($Z = 92$) having $A = 238$, and hence $N = A - Z = 146$. After the emission of an α-ray and two β-rays, we arrive in Table 7.2 again at the uranium atom. Given that an α-ray is a helium nucleus consisting of 2 protons and 2 neutrons, and that the β-rays are electrons resulting from the process $n \longrightarrow p + e + \bar{\nu}_e$, obtain the values of N and A appropriate to this second atom of uranium. Verify that $Z = 92$. Work through Table 7.2, obtaining values of N, Z, A, throughout.

The values of Z and N for a nucleus cannot be chosen arbitrarily. If *for a specified Z* we were to choose N too large, the nucleus would simply disgorge neutrons until a suitably smaller value of N were attained. And if N were chosen too small, protons or helium nuclei would similarly be disgorged. Nevertheless, for a specified Z, many values of N are in general permitted; i.e., many isotopes of the element of specified Z can exist. Yet this does not mean that all such permitted isotopes are permanently *stable*. For any one isotope, we may have $n \longrightarrow p + e + \bar{\nu}_e$, increasing Z by one unit and thereby changing the element, or we may have one or the other of $p \longrightarrow n + \nu_e + e^+$, $p + e \longrightarrow n + \nu_e$, decreasing Z by one unit. Only if these so-called β-processes do *not* occur is the isotope stable. An elegant way of considering this problem of stability is worth discussing.

Take, first, the case in which A is an odd number. It is possible to make up the same fixed value of A in many ways. Thus, if we begin by choosing Z, then N is determined by $N = A - Z$, since we begin from the fixed value of A. Suppose Z to be varied for all permissible nuclei, with A remaining fixed in this way. When the energies of these nuclei are calculated, it turns out that they can be plotted in the manner of Figure 7.4. Among the plotted points there will be a lowest one. This lowest one indicates the only stable nucleus at the value of A under consideration. For if we have a nucleus at point P of Figure 7.4, lying to the right of the lowest point, there will be a cascade of processes of the type $p \longrightarrow n + \nu_e + e^+$, or of $p + e \longrightarrow n + \nu_e$, which will go on until the bottom point is reached. And if we have a nucleus at the point Q, to the left of the lowest point, there will be a similar cascade arising from processes of the type $n \longrightarrow p + e + \bar{\nu}_e$.

Each step of such a cascade takes a finite time to occur. Unstable nuclei at a higher point of Figure 7.4 can therefore exist for a while, and if prepared artificially in the laboratory can be subjected to study. Unstable nuclei are not usually found in nature, however, since if they existed at one time they would

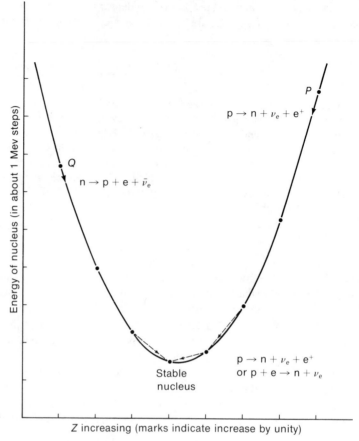

FIGURE 7.4.

Keeping the mass number A fixed, the energies of nuclei of different proton numbers Z fall, when A is odd, on a single U-shaped curve. Decays occur down either arm of the curve, until the bottom point is reached.

by now have undergone change. But there is one important exception to this statement. It may happen that the bottom point of our figure is only slightly lower than an adjacent point, as in Figure 7.5, where point 1 is only slightly lower than point 2. A nucleus starting at a point to the left of point 1 falls comparatively quickly down to point 1, but a nucleus starting at a point to the right of point 2 falls quickly to point 2, where it may remain for a very long time, because the energy step to point 1 is now very small. An example of this situation is given by the isotope $N = 112$ of the element rhenium ($Z = 75$), which has a slow decay in a characteristic time of about 4×10^{10} years to the isotope $N = 111$ of the element osmium ($Z = 76$), $^{187}\text{Re} \longrightarrow \, ^{187}\text{Os}$.

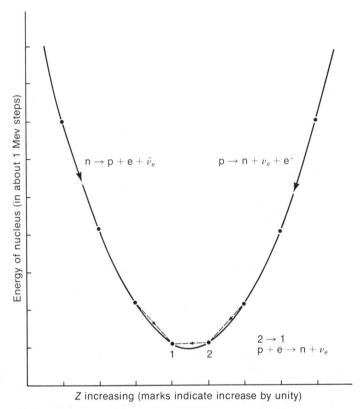

FIGURE 7.5.
Occasionally the last downward step may be very shallow, in which case the slight drop down this last step may take a long time to occur, longer even than the age of the Earth. Here A is odd.

Next we consider the case where the fixed value of A is an even number. We proceed as before, but there is now the interesting complication that, as Z is varied, the energy values fall alternately on two smooth curves. A situation of the kind shown in Figure 7.6 can then occur, with *two* stable possibilities, at points 1 and 2. A nucleus starting at A in this figure bounces downward from curve to curve until it arrives at point 1, the processes being either $p \longrightarrow n + e^+ + \nu_e$ or $p + e \longrightarrow n + \nu_e$. A nucleus starting at B falls from curve to curve until it arrives at point 2, the processes being $n \longrightarrow p + e + \bar{\nu}_e$. There is the curiosity that a nucleus starting at C can go either way, to point 1 or to point 2. If such a nucleus goes to point 1, then Z increases by one unit. If it goes to point 2, then Z decreases by one unit. This is the situation for the radioactive isotope of potassium, which can decay either to calcium (Z increasing) or to argon (Z decreasing).

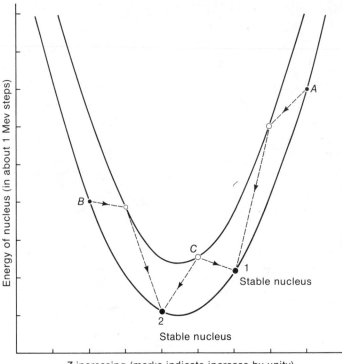

FIGURE 7.6.
The situation is more complicated when the mass number is even. Instead of
the single U-shaped curve of Figure 7.4, there are now two such curves. Decays occur down both the left-side and the right-side channels, which may
lead to different stable nuclei.

PROBLEMS:

1. The value of A for the radioactive isotope of potassium is 40. Does this
nucleus satisfy the condition, appropriate to point C of Figure 7.6, that both
N and Z are odd numbers? [Hint: From Table 7.1, find the value of Z for
potassium. Then use $A - Z$ for N.]

2. Except for two isotopes of uranium, ^{235}U, ^{238}U, and one isotope of thorium,
^{232}Th, nuclei with A greater than 209 quickly lose bits of themselves, until A
is reduced below 210. Estimate the total number of stable nuclei. [Hint: Use
Figure 7.4 when A is odd, and Figure 7.6 when A is even. Notice, however,
that your estimate, while giving a tolerable answer, is not accurate, because point
1 or point 2 of Figure 7.6 is sometimes placed above point C, in which case
there is only one stable nucleus, not two. This is especially the case for the
smaller values of A.]

321

§7.7. Nuclear Energy
and the Energy
of the Stars

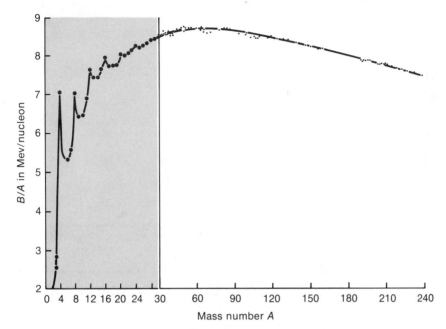

FIGURE 7.7.
The average energy yield per unit mass number plotted for all the naturally occurring nuclei.

It is of interest in relation to Problem (2) that every stable nucleus is actually found among naturally occurring materials. A complete list of all the isotopes of the elements found in nature is given in Table III.1 of Appendix III.2.

§7.7. Nuclear Energy and the Energy of the Stars

If we imagine protons and neutrons to be added together to form a nucleus, energy will be yielded up during the aggregation process, because the nuclear force pulls the particles together once they are within close-enough range of each other. Write B for this energy of formation, and consider the quantity B/A. This average energy yield per neutron or proton will in general be different from one nucleus to another. Figure 7.7 shows B/A plotted against A for all the naturally occurring nuclei.*

For comparison with Figure 7.7, the relative abundances of the naturally occurring nuclei are plotted in Figure 7.8, a schematic representation of Figure

---

*The unit of energy used in this figure is not of great importance in our discussion, but is one that is frequently used in physics: the energy acquired by an electron in moving through a static electric field with a potential of one volt is called the *electron volt* (eV). An electron with an energy of 1 eV has a speed of about 600 km per second. A million electron volts (MeV) is the unit in Figure 7.7.

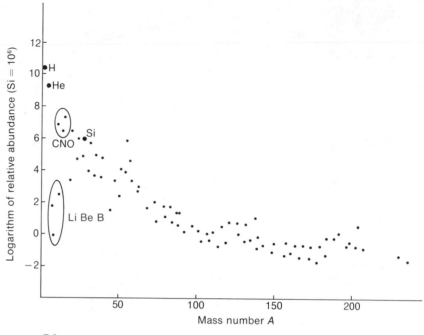

FIGURE 7.8.
The relative abundances of the naturally occurring nuclei. The abundances are highest for
hydrogen and helium. Notice from Figure 7.7 that it is just these light nuclei which can yield
the greatest energy.

7.8 being given in Figure 7.9. The curve marked α in Figure 7.9 is drawn
through the abundance values for $A = 28, 32, 36, 40$. The nuclei at these values
of A can be thought of as being made up of helium nuclei, i.e., made up of
α particles, and hence the designation α. The α curve of Figure 7.9 is seen
to lie well above the curve joining the abundances of the non-α nuclei. Comparison with Figure 7.7 shows that the curve marked "He-burning" passes
through the nuclei at the sharply peaked maxima of the B/A curve, a connection
which can hardly be accidental. Then again, Figure 7.9 has a peak, often called
the "iron peak," since the element iron sits at the top of it, which is close to
the general broad maximum of the B/A curve. These connections show a
relation between the physical properties of the nuclei and their naturally
occurring abundances, indicating an origin of a physical kind. Just as plants
and animals have been produced by evolution, not by special creation, so the
nuclei seem to have been produced by some form of physical evolution, not
by special creation.

We shall be returning to Figure 7.8 later, but before leaving it here let us
notice once again how very small the abundances, *logarithmically plotted,* are
for nuclei of large A. This is why the radioactive decay of elements of large

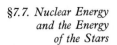

323

§7.7. Nuclear Energy
and the Energy
of the Stars

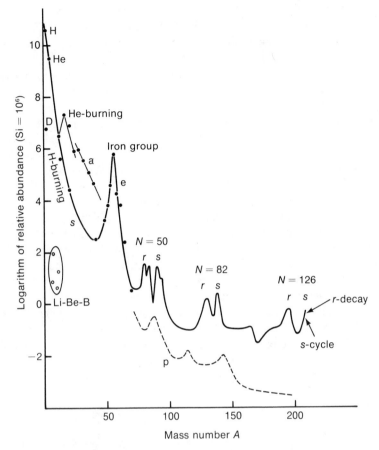

FIGURE 7.9.
The relative abundances of Figure 7.8 plotted in a schematic form.

A, like uranium and thorium, cannot be responsible for the energy produced by the stars; there is far too little of these materials.

The curve of Figure 7.7 shows that energy can be obtained either by adding nuclei of small A, so that the mass number of the product nucleus increases towards the broad flat maximum at A about 60, or by removing pieces from nuclei of very large mass number, as by the emission of α-particles from uranium. Energy could only be lost, however, by attempting either to add or to divide nuclei with mass numbers near 60. The atoms corresponding to such nuclei are just the well-known metals, iron, nickel, chromium, manganese, cobalt, copper, zinc, titanium, and vanadium. Although these metals play a dominant role in the economy of modern society, it must be recognized that, in terms of energy, they are nothing but the "ashes" into which lighter nuclei can be "burned," or into which heavier nuclei can be broken down. The "iron peak" of Figure 7.9 is indeed just a pile of cosmic ashes that has resulted from the burning of lighter nuclei. In the remainder of this chapter, we will trace the details of the way this has happened.

The curve of Figure 7.7 rises particularly sharply from hydrogen at $A = 1$ to helium at $A = 4$. Thus the largest energy yield comes from adding the simplest *and the most abundant* atoms, in just the process suggested by Eddington, namely $4\ ^1\mathrm{H} \longrightarrow\ ^4\mathrm{He}$. (We are now adopting the convention of writing the value of A as a superscript in front of the chemical symbol. Since the chemical symbol defines Z, the neutron number N is given immediately by the subtraction $A - Z$. This was the situation above when we wrote $^{87}\mathrm{Rb}$, $^{87}\mathrm{Sr}$, $^{40}\mathrm{K}$.)

Unlike the situation in 1920, at the time of Eddington's suggestion, we know today of many ways to convert hydrogen to helium, of which two are of special importance inside stars. A complex of nuclear reactions known as *the proton-proton chain*, shown in detail in Figure 7.10, is mainly responsible for the conversion of hydrogen to helium inside stars of comparatively small mass like the Sun. Another complex of reactions, given in Figure 7.11 and known as the *carbon-nitrogen cycle*, is of most importance inside main-sequence stars of larger mass. The proton-proton chain was first outlined in 1938 by H. A. Bethe and C. L. Critchfield, and the carbon-nitrogen cycle was proposed by Bethe in 1939. These first steps have been considerably extended and developed by later workers, notably by C. C. Lauritsen and William A. Fowler at the California Institute of Technology.

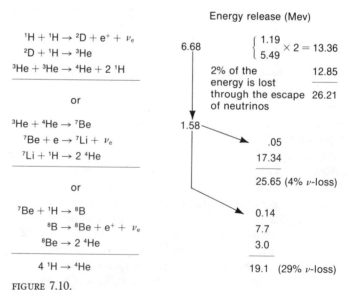

FIGURE 7.10.
The proton-proton chain. The MeV unit is equal to 1.6×10^{-16} kilowatt seconds.

§7.7. Nuclear Energy
and the Energy
of the Stars

$^{12}C + {}^{1}H \rightarrow {}^{13}N$	1.95
$^{13}N \rightarrow {}^{13}C + e^+ + \nu_e$	1.50
$^{13}C + {}^{1}H \rightarrow {}^{14}N$	7.54
$^{14}N + {}^{1}H \rightarrow {}^{15}O$	7.35
$^{15}O \rightarrow {}^{15}N + e^+ + \nu_e$	1.73
$^{15}N + {}^{1}H \rightarrow {}^{12}C + {}^{4}He$	4.96

25.03 Mev

(6% of the energy
is lost through the
escape of neutrinos)

In 1 case in 1000 there
are the alternative reactions:

$^{15}N + {}^{1}H \rightarrow {}^{16}O$ $^{17}O + {}^{1}H \rightarrow {}^{18}F$

$^{16}O + {}^{1}H \rightarrow {}^{17}F$ $^{18}F \rightarrow {}^{18}O + e^+ + \nu_e$

$^{17}F \rightarrow {}^{17}O + e^+ + \nu_e$ $^{18}O + {}^{1}H \rightarrow {}^{15}N + {}^{4}He$

$^{17}O + {}^{1}H \rightarrow {}^{14}N + {}^{4}He$ or

FIGURE 7.11.
The carbon-nitrogen cycle. The MeV unit is equal to
1.6×10^{-16} kilowatt seconds.

There is a general rule to the effect that reactions involving nuclei of larger positive electric charge—i.e., with more protons—go slower than reactions involving nuclei of smaller positive charge. This is because charges of the same sign repel each other, and the repulsion is greater when the nuclear charge is larger. Higher speeds of approach of the particles toward each other, involving higher temperature, are therefore needed when the charge is large than when it is small. According to the system of reactions of Figure 7.10, it is possible to generate helium from hydrogen without involving any nucleus that has a charge of more than two units, $Z = 2$ for ^3He, whereas in the system of Figure 7.11, helium production involves ^{15}N with $Z = 7$. Consequently we expect the system of Figure 7.10 to operate at a lower temperature than that of Figure 7.11, and this is correct. Thus the reactions of Figure 7.10 control the energy production in stars with central temperatures up to about $2 \times 10^{7}\,^{\circ}K$, whereas the reactions of Figure 7.11 control the situations at temperatures higher than this. It is indeed just this difference of behavior with respect to temperature that causes the processes in Figure 7.10 to be operative in main-sequence stars of small mass, and those in Figure 7.11 to be operative in main-sequence stars of large mass.

The difference here turns out to lead to an interesting difference of structure between stars like the Sun and main-sequence stars of larger mass. This is illustrated in Figure 7.12. In the shaded parts of each star energy is transported outward mainly by convection, i.e., by a boiling motion of the stellar material, whereas in the unshaded parts transport of energy is by radiation. The structures

FIGURE 7.12.
Left: Star with mass similar to that of the Sun.
Right: More massive star, on upper part of main
sequence. In each case shading denotes region
where energy is transported mainly by convection.
In unshaded region transport is by radiation.

of the two kinds of stars are completely opposite in character. In massive
main-sequence stars, we have convection near the center and radiation outside,
whereas in solar-type stars we have radiation carrying the energy throughout
the inner portions and convection on the outside. It is this transport of energy
by convection in the outer regions of the Sun that probably accounts for the
highly complicated behavior of the gases of the solar atmosphere, the behavior
we discussed in Chapter 2.

The system of reactions of Figure 7.10 would operate even faster at comparatively low stellar temperature if it were not for a peculiar feature of the first
reaction. Two protons approach each other, which they can do with ease,
because only $Z = 1$ is involved. However, the two protons simply come apart
again, $p + p \longrightarrow p + p$, unless one of them switches into a neutron, through
$p \longrightarrow n + e^+ + \nu_e$. The neutron and proton then hold together as a *deuteron*,
denoted in Figure 7.10 by the symbol D.

According to Figure 7.10, there are two possible branch points. Our objective
of building ^{4}He can be attained directly from ^{3}He $+ ^3$He $\longrightarrow ^4$He $+ 2p$, or
we can have ^{3}He $+ ^4$He $\longrightarrow ^7$Be $+$ energy. The relative probability of these
two routes is well-known from laboratory data. Then, in the route via ^{7}Be,
there is a further branch, according to whether ^{7}Be first decays to ^{7}Li or first
acquires a proton, thereby forming ^{8}B. The probabilities at this second branch
are also known from laboratory data. No matter which branch is followed, the
outcome is always to build ^{4}He. It has taken us some time and effort to arrive
at Figures 7.10 and 7.11, but the problem is not inherently simple, as can be
seen from the fact that man-made fusion reactors capable of building ^{4}He are
still a dream for the future. In the above work, we have shown how a process
that cannot yet be carried out terrestrially takes place in the stars.

§7.8. The Evolution of Stars

The generation of energy inside stars has an inevitable quality about it. We
saw in Chapter 6 that protostars are drawn together by gravitation, until
compression raises their internal temperatures high enough for nuclear reactions

to produce energy on a scale sufficient to compensate for the radiation that is being lost continuously from their surfaces. At this stage newly formed stars are said to populate the zero-age main sequence. In this chapter, especially in Figures 7.10 and 7.11, we have formulated the detailed nuclear reactions responsible for this energy generation.

In Chapter 6 we also saw that the amount of time for which hydrogen-to-helium conversion can generate energy depends markedly on the mass of the star. For solar-type stars energy will be generated by this process for at least 10^{10} years, but for stars of large mass the time-scale is much shorter, only 10^6 years for a mass of 20 $\odot$. The latter interval is very short compared to the terrestrial geological time-scale, and is therefore extremely short compared to the age of our galaxy. It follows that we must consider what happens next.

Suppose we return to the luminosity-color diagram—the Hertzsprung-Russell diagram—as in Figure 7.13, with the zero-age main sequence going

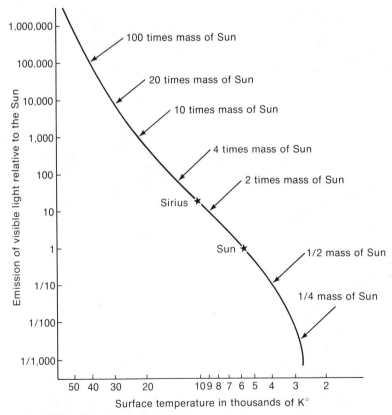

FIGURE 7.13.
Newly formed stars lie in the Hertzsprung-Russell diagram on a "zero-age main sequence," their position on this sequence being determined by mass. Notice that luminosities with respect to *visual light* are plotted here.

from lower right to upper left. It is natural to ask how the stars behave in this diagram as hydrogen-burning proceeds. The answer to this question, determined by extensive calculations, is shown schematically in Figure 7.14. Stars with masses comparable to the Sun evolve to the right and markedly upward, while stars of large mass evolve mainly toward the right. The effect is to produce a kind of funneling of stars into a certain region of the diagram. This region is characterized by red color and by quite high luminosity, a combination that requires such evolving stars to have radii much greater than that of the Sun, usually from 10 to 100 times the Sun's radius, and in some cases even more, as can be seen from Figure 6.23. It is indeed because of their large radii that stars in this part of the diagram are called *giants*.

It is interesting to compare these expectations with the observations of Figure 7.15, which gives a logarithmic plot of visual luminosity versus surface temperature—i.e., color—for the nearest stars, denoted by crosses, and for the brightest stars, denoted by circles. The remarkable feature of Figure 7.15 is that, whereas the nearest stars are either comparable to, or fainter than, the Sun, the brightest stars as they appear in the sky are intrinsically more luminous and are undergoing evolution toward the giant region of the diagram, exactly as was predicted in the schematic drawing of Figure 7.14. Many of the brightest stars are evolving away from the main sequence, and are therefore already approaching hydrogen exhaustion.

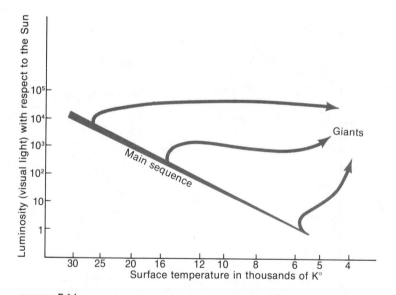

FIGURE 7.14.

As more and more hydrogen is converted into helium, stars evolve from the main-sequence toward the right of the Hertzsprung-Russell diagram, and they do so in such a way that they all arrive at what is known as the "giant" region of the diagram, so-called because the radii of stars in this region are large, comparable in scale to the radius of the Earth's orbit around the Sun.

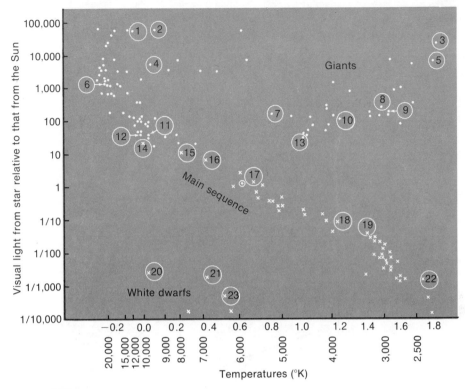

FIGURE 7.15.
The positions in the Hertzsprung-Russell diagram of the nearest stars (x), and of the stars which appear brightest to the eye (•). Key: 1, Rigel; 2, Deneb; 3, Betelgeuse; 4, Polaris; 5, Antares; 6, Spica; 7, Capella; 8, Mira; 9, Aldebaran; 10, Arcturus; 11, Castor; 12, Vega; 13, Pollux; 14, Sirius A; 15, Altair; 16, Procyon A; 17, Alpha Centauri; 18, 61 Cygnus A; 19, 61 Cygnus B; 20, 40 Eridani B; 21, Sirius B; 22, Barnard's Star; 23, Procyon B.

Two points concerning the observations of Figure 7.15 should be noted. In our previous representations of the luminosity-color diagram, total luminosities involving radiation of all frequencies were plotted. In Figure 7.15, however, the luminosities are given for visual frequencies only. The difference is not appreciable for stars with colors in the green-yellow region, but for blue and for red stars the difference can be very significant, since the ultraviolet and the infrared ranges of frequency are both excluded from the visual luminosities. Contours of equal radius, using the solar radius as unit, are shown in the visual-luminosity plot of Figure 7.16. A comparison with Figure 6.23 makes clear the effect of changing from total luminosities to visual luminosities.

The second point concerns the distances of the stars given in Figure 7.15. Distances must be known in order for observed fluxes to be converted into intrinsic luminosities. The distances were mostly determined by the trigonometric method already described in the early part of Chapter 2, but some

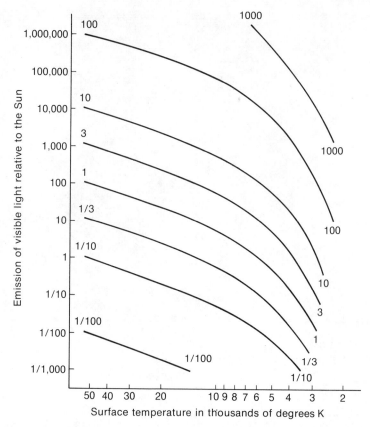

FIGURE 7.16.

Contours of equal stellar radii with the solar radius as unit. These contours are
different from those of Figure 6.23 because the lefthand scale is for radiation at
visual frequencies only, but for radiation of all frequencies in Figure 6.23.

were obtained by other methods, a discussion of which will be the main subject
of Chapter 8. If we anticipate that discussion a bit more, even more striking
evidence for the evolution of stars away from the main sequence, and into the
region of the giants, can be given. Figure 7.17 shows the result of superimposing
the different regions of the luminosity-color diagram occupied by the stars of
several open clusters. The effect of evolutionary funneling into the giant region
is clearly shown from this side-by-side comparison. The clusters represented
here are not by any means of the same age, as can be seen from the approximate
ages marked at the right-hand margin of the diagram.

It should be noted that the luminosities of the giant members of the cluster
h and χ Persei are less than those of the members that lie near the main sequence
because here we are using visual luminosities; most of the radiation from these
giants lies in the infrared and is not included in the visual luminosities.

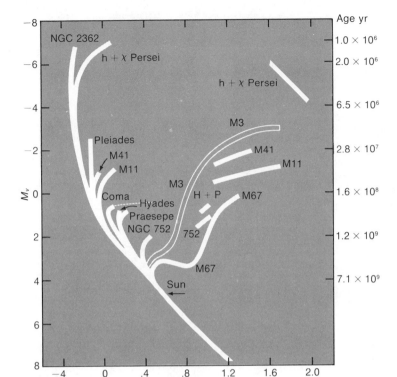

FIGURE 7.17.
A composite representation of stars belonging to a number of open clusters (solid lines) and to the globular cluster M3. (From A. R. Sandage, *Astrophysical Journal*, vol. 125, 1957, p. 436. University of Chicago Press.)

§7.9. *The Ultimate Stellar Problem*

By the time a star evolves away from the main sequence, it is already in its death throes. Although it is true that the general trend of evolution is from the main sequence and toward the giant region of the color-luminosity diagram, the detailed behavior of an individual star can be very complex, with sudden movements both up and down and left and right, as if the star were searching for some resting place. Figure 7.18 shows in schematic form a part of the path followed by a solar-type star.

If we return to Figure 7.7, which gave the energy yield per nuclear particle, B/A, plotted against the mass number A, it is again important to notice that much the largest possible yield of nuclear energy comes from the conversion of hydrogen to helium. Once a star begins to exhaust this possibility, most of its energy availability has gone. So long as the star contains nuclei with mass numbers appreciably less than 60, the region of the broad maximum of

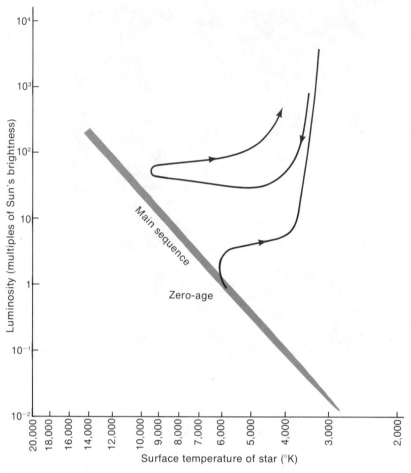

FIGURE 7.18.
The schematic track of a solar mass star.

Figure 7.7, it can generate energy in other ways, but they are more complex and less effective than the simple conversion of hydrogen to helium. What stars are doing in following paths like that of Figure 7.18 or like that of Figure 7.19 for stars of 7 ⊙ and 9 ⊙, is seeking these alternative possibilities. In Table 7.3 these further energy-producing processes, and typical temperatures at which they occur in the stars, are outlined. At the end of the table we have arrived at $A = 56$, which lies close to the maximum of Figure 7.7, and so indeed we have now arrived at the end of the road for nuclear-energy availability. At the end of Table 7.3 the material has been converted into nuclear "ashes."

It will also be seen that the temperature values in the righthand column of Table 7.3 rise as nuclei of larger and larger mass numbers are involved. Higher

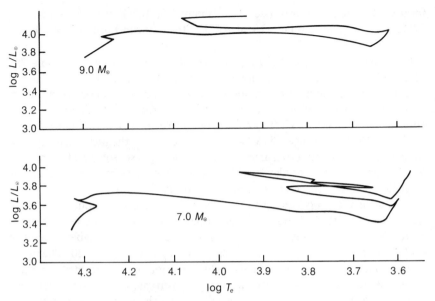

FIGURE 7.19.
Schematic tracks of massive stars. (The track for 9 ⊙ is from I. Iben, Jr., *Astrophysical Journal,* vol. 140, 1964, p. 1631. University of Chicago Press. The track for 7 ⊙ is from E. Hofmeister, R. Kippenhahn, and A. Weigert, *Stellar Evolution,* Stein and Cameron, eds., New York, Plenum Press, 1966.)

TABLE 7.3.
Burning of light elements

Process	Typical temperature of operation (°K)
$3\ ^4He \longrightarrow\ ^{12}C$	2×10^8
$^{12}C +\ ^4He \longrightarrow\ ^{16}O$	2×10^8
$2\ ^{12}C \longrightarrow\ ^4He,\ ^{20}Ne,\ ^{24}Mg$	8×10^8
$2\ ^{16}O \longrightarrow\ ^4He,\ ^{28}Si,\ ^{32}S$	1.5×10^9
$2\ ^{28}Si \longrightarrow\ ^{56}Ni$	3.5×10^9

speeds of motion are required, since the electrical forces increase with the mass number, in order that the nuclei can approach close enough to each other for the short-range nuclear forces to produce a general reordering of the protons and neutrons of which they are composed. For example, in the reaction $^{12}C +\ ^{12}C$, we have two similar nuclei approaching each other containing six protons, whereas for $^{16}O +\ ^{16}O$ the nuclei each have eight protons. The electrical forces are greater therefore for "oxygen-burning" than for "carbon-burning," and a higher temperature is needed to give higher speeds for the former. The temperatures rise from about $2 \times 10^8 \, ^\circ K$ for helium-burning, $3 \, ^4He \longrightarrow \, ^{12}C$, to about $3.5 \times 10^9 \, ^\circ K$ for silicon-burning, $2 \, ^{28}Si \longrightarrow \, ^{56}Ni$.

Only the main effects of these processes of fusion are shown in Table 7.3. Many other reactions of lesser importance also take place. Particularly during silicon-burning, many of the elements ranging in mass number from sulfur to nickel are produced. We shall return to such processes of *element synthesis* later in this chapter. Here we simply notice that by the two decays, $^{56}Ni \longrightarrow \, ^{56}Co + e^+ + \nu_e$, $^{56}Co \longrightarrow \, ^{56}Fe + e^+ + \nu_e$, the nickel arising from silicon-burning is converted into ^{56}Fe, the commonest isotope of iron. The iron we find in our everyday world is believed to have been formed in this way, in a stellar furnace at a temperature of $3.5 \times 10^9 \, ^\circ K$.

Several of the processes set out in Table 7.3 can occur simultaneously in the same star, with the process of highest temperature occurring nearest to the center. A situation involving all the stages of Table 7.3, applicable to the late evolution of a star of large mass, say 20 ⊙, is shown in Figure 7.20. The figure is schematic, the radii of the various shells not being drawn accurately to scale, since it is the outer low-density envelope of hydrogen and helium, enormous compared to the inner dense core of silicon, which causes such a star to lie in the giant region of the luminosity-color diagram.

Silicon-burning occurs at the center of the core. Other burning processes occur at the interfaces between one zone and another. Working inward from the surface, hydrogen-burning can occur at the interface between the first and second shells, i.e., between the zones marked H + He, He; helium-burning can occur at the interface between the zones marked He and C + O; carbon-burning at the interface between the C + O and O + Ne + Mg zones; and oxygen-burning at the interface between the latter zone and the core. The different interfaces can vary markedly in their contributions to the nuclear-energy

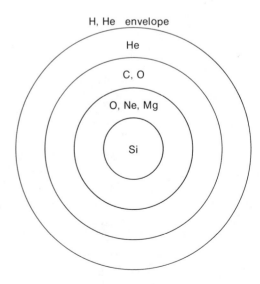

FIGURE 7.20.
The layered structure and varied chemical composition of a star of mass 20 ⊙ at an advanced evolutionary stage.

production, and indeed it is just such variations that cause the star to make sudden changes in the path which it follows in the luminosity-color diagram.

When material burns at an interface between two shells, there is an evolution in which the products of the nuclear reactions become added to the inner shell. Each shell gains material at its outer surface and loses material at its inner surface. Depending on the relative rates of burning at its inner and outer surfaces, a shell may either gain or lose mass. The investigation of such a complex nuclear evolution demands careful and extended calculations which can only be performed with the aid of a powerful digital computer. One cannot easily judge by general intuition how the evolution will develop, or exactly what move the star will next make in the luminosity-color diagram.

But all this "sound and fury" is of no ultimate avail. However complicated the structure a star may adopt, there is no gainsaying the fact that its energy sources are limited, that it tends to exhaust them at an increasing rate, and that the nuclear-energy production must eventually fail to make good the losses from the star into outer space. This lack of finality in the evolution along complex paths like those of Figures 7.18 and 7.19 leads us to ask if there can be any final resting state for the star, any ultimate graveyard in the luminosity-color diagram.

There can be no final static state of this kind so long as a star goes on losing energy into space; so to achieve such a state radiation must stop—the luminosity must die to zero—which means that the temperature of the surface must go to zero. This is an impossible condition to achieve so long as the interior of the star is hot, for then there must always be a steady flow of energy from the interior to the surface, where it will be lost to outer space—contradicting our requirement. It is necessary therefore that the temperature of the interior shall also go to zero. But then how is the interior material to develop a pressure adequate to support the weight of the star? Only if we can resolve this dilemma can there be any final resting place for the star.

From the point of view of classical physics this dilemma was unresolvable, but with the development of modern quantum mechanics (Chapter 4) it was realized that a new form of pressure would arise in cold matter of very high density, of the order of 10^6 grams per cm^3, a ton to a cube the size of a sugar lump. So if a star were to contract to this very high density, a pressure adequate to support the weight of the star might be possible. Such highly compressed stars do in fact exist and are known as *white dwarfs*. Several of them occur among the nearest stars, and were already plotted in Figure 7.15. Here was the graveyard we were seeking (at any rate, one of the graveyards).

PROBLEM:
There are seven white dwarfs among the nearest 100 stars. Explain why, if these 100 stars are considered to be a typical sample of the more numerous kinds of star, there must be on the order of 10^{10} white dwarfs in the whole galaxy.

It is tempting now to argue that, in its final stages of evolution, a star consumes its last nuclear-energy resources and then contracts to a white-dwarf state, cooling gradually, emitting less and less radiation into space, and ultimately ending its life as a cold inert body of exceedingly high density. This line of argument is destroyed, however, by the circumstance that the new form of pressure existing in a white dwarf is not adequate to support the weight if the mass of the star exceeds 1.3 to 1.4 ⊙ (the exact limit here depends on the precise nuclear composition of the star). Although such a scheme could indeed be applied to a solar-mass star, it cannot be used for a star of large mass, like the stars with the evolutionary tracks shown in Figure 7.19.

For many years astronomers thought the *only* way to meet this difficulty would be for a massive star to eject a large fraction of its material back into space, where the material would eventually join the clouds of gas that exist along the plane of the Milky Way. Two arguments could be advanced to support this idea, one theoretical, the other practical. It is a curious fact that the nuclear reactions yielding the most energy, those that convert hydrogen to helium, never lead to an explosive disintegration of a star. Later stages of the nuclear evolution, like the $^{16}O + ^{16}O$ reaction of Table 7.3, are potentially explosive, however. Although delivering less energy in total than the hydrogen-to-helium conversion, oxygen-burning, if it should become unstable, would be capable of suddenly lifting off the outer shells of a star with the structure of Figure 7.20. This would suddenly reduce the mass of the star, perhaps to less than the range 1.3 to 1.4 ⊙, in which case the remaining residue could at last settle down to becoming a white dwarf star.

Practical support for this idea came from observations showing clearly that stars do lose mass by violent ejection processes. This is even true for solar-type

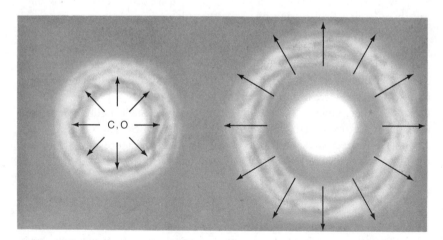

FIGURE 7.21.
A schematic representation of the ejection of the outer envelope of a highly evolved star of solar mass.

stars, the ones that might have reached the white-dwarf state without the need for mass ejection. Solar-type stars go to the white-dwarf state without completing the nuclear evolution of Table 7.3. After a star has zigzagged for a while in the giant region of the luminosity-color diagram, a situation is reached where the central regions consist largely of carbon and oxygen as in Figure 7.21, and where the outer envelope of hydrogen and helium is expelled away from the star. The hot exposed core cools rapidly, following a track in the luminosity-color diagram of the form shown in Figure 7.22. The expulsion of the hydrogen and helium would seem to occur in a number of puffs, rather than as a single explosion. Such a puff of expanding gas is susceptible to observation, since it shines by absorbing light from the parent star. The *planetary nebulae*, of which two examples are shown in Figures 7.23 and 7.24, are thought to arise in this way. Thus the planetary nebulae are a phenomenon

For FIGURE 7.24, see PLATE IX.

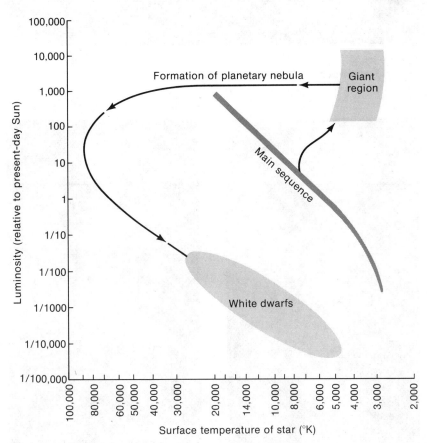

FIGURE 7.22.
Evolution of a star of solar mass to the white-dwarf region of the Hertzsprung-Russell diagram. Planetary nebulae are believed to be formed during the final sweep of the evolutionary track.

338

FIGURE 7.23.
The planetary nebula NGC 7293, about 2 light years in diameter (courtesy of Hale Observatories).

June 1959

May 1972

FIGURE 7.25. and FIGURE 7.26.
Notice the appearance of the bright supernova, believed to be associated with the galaxy (courtesy of Hale Observatories).

associated with the death throes of stars of comparatively small mass. Although spectacular in appearance, they contain rather little gas, a small fraction of $\odot$, and so are not to be thought of as being in any way similar to the bright nebulae that were studied in the preceding chapter, which contained masses of gas exceeding $10^3 \odot$.

Stars of solar type evolve as in Figure 7.22 to a white-dwarf state before oxygen-burning occurs, and so avoid the kind of violently explosive situation that can arise from $^{16}O + ^{16}O$. Stars of much larger mass reach the oxygen-burning phase, however, and are thus exposed to the possibility of a bomblike detonation. Stellar outbursts are in fact observed, as can be seen from the dramatic comparison of Figures 7.25 and 7.26. Such explosions are called *supernovae*. The supernova seen in Figure 7.26, but not in Figure 7.25, is comparable in brightness to the whole of the parent galaxy. This is plainly an outburst of very great intensity, requiring a nuclear instability affecting the whole of one, or more, of the shells of Figure 7.20. A mass of the order of $10 \odot$ has to be concerned in such explosions, the scale of which exceeds by 10^{27} the energy released in a man-made hydrogen bomb. Stars in their death throes are capable of a degree of violence quite outside our terrestrial experience.

This was the basis on which it was argued that explosions, arising out of nuclear instability, enabled stars of initially large mass to reduce their remaining residues to solar order, which then evolved by cooling into white dwarfs. Yet there were causes for disquiet. How were unstable stars able to judge the intensity of explosion with sufficient precision to ensure that the residues always had masses below the white-dwarf limit of 1.3–1.4 $\odot$? Could there be cases in which the star experienced total disintegration, no residue being left at all?

Because of these questions it became imperative to investigate this whole problem by careful calculations using powerful digital computers. Although this program of investigation is still not complete, several conclusions relating to our questions have emerged. After some controversy, it now seems agreed that the picture discussed above is indeed valid for stars of initial mass about 5 $\odot$. A white-dwarf residue seems to be left over after an explosion which adjusts itself appropriately so that the white-dwarf limit of 1.3–1.4 $\odot$ is not exceeded. However, the situation for stars of much larger initial mass, say 20 $\odot$, has turned out to be quite different from the older picture.

The difference for stars of initially large mass depends on a circumstance that was already suspected by 1960, namely, that the core of such a star—the silicon core of Figure 7.20—is too massive and too dense for it ever to become a white dwarf. So even if the outer shells of the star were entirely lost by an explosion, the remaining core could not end its life in the graveyard of the white dwarfs. What has now been found is that such a core evolves to a quite different graveyard, one of a dramatic and remarkable kind.

It will be recalled that the pressure responsible for supporting the weight of a white dwarf cannot be understood in terms of classical physics, because in classical physics there is no motion to supply pressure when the temperature is zero. In quantum physics, however, there is always motion, even though the temperature is zero, because the particles do not follow unique paths—there are always paths having appreciable probability besides the paths corresponding to zero motion. And as the density increases, the paths of one particle correlate with paths of other particles in a way that gives higher and higher probability to those which represent rapid motion. In short, high density forces the particles to move fast, and it is this which generates the pressure needed to support the weight of the star.

In white dwarfs, at densities of order 10^6 grams per cm^3, the fastest-moving particles are electrons, and it is electrons therefore that are mainly responsible for the pressure in these stars. Suppose now we increase the density substantially above 10^6 grams per cm^3. What happens? From what has just been said, we might expect still more rapid electron motions, and consequently a still more marked rise of pressure. This would indeed be correct if the total number of electrons remained always the same, but because of $p + e \longrightarrow n + \nu_e$ this need not be so, and in fact it is not so. For densities rising into the range 10^9 to 10^{10} grams per cm^3, protons and electrons are squeezed into neutrons. It follows that the pressure supplied by the electrons therefore tends to disappear as the density rises. Does the pressure now disappear entirely? No, because exactly the same argument applies also to the motions of the neutrons. The neutrons also supply pressure, although for them to do so effectively requires a very much higher density than 10^6 grams per cm^3. Effective pressure from neutrons, sufficient to withstand the weight of a star, requires densities of the

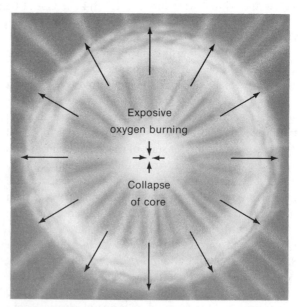

FIGURE 7.27.
Schematic representation of the supernova outburst of a massive star.

fantastic order of 10^{14} grams per cm³; that is, 10^8 tons to a cube the size of a sugar lump.

This excursion into physics gives the clue to what happens ultimately in the problem of a star of large initial mass. The core becomes too massive and of too high a density for evolution to the white-dwarf state to be possible. Instead, after explosion has driven off the outer shells of the star, in the manner of Figure 7.27, the core shrinks very markedly until the density becomes sufficient for the neutron pressure to become important, i.e., sufficient to support the weight of the remaining residue.* When this stage is reached the radius has become tiny—only a few per cent of the radius of the Earth, as indeed was shown in Figure 7.2. The core is said to have then become a *neutron star*.

The concept of a neutron star had already been formulated on theoretical grounds by 1932. The idea that such stars might exist remained a speculation, however, until objects called *pulsars* were discovered in 1967. The pulsars, which we will study later, have turned out to be neutron stars, and are thought to be residues of exploding stars. The pulsar is also a graveyard, but a less placid one—as we will see—than that of the white dwarfs.

§7.10. The History of Matter

At the outset of Chapter 6, it was remarked that the material of the galaxy is believed by astronomers to have been composed initially of hydrogen (and perhaps helium) only. The evidence for this view is that old stars contain much lower concentrations of elements having Z greater than 2 than do younger stars. How have elements heavier than helium been produced? By nuclear processes occurring inside stars, which act to produce fusion of lighter nuclei into heavier ones in accordance with the scheme laid out in Table 7.3.

Subsequent to their production inside stars, the elements are broadcast into space by stellar explosions. We have the cyclical picture of Figure 7.28, in which material is transferred backward and forward between the interstellar gas and the stars. Gas condenses into stars, as we saw in Chapter 6, but material is later returned from the stars to the interstellar gas by the processes we have just been studying.

It is of interest to estimate the amount of material circulated around the loop of Figure 7.28. For this estimate we need to know how often supernovae occur. It has been found that there is about one per 30 years per galaxy, a result coming from the observation of supernovae in many galaxies. Supernovae are very bright indeed, as can be seen from Figure 7.26, so they can be detected even in quite distant galaxies. Hence their average frequency can be determined from a day-by-day patrol of several hundred galaxies. Since there is no reason to think our galaxy is any different from the others in its supernova rate, we can take one per 30 years to be a satisfactory basis for calculating the total number of supernovae that have occurred during the whole life of our galaxy.

---

*It is also possible that the core may collapse into a *black hole*, as the core in the system of Cygnus X-1 may have done (see Appendix VI.3).

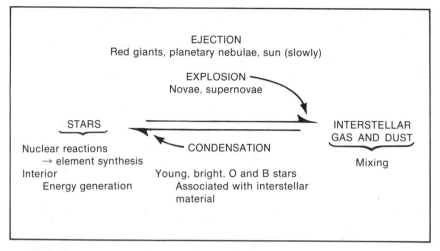

FIGURE 7.28.

A cyclical process in which matter is transferred backward and forward between stars and the interstellar gas.

The age of our galaxy cannot be much different from 10^{10} years, as can be seen from the ages of the oldest clusters of stars plotted in Figure 7.17. Consequently the total number of supernovae that have occurred must be of order $10^{10} \div 30$. If we take $6 \odot$ as a reasonable estimate for the mass ejected by the average supernova, the total mass returned to the interstellar gas then comes out at about $2 \times 10^9 \odot$, which is only 1 or 2 per cent of the mass of the whole galaxy. Our result therefore is that of the order of 2 per cent of the whole galaxy has been circulated by supernovae around the loop of Figure 7.28.

PROBLEMS:

1. The composition of the material ejected by a supernova must be dominated by the outer shells of Figure 7.20. Thus the ejected material must be mostly hydrogen, helium, carbon and oxygen. Taking half of it to be carbon and oxygen, and taking 2 per cent of the galaxy to have been circulated around the loop of Figure 7.28, the total mass of carbon and oxygen produced by supernovae should then be about 1 per cent of the whole mass of the galaxy. From the abundances of the elements given in Table 7.1, show this result to be in agreement with the concentration of carbon and oxygen in the Sun. [Hint: The abundances of Table 7.1 are with respect to the numbers of atoms. Remember to multiply by A in order to obtain abundances with respect to mass.]

2. Some silicon and some of the products of silicon-burning will also be expelled by supernovae, especially from supernovae of large mass. Taking the average ejection of these heavier elements to be one-tenth of the ejection of carbon and oxygen, in total about 0.2 per cent of the mass of the galaxy, show this result also to be in satisfactory agreement with the abundances given in Table 7.1.

The implication of these considerations is that the common materials of our daily world, the carbon that is the basis of life, the oxygen we breathe, the metals we use, all have experienced the cycle of Figure 7.28. They were produced in stellar furnaces at temperatures upward of a billion degrees, and they had already been flung violently into space *before the Sun and planets of our system were formed.* The constituents of our daily world have not been fashioned here *in situ* within the solar system itself, but they were fashioned many aeons ago within the galaxy. The parent stars are by now faint white dwarfs or superdense neutron stars which we have no means of identifying. So here we have the answer to the question in Blake's poem, "The Tyger": "In what furnace was thy brain?" And Voltaire was right when he wrote of "sermons in stones," but for a reason he could not have appreciated. The silicon, magnesium, and oxygen that make up a commonplace piece of stone have a remarkable, and even fantastic, history behind them.

General Problems and Questions

1. Explain the logical reasons why stars must possess energy sources of a nongravitational and nonchemical nature. Why did the views of the scientists of the nineteenth century impede the search for this energy source?

2. What are the elements, and how are they ordered? How many different elements are known?

3. What is radioactivity and how was it discovered?

4. What is an α-particle? What is a β-ray? What is a γ-ray?

5. What was Eddington's suggestion for the main source of the energy of the stars?

6. Describe the electronic shell structure of atoms, and discuss the relation of shell structure to the periodic table of Mendeleev.

7. How does the idea of shell structure help to explain the properties of molecules?

8. What is the relation of the number of electrons in an atom to the number of protons?

9. Why does electrical repulsion not cause the protons in the nucleus of an atom to burst apart? What is the general size relationship of the nucleus to the scale of the electron shells?

10. What is the neutron?

11. Discuss the interrelations, usually referred to as β-processes, between the neutron, proton, electron, and neutrino. Give a general idea of the mass ratios of these particles.

12. What are antiparticles?

13. What are the isotopes of an element?

14. You are given the mass number A of a nucleus. What are the criteria which determine the number of protons Z with respect to which the nucleus is stable? Discuss the cases of A odd and of A even.

15. What is a long-lived β-decay? Give an example.

16. Explain why, for small mass numbers, it is necessary to fuse nuclei to obtain energy, whereas the opposite process of fission is necessary for very large mass numbers.

17. Compare in general terms the energy yield of a nuclear fuel with that of a chemical fuel.

18. Write down in detail the nuclear reactions of the proton-proton chain.

19. Write down in detail the nuclear reactions of the carbon-nitrogen cycle.

20. What criterion in main-sequence stars determines whether the proton-proton chain or the carbon-nitrogen cycle is the more important? How does the over-all structure of a main-sequence star depend on this issue?

21. What is meant by the evolution of a star?

22. Draw a sketch of the Hertzsprung-Russell diagram to illustrate how stars evolve away from the main-sequence. Why is a giant star so named?

23. Give an account of the nuclear reactions which occur in stars after hydrogen has become exhausted in the central regions. What is the general effect of these further nuclear reactions on the evolution of a star?

24. How high can the temperature rise in an evolving star of large mass?

25. What is a white dwarf star, and how does the pressure inside it differ from the form of pressure within the Sun?

26. Describe supernovae, and discuss a sequence of events that could lead to their occurrence.

27. How is it thought that planetary nebulae are formed?

28. How does a neutron star differ from a white dwarf?

29. Write an essay on the history of matter.

Chapter 8:
The Measurement of
Astronomical Distances

The flux f from a luminous object will be taken throughout this chapter to be measurable, provided it is not too small, i.e., is above a certain "threshold." The modern electronic methods for making such measurements were considered in Appendix II.3.

If an object is obscured appreciably by interstellar dust (Chapter 6), we will suppose that we have a suitable way to correct the observed flux, and that this correction has been made in the specification of f. The flux is related to the intrinsic luminosity L and to the distance d by the equation $f = L/(4\pi d^2)$. If both L and d are unknown, not much is to be gained from this equation, but if either L or d happens to be known *from other considerations*, then the equation is immediately useful. If we happen to know d, the luminosity is determined from the observation of f, by $L = (4\pi d^2) \cdot f$. And if we know L, the distance is determined by $d = \sqrt{L/(4\pi f)}$.

§8.1. The Use of the Main Sequence

In Figure 7.15, the nearest stars were denoted by crosses. Their L values plotted in this figure were determined from $L = (4\pi d^2) \cdot f$, with d calculated by the trigonometric method of Chapter 2 (see Figure 2.4), and the color temperatures were obtained by subjecting the light from each star to a frequency analysis, using a spectrograph (Chapter 5, and Appendix II.9). The brightest stars, plotted as circles in Figure 7.15, had their colors determined in the same way. For about half of them, the *L* values were also obtained by using the trigonometric method of Chapter 2 to calculate *d*. The *L* values for the other half were obtained by the different method described below.

TABLE 8.1.
Relation of luminosity and magnitude

Magnitude	Luminosity in solar units
−0.38	100.00
2.12	10.00
4.62	1.00
7.12	.10
9.62	.01

With the main sequence thus delineated, suppose we observe the color and the flux *f* of a star whose distance is unknown. From the color, and also from the discrete absorption lines revealed by the spectroscopic frequency analysis, we may be able to decide that the star belongs to the main sequence. If we can, then we can easily use the observed color to decide where on the main sequence the star must be placed, as indicated in Figure 8.1. The luminosity *L* can now be read off on the lefthand scale, and the distance of the star is then immediately given by $d = \sqrt{L/(4\pi f)}$. The distance in such a case is said to be *spectroscopically* determined. With some care, we can extend this method to stars not of main-sequence type, as has been done for some of the brighter stars of Figure 7.15.

Figure 7.17 showed the luminosity-color distribution for the stars of some open clusters. The colors were again obtained by analyzing the frequency distribution of the light from each individual star. The *L* values were obtained with the aid of the main sequence, but in a somewhat different way from that just discussed. The stars of a cluster all have effectively the same distance *d*, so that their observed flux values immediately give their relative luminosities. This fixes the shape of the star distribution in the luminosity-color diagram, but an ambiguity in the absolute luminosity still remains. The ambiguity can be removed, however, in the manner of Figure 8.2, if we adjust the absolute luminosity level so that the main-sequence part of the cluster distribution falls on the known main sequence, for example, on the main sequence of Figure 7.15. With *L* thus known for every star in the cluster, the distance follows from $d = \sqrt{L/(4\pi f)}$. A similar procedure has been followed for each of the clusters of Figure 7.17.

Here we should consider a problem that arises with the foregoing procedure. Much of the data that makes up the main-sequence diagram of Figure 7.15

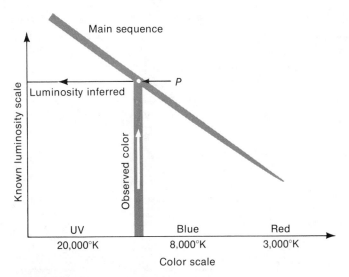

FIGURE 8.1.
The observed color of a main-sequence star is used to determine its
luminosity. A star at point P on the main sequence has the observed
color.

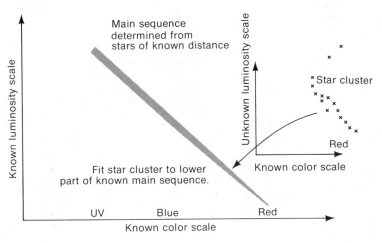

FIGURE 8.2.
The scale of the luminosity distribution of the stars of a cluster is determined by
fitting the lower part of the distribution to the zero-age main sequence.

is subject to some uncertainty, and as a result there is a significant amount
of "play" in the way the curves could fit when one superimposes the main-
sequence portion of a cluster distribution onto Figure 7.15. The imprecision
of Figure 7.15 results from the great ages of many of the nearby stars. Evolution
is somewhat affecting their positions in the luminosity-color diagram—they

are approaching the stage where they will begin evolving away from the main sequence (Chapter 7). Strictly, then, Figure 7.15 does not give the *zero-age* main sequence of Chapter 6. On the other hand, many of the clusters of Figure 7.17 are indeed comparatively young objects. Consequently their main-sequence distributions, at any rate in the lower righthand parts, are quite close to the zero-age condition. Hence the procedure described above does not compare "like with like" in a satisfactorily strict way.

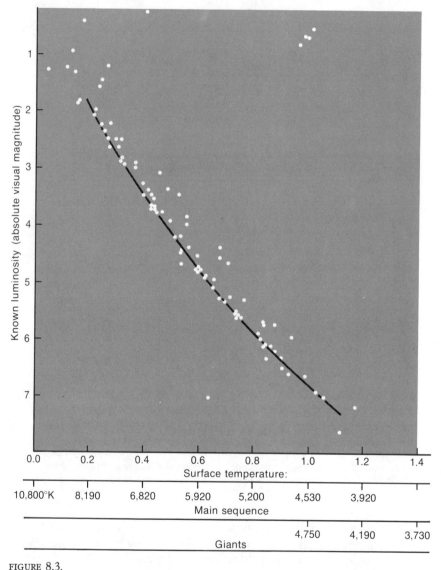

FIGURE 8.3.
The Hyades stars, showing the well-defined lower envelope. This envelope is taken to determine the zero-age main sequence. On the vertical luminosity scale, see Table 8.1.

Astronomers, fortunately, can replace Figure 7.15 by the main-sequence distribution of a group of stars that have a uniform age of about 10^9 years, which is not considered old enough for the lower righthand part of their distribution to have been much affected by evolution. The Sun is luckily close to several comoving groups of stars, of which one—the Hyades—is particularly useful for this purpose. For these stars we can determine distance values by a method analogous to the trigonometric method, and then use $L = (4\pi d^2) \cdot f$ to obtain luminosities. The way in which this is done is explained in Appendix III.1. Here we simply use the main-sequence distribution itself, which is shown in Figure 8.3. It is seen that a lower envelope to the Hyades stars is indeed quite well-defined. The line drawn in Figure 8.3 can now be used to superimpose onto the open-cluster distributions. It is in this way that Figure 7.17 was obtained.

We consider next how Figure 7.17 can be used—i.e., how clusters can be used—in connection with a distance indicator of a very different and remarkable kind. The ideas, which will take a while to develop, can perhaps best be introduced by referring back to the discussion of stellar evolution in Chapter 7.

§8.2. The Cepheid Variables

Evolving stars of appreciable mass, say from 3 ⊙ to 20 ⊙, follow evolutionary paths that zigzag back and forth in the giant region of the luminosity-color diagram, as we saw in Figure 7.19. During these wanderings, stars occasionally develop pulsational instability. This is a dynamic condition in which the radius of the star goes through a long series of periodic oscillations. Such oscillations were mentioned briefly in Chapter 2, and were illustrated in Figure 2.22, repeated here as Figure 8.4.

The outer regions swell in size for a while and then fall back to a condition of minimum radius, after which the outward motion occurs again. Although for some stars the cycles of expansion and contraction are irregular, for other stars the cycles are regular, the time and amount of expansion and contraction being exactly repeated. Although the physical reasons for these pulsations have been understood for about thirty years, it has only been within the last few years, with the aid of computers, that astronomers have been able to examine

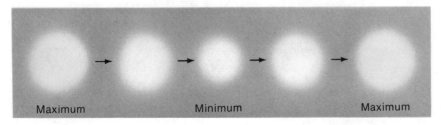

Maximum   Minimum   Maximum

FIGURE 8.4.
Oscillation in size of a Cepheid.

the situation at all thoroughly. The subtleties that turn out to be involved reflect the unusual nature of the whole phenomenon. It is commonly the case that dramatic phenomena of infrequent occurrence are highly complex in their nature, and therefore awkward to analyze in detail.

An important class of stars that show regular pulsations is known as the *Cepheid variables*. These stars play a critical role in astronomical distance measurements, since the Cepheids bridge the gap between measurements of comparatively small distances of a few thousand light years and the first stage of measuring a really large distance, that to the galaxy M31 shown in Figure 6.29. If it had happened that the Cepheids had not existed, an alternative way to bridge this gap could certainly have been found, but it would not have been as accurate as the method of the Cepheids.

During a pulsation, the radius of a Cepheid varies by about 10 per cent, as shown in Figure 8.4. This is considerably less than the range of pulsation of certain other kinds of variable star. The irregularly pulsating star Mira, for example, varies in radius by about 20 per cent. If we draw a diagram like Figure 8.4 for Mira, it would contain the first four planetary orbits of the solar system, as in Figure 8.5. The visual light emitted by Mira changes enormously during its cycle. At its brightest, Mira appears as a red star of 2nd magnitude; at its faintest it cannot be seen at all by the unaided eye. To early astronomers, Mira was an astonishing phenomenon, a star appearing regularly in the sky every eleven months, and then disappearing again! The name itself, Mira, means "The Wonderful." The fact that Mira was so well-known to early atronomers shows the care with which they watched the sky.

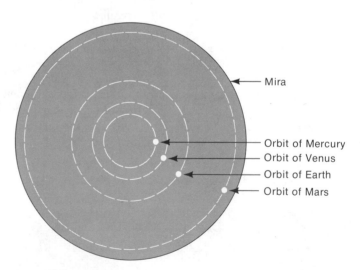

FIGURE 8.5.
The oscillating star Mira at maximum size would contain the orbits of the four inner planets of our solar system, drawn here schematically.

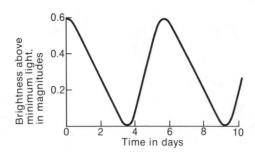

FIGURE 8.6.
The light curve of δ Cephei, with period 5.37
days, measured in yellow light. (From J. C.
Brandt and S. P. Maran, *New Horizons in Astronomy.* W. H. Freeman and Company. Copyright ©
1972.)

Figure 8.6 shows the way in which the light of the Cepheids varies with the oscillation of Figure 8.4. It is from the light variation that Cepheid variables are discovered. Astronomers do not examine stars one by one until a star that shows variation is discovered, however. Two photographic plates of the same region of the sky are taken at different times. Stars with light variation will in general have changed from one time to the other, and will show up immediately to the eye if the two plates are viewed in rapid alternation with each other. Not all the variable stars discovered in this way will turn out to be Cepheids; others will also be found by this simple technique. The Cepheids are those which show regular light variations of the form of Figure 8.6. The time required for a complete oscillation of the light can be measured with considerable accuracy, and is called the *period* of the Cepheid. The period of the star shown in Figure 8.6 is 5.37 days. This is the star δ Cephei, the first to be found, by John Goodricke in 1784. The periods of Cepheid variables range from 1 to 2 days at the short end to more than 100 days at the long end.

Early in this century an epoch-making discovery was made, relating the periods of the Cepheid variables to their intrinsic luminosities. It was found that the longer was the period *P*, the greater was the luminosity *L*. The discovery was made in 1912 by Henrietta Leavitt, an astronomer of the Harvard College Observatory working in South Africa. Two great clouds of stars can be seen in the southern skies, known as the Magellanic Clouds, the "Small Cloud" and the "Large Cloud," shown in Figures 8.7 and 8.8. These objects are not ordinary clusters of stars, but small galaxies in their own right. We know today that both Clouds are situated outside our galaxy, although they are associated with it, as can be seen in Figure 1.28.

The distances of the Magellanic Clouds were not known to Miss Leavitt, however. What was relevant to her observations was that both Clouds contain many Cepheid variables. In each Cloud the variables could all be considered

FIGURE 8.7.
The Small Magellanic Cloud, about 10,000 light years in diameter (courtesy of Dr. V. C. Reddish, Science Research Council).

as being at essentially the same distance away from us. And since there was very little dust to produce obscuration in the Small Cloud (but not in the Large Cloud, which has a good deal of dust), the fluxes of the Small Cloud variables therefore gave the *relative values* of their luminosities. Thus the fact that Miss Leavitt found a relation between the periods P and the fluxes f meant there had to be a relation between P and the intrinsic luminosities L.

Harlow Shapley then realized that, if L could be determined *in an independent way* for any one Cepheid, the whole of the relation discovered by Miss Leavitt could be represented as a relation between P and L. This would mean that an observational determination of P for any Cepheid, whether in the Magellanic Clouds or not, would give L, and hence that the distance d of the Cepheid would follow immediately from $d = \sqrt{L/(4\pi f)}$. The critical problem therefore was to make an independent determination of L for at least one Cepheid, and preferably, of course, for several Cepheids.

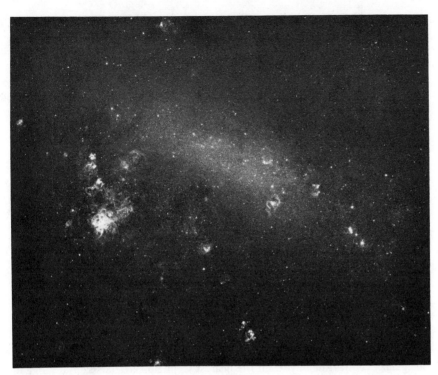

FIGURE 8.8.
The Large Magellanic Cloud, about 30,000 light years in diameter (courtesy of Dr. V. M. Blanco, Cerro Tololo Inter-American Observatory).

Currently, we have found several Cepheids in open clusters, and the distances of these clusters can be determined by using the main sequence, the method shown schematically in Figure 8.2. With the cluster distances d thus known, the L values of the Cepheids follow from $L = (4\pi d^2) \cdot f$, and the problem is solved. The resulting relation between periods and intrinsic luminosities is shown in Figure 8.9. We can therefore calculate the distance of any stellar group that contains a Cepheid variable, provided the group is not so distant that the flux f is too small to permit accurate measurement. The procedure is very simple in principle. Measure f and P. Read off L from Figure 8.9, and work out $d = \sqrt{L/(4\pi f)}$. The logical steps used in establishing this method are:

Lower main sequence (Hyades)
$\longrightarrow$ Open clusters
$\longrightarrow$ Cepheids

The Cepheids used nowadays for establishing the "calibration" of the P-L relation, those in clusters, were not known until the 1950's, when the first of them was discovered by J. B. Irwin. Shapley had to use the trigonometric method to calculate the distances of the nearest few Cepheids to be found in

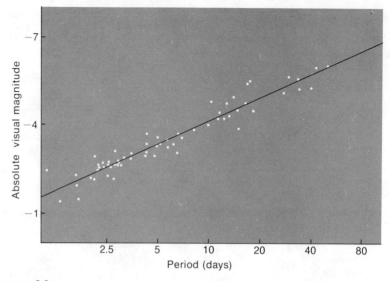

FIGURE 8.9.

The observed relation between the periods and average visual magnitudes of Cepheids. (Adapted from H. Arp, *Astronomical Journal,* vol. 65, 1960, p. 426.)

our galaxy. Unfortunately, even the nearest Cepheids are outside the range of two or three hundred light years within which the trigonometric method can be used with satisfactory accuracy. Nevertheless, in spite of this difficulty, the method proved so powerful that Shapley was able to establish a result of far-reaching importance. He was able to show that our galaxy is very much larger than it had been thought to be. Most astronomers at that time, 1917, thought that our galaxy was only a few thousand light years in size, and that the solar system lay quite near the center, instead of being about 30,000 light years out from the center, as we saw already in Chapter 1. This old erroneous picture came from the refusal of orthodox astronomical opinion to recognize that obscuration by dust occurs along the plane of the galaxy, that is, from supposing that our local "swimming hole" was the entire galaxy.

Within a few years of Shapley's work, another and even greater astronomical revolution occurred. Hubble discovered Cepheids in M31, M33, and NGC 6822, and so was able to calculate distances to galaxies outside our own. This settled, once and for all, a controversy which had developed in the nineteenth century and had raged throughout the first two decades of this century. When Messier compiled his catalogue more than 150 years ago, all prominent "diffuse objects" were listed together, as they were when the New General Catalogue (NGC) was compiled. Thus M42 (NGC 1976) was the Orion Nebula, shown in Figure 6.1, an object within our own galaxy, whereas M31 (NGC 224) was the Andromeda Nebula, a giant galaxy outside our own. This meant that gas clouds illuminated by stars inside our own galaxy were listed alongside external galaxies. All were classified as "nebulae." Although it had been clear from observations made in the latter part of the nineteenth century that a class of nebulae with remarkable spiral forms, like that of Figure 3.10 (M33), could be distinguished among the Messier catalogue of "nebulae," most astronomers continued to think of all such diffuse objects as being simply clouds within our own galaxy. There were a few strong protagonists for the opposite view, that the spiral "nebulae" were really very distant systems like our own Milky Way, notably the Englishman R. A. Proctor in the last century, and H. D. Curtis of the Lick Observatory in this century, but many distinguished men were ranged against them, including Shapley himself. After his pioneering and imaginative work on the scale of our own galaxy, it is curious to find Shapley taking a pedestrian point of view in this other controversy—perhaps, having shown our galaxy to be very large, he had the subconscious hope that everything belonged to it!

PROBLEM:

The distribution of spiral nebulae on the sky shows a zone of avoidance along the plane of the galaxy. This fact was used to support the view that the spiral nebulae were associated with our galaxy, and hence are a part of our galaxy. Where was this argument wrong?

Hubble first announced his proof of the existence of galaxies outside our own in the Halley Lecture for 1925, delivered at Oxford University. An account of the lecture appeared shortly thereafter in the *Observatory* magazine. It is an interesting indication of the leisurely pace of science in those days that Hubble did not trouble to publish the details of his work on M31 until 1929.*

PROBLEM:

A Cepheid variable in a certain "object" has a period of 40 days and an average flux that is 1 part in 10^{18} of that of the Sun, whose absolute visual magnitude is $+4.7$. What is the distance of the object in terms of the Earth's distance from the Sun (which is 1.496×10^{13} cm)? The light year is 9.460×10^{17} cm. What is the distance in light years? What might the object be?

---

*Things are very different nowadays. Instead of delaying the publication of a critical discovery, we seek nowadays to publish instantly, almost before the discovery itself is made. The change of the past fifty years is often attributed to personal ambition: the scientist today faces tougher competition, and so must seek recognition more fiercely than was necessary in the first quarter of the century. From my own experience, I do not believe this to be true. When I was a young assistant professor, without tenure, with a young family to support and very much in need of recognition, I was more relaxed about publishing than I became 20 years later, although by then I had attained the dignity of a full professorship, with tenure and no longer with serious family responsibilities. My own experience suggests that the phenomenon is social rather than personal, and I believe that it results from a deterioration in the way science is financed.

Hubble found that both the galaxies M31 (Figure 6.29) and M33 (Figure 3.10) contained stars that were significantly brighter than the brightest Cepheids. With the distance d of either galaxy known from the Cepheids, he was able to calculate the L value for the brightest stars, using $L = (4\pi d^2) \cdot f$. Observing what seemed to be stars in more distant galaxies, and assuming that the brightest stars in these galaxies would have the same intrinsic L values as the brightest stars in M31 and M33, he was able to calculate their distances from $d = \sqrt{L/(4\pi f)}$. Starting from the Cepheids, the logical chain was therefore:

Cepheids
$\longrightarrow$ Nearest galaxies
$\longrightarrow$ Brightest stars
$\longrightarrow$ More distant galaxies.

The class of "more distant galaxy" reached in this way actually contained about a thousand galaxies, including a cluster of galaxies in the constellation of Virgo. This sample of "more distant galaxy" was then assumed to be large enough that one could pick out a typically "brightest galaxy." The brightest galaxies turned out to be the giant ellipticals discussed in Chapter 3, of the form of M87, shown in Figure 8.10. The brightest galaxies in much more distant rich clusters were then taken to be intrinsically similar to M87 and to the other giant ellipticals of the Virgo cluster. Once again there was an inversion in the use of the flux equation. First, with the distance d obtained to the Virgo cluster of galaxies, and with M87 a member of this cluster, the intrinsic luminosity of M87 was calculated from $L = (4\pi d^2) \cdot f$. Then with the intrinsic luminosity of a still more remote elliptical galaxy taken to be the same as M87, the distance was calculated from $d = \sqrt{L/(4\pi f)}$. Starting from the Cepheids, the chain of logic had become:

Cepheids
$\longrightarrow$ Nearest galaxies
$\longrightarrow$ Brightest stars
$\longrightarrow$ More distant galaxies
$\longrightarrow$ Giant ellipticals
$\longrightarrow$ Still more distant clusters of galaxies.

There has turned out to be a snag in this otherwise extremely imaginative and effective scheme. The objects that Hubble distinguished in the Virgo cluster of galaxies, and which he took to be stars like the brightest stars of M31 and M33, have turned out instead to be bright nebulae, objects rather like the whole

FIGURE 8.10.
The brightest of the galaxies have globular or elliptic forms, like
the well-known galaxy M87 (NGC 4486) in Virgo. Note the
many clusters in the outer regions of M87. (courtesy of Hale
Observatories).

of the Orion Nebula (Figure 6.1). Such gas clouds emit light by absorbing
the radiation, not from a single bright star, but from a whole group of bright
stars contained within them. Hence the objects distinguished by Hubble had
intrinsic luminosities that were more characteristic of a group of stars than
of a single star. As a result, Hubble had underestimated the distance of the
Virgo cluster, and hence all the distances in the final links of the chain of
measurement.

To overcome this difficulty, A. R. Sandage decided to work with bright
nebulae, known technically as H II regions (because the gas within them is
so hot that hydrogen atoms are mostly ionized, the electrons from the hydrogen
atoms are set free), instead of with the brightest stars. He found that M31
and particularly M33 also contained H II regions, and he calculated their
intrinsic luminosities, again from $L = (4\pi d^2) \cdot f$, with d known from the

Cepheids. Assuming H II regions in the more distant galaxies of the Virgo cluster to be similar, Sandage in his early work modified Hubble's scheme to the form:

Cepheids
⟶ Nearest galaxies
⟶ H II regions
⟶ Virgo cluster
⟶ Giant ellipticals of Virgo cluster
⟶ More distant clusters containing elliptical galaxies.

More recently, Sandage has adopted what he feels to be a still better procedure. It has turned out that the brightest members of the Sc class of spiral galaxy (discussed in Chapter 3) are remarkably similar to each other. An example of this class of galaxy is shown in Figure 8.11. With the help of these giant Sc galaxies, one can reach a wider sample of galaxies than the Virgo cluster

FIGURE 8.11.
The galaxy M101 (NGC 5457), a giant Sc about 150,000 light years in diameter (courtesy of Kitt Peak National Observatory).

before determining L values for the brightest giant ellipticals. With this modification included, we now write down the whole logical chain whereby astronomical distances are measured:

Lower main sequence (Hyades)
$\longrightarrow$ Open clusters
$\longrightarrow$ Cepheids
$\longrightarrow$ Nearby galaxies
$\longrightarrow$ H II regions
$\longrightarrow$ Bright Sc galaxies
$\longrightarrow$ Moderately distant galaxies
$\longrightarrow$ Giant ellipticals
$\longrightarrow$ Very distant clusters containing giant ellipticals.

When astronomers assert that such-and-such a galaxy is at a distance of several thousand million light years, their statement rests on the strength of this chain.

One is compelled to ask: how strong is it? There is no generally agreed-on answer to this question. My own impression is that each of the links is quite accurate, by which I mean that I would expect some error, but perhaps not of more than 10 per cent at any one link. Yet with so many links in the whole chain, a cumulative error of 30 per cent might still be present in the final outcome. If this assessment of the situation is not overoptimistic, the achievement of establishing the great distances in the universe to such a degree of accuracy must be judged a very major one indeed.

General Problems and Questions

1. The observed flux f from an astronomical object is related to its distance d and to its luminosity L by the equation $f = L/4\pi d^2$. If L were also known, how would you use this equation to calculate d? Alternatively, if d were known, how would you use the equation to calculate L?

2. Satisfy yourself (by referring back if necessary to Section I) that you understand why a sphere of radius r, situated at a distance d that is large compared to r, has an apparent angular size θ given in degrees by the formula $\theta = 360\, r/(\pi d)$.

3. Use the result of Problem 2 to explain how the Earth's annual motion around the Sun can be used to calculate the distances of nearby stars. At what order of distance does this "trigonometric" method of distance determination become seriously inaccurate?

4. How can the zero-age main sequence be used to calculate distances that go well beyond the range of the trigonometric method?

Appendixes to Section III

Appendix III.1. The Hyades Main Sequence

The reader wishing to understand how to obtain the main sequence of the Hyades stars should recall from Appendix II.8 the meaning of the Doppler shift of the frequencies of the spectrum lines of a star. In a discussion (p. 240) of moving groups of stars, such as the Hyades, it was pointed out that the Doppler shift, together with a measurement of the proper motions of the stars, can be used to determine how such a group is moving with respect to the Sun.

To understand the importance of the measurement of proper motions, consider first the special case in which the line of sight from the solar system is at right angles to the motion of a star, as in Figure III.1. For the moment, let us suppose the speed V to be known. Then the star in question moves through a distance tV in a specified time t, and the line of sight changes, as in Figure III.2, by the small angle $\theta = 180\ tV/(\pi d)$ in degrees, where d is the distance of the star. To see that this formula is correct, imagine the distance

FIGURE III.1.
The special case in which the direction of motion of a star
happens to be at right angles to the line of sight to the Earth.

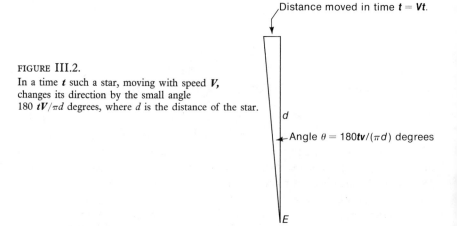

FIGURE III.2.
In a time t such a star, moving with speed V,
changes its direction by the small angle
180 $tV/\pi d$ degrees, where d is the distance of the star.

tV to form a small element along the circumference of a circle of radius d,
as in Figure III.3. The total length of the circumference is $2\pi d$; so the distance
tV is a fraction $tV/(2\pi d)$ of the whole circumference. This means that the
angle θ we are seeking must be a fraction $tV/(2\pi d)$ of the whole angle around
a circle, which is 360°. Hence $\theta = 360\ tV/(2\pi d)$ in degrees. Now suppose θ

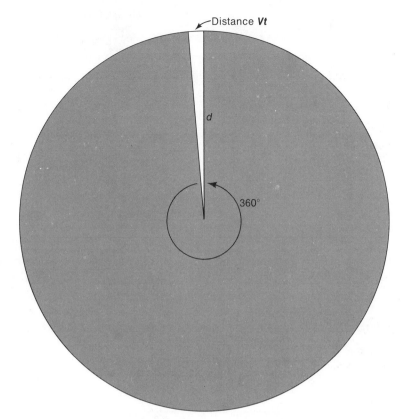

Distance **Vt**

d

360°

FIGURE III.3.
With the distance $t\boldsymbol{V}$ being a fraction $\boldsymbol{V}t/(2\pi d)$ of the circumference of a circle of radius d, the angle subtended by the element at the center of the circle must be $360\ t\boldsymbol{V}/(2\pi d)$ degrees.

is measured, so that we can write $\boldsymbol{\theta} = 180\ t\boldsymbol{V}/(\pi d)$. This immediately determines d, namely, $d = 180\ t\boldsymbol{V}/(\pi\boldsymbol{\theta})$. By measuring the flux $\boldsymbol{f}$ of the star in question, we then have $L = (4\pi\boldsymbol{d}^2)\cdot\boldsymbol{f}$, and the luminosity is obtained.

Only rarely can we expect to find a case in which the line of sight to the solar system happens to be essentially at right angles to the direction of motion of the star. In general, we will have the situation of Figure III.4, with some other angle between the line of sight and the direction of motion. It is important to notice that this angle is always known, however, and therefore that the ratio EN/ES of the lengths EN and ES of Figure III.4 is known. Provided we then replace $\boldsymbol{V}$ in the formula of the previous paragraph by $\boldsymbol{V}(EN/ES)$, the same result holds good, namely, $d = 180\ t\boldsymbol{V}(EN/ES)/\pi\boldsymbol{\theta}$, and the required luminosity is obtained from $L = (4\pi\boldsymbol{d}^2)\cdot\boldsymbol{f}$. This result is not hard to prove by mathematical reasoning, an example of the usefulness of mathematics.

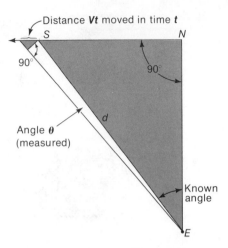

FIGURE III.4.
In general, the line of sight from the Earth is *not* at right angles to the direction of the star's motion. The small heavily marked triangle has the same shape as the large heavily marked triangle.

The speed V was assumed known in the preceding discussion. The Doppler-shift method of Appendix II.8 determines the product of V and the ratio NS/ES. Since this ratio is known, as well as EN/ES, we are therefore justified in considering V to be known. Used for the Hyades group of stars, this method leads to the results of Figure 8.3.

PROBLEM:
Prove the result just stated concerning the ratio EN/ES. [Hint: Use the fact that the small triangle of Figure III.4 has the same shape as the triangle ENS.]

Appendix III.2. The Fission of Heavy Nuclei

Table III.1 gives the abundances of the isotopes of the elements. From this table it can readily be seen that up to $A = 40$ the isotopes found most commonly in nature are: ^{1}H(1p), ^{4}He(2p,2n), ^{7}Li(3p,4n), ^{9}Be(4p,5n), ^{11}B(5p,6n), ^{12}C(6p,6n), ^{14}N(7p,7n), ^{16}O(8p,8n), ^{19}F(9p,10n), ^{20}Ne(10p,10n), ^{23}Na(11p,12n), ^{24}Mg(12p,12n), ^{27}Al(13p,14n), ^{28}Si(14p,14n), ^{31}P(15p,16n), ^{32}S(16p,16n), ^{35}Cl(17p,18n), ^{36}A(18p,18n), ^{39}K(19p,20n), and ^{40}Ca(20p,20n). With the exception of hydrogen, the pattern is the same throughout this list. When the mass number is odd, there is one more neutron than proton. When the atomic number is even, the numbers of neutrons n and of protons p are the same.

The mass of the neutron is about $\frac{1}{7}$ of a per cent greater than that of the proton. Since mass and energy are equivalent quantities, and since nuclei evolve to their state of lowest energy, we might at first sight expect the greater mass of the neutron would cause nuclei to have more protons than neutrons. This expectation is not borne out, however, because there is an electrical effect which overcompensates this mass effect. For example, in ^{19}Ne(10p,9n) the 10 protons have more electrical energy than the 9 protons of ^{19}F, since 10 protons generate a stronger electrical field than 9 protons do. So ^{19}Ne actually has more energy than ^{19}F, even though the extra neutron in ^{19}F has more mass energy than the extra proton in ^{19}Ne. From the curve of Figure 7.4, we see that ^{19}Ne must therefore decay to ^{19}F,

$$^{19}\text{Ne}(10p,9n) \longrightarrow \, ^{19}\text{F}(9p,10n) + e^+ + \nu_e.$$

The characteristic time for one half the ^{19}Ne to decay is about 18 seconds, very much faster than the times of the naturally occurring decays of ^{40}K and ^{87}Rb, meaning that the points on the curve of Figure 7.4 representing ^{19}Ne and ^{19}F have quite a wide energy spacing.

The effect of the energy of the electrical field generated by the protons becomes more and more marked as the mass number A increases above 40, causing the difference between the number Z of protons and the number $A - Z$ of neutrons to increase. For the commonest isotope of iron, ^{56}Fe, $Z = 26$, $N = 30$, whereas for the isotope ^{238}U of uranium, $Z = 92$, $N = 146$. Thus the electrical effects become severe for elements at the upper end of the periodic table. It is indeed the effect of the electrical field of the protons that causes the B/A curve of Figure 7.7 to possess the broad maximum at about $A = 60$. If it were not for the electrical effects, this curve would go on rising slowly with increasing A, in which case it would be possible to obtain further energy by fusion processes going past mass number 50 or 60. The "burning" of light nuclei by fusion to nuclear "ashes" in the iron peak, by processes discussed in Chapter 7, is a consequence therefore of the electrical fields of the protons in nuclei, in the sense that without the energy of these fields there would not be an iron peak, and our everyday world would not contain its large concentrations of iron, cobalt, copper, nickel, chromium, manganese, and zinc. Instead, we would have much more of the heavier elements, with the consequence that platinum and gold would not have their present-day rarity.

Omitting details among the light elements, the curve of Figure 7.7 is shown in schematic form in Figure III.5. A nucleus of very large A can evidently be made to yield energy by breakup into two nuclei of approximately comparable mass numbers, say, A_1 and A_2, with $A_1 + A_2 = A$. Such a process is called *nuclear fission*. All sufficiently heavy nuclei are unstable and undergo fission, as can be seen from Figure III.6, which plots on a logarithmic scale the characteristic times for fissional breakup of nuclei with mass numbers between 232 and 256. Of these, ^{232}Th, ^{235}U, ^{238}U, occur naturally on the Earth.* The others plotted in Figure III.6 were produced experimentally in the laboratory. The fissional time-scales for the three naturally occurring nuclei are much longer than the decay times for the emission of α particles (helium nuclei), which are 1.41×10^{10} years for ^{232}Th, 7.1×10^8 years for ^{235}U, and 4.51×10^9 years for ^{238}U. However, for ^{250}Cm and ^{254}Cf the situation is reversed; the fission time-scales, 1.7×10^4 years and 60.5 days, respectively, are

---

*A trace of ^{244}Pu also occurs naturally.

366

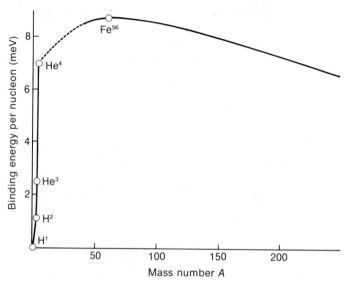

FIGURE III.5.
The energy yield curve of Figure 7.7 reproduced in schematic form.

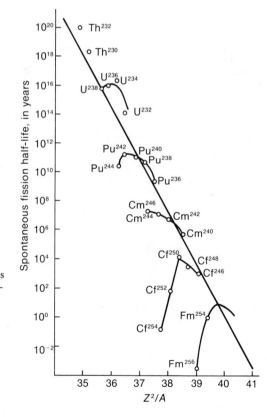

FIGURE III.6.
The fissional decay times for nuclei with atomic masses between 230 and 255. (From E. Segré, *Nuclei and Particles*. Copyright © 1964 by W. A. Benjamin, Inc., Menlo Park, Calif.)

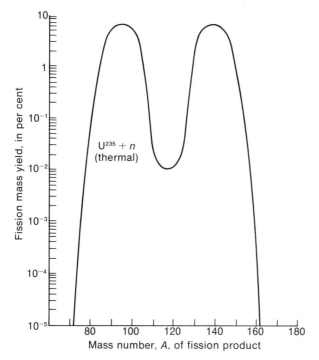

FIGURE III.7.
The distribution of fission products from the decay of ^{236}U.

shorter than the decay times for the emission of α particles. And this is believed to be the general rule for nuclei with A greater than about 260. Such very heavy nuclei decay in general by fission rather than by the emission of α particles.

Although under natural conditions fission is unimportant for ^{232}Th, ^{235}U, ^{238}U, these nuclei can be made to undergo fission by adding a further neutron to them, ^{232}Th $+$ n $\longrightarrow$ ^{233}Th, ^{235}U $+$ n $\longrightarrow$ ^{236}U, ^{238}U $+$ n $\longrightarrow$ ^{239}U, all leading to essentially immediate fissional instability, provided the added neutron is fast-moving. For slowly moving neutrons, on the other hand, only ^{235}U $+$ n $\longrightarrow$ ^{236}U gives fission. For this last process, the distribution of fission products has the form shown in Figure III.7, from which it is clear that fission does not occur in one special way; many breakup modes are possible, although the mass numbers A_1 and A_2 of the fission fragments must always add to 236.

A very critical point concerning the immediate fragments A_1 and A_2 is that, in general, they are themselves unstable and emit one or more neutrons: $A_1 \longrightarrow (A_1 - 1) + $ n, $A_2 \longrightarrow (A_2 - 1) + $ n, $(A_1 - 1) \longrightarrow (A_1 - 2) + $ n, $(A_2 - 1) \longrightarrow (A_2 - 2) + $ n, where $(A_1 - 1)$ is a nucleus of mass number $A_1 - 1$, and so on. On the average about 2.5 neutrons are released in the fission of an atom of ^{236}U. This neutron emission is illustrated in Figure III.8.

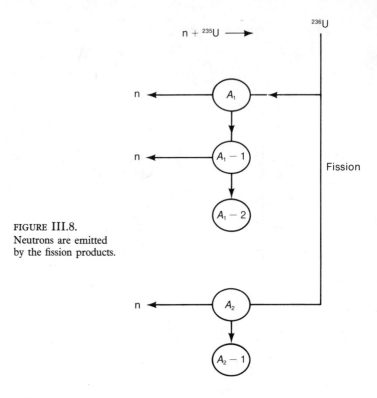

FIGURE III.8.
Neutrons are emitted
by the fission products.

If now these further neutrons are used to induce more fission, as in Figure III.9, the possibility exists for an amplifying cycle. Such an amplifying cycle occurs in the explosion of a uranium bomb. If, instead of the unrestrained situation of Figure III.9, the neutrons are first slowed down (moderated), and their supply to the ^{235}U is accurately controlled, as is indicated in Figure III.10, the cycle can be arranged to be precisely self-maintaining, neither growing explosively nor dying away, in which case we have the principle of the *nuclear reactor*. The fragments A_1 and A_2 are fast-moving, as also are the neutrons from $A_1 \longrightarrow (A_1 - 1) + n$, etc., and the energy of this motion is converted to heat when the fragments and the neutrons slow down, as they eventually do because of collisions with other particles. Consequently fission yields energy in the form of heat.

Because ^{232}Th, ^{238}U, are not made fissionably unstable by slow-moving neutrons, the energy-producing scheme of Figure III.10 cannot be used for these other two naturally occurring nuclei. Table III.1 shows that ^{235}U comprises less than 1 per cent of naturally occurring uranium. The characteristic decay time of ^{235}U for the emission of an α particle, 7.1×10^8 years, is significantly less than the corresponding decay time of ^{238}U, 4.51×10^9 years, whereas the decay time of ^{232}Th is the longest of all, 1.41×10^{10} years. Because of its shorter decay time, most of the ^{235}U initially present on the Earth has by now disappeared, whereas most of the ^{232}Th, and much of the ^{238}U, remains.

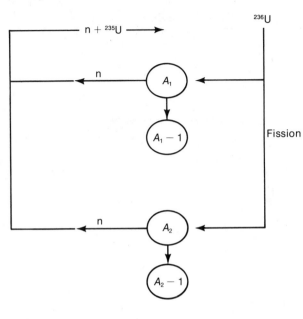

FIGURE III.9.
The neutrons from the fission products of ^{236}U can be used to induce more fission, leading to an amplifying cycle.

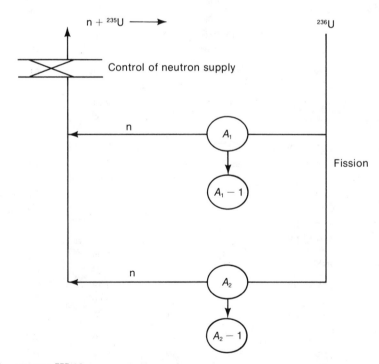

FIGURE III.10.
By introducing control of the neutrons from the fission products of ^{236}U, a non-explosive cycle can be maintained.

TABLE III.1.
Abundances of nuclei

Element	A	% Abundance	Abundance
1 H	1	~ 100	3.18×10^{10}
	2		5.2×10^5
2 He	3		$\sim 3.7 \times 10^5$
	4	~ 100	2.21×10^9
3 Li	6	7.42	3.67
	7	92.58	45.8
4 Be	9	100	0.81
5 B	10	19.64	68.7
	11	80.36	281.3
6 C	12	98.89	1.17×10^7
	13	1.11	1.31×10^5
7 N	14	99.634	3.63×10^6
	15	0.366	1.33×10^4
8 O	16	99.759	2.14×10^7
	17	0.0374	8040
	18	0.2039	4.38×10^4
9 F	19	100	2450
10 Ne	20	(88.89)	3.06×10^6
	21	(0.27)	9290
	22	(10.84)	3.73×10^5
11 Na	23	100	6.0×10^4
12 Mg	24	78.70	8.35×10^5
	25	10.13	1.07×10^5
	26	11.17	1.19×10^5
13 Al	27	100	8.5×10^5
14 Si	28	92.21	9.22×10^5
	29	4.70	4.70×10^4
	30	3.09	3.09×10^4
15 P	31	100	9600
16 S	32	95.0	4.75×10^5
	33	0.760	3800
	34	4.22	2.11×10^4
	36	0.0136	68
17 Cl	35	75.529	4310
	37	24.471	1390

TABLE III.1. (*continued*)

Element	A	% Abundance	Abundance
18 Ar	36	84.2	9.87×10^4
	38	15.8	1.85×10^4
	40		~ 20 ?
19 K	39	93.10	3910
	40		5.76
	41	6.88	289
20 Ca	40	96.97	6.99×10^4
	42	0.64	461
	43	0.145	105
	44	2.06	1490
	46	0.0033	2.38
	48	0.185	133
21 Sc	45	100	35
22 Ti	46	7.93	220
	47	7.28	202
	48	73.94	2050
	49	5.51	153
	50	5.34	148
23 V	50	0.24	0.63
	51	99.76	261
24 Cr	50	4.31	547
	52	83.7	1.06×10^4
	53	9.55	1210
	54	2.38	302
25 Mn	55	100	9300
26 Fe	54	5.82	4.83×10^4
	56	91.66	7.61×10^5
	57	2.19	1.82×10^4
	58	0.33	2740
27 Co	59	100	2210
28 Ni	58	67.88	3.26×10^4
	60	26.23	1.26×10^4
	61	1.19	571
	62	3.66	1760
	64	1.08	518
29 Cu	63	69.09	373
	65	30.91	167

TABLE III.1. (*continued*)

Element	A	% Abundance	Abundance
30 Zn	64	48.89	608
	66	27.81	346
	67	4.11	51.1
	68	18.57	231
	70	0.62	7.71
31 Ga	69	60.4	29.0
	71	39.6	19.0
32 Ge	70	20.52	23.6
	72	27.43	31.5
	73	7.76	8.92
	74	36.54	42.0
	76	7.76	8.92
33 As	75	100	6.6
34 Se	74	0.87	0.58
	76	9.02	6.06
	77	7.58	5.09
	78	23.52	15.8
	80	49.82	33.5
	82	9.19	6.18
35 Br	79	50.537	6.82
	81	49.463	6.68
36 Kr	78	0.354	0.166
	80	2.27	1.06
	82	11.56	5.41
	83	11.55	5.41
	84	56.90	26.6
	86	17.37	8.13
37 Rb	85	72.15	4.16
	87		1.72
38 Sr	84	0.56	0.151
	86	9.86	2.65
	87		1.77
	88	82.56	22.2
39 Y	89	100	4.8
40 Zr	90	51.46	14.4
	91	11.23	3.14
	92	17.11	4.79
	94	17.40	4.87
	96	2.80	0.784

TABLE III.1. (*continued*)

Element	A	% Abundance	Abundance
41 Nb	93	100	1.4
42 Mo	92	15.84	0.634
	94	9.04	0.362
	95	15.72	0.629
	96	16.53	0.661
	97	9.46	0.378
	98	23.78	0.951
	100	9.63	0.385
44 Ru	96	5.51	0.105
	98	1.87	0.0355
	99	12.72	0.242
	100	12.62	0.240
	101	17.07	0.324
	102	31.61	0.601
	104	18.58	0.353
45 Rh	103	100	0.4
46 Pd	102	0.96	0.0125
	104	10.97	0.143
	105	22.23	0.289
	106	27.33	0.355
	108	26.71	0.347
	110	11.81	0.154
47 Ag	107	51.35	0.231
	109	48.65	0.219
48 Cd	106	1.215	0.0180
	108	0.875	0.0130
	110	12.39	0.124
	111	12.75	0.189
	112	24.07	0.356
	113	12.26	0.181
	114	28.86	0.427
	116	7.58	0.112
49 In	113	4.28	0.008
	115	95.72	0.181
50 Sn	112	0.96	0.0346
	114	0.66	0.0238
	115	0.35	0.0126
	116	14.30	0.515
	117	7.61	0.274
	118	24.03	0.865

TABLE III.1. (*continued*)

Element	A	% Abundance	Abundance
	119	8.58	0.309
	120	32.85	1.18
	122	4.72	0.170
	124	5.94	0.214
51 Sb	121	57.25	0.181
	123	42.75	0.135
52 Te	120	0.089	0.0057
	122	2.46	0.158
	123	0.87	0.056
	124	4.61	0.296
	125	6.99	0.449
	126	18.71	1.20
	128	31.79	2.04
	130	34.48	2.21
53 I	127	100	1.09
54 Xe	124	0.126	0.00678
	126	0.115	0.00619
	128	2.17	0.117
	129	27.5	1.48
	130	4.26	0.229
	131	21.4	1.15
	132	26.0	1.40
	134	10.17	0.547
	136	8.39	0.451
55 Cs	133	100	0.387
56 Ba	130	0.101	0.00485
	132	0.097	0.00466
	134	2.42	0.116
	135	6.59	0.316
	136	7.81	0.375
	137	11.32	0.543
	138	71.66	3.44
57 La	138		0.00041
	139	99.911	0.445
58 Ce	136	0.193	0.00228
	138	0.250	0.00295
	140	88.48	1.04
	142	11.07	0.131
59 Pr	141	100	0.149

TABLE III.1. (*continued*)

Element	A	% Abundance	Abundance
60 Nd	142	27.11	0.211
	143	12.17	0.0949
	144	23.85	0.186
	145	8.30	0.0647
	146	17.22	0.134
	148	5.73	0.0447
	150	5.62	0.0438
62 Sm	144	3.09	0.00698
	147		0.0349
	148	11.24	0.0254
	149	13.83	0.0313
	150	7.44	0.0168
	152	26.72	0.0604
	154	22.71	0.0513
63 Eu	151	47.82	0.0406
	153	52.18	0.0444
64 Gd	152	0.200	0.000594
	154	2.15	0.00639
	155	14.73	0.0437
	156	20.47	0.0608
	157	15.68	0.0466
	158	24.87	0.0739
	160	21.90	0.0650
65 Tb	159	100	0.055
66 Dy	156	0.0524	0.000189
	158	0.0902	0.000325
	160	2.294	0.00826
	161	18.88	0.0680
	162	25.53	0.0919
	163	24.97	0.08099
	164	28.18	0.101
67 Ho	165	100	0.079
68 Er	162	0.136	0.000306
	164	1.56	0.00351
	166	33.41	0.0752
	167	22.94	0.516
	168	27.07	0.0609
	170	14.88	0.0335
69 Tm	169	100	0.034

TABLE III.1. (*continued*)

Element	A	% Abundance	Abundance
70 Yb	168	0.135	0.000292
	170	3.03	0.00654
	171	14.31	0.0309
	172	21.82	0.0471
	173	16.13	0.0348
	174	31.84	0.0688
	176	12.73	0.0275
71 Lu	175	97.41	0.0351
	176		0.00108
72 Hf	174	0.18	0.00038
	176	5.20	0.0109
	177	18.50	0.0389
	178	27.14	0.0570
	179	13.75	0.0289
	180	35.24	0.0740
73 Ta	180	0.0123	0.00000258
	181	99.9877	0.0210
74 W	180	0.135	0.000216
	182	26.41	0.0422
	183	14.40	0.0230
	184	30.64	0.0490
	186	28.41	0.0454
75 Re	185	37.07	0.0185
	187		0.0341
76 Os	184	0.018	0.000135
	186	1.29	0.00968
	187		0.0088
	188	13.3	0.0998
	189	16.1	0.121
	190	26.4	0.198
	192	41.0	0.308

TABLE III.1. (*continued*)

Element	A	% Abundance	Abundance
77 Ir	191	37.3	0.267
	193	62.7	0.450
78 Pt	190	0.0127	0.000178
	192	0.78	0.0109
	194	32.9	0.461
	195	33.8	0.473
	196	25.3	0.354
	198	7.21	0.101
79 Au	197	100	0.202
80 Hg	196	0.146	0.000584
	198	10.2	0.0408
	199	16.84	0.0674
	200	23.13	0.0925
	201	13.22	0.0529
	202	29.80	0.119
	204	6.85	0.0274
81 Tl	203	29.50	0.0567
	205	70.50	0.135
82 Pb	204	1.97	0.0788
	206	18.83	0.753
	207	20.60	0.824
	208	58.55	2.34
83 Bi	209	100	0.143
90 Th	232	100	0.058
92 U	235		0.0063
	238		0.0199

This table is taken from A. G. W. Cameron, *Space Science Reviews* 15 (1970), 121.

It is unfortunate, economically, that the most convenient of the three is also the one that is least abundant. To this we must add the further awkward point that it is technically difficult to separate ^{235}U from the much more abundant ^{238}U.

Possibilities exist for overcoming these difficulties. If the neutrons from A_1, A_2, were not slowed, ^{238}U would be fissionally unstable. However, fast-moving neutrons are hard to control, so alternatively we might seek to take

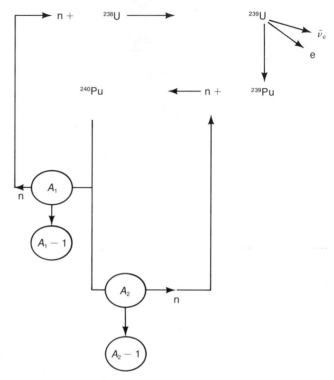

FIGURE III.11.
A fission cycle based on ^{238}U.

advantage of the fact that ^{239}U, produced from ^{238}U $+$ n $\longrightarrow$ ^{239}U, itself undergoes decay by emitting a β-ray; ^{239}U $\longrightarrow$ ^{239}Pu $+$ e $+$ $\bar{\nu}_e$. The importance of this transition is that the nuclear properties of ^{239}Pu are similar to those of ^{235}U. Thus the ^{240}Pu produced by *slow* neutrons, n $+$ ^{239}Pu $\longrightarrow$ ^{240}Pu, undergoes fission like ^{236}U. Hence we can contemplate the scheme of Figure III.11. Notice that two neutrons are required to maintain the cycle of Figure III.11, compared to only one neutron for that of Figure III.10. A fission cycle based on ^{235}U yields spare neutrons, whereas there are little to spare in the cycle of Figure III.11.

How much energy can be produced by these processes? A reactor based on Figure III.10 yields about 7×10^7 kilowatts from the consumption of one gram per second of ^{235}U. The energy requirement of modern industrial society is about 5×10^9 kilowatts. Hence a consumption of some 70 grams of ^{235}U per second would be needed to supply the raw energy demands of society. In the course of a year about 2,000 tons of ^{235}U would therefore be required. Since rather less than 1 per cent of all uranium consists of ^{235}U, total uranium production would need to be about 200,000 tons per year, if only the ^{235}U were used in the manner of Figure III.10. Such a uranium supply is outside the mining capability of the Earth to maintain for more than a few years. On the other hand, for a reactor based on ^{238}U, the uranium requirement would be reduced back to 2,000 tons per year, which could be sustained from high-grade uranium ores for several centuries.

In principle, a supply of 2,000 tons of uranium per year could be maintained for an essentially unlimited future period if the uranium were extracted from ordinary rock rather than from high-grade ores. Uranium makes up about one or two parts per million of the granite rocks of the whole of the Earth's crust; so mining some 10^9 tons per year of granite could yield the needed amount of uranium. Strip mining a few square kilometers a year to a depth of 100 meters would be adequate, and would hardly create a serious environmental problem. The main technological difficulty would lie in extracting the small uranium content from a considerable mass of rock.

Perhaps the main problem with such an energy scheme lies in potential long-term side effects. Large quantities of radioactive wastes in the form of the nuclei A_1, A_2, and of their products after neutron emission, $A_1 \longrightarrow (A_1 - 1) + n$, etc., must necessarily be produced, of the order of 1,000 tons per year of it. Most of these resulting nuclei are unstable and emit β-rays, $n \longrightarrow p + e + \bar{\nu}_e$ occurring within the nuclei. Usually such β-processes are completed within a few years, and if all of them were, the problem of the disposal of radioactive wastes would probably not be serious. However, for some of the nuclei the situation illustrated in Figure 7.5 occurs, with the characteristic decay time for emission of β-rays being thousands, or even hundreds of thousands, of years. A real social danger could lie in a progressive accumulation year by year of these long-lived unstable nuclei. At first the danger would be small, but as millenium followed millenium the amount of potentially harmful material would pile up inexorably in larger and larger amounts, possibly to a point of final disaster for all life on the Earth.

An interesting suggestion for overcoming this difficulty is, first, to separate out the dangerous long-lived decay products from the short-lived ones, and then to blast the long-lived stuff out from the Earth into distant regions of the solar system. This could conceivably be an acceptable solution to the problem, and indeed could eventually turn out to be the most important application of space technology. So far as I can see, there is no reasonable environmental objection to this proposal, since dangerous radiation of natural origin already exists outside the Earth, and in an intensity that Man could not remotely equal.

TOPICS FOR DISCUSSION:
Compare these difficulties with those involved in obtaining energy from the fusion of light nuclei. What is the weight of coal and oil burned per year? What is the amount of iron ore mined per year?

The radioactive series starting with ^{238}U and ending at ^{206}Pb was set out in detail in Table 7.2. There are very short-lived stages in this series. The characteristic time for half of all radon nuclei, ^{222}Rn, to decay to the polonium isotope ^{218}Po is 3.8 days, but half the ^{218}Po will decay in a characteristic time of only 3 minutes. It may be wondered how in these circumstances elements like uranium came to be formed in the first place. Suppose that we painstakingly tried to add together individual protons and neutrons, proceeding along the series of Table 7.2, but in a reversed sense. A difficulty would arise at nuclei like ^{218}Po, because they would break down essentially as soon as they were formed. It would be better to seek an explanation in terms of an inversion of the fission process, by joining nuclei with mass numbers A_1, A_2, of order 120, to produce a compound nucleus with mass number $A = A_1 + A_2$ of order 240. However, the electrical repulsion between two such nuclei is so strong that this kind of large mass fusion cannot occur in any astrophysical environment that can reasonably be contemplated. We must look therefore to a different form of process.

Let us begin by considering the addition of a neutron to the most abundant isotope of carbon, ^{12}C + n $\longrightarrow$ ^{13}C, thereby producing the less abundant isotope ^{13}C. Then add a further neutron, ^{13}C + n $\longrightarrow$ ^{14}C. It makes no difference how quickly these two neutrons are added, because both ^{12}C and ^{13}C are stable—they will stay around for any length of time waiting for the neutrons to come along. But the time does matter if we attempt to add a third neutron to ^{14}C, since ^{14}C is not stable; it decays by ^{14}C $\longrightarrow$ ^{14}N + e + $\bar{\nu}_e$ in a characteristic time of about 5,700 years. If a third neutron were made available in a time appreciably less than 5,700 years, the ^{14}C will still be around and then indeed we will have ^{14}C + n $\longrightarrow$ ^{15}C. But if the third neutron were not to come along for much more than 5,700 years, the ^{14}C would have decayed to ^{14}N. What then happens is that the third neutron is "wasted," in the sense that mass number 15 is not formed. Instead we have ^{14}N + n $\longrightarrow$ ^{14}C + p.

This example brings out the important point that if we attempt to build nuclei of increasing A simply by repeatedly adding neutrons, and so avoiding electrical effects, since the neutron is uncharged, we must specify the time-scale within which the neutrons are made available. The outcome of such an addition depends on the time-scale of the addition. The same point arises in a somewhat different way if, instead of choosing ^{12}C as our starting nucleus, we choose ^{56}Fe, the most abundant nucleus in the so-called iron peak of the abundance curve of Figure 7.9.

The first three steps in this second example are ^{56}Fe + n $\longrightarrow$ ^{57}Fe, ^{57}Fe + n $\longrightarrow$ ^{58}Fe, ^{58}Fe + n $\longrightarrow$ ^{59}Fe, and these steps occur whatever the neutron time-scale, because ^{56}Fe, ^{57}Fe, ^{58}Fe, are all stable. At ^{59}Fe, however, there are alternatives. If the time-scale for neutron addition is appreciably less than 45 days, we have ^{59}Fe + n $\longrightarrow$ ^{60}Fe, but for a time-scale appreciably greater than 45 days, the ^{59}Fe emits a β-ray, ^{59}Fe $\longrightarrow$ ^{59}Co + e + $\bar{\nu}_e$, and the neutron is then added to the cobalt, ^{59}Co + n $\longrightarrow$ ^{60}Co. For a time-scale of about 45 days, with a fraction of the ^{59}Fe undergoing decay, both possibilities occur. Supposing the time-scale to be appreciably less than 45 days, ^{60}Fe is formed and is effectively stable, since the time for ^{60}Fe $\longrightarrow$ ^{60}Co + e + $\bar{\nu}_e$ is actually very long, 3×10^5 years. A further neutron therefore gives ^{60}Fe + n $\longrightarrow$ ^{61}Fe. But ^{61}Fe now decays to ^{61}Co in 6 minutes. If a neutron comes along in appreciably less than 6 minutes, and so gives ^{61}Fe + n $\longrightarrow$ ^{62}Fe, the decay problem is rendered still more acute, because ^{62}Fe emits a β-ray, ^{62}Fe $\longrightarrow$ ^{62}Co + e + $\bar{\nu}_e$, in a characteristic time of only 0.19 seconds. Hence the neutrons are required to come along in shorter and shorter times if we are to proceed to heavier and heavier isotopes of iron, otherwise decay into cobalt takes place. After a decay into cobalt, a similar sequence of events would be repeated. Neutrons would be added for a while, but unless the time-scale for their availability were made shorter and shorter, there would be a further decay through the emission of a β-ray, the result of this second decay being an isotope of nickel. Again the sequence of events would be repeated, the element formed at the next decay being copper.

The pattern we have been considering is shown in Figure III.12. In this figure the mass number A is plotted against the proton number Z. The addition of a neutron, as in ^{56}Fe + n $\longrightarrow$ ^{57}Fe, produces a step to the right, while the emission of a β-ray produces a step upward. With the addition of a neutron

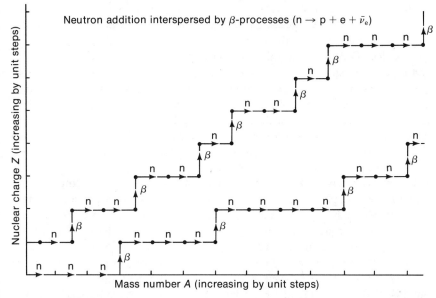

FIGURE III.12.
The addition of a neutron produces a step to the right in this diagram, whereas the emission of a β-ray produces a step upward. The more rapidly neutrons are made available, the more the path taken by a growing nucleus lies to the right of the diagram.

A increases by 1 with Z staying the same; with the emission of a β-ray A stays the same, but Z increases by 1. The track followed in this diagram by a nucleus depends on the rate at which neutrons are made available. The more rapid the supply of neutrons, the more the track lies to the right in Figure III.12.

In Chapter 7 we considered nuclear fusion processes within stars in terms of how much energy they produced. Many other reactions, less important for energy production, also occur. One of these is of special significance here, since it leads to neutron availability. The process is $^{13}C + {}^{4}He \longrightarrow {}^{16}O + n$. This reaction yields enough neutrons to build nuclei of the iron peak, like ^{56}Fe, into all the elements of Table 7.1 that are heavier than nickel. Usually ^{13}C is some three to ten times more abundant than ^{56}Fe, so typically iron is converted by neutron addition into elements like copper and zinc. Exceptionally, however, the carbon-to-iron ratio can be much larger, yielding on the order of 100 neutrons per iron nucleus. It is in such exceptional situations that the heaviest elements are produced. It may be noted here that the iron-peak elements have a considerably greater affinity for neutrons than has ^{13}C or ^{16}O, so that neutrons from $^{13}C + {}^{4}He \longrightarrow {}^{16}O + n$ tend to attach themselves to the heavier elements rather than to the carbon or oxygen or helium.

Suppose we start with ^{56}Fe and add neutrons progressively, without worrying about the neutron supply ceasing. Suppose further, that we consider two quite different time-scales for the process: a time scale s associated with the evolutionary peregrinations of a massive evolving star as it zigzags in the giant region of the luminosity-color diagram; and a time-scale r associated with the ultimate explosive outburst of such a star. Typically, we think of the s time-scale as being of order 100,000 years, and of the r time-scale as being on the order of a few seconds. What will be the tracks in the Z, A, diagram for these two time-scales? The answer to this question is shown in Figure III.13. The track farther to the right, for the r time-scale, goes to values of A of order 260, where fission occurs. The fission fragments would themselves then grow until A again reached about 260, when further fission would occur, and so on repeatedly around a large-scale cycle ranging from A about 130 to A about 260. For the s time-scale, on the other hand, the mass number A grows until the nucleus enters a small-scale two-part cycle between the elements lead and bismuth, as shown in Figure III.14.

Figure III.13 shows the s and r tracks of a nucleus on the assumption that an unlimited supply of neutrons is made available. Only a portion of these tracks will usually be followed, however, since sometimes only 10 neutrons per ^{56}Fe may be available, or 30 neutrons per ^{56}Fe, or 100–200 neutrons, according to the stellar circumstances. Yet whatever the circumstances there must be a strong tendency for the nuclei to go to one or another of the slanting lines of Figure III.13 marked at the values 50, 82, and 126 of the neutron number $N = A - Z$, because a nucleus reaching one of these lines is less ready to accept a further neutron than a nucleus not on one of these lines. Experimental data showing the ability of nuclei to accept further neutrons are given in Figure III.15. Pronounced minima are seen to occur for $N = 50$, $N = 82$, $N = 126$. This nuclear property is connected with the formation of the neutrons

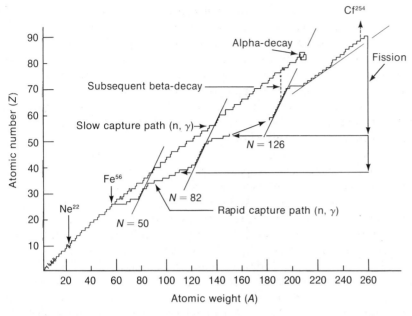

FIGURE III.13.
The two paths corresponding to the slow addition and the fast addition of neutrons. The "magic" neutron numbers at $N = 50, 82, 126$, produce a kind of staircase effect on the paths.

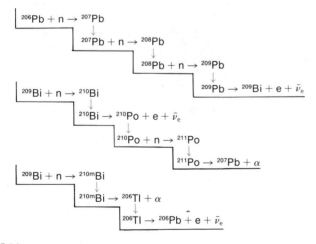

FIGURE III.14.
The two-part cycle between Pb and Bi, neutrons being added on a time-scale of order 10^6 years. The alternative paths from Bi to Pb depend on whether $A = 210$, $Z = 83$, is formed in its normal form, ^{210}Bi, which decays to ^{210}Po in a characteristic time of five days, or whether the isomeric form, ^{210m}Bi, is produced. This isomeric form undergoes α decay to ^{206}Tl in a characteristic time of 3×10^6 years, being stable against decay to ^{210}Po. What would happen if neutrons were made available in a time appreciably less than 3×10^6 years?

into shells. The protons also form similar shells. In Chapter 7 we studied the shells of electrons, and saw that they occur for electron numbers equal to 2, 10, 18, 36, 54, and 86. Just as the reason for these specific electron numbers is to be sought in the application of quantum mechanics to the electrical forces binding the electrons to the nucleus, so the reason for the numbers 50, 82, and 126 is to be sought in the application of quantum mechanics to the forces that bind the neutrons and protons together. Because these *nuclear forces* are different from electrical forces, the "magic numbers" of 50, 82, and 126 are different from the electron shell numbers.

The above considerations have an important relation to the observed abundances of the heavy elements, plotted in detail in Figure III.16. Notice first the double peak, one at A about 125, the other at A about 135. These two peaks correspond very closely to the places where the $N = 82$ line of Figure III.13 meets the r and s tracks, respectively. A similar double peak occurs at A about 195 and at A about 210 (although not fully formed at Bi), corresponding to the places where the $N = 126$ line meets the r and s tracks. Two peaks for $N = 50$ can also be discovered in the abundance data of Figure III.16, although the separation of these peaks is less—and should be less—than for the other two double peaks. The comparatively high abundances for A in the range from about 80 to 90 shows the limitations imposed by the availability of neutrons, which has not usually been sufficient to push the growing nuclei beyond the $N = 50$ line. This accords with what was said above—we saw that the number of neutrons available per ^{56}Fe nucleus would mostly be in the range 3 to 10, less often in the range 10 to 30, and still less often above 100.

The problem of the origin of uranium and thorium is now solved. These elements were produced along the r track of Figure III.13 under exceptional conditions of neutron availability: 100–200 per ^{56}Fe. Very heavy elements like uranium and thorium are built only by rapid neutron addition, which keeps the tracks in the Z, A, diagram far away from the places occupied by nuclei like ^{218}Po and ^{222}Rn that are exposed to incipient breakdown, as can be seen from Figure III.17.

Nuclei along the r track of Figure III.13, when placed in one or the other of the energy diagrams of Figures 7.4 or 7.6, according to whether the mass

FIGURE III.15.
The lefthand scale is a measure of the ability of nuclei to accept neutrons. These are natural-element neutron-capture cross sections near 30 keV. The data are from Oak Ridge.

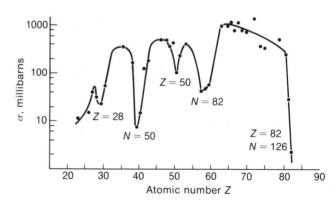

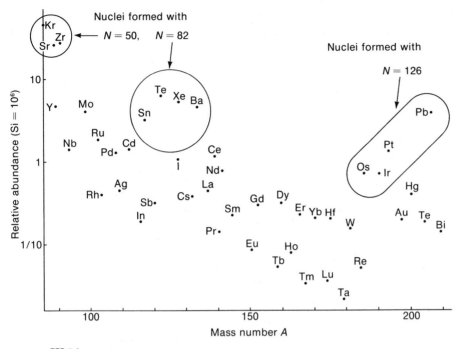

FIGURE III.16.
The observed relative abundances follow a varied distribution, having peaks that are related to the minima of Figure III.15.

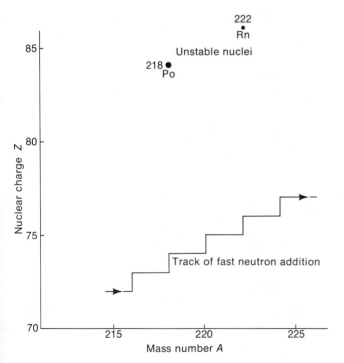

FIGURE III.17.
Detail from Figure III.13, showing how the track followed when neutrons are rapidly available avoids unstable nuclei like ^{218}Po and ^{222}Rn.

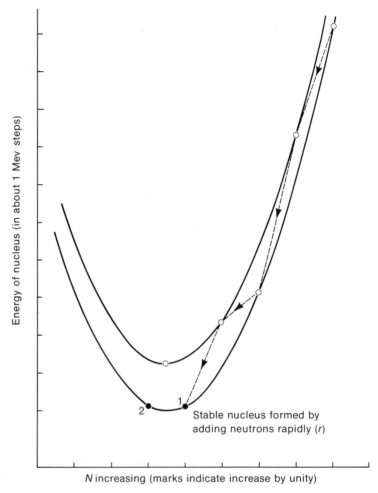

FIGURE III.18.
Once neutrons are no longer available, decays by the emission of β-rays
occur, until a stable nucleus is eventually reached. Note that the stable nucleus on the left is *not* formed by adding neutrons rapidly.

number A is odd or even, lie high on the righthand side, as is illustrated for
an even A in Figure III.18. Once neutrons cease to be available, as happens
in a time-scale of a few seconds, such nuclei then decay downward in the manner
of Figure III.18. Very heavy nuclei, with the mass number A greater than 240,
can emit α particles, or even undergo fission, before the cascade process of
Figure III.18 reaches its bottom point. However, less heavy nuclei reach point
1 in Figure III.18, which is one of the two points of stability, point 2 being
the other. Nuclei are always found at point 2 in nature (as well as at point
1), but reach it by the s track, never by the r track.

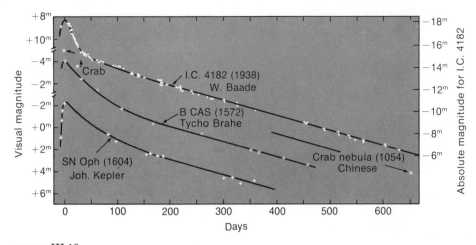

FIGURE III.19.
The behavior of the supernova that occurred in 1938 in the galaxy I.C. 4182 is shown at the top of the figure. The corresponding behavior of the supernovae observed by Tycho Brahé, by Kepler, and by Chinese astronomers (A.D. 1054), as reconstructed from relative brightness observations, are also shown.

It is of interest to relate these ideas to the light emitted by supernovae. In Chapter 7 we saw that stars exploding because of nuclear instability are observed as supernovae. The light curves observed for several supernovae are shown in Figure III.19. The early rise of these curves is associated with the energy released by the nuclear instability: as the debris hurled outward from the explosion spreads into space, the light first rises and then falls away, the rise and fall being accomplished in some 20 to 30 days. Thereafter the light declines more slowly during a period of several hundred days. The energy emitted during this slower decline is believed to be due to the r-process nuclei, for example, to the β-decay of ^{59}Fe in a characteristic time of 45 days and to the fission of ^{254}Cf in a characteristic time of 60 days, to mention two such energy sources.

Appendix III.4. Hydrogen-Burning and the Puzzle of the Helium Abundance

In Chapter 7 it was seen that the source of energy in main-sequence stars is hydrogen-to-helium conversion, and the details of the two modes of conversion, the proton-proton chain and the carbon-nitrogen cycle, were given in Figures 7.10 and 7.11. The relative importance of these two modes is sensitive to the temperature T. Writing ϵ_{pp} for the rate of energy production by the proton-proton chain, and ϵ_{CN} for that by the carbon-nitrogen cycle, the ratio $\epsilon_{CN}/\epsilon_{pp}$ varies about as T^{13}. With the abundances of CNO given in Table 7.1 taken for the initial composition of a main-sequence star (see also the abundances in Table III.1), the two modes of energy production are equal at about $T = 1.8 \times 10^7 \,°\text{K}$. It follows that if, instead of measuring T in $°\text{K}$, we use $10^6 \,°\text{K}$ as the unit of T, then $\epsilon_{\text{CN}}/\epsilon_{\text{pp}} = (T_6/18)^{13}$, the notation T_6 being used here to denote such a choice of the temperature unit. This equation for $\epsilon_{\text{CN}}/\epsilon_{\text{pp}}$ is represented graphically in Figure III.20. The two modes of energy production become equal in their total output for a main-sequence star with central temperature about $T_6 = 20$, a value somewhat higher than $T_6 = 18$ being necessary since most of the material of a star is cooler than the material at its center. In Table III.2 there is a selection of values of the central temperature and central density for main-sequence stars of variable mass M. A changeover

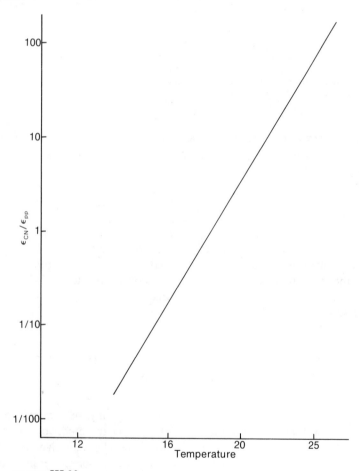

FIGURE III.20.
The behavior of the ratio $\epsilon_{\text{CN}}/\epsilon_{\text{PP}}$ with temperature. Energy generation from the carbon-nitrogen cycle equals that from the proton-proton chain at a temperature of about 18 times $10^6 \,°\text{K}$. The unit of temperature is $10^6 \,°\text{K}$.

TABLE III.2.
Temperature and density in some main-sequence stars.[a]

Mass (in terms of solar mass)	Central temperature (°K)	Central density (gm cm^{-3})	Radius of star (in terms of solar radius)
0.6	8.05×10^6	65.02	0.64
1	12.00×10^6	76.92	1.02
2.5	19.82×10^6	48.31	1.59
10	27.81×10^6	7.79	3.63

[a] From M. Schwarzschild, *Structure and Evolution of the Stars*, Princeton, 1958. The values here are of course calculated, not observed. The calculations depend on taking convenient averages of certain rather complex physical quantities. Other authors, using different averages, have calculated central temperatures that are somewhat higher than those of Schwarzschild. Thus, for a star of mass 10 ⊙, Anders and Peterson (*Stellar Evolution*, Plenum Press, 1966) obtain a central temperature of 30×10^6°K. Haselgrove and Hoyle (Monthly Notices Royal Astronomical Society, Vol. 119, page 112) also obtained rather high values for the central temperature.

from dominance of the energy production by the proton-proton chain to dominance by the CN cycle occurs as M increases through a value of about 2.5 ⊙.

In Chapter 7 we also saw that the intrinsic luminosity L of a main-sequence star, measured in terms of the solar luminosity as unit, was given approximately by $(M/\odot)^4$. More simply, if M is measured in terms of the solar mass ⊙, we have $L = M^4$, approximately. The relation between L and M is shown more precisely in Figure III.21, from which one sees that L varies more as M^5 for masses of solar order, but more as M^3 for masses around 10 ⊙. The L dependence on M becomes still less steep for masses larger than 30 ⊙, becoming eventually proportional to M itself.

The ratio L/M can be taken as a measure of the efficiency of the production of radiation by a star. With L behaving about as M^4, provided M is not too large, this ratio is proportional to M^3, giving a rather marked increase of efficiency with mass. The actual efficiency averaged over the whole of our galaxy is about the same as that of a star with $M = \frac{2}{3}$ ⊙. On the average, the stars of our galaxy are not therefore as efficient as the Sun in their production of light. The average efficiency is thus rather low, although not so low as in some galaxies—the efficiency is particularly low for elliptical galaxies. On the other hand, the efficiency is much higher for young open clusters of stars.

The conversion of one gram per second of hydrogen to helium yields a power production of 6×10^8 kilowatts. The total output of radiation of all forms by our galaxy is believed to be about 6×10^{33} kilowatts, an output which can be maintained therefore by the conversion to helium of 10^{25} grams of hydrogen per second. Now, the ages of the oldest star clusters in Figure 7.17 were shown to be about 10^{10} years, about 3×10^{17} seconds. Consequently the radiative output of our galaxy could have been maintained during a lifetime of 10^{10} years

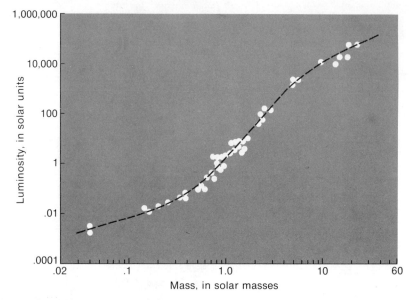

FIGURE III.21.

The empirical relation for main-sequence stars between luminosity (all radiation frequencies) and mass. The unit of luminosity is that of the Sun, and the unit of mass is the solar mass.

by the conversion to helium of 3×10^{42} grams of hydrogen. Since the total mass of our galaxy is believed to be about 3×10^{44} grams, we see that only a 1 per cent conversion of hydrogen to helium would have been needed to supply a steady radiative output *at the present-day level* throughout the lifetime of the oldest star clusters.

It is of interest to compare this result with the amount of helium found in naturally occurring material. According to Table 7.1, or to Table III.1, hydrogen atoms outnumber helium atoms in a ratio of about 10 to 1. Since the mass of the helium atom is some four times that of the hydrogen atom, the naturally-occurring mass ratio is therefore 5 to 2, which implies that a fraction $\frac{2}{7}$ of the mass of the galaxy is helium. Evidently, then, with the total mass of the galaxy taken as 3×10^{44} grams, the mass of helium in the galaxy is nearly 10^{44} grams, very much more than the mass of 3×10^{42} grams estimated in the preceding paragraph to have been produced from hydrogen-burning.

It seems, then, that at least one of the following statements must be true.

1. Most of the helium found in the galaxy has *not* been produced by nuclear reactions occurring within stars.
2. Our galaxy emitted far more radiation at times in the past than it does now.
3. Our galaxy is actually much older than the age of 10^{10} years attributed to the oldest stars.

Of these possibilities, (1) is favored by many astronomers, and we will discuss it further in Section VI. The second possibility also requires serious consideration, since galaxies are known that do have exceedingly large radiative emissions; very large outputs of infrared radiation have recently been found to be coming from some. Our galaxy could be in a quiescent phase at present. The third possibility is not considered plausible by most astronomers, but it is mentioned here for the sake of completeness.

The puzzling aspect of this situation is that, whereas nuclear processes inside stars can be shown to be entirely adequate to explain the naturally occurring abundances of all common elements heavier than helium, the stars seem to be inadequate for explaining the origin of their simplest product, namely, the helium synthesized directly from hydrogen.

Appendix III.5. The Mystery of the Missing Solar Neutrinos

The details of the proton-proton chain set out in Figure 7.10 show that ^{4}He can be produced in the ways illustrated in Figure III.22. Either ^{4}He is formed directly from two nuclei of ^{3}He, by ^{3}He + ^{3}He $\longrightarrow$ ^{4}He + 2p, or an extended chain of reactions is set up through ^{3}He + ^{4}He $\longrightarrow$ ^{7}Be. The ^{7}Be produced in the latter reaction can itself follow two alternative paths, as is also shown in Figure III.22. We can have ^{7}Be $\longrightarrow$ ^{7}Li + e^+ + ν_e, followed by ^{7}Li + p $\longrightarrow$ ^{8}Be, ^{8}Be $\longrightarrow$ 2 ^{4}He, or the ^{7}Be can first acquire a proton, ^{7}Be + p $\longrightarrow$ ^{8}B, in which case

$$^8B \longrightarrow {}^8Be + e^+ + \nu_e$$

follows, and the ^{8}Be again undergoes spontaneous decay to 2 ^{4}He. For material of specified hydrogen and helium content, and for specified temperature and

$$^1H + {}^1H \rightarrow {}^2D + e^+ + \nu_e$$
$$^2D + {}^1H \rightarrow {}^3He$$
$$^3He + {}^3He \rightarrow {}^4He + 2\ {}^1H$$

---

or

$$^3He + {}^4He \rightarrow {}^7Be$$
$$^7Be + e \rightarrow {}^7Li + \nu_e$$
$$^7Li + {}^1H \rightarrow 2\ {}^4He$$

---

or

$$^7Be + {}^1H \rightarrow {}^8B$$
$$^8B \rightarrow {}^8Be + e^+ + \nu_e$$
$$^8Be \rightarrow 2\ {}^4He$$

FIGURE III.22.
The production of ^{4}He in the proton-proton chain. Three routes are possible.

density, the rates of all the reactions in Figure III.22 can be calculated. In particular, the rate at which neutrinos are produced by the decay of ^{8}B can be calculated using experimental nuclear data.

With these general ideas in mind, we can proceed to discuss one of the most baffling problems of modern astronomy. In brief, neutrinos from the decay of ^{8}B, which were expected to be generated inside the Sun, have not in fact been found.

A "model" is calculated for the Sun that satisfies a number of important requirements. The model must have the age which we think the Sun has, about 4.6×10^9 years. The initial composition of the material must be the same as the Sun is believed to have had, and the intrinsic luminosity must be the same as the Sun is now observed to have, 3.9×10^{23} kilowatts. The outcome is a model having the same essential properties as the Sun, with the temperature and density distributions everywhere known—i.e., known at all distances from the center. With the composition also known, it is then possible to calculate what the output of neutrinos from the decay of ^{8}B should be.

The puzzle is that an experiment, set up by Dr. R. Davis of the Brookhaven Laboratory to detect these neutrinos, shows no clear-cut positive result. The neutrinos are apparently not as numerous as the above argument requires. The method of detection is based on the reaction

$$^{37}\text{Cl} + \nu_e \longrightarrow {}^{37}\text{A} + e,$$

in which the nucleus ^{37}Cl(17p,20n) of chlorine is changed to the nucleus ^{37}A(18p,19n) of argon. Although the reaction is exceedingly weak and rare, the few times per month when it occurs can be detected by means of a highly sensitive technique for the detection of argon. The experiment consists in exposing a large quantity of fluid perchlorethelene, C_2Cl_4, for several months to the presumed solar neutrinos. Because the neutrinos are hardly absorbed at all in the body of the Earth, the tank containing C_2Cl_4 is placed deep in a mine, and is surrounded by a jacket of water. This prevents ^{37}A from being formed except by the neutrinos. The experimental arrangement is shown in Figure III.23.

For FIGURE III.23,
see PLATE XXIII.

The failure of this experiment to detect the expected neutrinos from the Sun sets a problem which has not yet been solved. Several ways of resolving the discrepancy have been considered.

1. We have to rely on calculation for determining the rate of $p + p \longrightarrow D + e^+ + \nu_e$. If there were a mistake in the theoretical basis of the calculation, in the sense that too slow a rate were being estimated, then the actual production of ^{8}B, and hence of neutrinos, would be less than the calculated value, perhaps sufficiently less to be consistent with the experiment described above.

2. The rates of certain of the other reactions of Figure III.22 also affect the production of ^{8}B. If ^{3}He + ^{3}He $\longrightarrow$ ^{4}He + 2p were faster than we think

it is, nuclei of ^{3}He would "go" more by this route than by ^{3}He + ^{4}He $\longrightarrow$ ^{7}Be, and there would then be less ^{8}B produced by ^{7}Be + p $\longrightarrow$ ^{8}B, and hence fewer neutrinos.

3. There could be something wrong in the astrophysical calculations leading to the distributions of density and temperature in the model Sun.
4. The basic assumption, that the radiative loss from the surface of the Sun is balanced by nuclear energy production, could be wrong.
5. Something could happen to the neutrinos during their journey from the Sun to the Earth.

What can be done by way of examining these alternatives? Where calculations along well-understood lines are involved, as in 1 and 3, the calculations can be carefully repeated by several workers. Where experimental data are involved, as in 2, further experiments can be performed. The situation is that, after several years of concerted effort along these logically straightforward lines, the discrepancy remains unresolved.

For 5 to be possible, the neutrinos would have to decay into some other particle, not subject to detection, during the eight minutes required for their flight from the Sun to the Earth. For 4, the Sun would have to be out of short-term equilibrium. We would still need to say that, on a long-time average, there must be a balance between energy lost from the solar surface and the production of nuclear energy, but we might seek to argue that the balance need not apply at any one moment. This leads to the problem of *why* the Sun should thus be out of short-term equilibrium. Mixing of material from one part of the Sun to another might supply the explanation here, but this idea is far from being straightforward.

The picture of the Sun being out of "short-term" balance means on a time-scale less than ten million years, "short-term" being quite "long-term" from a human point of view. Such a picture could be connected with the fact that, during the past million years, the Earth has experienced a quite unusual sequence of Ice Ages. The present-day solar luminosity could conceivably be lower than its long-term average. Calculations along these lines suggest that the solar emission could indeed go considerably lower yet, implying that the Ice Age sequence is very far from being finished. The most extensive and devastating climatic disturbances could still lie in the future.

Possibility 5 would seem to have no great significance for the Sun, but it would have considerable importance for physics.

What then are we to believe? A mistake of detail somewhere, or an issue of far-reaching principle? It is disturbing to find ourselves faced by such a choice. The Sun is apparently a very straightforward kind of star. If such complexities arise for the Sun, what might be the situation for other, obviously more complex, stars? Eddington once said, "it is reasonable to hope that in the not too distant future we shall be competent to understand so simple a thing as a star." His prediction here has proved less accurate than his hopes for the process ^{4}H $\longrightarrow$ ^{4}He. More than thirty years after his death, we still seem far from fully understanding even the simplest stars.

Appendix III.6. Nuclear Explosions, Man-Made and Stellar

Returning to the fission chain of Figure III.9 on which man-made nuclear explosions are based, one can ask: What sets the explosion going? How does the explosion keep going? Why doesn't the heat released simply blow the uranium apart as a chemical explosive would do?

The answer to the first of these questions turns on whether enough of the neutrons released by the fission fragments A_1, A_2, "find" more ^{235}U. In Figure III.9 the neutrons were shown moving directly toward the ^{235}U. But neutrons would also be emitted in all directions, as in Figure III.24, and so some would be lost. This loss could be avoided by arranging for the uranium entirely to surround the source of the neutrons, i.e., the fragments A_1, A_2, derived from the uranium itself. To prevent loss we require uranium always to surround uranium, as indicated schematically in Figure III.25. Only by having an endless sequence of shells can such a condition of no loss be fully satisfied. However, since more neutrons are made available than are actually needed, some can be lost without a lump of ^{235}U necessarily becoming stable. We saw above that about 2.5 neutrons are produced on the average per fission, but as long as more than one neutron per fission generates further fission, the cycle will be amplified. About half the neutrons from the fission fragments can thus be lost, which means that the uranium does not need to be distributed in a way that supplies perfect neutron absorption.

Anyone seeking to produce an explosive situation must determine two configurations for the fissile material, which here is ^{235}U. One configuration must meet the condition we have just considered: it must be explosive. The

In the fission process
$$^{236}U \rightarrow A_1 + A_2$$
$$A_1 \rightarrow (A_1 - 1) + n$$
$$A_2 \rightarrow (A_2 - 1) + n$$

the neutrons may be emitted in any direction:

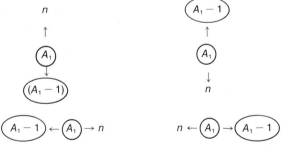

FIGURE III.24.
During the fission cycle, neutrons are emitted in all directions.

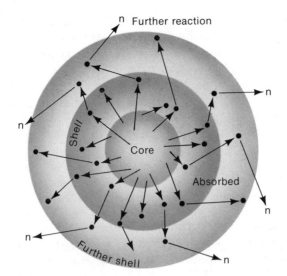

FIGURE III.25.
To prevent any neutrons from being lost during the
fission cycle, an infinite distribution of uranium
would be needed.

other must be nonexplosive, that is, too inefficient to absorb as much as one
neutron per fission. The explosion is then triggered by a switch from the
inefficient arrangement to the efficient one, the switch being made quickly by
moving the uranium with a chemical explosive; we could imagine two halves
of a sphere being brought suddenly together, as in Figure III.26.

A hot ball of gas blows apart, regardless of whether the gas has been heated
by nuclear fission or by a chemical explosion. As energy is released in a nuclear

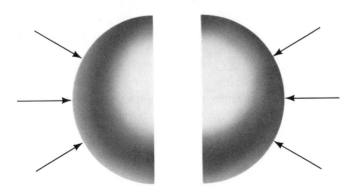

FIGURE III.26.
One way to trigger an amplifying fission cycle is to drive two halves of
a sphere of ^{235}U explosively together.

bomb, the originally solid uranium develops into a ball of hot gas which must thus tend to blow apart. If the expansion of the gas were quicker than the time required for quite a few amplifying cycles of the fission process, the release of nuclear energy would obviously soon be cut off. This does not happen, however, because the nuclear processes are in fact much quicker than the expansion time of the hot ball of uranium gas. At a temperature as high as 100 million °K, a ball, say, 10 cm across, would blow apart in about 10^{-7} seconds, but short as this may seem, the time-scale for neutron-induced fission is much shorter.

Similar considerations are needed for stellar explosions. Because in stars we are dealing with fusion, not fission, the problem of the geometrical arrangement of the explosive material is of less concern; the material is in any case arranged in spherical shells, as in Figure 7.20. The basic reaction in massive stars is $^{16}O + {}^{16}O \longrightarrow {}^{28}Si + {}^{4}He$. This reaction cannot proceed essentially to completion before the energy released by the reaction itself causes the oxygen to blow apart, unless the temperature rises rapidly enough and high enough, say, from 2×10^{9} °K to 3×10^{9} °K. The question now is what process could cause such a temperature change within a few seconds, which is the order of the time in which the oxygen blows apart.

To consider this issue, let us return to the picture given in Chapter 7 of the evolution of a massive star, say, a star of mass 20 $\odot$. As hydrogen-burning proceeds, the star evolves from the main sequence into the giant region of the luminosity-color diagram. The helium formed from hydrogen-burning, together with the initial helium, is burned to form carbon and oxygen. Further burnings then occur progressively toward the center, carbon burning to neon and magnesium, oxygen burning to silicon, silicon burning to iron, leading thereby to the multiple-shell model of Figure 7.20. During these burning phases, the star follows a quite complicated path in the luminosity-color diagram. Where does this successive burning of nuclear fuels end? What finally happens?

Radiation from the surface out into space continues, inevitably leading to changes in the interior and so causing evolution to proceed. Moreover, a new source of energy loss from the star eventually becomes a more important factor in determining evolution than the radiation emitted from the surface. As temperatures rise, neutrinos are emitted from the interior of the star in such great numbers that their energy loss becomes severe. Unlike radiation, the neutrinos have only a very weak interaction with the material of the star, and they are able to escape essentially freely as soon as they are produced. The process occurs in two stages. Radiation is first converted into pairs of electrons and positrons,

$$\text{Radiation} \longrightarrow e^{+} + e,$$

and then the electron-positron pairs go into neutrino-antineutrino pairs,

$$e^{+} + e \longrightarrow \nu_{e} + \bar{\nu}_{e}.$$

Normally we think of stars evolving during millions of years, for example, in the funneling process discussed in Chapter 7, whereby stars go from the main sequence to the giant region of the luminosity-color diagram, as in Figure 7.17. Very much shorter evolution times occur, however, when the central temperature of a star rises above 10^9 °K. Indeed, when the central temperature reaches 3×10^9 °K, the energy lost by neutrino emission becomes so exceedingly severe that evolution then proceeds in only a few thousand seconds. In this short time, the layered structure of Figure 7.20 changes appreciably.

Evolution because of neutrino loss takes place most rapidly in the core, because the temperature is highest there. Calculation shows that the core draws away more and more from the other shells of Figure 7.20. That is to say, the core shrinks inwards on itself. The heat generated by the compression raises the temperature still further, increasing the neutrino losses, making matters worse. The ultimate result is a total collapse of the core. The strong gravitational forces pull the material inward until the superdense condition of the neutron stars is reached.

We saw already in Chapter 7 that neutron stars are of extremely small size, much smaller the Earth. Just as in a normal star, in a neutron star gravity, which is exceedingly strong, must be resisted by internal pressure. Also in Chapter 7 we saw how such a pressure can arise from the application of quantum mechanics to the neutrons which are formed from $p + e \longrightarrow n + \nu_e$.

The core is still in a state of rapid infall when this form of pressure, requiring the enormous density of 10^{38} neutrons per cm^3, comes into play. Before any nondynamic state can be set up, the pressure developed within the neutrons must operate therefore to stop the rapid infall of the core. This leads to what is usually referred to as a "bounce." To stop the infall, the pressure must become even greater than would be needed in a static neutron star. Once the infall is halted, this excess pressure generates an outward motion of the core. Thus the core can only attain a static neutron-star condition after it has gone through a sequence of rapid radial oscillations. These oscillations do not occupy much time, probably not more than a few seconds, but they can have an important effect on the outer shells of the star.

Although the core has thus undergone a dramatic change of structure, the outer shells of the star have not yet changed appreciably. With pressure support from the core removed, the outer shells tend to fall inward, but the time-scale for their collapse is much longer than that of the core. Consequently the core undergoes its first bounce before the shells have much changed. At its bounce, the core transmits what is known as a *shock wave* into the outer shells, causing sudden compression of them with a resulting sharp rise of temperature.* These are the shells still containing helium, carbon, and the potentially explosive oxygen. With the sharp rise of temperature, nuclear energy, particularly from $^{16}O + {}^{16}O \longrightarrow {}^{28}Si + {}^4He$, is released catastrophically on a time-scale of a few seconds. This energy release enables the outer layers to lift themselves

---

*An interesting idea now being investigated is that the shock wave arises from the absorption, in material immediately outside the core, of neutrinos emitted by the core. It is uncertain whether there could be enough absorption to cause the shock wave.

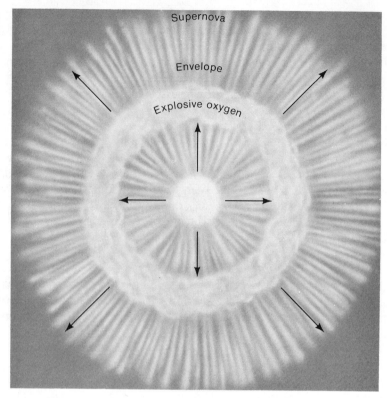

FIGURE III.27.
Sequence of processes leading to the outburst of a supernova. The core transmits a shock wave into the unstable oxygen, causing it to be compressed and explode.

explosively off the star. Most of the initial mass of the star is lost in this way, being expelled outward at a speed of 5,000 to 10,000 km per second, and leaving behind the core, which after bouncing for a short while settles into the equilibrium structure of a neutron star. This sequence of processes is illustrated in Figure III.27, a sequence believed to constitute the explosive phenomena observed as *supernovae*, with properties discussed already both in Chapter 7 and in connection with Figure III.19.

Appendix III.7. *White-Dwarf Binaries and the Occurrence of Novae*

White dwarfs were considered in Chapter 7 to be dead stars, cooling as they pass to a final state. This picture remains valid provided the star has no further access to nuclear fuel, as would be expected for a white dwarf situated by itself. However, for a white dwarf that is one member of a binary system, the situation can be markedly otherwise. Under suitable circumstances the white dwarf can receive material from its companion star, especially if the companion star lies in the giant region of the luminosity-color diagram, for then the companion's

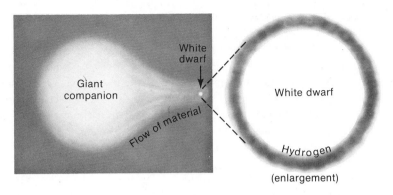

FIGURE III.28.
Interchange of material within a binary system can allow a white dwarf to acquire a supply of hydrogen at its surface.

radius is large and the gravity at its surface is weak, which permits material to pass without undue difficulty to the white dwarf, a situation illustrated in Figure III.28.

The transferred material in such a situation comes from the extreme outer regions of the giant companion, where it has not been subjected to nuclear fusion processes. The material consists largely of hydrogen. Compressed in the strong gravitational field of the white dwarf, the transferred material becomes heated, and the hydrogen begins to burn, within a skin covering the surface of the white dwarf. At first the hydrogen burns slowly, producing helium mainly by the carbon-nitrogen cycle. The energy released by the helium production remains largely trapped within the skin of transferred material, and so causes the temperature to rise. The process is cumulative. The more the temperature rises, the more energy is produced, and the temperature rises still further. The hydrogen-burning continues until the skin eventually becomes hot enough to burst outward even against the strong gravitational field which has hitherto clamped it onto the surface of the white dwarf. The violence of the process can be augmented by a mixing of the surface hydrogen layers with material from the lower-lying regions of the white dwarf itself. The latter material often consists of carbon and oxygen, which both increase the speed of the carbon-nitrogen cycle and are themselves explosive when the temperature is sharply lifted. The outcome is an ejection of material from the white dwarf, a violent shedding of its skin of acquired material, that has some similarity to the outburst of a supernova. The situation is similar in principle, but not in degree. Calculations indicate that the ejection will occur when the surface skin of acquired material on the white dwarf attains a mass of only about 10^{-3} $\odot$, much less than the masses involved in supernovae. Explosions of this kind appear to be the cause of ordinary *novae*.

We saw in Chapter 7 how planetary nebulae are produced during the final stages of a star's evolution from the giant region of the luminosity-color diagram to the region of the white dwarfs. Astronomers at one time believed that novae usually occur during this stage of final evolution of a single star; perhaps some among the different forms of novae can be explained in this way. Yet observation has shown the novae typically occur in binary systems, in agreement with the preceding discussion. Indeed, I recall Dr. Rudolf Minkowski, who had made a detailed study of planetary nebulae, insisting to me in a private conversation many years ago that there seemed to be no linking bridge between the class of planetary nebulae and the class of novae. His percipient remark was made long before the data on which this discussion is based were available.

Notice that the process described above is likely to be repetitive. The white dwarf, after violently throwing off its skin of acquired material, may be supplied with a further skin by its giant companion. This too will be rejected in the same way. Repetitive novae are indeed observed, in accordance with this expectation.

The hot surface of a white-dwarf member of a binary, developed at the time of an outburst, can be expected to emit radiation of very high frequency, x-rays. It is of interest that two x-ray binaries have indeed been recently detected by the satellite *Uhuru*. They are the two x-ray sources known as Centaurus X-3 and Hercules X-1. The situation for Centaurus X-3 is illustrated in Figure III.29. The period of the binary pair is about two days, with the large star

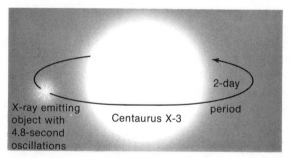

FIGURE III.29.
Schematic representation of the x-ray source Centaurus X-3.

of Figure III.29 eclipsing the x-ray-emitting star for about half a day on every revolution. The remarkable feature of the x-ray star is that its emission pulsates in the very short time of 4.8 seconds, as if the hot disturbed surface of a white dwarf were in rapid oscillation, as it might well be after a sudden ejection of material. The source Hercules X-1 has been identified with a previously known binary, HZ Herculis. The situation is similar to that of Centaurus X-3, with a binary motion of one star around the other in 1.7 days, and with the x-ray star pulsating in the even shorter time of 1.2 seconds.

It should be noted, however, that another possibility for understanding these remarkable x-ray oscillations would be to ascribe them to the rotation of *neutron stars* instead of to the pulsation of white dwarfs. We will consider the effect of the rotation of neutron stars in connection with the phenomenon of *pulsars,* at a later stage in Section V. Many astronomers favor the latter explanation, rather than that of the preceding paragraph.

Although by itself a white dwarf may be a dead star, when another star is close by a white dwarf is by no means dead. The remarkable processes which can then ensue are due at root to the strong gravitational forces which the white dwarf exerts at its own surface.

Appendix III.8. Nuclear States

In Chapter 4 we saw that the electrons of an atom can be considered to be in one or another of a number of states which were denoted symbolically by $\psi_1, \psi_2, \ldots$. Transitions can occur between one state and another. If we order the sequence $\psi_1, \psi_2, \ldots$, in a suitable way, a leftward transition gives emission of radiation, a rightward one absorption of radiation. Diagrams can be drawn, as in Figure 4.30, in which the states are assigned various levels, with ψ_1 the bottom level. The level to be assigned to any other state, say, ψ_r, is determined by the frequency of the radiation emitted in the transition $\psi_r \longrightarrow \psi_1$. Such a way of representing the states is known as an *energy-level diagram.*

In a similar way, energy-level diagrams can be constructed for the *nuclei* of atoms. The protons and neutrons making up a nucleus have bundles of paths which can be denoted by $\varphi_1, \varphi_2, \ldots$, such that a leftward transition in the sequence gives emission of radiation, a rightward one absorption of radiation. With φ_1 forming the bottom level, we again represent the *nuclear states* φ_2, $\varphi_3, \ldots$, by the frequency of the radiation emitted in the transition $\varphi_r \longrightarrow \varphi_1$, where φ_r may be any one of $\varphi_2, \varphi_3, \ldots$. Such an energy diagram is shown for the nucleus ^{12}C in Figure III.30.

The frequencies involved in transitions between nuclear-energy levels are on the order of a million times larger than those given by the transitions between electronic energy levels. Typically, electronic transitions yield radiation frequencies on the order of 10^{15} oscillations per second—i.e., light, whether visible, ultraviolet, or infrared—whereas nuclear transitions give radiation frequencies on the order of 10^{21} oscillations per second. Such very high-frequency radiation is described as γ-radiation, or more briefly as γ-rays. Indeed, it is from nuclear transitions that γ-rays are mainly generated. The damaging γ-rays from a nuclear weapon arise mainly in this way.

We have so far been writing the reactions between protons, neutrons, and nuclei without explicitly mentioning the γ-rays which the various reactions generate. For example, we wrote ^{12}C $+ p \longrightarrow {}^{13}$N, ^{56}Fe $+ n \longrightarrow {}^{57}$Fe. In both these reactions γ-rays are emitted, which may be denoted by writing

$$^{12}\text{C} + p \longrightarrow {}^{13}\text{N} + \gamma, \, {}^{56}\text{Fe} + n \longrightarrow {}^{57}\text{Fe} + \gamma.$$

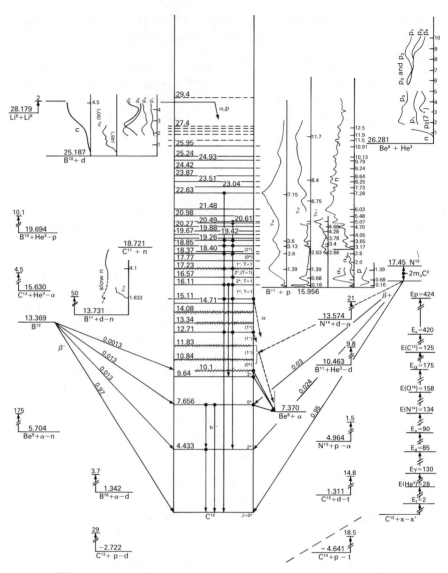

FIGURE III.30.

The energy-level diagram of the nucleus ^{12}C. Energy values are in MeV, one MeV being 1.6 times 10^{-16} kilowatt seconds. (Courtesy, W. A. Fowler, California Institute of Technology.)

Physicists usually take the trouble to write nuclear reactions in this more informative way.

Although diagrams such as Figure III.30 may seem complex, it should be noted that the electronic energy-level diagrams for atoms with as many electrons as the nuclei have protons and neutrons are essentially just as complex, as can be seen from Figure III.31.

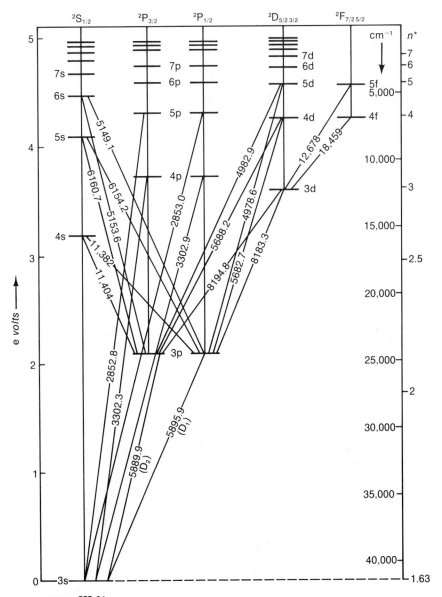

FIGURE III.31.
The energy-level diagram of the sodium atom (Na). Notice the two transitions marked D₁, D₂. Radiation from these transitions is of a yellow color. It is this radiation which gives the familiar yellow color from sodium street lamps. (From H. G. Kuhn, *Atomic Spectra*, Academic Press, 1962. Longman Group, Ltd.)

Appendix III.9. Nuclear Reaction Rates

We saw in Chapter 7 that the protons in one nucleus, say, A, repel electrically those in another nucleus, say, B. This repulsion will keep A and B apart, unless they are projected toward each other at high speed. However, should A and B thus come close together, *nuclear forces* of attraction become operative, permitting A and B to join, forming a combined nucleus, say, G; $A + B \longrightarrow G$. Such a reaction proceeds the more rapidly as the speed of approach of A to B increases. In stellar material this circumstance implies, since high temperature means high speeds, that the higher the temperature is, the faster the nuclear reactions take place within the material.

What other factors affect the rate of $A + B \longrightarrow G$? Referring back to Figure 7.7, the fusion of A and B to form G requires the mass numbers of the nuclei to be to the left of the broad maximum of the curve shown in this figure. With A and B sufficiently to the left of this maximum, we then have

$$A + B \longrightarrow G + \text{energy yield,}$$

the energy yield commonly occurring in the form of γ-rays.

The energy yield depends on the motions of A and B, as well as on the intrinsic properties of the nuclei themselves. Because the energy of the motions varies from one reaction, $A + B \longrightarrow G$, to another, it is convenient to separate the total yield into a fixed part, depending only on the intrinsic properties of A, B, G, and a variable part due to motion,

$$A + B \longrightarrow G + \text{fixed yield} + \text{energy of motions.}$$

The fixed yield here determines a level for the sum $A + B$ in relation to the energy-level diagram of G. Suppose, for simplicity, that G is produced by the reaction in its lowest level, φ_1 in the above notation. Then the fixed-energy yield determines the amount by which $A + B$ must be placed above the lowest level of G. Notice, as an example, the level of $^8\text{Be} + \alpha$ in Figure III.30. Since ^8Be can be thought of as $\alpha + \alpha \longrightarrow {}^8\text{Be}$, this reaction is important in the formation of ^{12}C, by $^8\text{Be} + \alpha \longrightarrow {}^{12}\text{C}$, which is equivalent to $3\alpha \longrightarrow {}^{12}\text{C}$.

We come now to a factor that can significantly increase the rate of $A + B \longrightarrow G$. Should the energy of motion of A and B, when added to the level of $A + B$, happen to be close to some level of G, the reaction is said to be *resonant* with respect to this level of G. Reactions that are resonant proceed much more rapidly than they do otherwise. The situation for resonance is illustrated schematically in Figure III.32. Resonance has the important effect of overcoming the inhibiting effect of the electrical repulsion between A and B.

Any reaction can be made resonant in accordance with Figure III.32, by taking the energy of motion of A and B to be just such as will raise the level

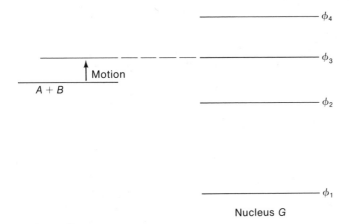

FIGURE III.32.

The situation for a resonant nuclear reaction. The energy of the motions of nuclei A and B, when added to the energy level of $A + B$, gives the same level as one of the states (here, ϕ_3) of the product nucleus G.

of $A + B$ to the next higher level of G. However, the next level of G may be so much higher than the level of $A + B$ that a large contribution from the motions of A and B would thus be required, and very often the temperature of stellar material is inadequate to supply such a large contribution. Except for stars in which the temperature is unusually high—i.e., except for massive stars nearing the end of their lives—resonance is not common in stellar material.

There is one important situation in which resonance plays a crucial role, however. From Figure III.33 it can be seen that the nucleus ^{12}C possesses a level, marked 7.6549 MeV (1 MeV is an energy equal to 1.6×10^{-16} kilowatt seconds), that lies only moderately above $^8Be + \alpha$. Resonance in this level is important for the formation of ^{12}C. In Figure III.33 it can also be seen that the level marked 8.872 (MeV) lies quite far above that of $^{12}C + \alpha$, and resonance in this level does *not* occur to any important degree under the temperature conditions in which carbon is formed within stars. This is fortunate indeed, since otherwise all the carbon produced by $^8Be + \alpha \longrightarrow {}^{12}C$ would be quickly converted to oxygen by $^{12}C + \alpha \longrightarrow {}^{16}O$, with the consequence that the world would be substantially without carbon, and hence without life as we know it.

A CURIOUS SPECULATION

The ^{16}O nucleus has a level marked 7.1187 (MeV) in Figure III.33 lying only slightly below the level of $^{12}C + \alpha$. Resonance cannot occur in this 7.1187 state, however, because the energy of motion of ^{12}C and α must always be a positive quantity, thereby always increasing the total energy of $^{12}C + \alpha$

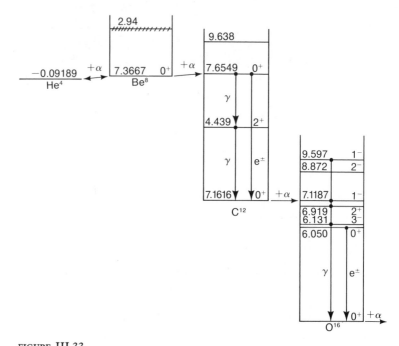

FIGURE III.33.
Details of the lower parts of the energy-level diagrams of ^{12}C, ^{16}O, shown in relation to each other, and also in relation to ^{8}Be + ^{4}He. (Courtesy of W. A. Fowler, California Institute of Technology.)

upwards and *away* from the 7.1187 MeV level of ^{16}O. Yet if this level had been only slightly higher in the diagram, resonance would have been possible, with the unfortunate consequences described above. This prompts the question: Is the positioning of the 7.6549 level of ^{12}C *above* ^{8}Be + α, thereby permitting resonance for ^{8}Be + $\alpha \longrightarrow$ ^{12}C, and the positioning of the 7.1187 level of ^{16}O *below* ^{12}C + α, thereby *not* permitting resonance for ^{12}C + $\alpha \longrightarrow$ ^{16}O, just a happy accident? Or can we find some reason why this should be so?

The positioning of these levels depends on the electrical repulsion between protons, and on the strength of the nuclear forces that bind together the protons and neutrons within the nuclei. Change these two opposing effects only slightly, and the levels in ^{12}C and ^{16}O could be changed by an amount that would produce a world essentially without carbon, and hence without life as we know it. Is the balance between these two effects immutable? Might it vary from one region of the universe to other, very distant regions? If so, in some places we would then expect the balance to be "right," as in Figure III.33, producing our form of environment, with approximately equal quantities of carbon and oxygen. In other places, however, the situation could be otherwise, with very little carbon, and so without life forms of the terrestrial kind.

The families of particles mentioned at the end of §7.5 are set out, together with their main properties, in Table III.3.

TABLE III.3. *Families of particles*

Family	Electric charge	Mass[a]	Mean lifetime (seconds)	Common decay products	Antiparticle
LEPTONS					
μ (mu minus)	$-e$	106	2.2×10^{-6}	$e\,\nu_\mu\,\bar{\nu}_e$	μ^+ (mu plus)
e (electron)	$-e$	0.511	stable		e^+ (positron)
ν_e (neutrino)	0	0	stable		$\bar{\nu}_e$ (antineutrino)
ν_μ (mu neutrino)	0	0	stable		$\bar{\nu}_\mu$ (mu antineutrino)
BARYONS					
p (proton)	$+e$	938.26	stable		$\bar{p}$ (antiproton)
n (neutron)	0	939.55	930	$p\,e\,\bar{\nu}_e$	$\bar{n}$ (antineutron)
λ (lambda)	0	1115.6	2.5×10^{-10}	$p\,\pi^-$ $n\,\pi^0$	$\bar{\lambda}$ (antilambda)
Σ^+ (sigma plus)	$+e$	1189.4	8.0×10^{-11}	$p\,\pi^0$ $n\,\pi^+$	$\bar{\Sigma}^-$ (anti sigma minus)
Σ^0 (sigma zero)	0	1192.5	less than 10^{-14}	λ + radiation	$\bar{\Sigma}^0$ (anti sigma zero)
Σ^- (sigma minus)	$-e$	1197.3	1.5×10^{-10}	$n\,\pi^-$	$\bar{\Sigma}^+$ (anti sigma plus)
Ξ^- (xi minus)	$-e$	1321.2	1.7×10^{-10}	$\lambda\,\pi^-$	$\bar{\Xi}^+$ (anti xi plus)
Ξ^0 (xi zero)	0	1314.7	3.0×10^{-10}	$\lambda\,\pi^0$	$\bar{\Xi}^0$ (anti xi zero)
MESONS					
π^+ (pi plus)	$+e$	139.6	2.6×10^{-8}	$\mu^+\,\nu_\mu$	π^- (pi minus)
π^- (pi minus)	$-e$	139.6	2.6×10^{-8}	$\mu\,\bar{\nu}_\mu$	π^+ (pi plus)
π^0 (pi zero)	0	135.0	10^{-16}	radiation	π^0 (pi zero)
K^+ (K plus)	$+e$	493.8	1.2×10^{-8}	$\mu^+\,\nu_\mu,\ \pi^+\,\pi^0$	K^- (K minus)
K^- (K minus)	$-e$	493.8	1.2×10^{-8}	$\mu\,\bar{\nu}_\mu,\ \pi^-\pi^0$	K^+ (K plus)
K^0 (K zero)	0	497.8	8.6×10^{-11} (Fast decay mode)	$\pi^+\,\pi^-,\ 2\pi^0$	$\bar{K}^0$ (anti-K zero)
			5.4×10^{-8} (Slow decay mode)	$3\pi^0,\ \pi^+\,\pi^-\,\pi^0,$ $\pi^+\,\mu\bar{\nu}_\mu,\ \pi^+\,e\,\bar{\nu}_e,$ $\pi^-\,\mu^+\,\nu_\mu,\ \pi^-\,e^+\,\nu_e$	
$\bar{K}^0$ (anti-K zero)	0	497.8	Same decays as K^0		K^0 (K zero)
η (eta)	0	548.8		$3\pi^0,\ \pi^0\,\pi^+\,\pi^-$ $\pi^+\,\pi^-$ + radiation radiation only	η (eta)

[a] The particle masses are given here in terms of 1 MeV (million electron volts) as the unit of energy. To relate this quantity to everyday units, note that 1 MeV is the same as a power of 1 kilowatt operating for 1.6×10^{-16} seconds. Note also that in all decays the products always have the same total electric charge as the parent particle. Can you find particle-antiparticle regularities among the decay products? The particles K^0, $\bar{K}^0$, are peculiar in having two decay modes. As a project, seek to discover a further peculiarity about these two particles.

1. How are the distances of the stars of the Hyades determined?

2. Discuss the process of fission. How can a self-amplifying fission cycle based on ^{235}U be set up? Why is such a cycle not so readily possible for ^{238}U?

3. Explain how a fission cycle based on ^{235}U can in principle be prevented from explosive amplification.

4. How may the elements uranium and thorium be produced by processes inside stars? How do these processes avoid the breakback instabilities of nuclei like ^{218}Po and ^{222}Rn?

5. Compare the explosion of a supernova with that of a man-made nuclear weapon. What important feature do the two have in common?

6. Describe the *r* and *s* processes of neutron addition. How may one conclude that *both* these processes must have been operative?

7. Give an argument to support the view that not all the helium found in the universe can have arisen through nuclear processes inside stars.

8. Is there any counter to the argument of Problem 7?

9. Describe the nuclear reactions which were expected to lead to an observable flux of neutrinos from the Sun.

10. What are the possible implications of the failure to observe these neutrinos?

11. In what way may the white-dwarf component in a binary system come "to life again," possibly displaying the characteristics of a nova?

12. Explain the emission of γ-rays in terms of transitions between nuclear states.

13. Under what circumstances is a nuclear reaction *resonant?* How does resonance affect the speed of a nuclear reaction?

References for Section III

For further reading, see R. Jastrow and M. H. Thompson, *Astronomy: Fundamentals and Frontiers* (New York: Wiley, 1972).

For the properties of nuclei, see H. A. Bethe and P. Morrison, *Elementary Nuclear Theory* (New York: Wiley, 1967).

For a discussion of the abundances of the elements, see A. G. W. Cameron, "A New Table of Abundances," and for the late stages of stellar evolution, see W. D. Arnett, "Some Quantitative Calculations of Final Stages of Stellar Evolution," both from D. N. Schramm and W. D. Arnett, eds., *Explosive Nucleosynthesis* (Austin: University of Texas, 1973).

For recent work on binary objects emitting x-rays, see E. Schreier, R. Levinson, H. Gursky, E. Kellogg, H. Tananbaum, and R. Giacconi, "Evidence for the Binary Nature of Centaurus X-3 from UHURU X-Ray Observations," *Astrophys. J.*, vol. 172 (1972), letter p. 79, and R. Giacconi, H. Gursky, E. Kellogg, R. Levinson, E. Schreier, and H. Tananbaum, "Further X-Ray Observations of Hercules X-1 from UHURU," *Astrophys. J.*, vol. 184 (1973), p. 227.

Section IV:
The Solar System

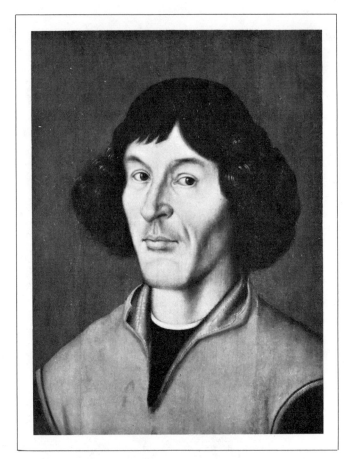

Nicolaus Copernicus, 1473–1543.
(Portrait courtesy of Muzeum Okregowe W Toruniu.)

Chapter 9:
The Pioneers of Astronomy

§9.1. The Greek Astronomers

Nowadays we know of the nine planets shown schematically in Figure 9.1. They vary greatly in their physical characteristics, as we will study in succeeding chapters. They vary also in their distances from the Sun, a feature that will assume great significance later. What they do have in common is that their orbits around the Sun are all very nearly circles. Of the six known in ancient times—Mercury, Venus, Earth, Mars, Jupiter, and Saturn—the greatest deviation from circularity occurs for Mercury, and is only about 2 per cent. The form of Mercury's orbit is shown in Figure 9.2, and it can be seen that the eye has difficulty in detecting any difference from a circle.

Not only is there a regularity about the shapes of the orbits, but all the orbits lie nearly in a plane. If one chooses the orbit of the largest planet, Jupiter, as a standard, the orbits of all the others lie in this plane to within an accuracy of a degree or two.

These properties much simplified the problems which faced astronomers of earlier centuries in their attempts to discover the forms of the planetary orbits, and also in seeking to give a physical explanation for the forms that were discovered. If the orbits had been as varied as those of comets, of which a

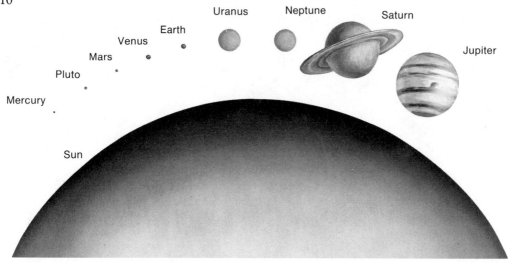

FIGURE 9.1.
The nine planets of the solar system in scale relative to the Sun. The diameter of the Sun is about 1.4×10^6 kilometers. (From J. Gilluly, A. Waters, and A. Woodford, *Principles of Geology*, 3d ed. W. H. Freeman and Company. Copyright © 1968.)

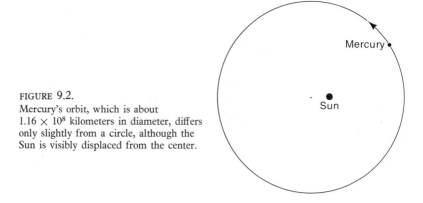

FIGURE 9.2.
Mercury's orbit, which is about 1.16×10^8 kilometers in diameter, differs only slightly from a circle, although the Sun is visibly displaced from the center.

selection is shown in Figure 9.3, these problems would have been well-nigh intractable, even though the physical laws that determine the cometary orbits have turned out to be exactly the same ones that determine the planetary orbits. Had it not been possible to solve the problem of the planetary motions, science would quite likely not have developed at all in the modern sense. Modern science, modern technology, modern civilization, might well not have happened. There may be some who feel that would have been a good thing, but their

FIGURE 9.3.
If the orbits of the planets had been as complex as those of short-period comets, the problem
of understanding the planetary motions would probably have been quite intractable. (From
N. B. Richter, *The Nature of Comets,* Methuen, 1963, by permission of Barth Verlag, Leipzig,
Germany.)

views would probably change sharply should they ever suffer a serious illness
of a kind that can be cured only by modern medical science. Modern civilization
has many unpleasant factors to it, but these have to be faced, not avoided.

Even though the problem of disentangling the planetary motions was much
simpler than it might have been, it was still quite complicated enough. Figure
9.4 shows a sample of how the planetary movements on the sky appear to a
terrestrial observer. In ancient times such an observer had to make his observations of these motions by naked eye, and then had to imagine a model for
the motions whose operations would lead to results that were consistent with
the observations. This problem demanded an ability to think geometrically.
Of all the ancient peoples, the Greeks seem to have been the first who had

For FIGURE 9.4,
see PLATE X.

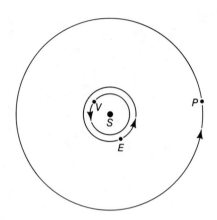

FIGURE 9.5.
In the heliocentric theory of Aristarchus of
Samos, the planets were required to move
around the Sun in simple circles (examples:
$E \equiv$ Earth, $V \equiv$ Venus, P schematic for outer
planet).

enough geometrical skill to make a reasonable attack on the problem. Indeed,
Aristarchus of Samos in the third century B.C. arrived at a picture strikingly
in accord with our modern view, the heliocentric picture shown in Figure 9.5.

Not only was it supposed in the picture of Figure 9.5 that the orbits are
simple circles, but the planets were supposed to move around the circles at
uniform rates. The rates were not the same from planet to planet, and had
to be determined from a comparison with the observations, as did the relative
sizes of the orbits. Since the observations were only concerned with the directions of the planets in the sky, it was not necessary to know the absolute scale
of the solar system. Nevertheless, Aristarchus made an ingenious attempt to
determine the distance from the Earth to the Sun, a method illustrated in Figure
9.6. He arrived at a result of about four million miles, considerably less than
the actual value of some 93 million miles, but at least showing the scale of
the solar system to be large judged by terrestrial standards.

It is remarkable that, even though the actual orbits of the planets are but
little different from circles, as we have seen in Figure 9.2, the slight deviations
produce quite large errors when we attempt to use the Aristarchus model of

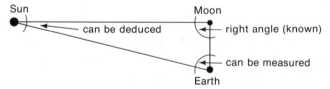

FIGURE 9.6.
When the Moon is halfway from being new to being full, the lines
from the Moon to the Sun and to the Earth form a right angle. The
angle between the lines from the Earth to the Sun and Moon can be
measured. Thus the shape of the Sun-Moon-Earth triangle is then
known, and the ratio of the Sun's distance from the Earth to that of
the Moon can be determined.

Figure 9.5 to predict ahead of time where a planet is going to be. Suppose we have been lucky enough to fix on the correct periods of revolution of the planets in their orbits, and on an appropriate relative scale for the circles of Figure 9.5. The errors of prediction will still far exceed the observational errors, even for observations with the naked eye, especially for the planets Mercury and Mars, whose orbits differ the most from circularity among those of the six planets known in ancient times. The predicted position of Mars could differ from the observed position in the sky by as much as 15°, which is some three hundred times greater than errors of observations made with the naked eye. Such a discrepancy could not be accepted by the astronomers of the ancient world, even though the theory of Aristarchus explained the main features of the planetary motions in an elegant style. It has been a characteristic of scientists in all ages that they will abandon an otherwise attractive theory if it has consequences that are found to disagree with observation or experiment.

What was to be done to improve the situation? The correct method, as we see today, would have been to ask the direct question: What are the actual orbits? Strangely enough, the determination to seek an answer to this uninhibited query had to wait fifteen hundred years, until Kepler. The Greeks felt convinced that orbits had somehow to be circles. So they forced their enquiry into a constrained form: "How can one represent the planetary orbits by a combination of circles that will reproduce the observed positions of the planets?" The concept that the orbits must be built from circles came from Greek ideas of art and of form. For many centuries the Greeks had maintained what today we would call their "international balance of payments" largely by the export of skillfully painted pottery. This activity may well have been responsible for stimulating the geometrical basis of Greek thinking, thereby giving a strong weapon of attack on the problem of the planetary motions. On the other hand, it suggested certain ideas of form and shape which then impeded a full resolution of the problem. So a cultural trait can both help and hinder. Our own modern culture probably has similar characteristics.

A major attack on the problem, using circles only, was made in the second century A.D. by Claudius Ptolemy, a skilled geometer fully conversant with the discoveries that had been made during the four centuries which separated him from Aristarchus. Ptolemy's theory succeeded in predicting planetary positions to within a maximum error of about 1°. It contained two major changes from the theory of Aristarchus.

The first change is illustrated in Figure 9.7. The planetary orbit is still a circle, but the Sun, represented by the point S, has now been offset by a small amount from the center C of the circle. Not only that but a point A, related to C and S by two conditions—that all three points are collinear, and that $AC = CS$—has the following curious property. The line from A to the planet P turns at a uniform rate, whereas the line to the planet from the center C of the circle does not. It was only this first change that improved the accuracy of the theory.

FIGURE 9.7.
The basic construction of Ptolemy. The planet
P moves around a circle of center *C*, but it is
the direction from *A* to *P* which turns round
at a uniform rate. The Sun *S* is placed so that
the distance *CS* is equal to *AC*. This is the
heliocentric form of Ptolemy's construction.

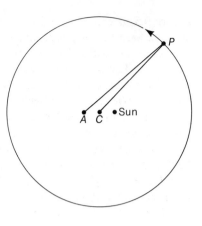

FIGURE 9.8.
The geocentric form of Ptolemy's theory was
obtained by adding an epicycle to Figure 9.7,
and by replacing the Sun *S* by the Earth *E*.

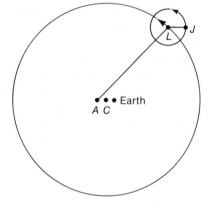

The second change was to refer the motion of each planet to the Earth instead
of to the Sun. This was done by replacing the Sun in Figure 9.7 by the Earth
and by also adding an epicycle. This is shown in Figure 9.8, drawn for the
planet Jupiter.

Ptolemy was free to choose several quantities in his theory. He could choose
the relative scale of the two circles of Figure 9.8, the rates of motion in these
circles, and also the amount of the offset from *C* to *E*. From our modern point
of view, it is remarkable that his choices for all these quantities were very nearly
optimum for obtaining the best fit of his theory to the actually observed
planetary motions.

The second of Ptolemy's changes away from the theory of Aristarchus,
namely, his preference for a *geocentric* theory rather than a *heliocentric* one,
was in no way helpful to accuracy. However, for convenience in calculation,
the geocentric theory had an advantage. In a heliocentric theory, planetary
positions are predicted with respect to the Sun, not to the Earth. Since all

observations are made from the Earth, however, a further calculation would be required in a heliocentric theory before comparison with observation would become possible. The need for this further calculation is obviated by passing from Figure 9.7 to Figure 9.8.

History has it that Ptolemy used Figure 9.8 instead of Figure 9.7 because he believed the planets went around the Earth, not around the Sun. But I suspect all we really know is that the *followers* of Ptolemy adopted this point of view. It seems entirely possible that convenience of calculation was Ptolemy's main motive for adopting the geocentric scheme, rather than any deep feelings that it represented a more correct view than the heliocentric picture. Today we understand that the two theories are essentially equivalent to one another. The difference is one of relative motion only.

§9.2. Copernicus

We must come forward almost fourteen hundred years from the age of Ptolemy (A.D. 150) to that of Nicolaus Copernicus (1473–1543) to find effective new ideas appearing on the problem. Copernicus greatly disliked the idea that a planet revolved at a uniform rate about a point which was not the center of a circle, the point A of Figure 9.7. He discovered that Figure 9.7 is mathematically equivalent to Figure 9.9, in which the Sun is still offset from the center of the main circle, but in which the planetary motion is represented by two uniform circular motions, that of L about K, and that of P about L. To obtain the mathematical equivalence just mentioned, Copernicus found that the second rate of turning (with respect to a fixed direction) had to be exactly twice the rate of turning of the line from K to L.

It will be seen that Copernicus was thereby forced into an *epicyclic* theory, even when planetary positions were reckoned with respect to the Sun. To have then proceeded to the second of Ptolemy's changes, to have referred all motions

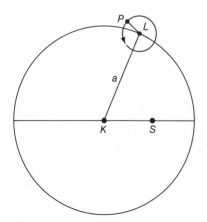

FIGURE 9.9.
The heliocentric theory of Copernicus also involved an epicycle. The line KL turns at a uniform rate. The line from L to the planet P turns at twice this rate, and the distance L to P is $\frac{1}{3}$ of the distance from K to the Sun S.

to the Earth in a geocentric theory, would have involved Copernicus in a double system of epicycles. It may well have been a dislike of the unpleasant complexity of such a scheme that caused Copernicus to reject the second step of Ptolemy. He continued to refer planetary positions to the Sun, not to the Earth. This turned out to be particularly fruitful for the later work of Kepler and Newton. Indeed, it is this aspect of the theory for which Copernicus is popularly esteemed. The view that Copernicus let the planets move around the Sun in simple circles is quite wrong, however. His system, like that of Ptolemy before him, contained epicycles. Copernicus, like Ptolemy, was overwhelmingly concerned with the accuracy of his system. He knew perfectly well that a theory based on simple circles was far too inaccurate. He said toward the end of his life that his ultimate hope would be to represent all the planetary motions to within an accuracy of $\frac{1}{6}°$. In this he did not succeed.

We now know that the difference between a heliocentric theory and a geocentric theory is one of relative motion only, and that such a difference has no physical significance. But such an understanding had to await Einstein's theory of gravitation in order to be fully clarified. The situation was not properly understood even in the nineteenth century. In earlier centuries it was believed that an important difference *in principle* existed between the two theories, and in the century following Copernicus a flood of inconsequential passion was released over the issue. In Catholic Europe those who "believed" in the Copernican theory came near to being burnt at the stake. The proscription placed on Galileo, for his advocacy of the Copernican point of view, springs immediately to mind.

This strong cultural reaction to Copernicus had two consequences. The first, chronologically, was to supply an impulse toward astronomical observation (we will save discussion of the second for §9.5). Copernicus had relied heavily on ancient Greek observations of the planets, and to a lesser extent on Arabic observations. European work in his day was comparatively meager. All this changed decisively in the half-century after his death. Tycho Brahe, lavishly supported in Denmark, made new observations of unprecedented accuracy. Quarreling with his Danish patrons, Tycho then moved helter-skelter, instruments and all, to Prague, where for a time he took on the young Johann Kepler as an assistant. Of these days, Kepler remarked, "Tycho possesses the best observations and consequently the material for the creation of a new structure. He also has workers and everything else one might desire. He lacks only the architect who uses all this according to a plan."

Tycho seems to have been impelled to make his observations by a strong desire to disprove Copernicus. This desire of the observer to disprove the theoretician is by no means confined to Tycho or to his moment in history. Most experienced theoreticians nowadays know that irritation is the best goad to focus the attention of observers on a particular problem. Nothing pleases the observer more than disproving the theoretician, and nothing worries him more than discovering something thoroughly new, since this opens up new fields of activity for the theoretician. The ideal situation for the observer is to find that everything fits the concepts which he happens to hold. Since the theoretician lives as a kind of parasite on the discoveries of the observer, this attitude can readily be understood. Actually, the work of both observer and theoretician is necessary to progress in science. Without the one, the other would soon stultify.

PROJECT:
Read pages 99 through 108 in Max Kaspar's *Kepler* (Abelard-Schuman, New York, 1959) for a remarkable insight into the differing outlooks of the observer and the theoretician.

---

§9.3. Kepler

The work of Tycho Brahe made it plain, particularly to Kepler (1571–1630), that the Copernican theory still contained unacceptable inaccuracies. Even though the worst of the discrepancies which could possibly happen did not occur for the planet Mars at the time of Tycho's observations, the discrepancies were still as large as $\frac{1}{6}°$, whereas the observations themselves were good to about $\frac{1}{100}°$. To Kepler, it therefore became imperative to search for an improved theory.

His first attempts were along exactly the same lines as those of Ptolemy and Copernicus. He accepted the constraint, imposed by the Greeks, that his theoretical orbits must be constructed out of circles. At first he was able to make some progress toward a better theory,* but after a very great deal of fruitless calculation, he finally abandoned the old system. He even abandoned for a while the ultimate aim of being able to predict correctly the position of a planet in the sky. Almost in desperation, he asked the critical geometrical question posed at the beginning of this chapter: "What *is* the orbit, anyway?" Partly because Kepler was willing to face an almost inordinate amount of calculation, and partly because the data necessary to answer this question had been accumulated by Tycho Brahe (and because Kepler had the good fortune to have access to it), he was at last able to arrive at an answer. To Kepler's chagrin, the answer turned out to be ludicrously simple. The orbit of Mars, the most difficult of all the planets to fit into a satisfactory scheme, was nothing but a squashed circle.

Figure 9.10 shows the kind of squashing of a circle required to produce an *ellipse*. One dimension of the circle is simply reduced relative to the other dimension. In Figure 9.10 the *y* dimension has been reduced by a squashing

---

*In fact, by using increasingly complicated patterns of circles one could, as we know today, approach more and more closely to the correct theory. We also can see today that it would hardly be possible, without first knowing the correct theory, to guess the complex systems that would ultimately be necessary to represent the motions in this way.

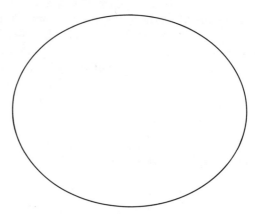

FIGURE 9.10.
An ellipse with eccentricity *e* equal to $\frac{3}{5}$.

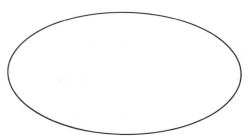

FIGURE 9.11.
An ellipse with the *y* dimension equal to $\frac{1}{2}$ of the *x*
dimension. What is the eccentricity?

factor of four-fifths. It would be possible to reduce the *y* dimension by other
amounts—to a half, for example, as in Figure 9.11—and another ellipse is then
obtained. The amount of squashing of the circle defines what is known as the
eccentricity of the ellipse. The eccentricity is usually denoted by *e*, and it is
calculated in the following way. Take the square of the squashing factor, $(\frac{4}{5})^2$
for Figure 9.10. Subtract this number from one, $1 - (\frac{4}{5})^2$; the result is $\frac{9}{25}$. Lastly,
take the square root of this result, to obtain $\frac{3}{5}$. The value of *e* for Figure 9.10
is therefore $\frac{3}{5}$.

The larger the value of *e*, the more squashed the circle becomes, as may
easily be verified from the rule that *e* is the square root of

$$1 - (\text{squashing factor})^2.$$

The values of *e* for the actual planetary orbits are all much less than 0.6, so
the actual orbits are much less flattened than Figure 9.10, as we saw already

in Figure 9.2. The value of *e* for the planet Mars is about 0.1, corresponding to a very slight measure of flattening, the squashing factor being about 0.995.

An ellipse has the important property shown in Figure 9.12. There are two points, symmetrically distributed on the long axis, known as foci and denoted by *S* and *K*. The sum of the distances *S* to *P* and *K* to *P*, where *P* is *any* point on the ellipse, is always the same. In fact, if one attaches a piece of string to two fixed points and moves a pencil in the manner of Figure 9.13, the point of the pencil traces out an ellipse. Different ellipses (i.e., with different values

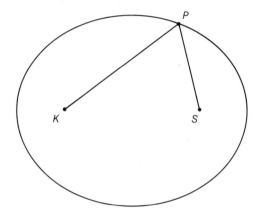

FIGURE 9.12.
There are two points on the long axis of an ellipse known as the foci, points *S* and *K* in this figure, with the property that the sum of the distances *S* to *P*, *K* to *P*, is the same whatever point on the ellipse may be chosen for *P*.

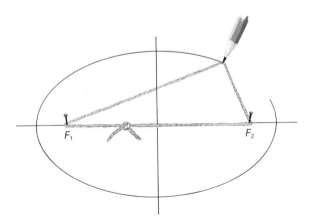

FIGURE 9.13.
The fact illustrated in Figure 9.12 permits the mode of construction of an ellipse shown here (see also Figure 1.9).

of the eccentricity) can be obtained in this way simply by varying the length of the string, as we saw at the outset in Chapter 1 (Figure 1.9).

For Mars, Kepler found not only that the orbit is an ellipse, but also that the Sun lies at one of the two focal points, an interesting and most elegant result. In theory the Sun could lie at either focus of an orbit, although in fact it must lie at one or the other of them. What *cannot* occur is an orbit in which the Sun is not at either focus.

It was necessary next for Kepler to return to the issues he had dismissed from consideration in his search for the *form* of the orbit. Particularly, how did a planet move in its elliptic orbit? An answer to this further question was essential for predicting the position of a planet at a given moment of time. Merely knowing the form of the orbit did not allow any prediction to be made.

The answer to this question also turned out to be exceedingly elegant. The planet did *not* move uniformly along the ellipse in the sense of moving equal distances in equal times. Instead, a line drawn from the Sun to the planet swept out *equal areas* in equal times. This rule is illustrated in Figure 9.14.

A little thought will probably soon convince the reader that Kepler's law of equal areas in equal times is not easy to apply in practice. How actually would one go about predicting where the planet will be located at some time *t* in the future? Suppose the planet is starting now from the point *P* of Figure 9.15, and suppose we know the total time, *T*, required for the planet to make a complete revolution around its orbit. Also, for simplicity, let *t* be less than *T*. Then to find the required position of the planet, we have to go round the orbit, starting from *P*, until we reach a point *X*, where the area bounded by the straight lines *SP* and *SX* and by the arc of the ellipse itself, that is, the shaded area of Figure 9.15, is a fraction *t/T* of the area of the whole ellipse. Although one can easily see that there will indeed be some point *X* with this

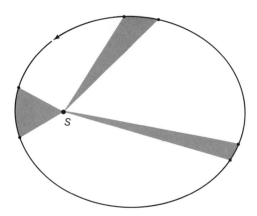

FIGURE 9.14.
Kepler's law of equal areas. A line drawn from the Sun to the planet sweeps out equal areas in equal times. The three shaded portions in this figure having equal areas, the planet takes equal time intervals to traverse the corresponding three segments of the orbit.

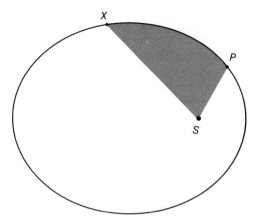

FIGURE 9.15.
Starting with the planet at the point *P*, we wish to use Kepler's law of equal areas to find where the planet will be after a time interval *t*. We have to find a point *X* such that the shaded area constitutes a fraction t/T of the whole area of the ellipse, where *T* is the time required to move around the whole ellipse back to the starting point. How is this problem to be solved?

property, to find it one must be able to calculate an actual numerical value for the shaded area of the figure. There are elementary graphical methods for doing this, but such methods would be exceedingly laborious to use many times over. The reader will probably be able to convince himself that there is no graphic method for solving this problem that is easy to apply. The need for an easily applied method was indeed one of the strong reasons why a new and powerful branch of mathematics was developed about three-quarters of a century after Kepler's discoveries. This new branch of mathematics became known as the *Calculus*. It was indeed because all the early astronomers from Aristarchus to Kepler lacked this branch of mathematics that they found it so hard to discover the orbits and predict the planetary motions.

All theories of the planetary motions must specify the relative sizes of the planetary orbits, as we noted for the simple circular model of Aristarchus shown in Figure 9.5. In place of the radii of these circles in the Aristarchus model, in modern work one usually specifies half of the long axis of the elliptic orbits. Taking the scale of the Earth's orbit around the Sun as a standard of reference, we can compare the values used by Ptolomy and by Copernicus with the modern values, as shown in Table 9.1.
The table shows that both Ptolemy and Copernicus were commendably accurate in their estimates. The values used by Copernicus were in fact based on those of Ptolemy, although he made some small changes in order to take into account observations which he had made himself.

Table 9.1 also gives the value for each planet of the quantity which above we called *T*, the time required for each planet to make a complete circuit of

TABLE 9.1.

The radii a_P and the periods T_P of the planetary orbits

Planet	a_P (unit is radius of the Earth's orbit)			T_P, modern value (unit is the tropical year)	$\dfrac{a_P{}^3}{T_P{}^2}$
	Ptolemy's value	Copernicus's value	Modern value		
Mercury	0.3708	0.3763	0.3871	0.2408	1.000
Venus	0.7194	0.7193	0.7233	0.6152	1.000
Earth	—	1.0000	1.0000	1.0000(4)[a]	1.000
Mars	1.5191	1.5198	1.5237	1.8809	1.000
Jupiter	5.2164	5.2192	5.2028	11.8622	1.000
Saturn	9.2336	9.1743	9.5388	29.4577	1.000

[a]Why is the period for the Earth not exactly unity?

its orbit. Kepler was apparently the first to notice a relationship between the sizes of the orbits and the values of T. The last column of the table gives the cube of the quantity a (one-half of the long axis of the elliptic orbit) divided by the square of T. It is seen that the resulting numbers are all equal to 1. Why?

This was a question that Kepler did not succeed in answering, for a reason connected with the difference between "kinematics" and "dynamics." By a kinematic description of the motion of a body, we mean a literal and correct description without any explanation of the *reason* for the motion. By a dynamic description of the motion, we mean that a physical reason is also given. Kepler's work was kinematic. He was not able to give a satisfactory physical explanation for the orbital motions which he himself had shown to be occurring in space. This was not for want of trying, but because the basic concepts necessary for a physical understanding of the motions were not available in Kepler's time. Nevertheless, Kepler's attempt on the problem is of interest and is worth considering.

The development of a physical understanding of the motions of bodies had been greatly impeded for almost two thousand years by an erroneous idea of the Greeks. Observing the motions of a heavy vehicle like an ox-cart, the Greeks thought it only a matter of common sense that motion required the action of a force on a body. Take away the force, and the body would cease to move, just as the cart did when the ox ceased to pull on it; so the view became widespread that motion implied force. Furthermore, it became natural to add that the motion was always in the same direction that the force was exerted—the cart moved in the direction in which the ox pulled, not sideways or backward.

Throughout the middle ages, thoughtful commentators expressed doubts from time to time about this idea. The bow and arrow became a more formidable weapon in medieval times than it had ever been before. The flight of an arrow became of more concern in men's minds: how to achieve greater distance and greater penetration into a target. Did it really require a constantly operating force to keep an arrow in flight after it had been released from the bow?

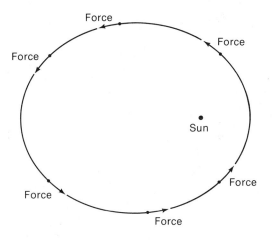

FIGURE 9.16.
Kepler's idea for keeping the planets moving in
their orbits. They were to be pushed continuously
by a force *pointing always in the direction of motion.*

Canonical opinion followed the teachings of Aristotle, which required a force
to be so operating, but the heterodox commentators remained skeptical.

As we see the question today, it does not require a force to keep an arrow
in flight. In fact, the forces which do exist between the arrow and the air actually
tend to impede the flight. The resolution of the ox-cart situation is, of course,
that the bearing surfaces between the wheels and shafts of the cart were so
crude in ancient times that the pull of the ox was required to overcome *friction*.
The cart stopped as soon as the ox ceased to pull because of friction. A car
driven on to a smooth icy lake does not stop when the wheels are locked and
the engine is stopped. The car continues to slide for a very long way. But the
experience of an unchecked skid of this kind was not available to the Greeks!

Kepler had not succeeded in ridding himself of this erroneous idea. He felt
some force must be present to push the planets along their orbits, as in Figure
9.16. He conceived of these forces as being magnetic. Although this notion
was entirely incorrect as an explanation of the ordinary planetary motions, it
has a remarkable similarity to modern ideas of the forces which played an
important role during the formation of the planets themselves. This we will
consider in Chapter 11.

§9.4. Galileo Galilei

It was Galileo (1564–1642) who first succeeded in escaping from the old
erroneous ideas and who thereby laid the foundation of the science of physics.
The concept he eventually arrived at was this:

A force is acting on a body only when its motion *changes.*

Notice that the motion of a body can change in two ways, and that both require
the operation of a force: the *speed* of the body can change, and the *direction*

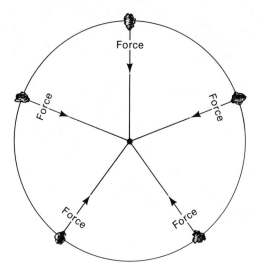

FIGURE 9.17.
To keep a stone whirling around in a circle, a force
must be applied to the stone, always in a direction
toward the center of the circle.

of its motion can change. We can imagine a stone whirled around in a sling
at a constant speed. The changing direction of the motion requires the operation
of an inward force, as in Figure 9.17. We are all familiar with this effect from
the experience of driving an automobile around a curve.

So the idea of Galileo amounted to this: Only for a body moving uniformly
in a straight line or for a body at rest is there no force. This concept is very
simple when one knows about it, but it was difficult to formulate, for the reason
that nobody has ever observed bodies which move uniformly in a straight line!
An arrow in free flight moves with an approximately uniform speed, but its
direction changes because of the gravitational force which the Earth exerts upon
it. An arrow discharged from a space vehicle would come nearer to meeting
the required condition, but even then the motion would deviate from a straight
line because of the gravitational effect of the Sun. The concept of the condition
for zero force is therefore quite idealized. It is of interest to see how Galileo
arrived at it.

The classic experiment of Galileo is illustrated in Figure 9.18. A heavy ball
rolled down a bowl rises on the far side of the bowl to the same height as
that at which it started. Notice this is something one can actually do, an actual
experiment.

In such an experiment the ball does not rise to quite the same height that
it started from, because there are small forces which tend to check the motion—
friction between the moving ball and the air, and friction at the small area
of contact of the ball and bowl. But Galileo, unlike the Greeks, was fully aware
of the effects of friction. Consequently it was not hard for him to understand
that, if the frictional forces had been absent, the ball really would have risen

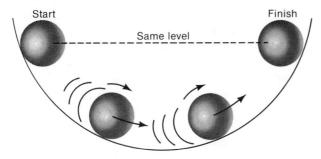

FIGURE 9.18.
A heavy ball rolled down a circular bowl rises on the far side of
the bowl to the height from which it started.

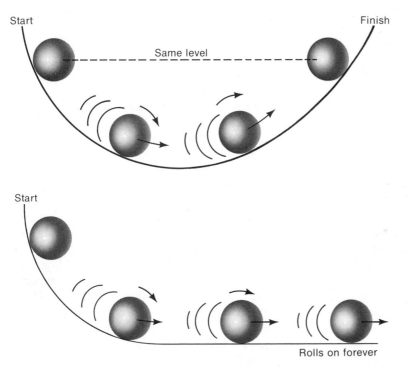

FIGURE 9.19.
Galileo noticed that this remains true for an elongated bowl.

to the height from which it started. Now change the shape of the bowl to the
form shown in Figure 9.19. Start the ball rolling down the steep side, and let
it roll up the gentle slope of the bowl. What happens? Just the same as before.
It rises to the height from which it started. Once again this is an experiment
that actually can be done. Proceed in the same way *a fortiori*. Lengthen the

bowl, making the slope on the one side more and more gentle. Always the ball rises to the height from which it started. And now at last we come to the step we cannot actually perform: imagine the bowl to be lengthened indefinitely. What would happen? Galileo's answer was that, provided there were no frictional forces, the ball would roll on and on forever. It would do so, seeking to rise to the height from which it started. The motion continues uniformly in a straight line if there is no force.

This simple argument broke the stranglehold of Greek ideas, opening the way to the modern development of the physical sciences. In seeking to dam back the flood of new ideas, the Roman Church placed Galileo in his later years under house arrest for advocating the Copernican theory. Ironically, it was precisely during the period of house arrest that these ideas occurred to him, ironic because in these ideas Galileo had a truly explosive effect on the course of science, an effect with far greater ultimate consequences than his advocacy of the Copernican theory could ever have had.

§9.5. Hooke

The hardening of the Roman Church against science and astronomy, centering around opposition to the Copernican theory, had another important cultural effect (the second of the two mentioned on p. 416). As the seventeenth century wore on, the development of new ideas was forced more and more into the Protestant lands of northwestern Europe, where, in the second half of the seventeenth century, the next big steps were taken.

Although to our modern way of thinking, the work of Kepler, perhaps more than that of Copernicus, was decisive in setting the direction in which progress was to be made, Kepler's discoveries were not regarded as central to the problem of dynamics except in northwestern Europe. They were especially highly regarded by the fellows of the newly formed Royal Society in London. Just as Kepler himself had achieved progress by not asking the hardest question first, so a group of English scientists, particularly Robert Hooke (1635–1703), put aside the most remarkable of Kepler's discoveries: the elliptic form of the planetary orbits. Instead they concentrated on the remarkable equality of the numbers in the last column of Table 9.1, an equality that became known as Kepler's "third law." What did it mean?

In attempting to answer this question, they decided to ignore the small deviation of the actual planetary orbits from circles for the moment. Denoting the radius of a planetary circle by a and the time for a revolution around the circle by T, why was T^2 proportional to a^3?

The first step in answering this question came from noticing that, since the planets are not in uniform motion in a straight line, they must be acted on by a force. Furthermore, we know from the circular motion that the force must act radially inwards, just as it takes a force directed radially inwards to hold a stone whirled around in a sling, as in Figure 9.17. The strength of the force that acts on the stone is found to depend on the following quantities: the size of the stone, the speed with which it is whirled, and the length of the sling. The relation between these quantities and the magnitude of the force was again a matter for experiment. It was discovered by Christian Huygens (1629–1695) that the force is given by the quantity mv^2/a, where m is the mass of the object, v its speed, and a the radius of the circle.

The force increases as the *square* of the speed—this is exactly why we are advised to slow down so much in driving an automobile around a sharp curve. The force also decreases as the radius a is increased, another fact with which we are familiar from our automobile experience; it is easier to drive around a gentle curve than around a sharp one. The mass m of a particle we have already encountered in Chapter 4. Mass is a measure of the quantity of material in our object. If we double the weight of a stone in a sling, we double both the mass and the force required to hold it in circular motion at a given speed.

The force required to whirl a stone around in a circle is supplied by the sling. The force which permits a car to be driven round a curve comes from friction between the tires and the road. What supplies the inwardly directed force required to maintain the planets in their nearly circular orbits? The important answer given to this question is that a force exists between the planet and the Sun, a force pulling the planet toward the Sun. This is said to be a *gravitational force*. Thus, regarding the planetary orbits as simple circles, we have

$$\text{Gravitational force} = \frac{mv^2}{a},$$

where m is now the mass of the planet, v its speed, and a the radius of its orbit.

The circumference of the circle is $2\pi a$, and the time for an orbital revolution, which we denoted by T, is given by dividing $2\pi a$ by the speed v,

$$T = \frac{2\pi a}{v},$$

which is the same as

$$v = \frac{2\pi a}{T}.$$

Substituting $2\pi a/T$ for v, we therefore get

$$\text{Gravitational force} = \frac{m}{a}\left(\frac{2\pi a}{T}\right)^2 = \frac{4\pi^2\, ma}{T^2}.$$

Now Kepler's third law showed that, going from one planet to another, T^2 varied like a^3. We can express this law by writing

$$T^2 = \text{(some constant)} \, a^3,$$

the constant that multiplies a^3 on the righthand side of this equation being the same for every planet. If we knew this constant, which we can denote by K, $T^2 = Ka^3$, we would then know the gravitational force:

$$\text{Gravitational force} = \frac{4\pi^2 \, ma}{T^2} = \frac{4\pi^2 \, ma}{Ka^3} = \left(\frac{4\pi^2 \, m}{K}\right) \cdot \frac{1}{a^2}.$$

But we do *not* know K from this argument alone, because (remembering that mentally we are still in the seventeenth century) we do not know the actual values of a for the planetary orbits. We know only relative values of a. Nevertheless, we have indeed deduced a crucial result. *The gravitational force is proportional to the inverse square of the distance from the planet to the Sun.*

Usually in astronomical texts, this inverse-square form of the gravitational force is stated as an axiom. Then the above argument is applied in reverse to deduce Kepler's third law. This is not the way the pioneers of astronomy proceeded, however. Starting from Kepler's law, they derived information about the gravitational force.

It is worth inserting at this point an argument of a kind that is typical of theoretical physics. The contestants in a tug-of-war soon became aware that both sides feel the tension in the rope between them. So we would think it peculiar if the planet in Figure 9.20 were to experience a gravitational force

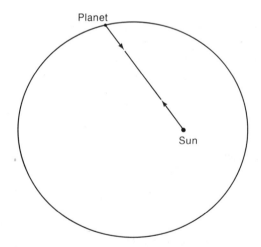

FIGURE 9.20.
The gravitational force which the Sun exerts on a plant is exactly balanced by the force which the planet exerts on the Sun.

from the Sun, and yet the Sun were not to experience a force from the planet. As in a tug-of-war, we expect the two bodies to feel similar forces, as indicated in Figure 9.20. If this argument is correct, the unknown quantity K must involve the mass $\odot$ of the Sun in such a way that we can write

$$\text{Gravitational force} = (\text{some constant})\, \frac{m\odot}{a^2}.$$

The constant here is not dependent on a or on the masses of the planet and of the Sun. This theoretical argument has not achieved much in practical terms, however, since we still do not know the constant which continues to appear in the formula. But at least to the eye of the physicist, the formula has been made to look a bit more tidy!

The constant here is usually written as G,

$$\text{Gravitational force} = G\frac{m\,\odot}{a^2}.$$

In the seventeenth century there was no way to discover the value of G. Two ways of calculating G were used in the eighteenth century. One was to determine the actual scale of the planetary orbits. Thus, with both T and a known in a particular case, $T^2 = Ka^3$ immediately determined K, $K = T^2/a^3$. The other way was to set up a laboratory experiment in which one large piece of matter was placed near another large piece and the gravitational force between them was actually measured. This was the method used by Henry Cavendish (1731–1810) in his famous experiment (see Figure 9.21).

FIGURE 9.21.
In the time of Newton, the distance of the Earth from the Sun was not well-known, so the constant G in Newton's gravitational formula was not known. One way to determine G was by an experiment in which the deflection of hanging pellets (x) toward the large known weights marked W was measured. Such an experiment was performed at the end of the eighteenth century, about 70 years after Newton's death, by Henry Cavendish.

We come now to a quite subtle point. Our argument has determined the form of the gravitational force, particularly the inverse-square dependence on distance, for a planet moving in a circular orbit around the Sun. Can we expect the same law to hold if the planet were moving in some other way? Since our formula does not contain the velocity of the planet, we may feel encouraged to answering the question affirmatively. But so far we have been dealing with a situation where the speed v of the planet does not change. We might imagine that a relation between the gravitational force and the velocity could exist for cases in which the speed was changing. So it would be wrong to be fully confident in the correctness of our formula in every situation. However, it is reasonable not to bring in such complexities unless we have clear evidence of a need for them. Rather does it seem sensible to first take a look at Kepler's other discoveries. Explicitly, given a gravitational force of the form set out above, can we explain:

1. Why planets actually move in elliptic orbits?
2. Why the motion is one in which equal areas are swept out in equal times?

The problem of the planetary motions was apparently first formulated in these terms by Robert Hooke. In astronomy books it is often said that the line of reasoning set out above was first worked through by Newton in the year 1665, during an intermission period from his youthful studies at Cambridge. This story was told by Newton in his old age, well after 1700, long after the events in question, after Hooke was dead. Letters between Hooke and Newton, written around 1679, which have recently been discovered, cast doubt on this claim, however. From these letters it seems clear that in 1679 Hooke understood the importance of the above form of the "law of gravitation" better than Newton.

§9.6. Newton

By normal standards, Robert Hooke was an exceedingly able scientist, but in mathematical skill he could not compare with Isaac Newton. Once Newton was on the right track in his understanding of the law of gravitation, Hooke's attempts to compete in solving the problem of the planetary motions were doomed to failure. Within a period of three or four years, Newton solved the Keplerian problems. Yes, a planet moving around the Sun under the law of gravitation would follow a path which in the general case was an ellipse with the Sun at one of the foci. Yes, the planet must sweep out equal areas in equal times. All this might have been expected from the first really powerful mathematical intellect to be brought to bear on the problem, once it was formulated in the correct terms. If Newton had not solved these problems of Kepler, their eventual solution could not have been long delayed. Some other considerable talent would soon have arisen, even if Hooke himself had not managed to discover a solution. What was unexpected was the shattering breadth of what Newton achieved during the years from 1680 to 1685. He solved far-reaching problems that might well have occupied astronomers and mathematicians for the next fifty years, and he opened up other problems which did in fact occupy mathematicians for a further century and a half. It took them until about 1850 to consolidate what Newton achieved in those five years of intensive effort.

Toward the close of the seventeenth century, Gottfried Wilhelm Leibniz (1646–1716) and Johann Bernoulli (1667–1748) devised between them a problem as a challenge to the mathematicians of all Europe:

Given two points such that the straight line joining them is neither horizontal nor vertical, to find how the curve joining them must be drawn, so that if a particle starts from the top end and falls under gravity, it shall reach the lower end in the least possible time. Notice that the simple suggestion of the straight line itself is not correct. A straight line gives the shortest way, not the quickest way. In the entry in her diary for January 29, 1697, Newton's niece gives the perfunctory statements: "Bernoulli sent problem. I.N. home at 4 P.M. Finished it by 4 A.M." For myself, I rather doubt the implication here that Newton entirely solved the problem at a single sitting. He may very well have written out the solution at a single sitting, but I suspect he spent quite a while beforehand thinking over his method of attack. For in his method of attack Newton invented in embryo a wholly new branch of mathematics, which nowadays we call the *calculus of variations*.

The last part of the story is perhaps the best. Newton allowed his friends at the Royal Society to publish his solution anonymously, an indication of his peculiar character. When Bernoulli saw it, he exclaimed, "Ah! I recognize the lion by his paw!"

It is interesting to add that the problem set by Bernoulli turned out to be much more than a brain teaser. It was a forerunner of the method used nowadays for stating the basic physical laws, the *principle of least action* as it is usually called. And the mathematical method we use today for dealing with this principle is the one already foreshadowed in Newton's solution of Bernoulli's problem.

Newton published his gravitational theory in 1687. As we have already noted, the book, *Philosophiae Naturalis Principia Mathematica*, contained much more than a solution of Kepler's problems. There is obviously nothing special about the gravitational force of the Sun, except that the Sun is much the most massive body in the solar system, and so produces the largest gravitational effects. But in a similar way one planet will also exert a gravitational force on another. How does one go about working out the motions of the planets taking account of all the mutual gravitational interactions that exist within the solar system? For the most part, the practical details of such a complex problem were not of any great consequence in Newton's time, except for one celestial body: the Moon. To have any hope of explaining the observed motion of the Moon, one must take into account both the Earth's gravitational pull on the Moon and the Sun's pull on it. So three bodies are involved in this case, the Moon itself and both the Earth and the Sun. The problem of the interactions and motions of three bodies has never been solved exactly, in the sense that Kepler's problems were solved exactly. All that can be done is to achieve an approximate solution. The problem which Newton tackled in his book was to find out how good he could make his approximate solution. The aim was to make the solution as accurate as the observations themselves. The problem is exceedingly difficult, and has occupied many mathematicians during the three centuries from Newton's time to our own day. Newton told his friends that it was the only problem which had ever made his head ache.

Generations of mathematicians subsequent to Newton have sought to improve the accuracy of the most general formulation of the problem of the planetary motions, a formulation in which all the mutual gravitational forces are included. There is one straightforward way of tackling this problem, however: by literal numerical calculation. Suppose we know where all the planets are right now. Then we can work out what the gravitational forces between them are right now. Provided we also know the present motions, we can work out by actual calculation where the planets will be, and what their motions will be, a short time hence, say, tomorrow. Then, knowing what the situation will be tomorrow, we can simply repeat the calculation, determining the positions and motions a further time ahead, say, the day after tomorrow. And so on, in principle in a never-ending sequence of steps.

The mathematicians of the nineteenth century rejected this simple approach, not only because the volume of calculation which it would require was vastly beyond their means, but also because there are always slight errors in a practical calculation, however carefully one performs it. These errors would not have mattered for a few steps of calculation, but they would have become progressively more important the longer the calculation proceeded. In the end, after very many steps, the errors would vitiate the result. For these reasons, all attacks on the problem proceeded in the nineteenth century in terms of what are called analytical methods, symbols being used instead of actual numbers. The use and development of these methods greatly influenced the teaching of mathematics. Particular tricks that were found useful in the planetary problem were thought "important," and became an essential part of a student's training. In retrospect we now see that such an intense concentration on particular mathematical forms was not really an advantage to science generally.

In recent times these complex old methods have largely been abandoned in favor of the straightforward numerical method, which has been made feasible by the invention and remarkable development of the digital computer, which is many hundreds of millions of times faster in performing actual calculations than even the most skillful human can ever be. Nor do we think it so useful nowadays to calculate where the planets are going to be located a long time ahead. We are content to calculate a few years ahead. Such calculations meet all the needs of the astronomer as well as those of the astronaut.

As an indication of the accuracy of modern numerical calculations, the position of Mercury can be estimated months ahead of time to within an

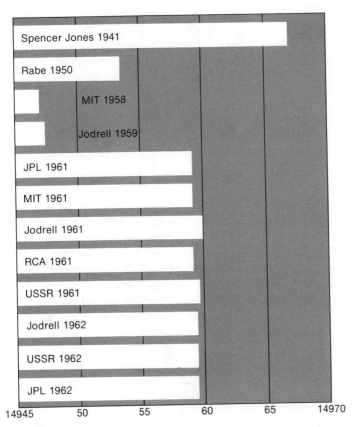

FIGURE 9.22.
Two optical determinations of the average distance of the Earth from the Sun, followed by two radar determinations, gave widely discrepant results. Since 1960, however, eight different radar determinations have given closely the same value, about 1.496×10^{13} centimeters. The unit of distance on the horizontal scale here is 10,000 kilometers.

accuracy of about a kilometer. Since the distance of Mercury from the Sun is about fifty million kilometers, this represents a level of accuracy of about one part in fifty million.

To end this chapter, let us return to the absolute scale of the solar system, which was not known in the time of Newton. The first tolerably accurate calculations of the distance of the Earth from the Sun were made in the eighteenth century. These were improved during the nineteenth century and into the present century. Two of the estimates made in this century, by Spencer-Jones and by Rabe, are shown in Figure 9.22, as are several made by using the quite new modern method of *radar*. The principle of this method is much easier to understand than the earlier ones. Radio waves are emitted from a transmitter on the Earth, in the form of short bursts. The bursts travel across space to some suitable planet, for example, to Venus, where they are partially reflected by the surface of the planet. Although the reflected waves are not directed wholly back to the Earth, enough return to Earth to be detected with the aid of a large radiotelescope, and the elapsed time-interval between the emission of a burst and its reception can thus be measured. (Subtleties are involved in distinguishing one burst from another, but this is a matter of technique, not of principle.) With the elapsed time determined, the distance of the planet in question is immediately given, since the speed of travel of the radio waves is known, being about 300,000 kilometers per second. Thus we have

$$\text{Distance} = (\text{elapsed time}) \times (\text{speed of waves}).$$

Once the actual distance of a single planet has been measured, the scale of the solar system is then established, because the relative scale of the planetary orbits is already well-known, as we have mentioned on several occasions.

By 1962, the radar technique had become well-established. The results from workers in many countries have agreed very closely with each other, as can be seen from Figure 9.22. The new method is plainly more accurate than the old, which illustrates yet again the crucial importance of improving our techniques of measurement. The average distance of the Earth from the Sun is seen to be nearly 149,600,000 kilometers.

General Problems and Questions

1. Describe the layout of the solar system, and relate the scale of the planetary orbits to the size of the Sun.

2. According to the theory of Eudoxus and Aristotle, each planet moves around the Earth at a certain fixed distance. By considering the variation in the brightness of Mars that occurs as Mars moves around its orbit, develop the criticism which caused this theory to be abandoned.

3. Why was the heliocentric theory of Aristarchus of Samos abandoned?

4. How did Aristarchus attempt to discover the distance of the Sun?

5. Illustrate by sketches the constructions of Ptolemy and Copernicus. Satisfy yourself that both lead to orbits of nearly elliptic forms.

6. Two ellipses have the same long axis. One has eccentricity $\frac{2}{3}$, the other has eccentricity $\frac{4}{5}$. What is the ratio of their shorter axes?

7. What discoveries concerning the planetary orbits were made by Kepler?

8. Discuss the Greek concept of force, and explain the reasoning by which Galileo arrived at the quite-different modern concept of force.

9. Write down the formula expressing the magnitude of the force required to keep an object of mass m moving in a circle of radius a with a steady speed v. In what direction must the force act?

10. You are traveling in a car driven around a circular curve at a steady speed. How much greater is the force experienced at 60 m.p.h. than at 30 m.p.h.? Does the force experienced depend on the mass of the car? On other persons in the car? On your own mass?

11. How was the answer to Problem 9 used, together with Kepler's third law, to arrive at a formula for the gravitational force between the Sun and a planet? What was this formula?

12. What was the gravitational constant, and how was it measured?

13. State the difference between a kinematic theory, like that of Kepler, and a dynamic theory, like that of Newton.

14. A circular disk rolls in a vertical plane along a horizontal table. Sketch the path of a point on the circumference of the disk. The answer to the problem of Bernoulli and Leibniz was a curve of this shape.

15. Why is the precise calculation of the actual orbit of a planet far more difficult than for the comparatively simple case of the Sun and planet alone?

16. Describe the method by which in recent years the average distance from the Earth to the Sun has been accurately measured.

Chapter 10:
The Physical Characteristics
of the Solar System

§10.1. The Discovery of New Planets

The relative sizes of the Sun and its nine planets are shown in Figure 10.1, and an attempt is made in Figure 10.2 to show the spatial arrangement of the planets. By comparing Figure 10.2 with a diagram given by Copernicus in his book *de Revolutionibus Orbium Caelestium* and reproduced in Figure 10.3, we can see that there are three additional planets in Figure 10.2, all lying beyond Saturn, which were not known to Copernicus. The first of these, Uranus, was discovered in 1781 by William Herschel (1738–1822). Herschel first noticed a wandering body which did not remain in the same position with respect to the stars. The nature of its motion soon made it clear that the new object was in orbit around the Sun, more or less in a circle, like the other known planets. However, its distance from the Sun was greater than for any of the then-known planets, as could be judged from Kepler's third law, T^2 proportional to a^3, where a represented the radius of the orbit and T the time for a revolution around the Sun. Since the motion of Uranus on the sky was soon found to be nearly three times slower than the motion of Saturn, it was immediately apparent that the value of a for Uranus had to be about twice that for Saturn. The actual values for a and T for all the planets are given in Table 10.1.

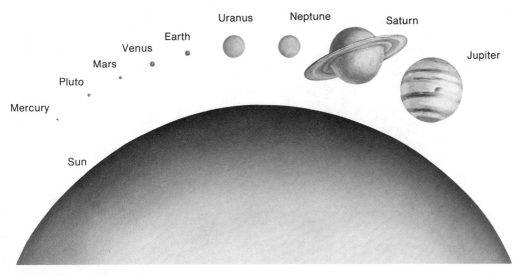

FIGURE 10.1.
The nine planets of the solar system in scale relative to the Sun, as in Figure 9.1 (From
J. Gilluly, A. Waters, and A. Woodford, *Principles of Geology*, 3d ed. W. H. Freeman and
Company. Copyright © 1968.)

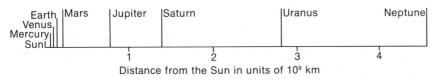

FIGURE 10.2.
The distances of the planets from the Sun.

This argument showed Herschel that Uranus was a planet. Besides, the new
object was seen not as a point of light, but as a disc like the other planets,
not very large, about 3 arc seconds in diameter, but clearly perceptible with
a reasonable telescope, even in as unfavorable a climate as that of England.
At first, because of its small apparent size, Herschel must have thought the
new planet to be only a small one, but when its motion was found to be so
slow, and hence its distance so large, he realized that it need not be small.
In fact, Uranus is a very substantial planet, with a diameter nearly four times
that of the Earth.

The discovery of a new planet had a great impact on popular opinion in
England, much as an actual observation of a UFO might have on us today.
It secured for Herschel an annual grant from King George III which enabled
him to quit his profession as a musician and to become a full-time astronomer,

TABLE 10.1. *Planetary data*

Planet	Half of long axis of orbit (km)	Eccentricity	Average orbital speed (km/sec)	Time around orbit (years)	Mass (gm)	Earth masses	Radius (km)	Density (gm/cm³)	Axial rotation period	Tilt of rotation axis to orbital plane
Mercury	5.791×10^7	0.206	47.90	0.2408	3.3×10^{26}	0.056	2439	5.4	58.7 days	7°
Venus	1.082×10^8	0.007	35.05	0.6152	4.9×10^{27}	0.81	6050	5.1	243 days	6°
Earth	1.496×10^8	0.017	29.80	1.0000(4)	6.0×10^{27}	1.00	6378	5.52	23h 56m	23.5°
Mars	2.279×10^8	0.093	24.14	1.8809	6.4×10^{26}	0.11	3394	3.97	24h 37m	24°
Jupiter	7.783×10^8	0.048	13.06	11.8622	1.9×10^{30}	318	71,880	1.33	9h 55m,a	3°
Saturn	1.427×10^9	0.056	9.65	29.4577	5.7×10^{29}	95	60,400	0.68	10h 38m,a	27°
Uranus	2.869×10^9	0.047	6.80	84.013	8.8×10^{28}	15	23,540	1.60	10h 49m	98°
Neptune	4.498×10^9	0.008	5.43	164.79	1.0×10^{29}	17	24,600	1.6	15h	29°
Pluto[b]	5.900×10^9	0.249	4.74	248.4	—	—	—	—	—	—

[a] Temperate zones
[b] Data for Pluto for last six columns is uncertain.

FIGURE 10.3.
This is Copernicus' own drawing, as it appears in his original manuscript. The figure in the printed book is different and inferior, and was probably changed without reference to Copernicus.

a process which no doubt was aided by the name which he gave to the new planet: *Georgium Sidus*. Fortunately, the name did not survive, being black-balled by popular acclaim. The name of Uranus was given to it by the German astronomer J. E. Bode, whom we shall meet again very shortly in a different connection.

Herschel much deserved his grant, for picking up a moving object as he had done, from out of the profusion of stars on the sky, demanded the meticulous skill of the true observer. It is also worth note that he built all his own telescopes, eventually achieving one with an aperture of 48″. Herschel's lifework lay in observing the heavens, to discover without fear or favor what was there. The catalogue of objects which he compiled might be described as the first sky survey. It is interesting that his largest telescope had just the same aperture as the 48″ Schmidt telescope on Palomar Mt., which was used to make our modern sky survey.

Long before the new planet had completed its first revolution about the Sun, it was found that Uranus was not moving quite in an ellipse. At first it was thought that the deviations could be explained as arising from the gravitational influences of Jupiter and Saturn. But by 1840 it had become clear that this was not so. There were deviations remaining even after the effects of Jupiter and Saturn had been allowed for. What could these be due to? Perhaps to still another unknown planet? If so, could the observed deviations be used to infer the whereabouts of the disturbing planet? Could a further new planet be found in this way?

Two young mathematicians began an assault on this problem, U. J. J. Le Verrier (1811–1877) in Paris and J. C. Adams (1819–1892) in Cambridge. Adams started his work a year or two ahead of Le Verrier, but had difficulty in persuading the authorities in charge of the Greenwich Observatory and of the Cambridge Observatory to make the comparatively small effort needed to verify whether the calculated planet was really there or not. Le Verrier may have had a similar problem in France, for instead of promoting a search in France he communicated his results in September 1846 to J. G. Galle in Berlin. It took Galle only a few hours of observation to find the predicted planet, Neptune. In this he was significantly aided by a young French student, H. L. d'Arrest. It will be seen from Table 10.1 that in size and mass Neptune is practically a twin of Uranus.

In our century, Percival Lowell (1855–1916), a New England businessman who became fascinated by astronomy, had dreams of repeating the nineteenth-century story of the discovery of Neptune. His plan was to use observations of the orbit of Neptune to discover a still more distant planet. And indeed, in February 1930, C. Tombaugh, working at the Observatory which Lowell himself had built at Flagstaff, Arizona, found the planet Pluto. So the story was in fact repeated.

Both these predicted discoveries contain peculiarities. That of Neptune has been examined in some depth in recent years, and the resulting hindsight conclusions are worth mentioning. Adams and Le Verrier were both concerned with predicting the direction of the new planet on the sky, in order that it could then be detected observationally. It was less important to them to determine the *distance* of the planet, however. Indeed, both investigators realized that the available data from Uranus would not suffice to calculate the distance of the planet. The direction on the sky was the best they could hope for, and then only if they made an *assumption* concerning the radius a of the orbit of the hypothetical planet. Both made the same assumption, namely, that a for Neptune was some 39 times greater than for the Earth. It is interesting to see how they arrived at this apparently peculiar number.

§10.2. The Titius-Bode "Law"

Measuring the values of a in terms of that of the Earth (often called the *astronomical unit*, a.u.), consider the values known to Adams and Le Verrier, given in Table 10.2.

TABLE 10.2.
Radii of the planetary orbits

Planet	a (astronomical units)
Mercury	0.39
Venus	0.72
Earth	1.00
Mars	1.52
Jupiter	5.20
Saturn	9.55
Uranus	19.2

Consider next the following method of arriving at a set of similar numbers, the method being known as Bode's law, although it appears first to have been noticed in 1766 by Titius of Wittenberg.

(i)	Start with	0.4	= 0.4.
(ii)	Add 0.3:	0.4 + 0.3 =	0.7.
(iii)	Add 0.3 × 2:	0.4 + 0.6 =	1.0.
(iv)	Add 0.3×2^2:	0.4 + 1.2 =	1.6.
(v)	Add 0.3×2^3:	0.4 + 2.4 =	2.8.
(vi)	Add 0.3×2^4:	0.4 + 4.8 =	5.2.
(vii)	Add 0.3×2^5:	0.4 + 9.6 = 10.0.	
(viii)	Add 0.3×2^6:	0.4 + 19.2 = 19.6.	

Apart from case (v), we have a set of numbers which are all quite close to the values of a in Table 10.2. This circumstance led astronomers, who all believed firmly in this numerology, to ask, "Why is case (v) missing from the known planets?" It was then found, by an intensive search for objects with a distance from the Sun of about 2.8 a.u., that indeed such objects do exist. They are small, smaller than the Moon even, but nevertheless they exist. The

TABLE 10.3.
The minor planets Ceres and Pallas

	Half of long axis of orbit (relative to Earth's orbit)	Eccentricity	Period in orbit (years)	Mass (gm)	Radius (km)
Ceres	2.767	0.079	4.60	6×10^{23}	350
Pallas	2.767	0.235	4.61	1.8×10^{23}	230

properties of the two largest, Ceres and Pallas, are shown in Table 10.3. Known as minor planets, or *asteroids,* their discovery appeared to confirm triumphantly the prediction of Bode's law, namely that some planetary-type object should be found at about 2.8 Au. The position that faced Adams and Le Verrier, before the discovery of Neptune, is summarized in Table 10.4. The 38.8 given in the third column comes from extending Bode's law to a hypothetical case (ix), namely, $0.4 + (0.3 \times 2^7) = 38.8$. Given the data known to Adams and Le Verrier, this seemed an entirely natural step for them to take. But fate can be cruel to those who put their trust in numerology. The eventual situation discovered by observation is shown by the two lines in Table 10.5.

Considering the quite large difference between the actual value of about 30 a.u. and the value of about 39 a.u. assumed by Adams and Le Verrier, one

TABLE 10.4.
Radii of planetary orbits and Bode's Law

Planet	a (observed)	Bode's Law
Mercury	0.39	0.4
Venus	0.72	0.7
Earth	1.00	1.0
Mars	1.52	1.6
Asteroids	2.8	2.8
Jupiter	5.20	5.2
Saturn	9.55	10.0
Uranus	19.2	19.6
Hypothetical planet		38.8

TABLE 10.5.
Radii and Bode's Law for Neptune and Pluto

Planet	a (observed)	Bode's Law
Neptune	30.1	38.8
Pluto	39.5	77.2

may wonder why their calculations were so spectacularly successful. It turns out that success depended on the particular configuration which Uranus happened to have in relation to Neptune *at that time*. At other times, for other configurations, the method used by Adams and Le Verrier would not have been successful. Success had about a fifty-fifty chance, something that old textbooks, bent on lauding their achievement, do not usually mention.

The calculation leading to the discovery of Pluto appears to have been even more suspect. Most modern experts in the field of planetary motions with whom I have discussed the matter seem to feel that the discovery of Pluto can only be ascribed to chance.

Leaving aside these historical questions, how good are we to consider Bode's "law" to be? Astronomers today differ markedly in the weight which they attach to it. Some have even made it a cornerstone in attempting to develop a theory of the origin of planets, but others feel it to be a mere set of coincidences. Let us consider the various cases in turn.

(i)   The fit of Bode's law for Mercury is not meaningful because we start with 0.4 AU, chosen to agree closely with the value of a for Mercury.

(ii) The fit for Venus is also not meaningful because the 0.3 used in Bode's law can be regarded as another "free choice."

(iii) The fit for the Earth, $0.4 + (0.3 \times 2) = 0.4 + 0.6 = 1.0$, is not particularly meaningful, because we might have elected to use a multiplier other than 2. For example, if the value of a for the Earth had happened to be 1.3 a.u., then we could have taken $0.4 + (0.3 \times 3) = 1.3$. There is perhaps a little to be argued from the fact that 2 is an especially simple multiplier.

(iv) Having fixed the numbers 0.4, 0.3, and the multiplier 2 for the first three cases, the agreement for Mars is a clear-cut success for the law.

(v)   The asteroids Ceres and Pallas seemed at first to provide another good success for the law, but by now many asteroids are known which disagree with it. Table 10.6 gives corresponding values for the asteroids Juno and Vesta. The position is thereby weakened.

(vi) Jupiter is a clear success.

(vii) Saturn is a clear success.

(viii) Uranus is a clear success.

(ix) Neptune is a serious failure.

(x)   Pluto is an even worse failure.

TABLE 10.6.
The minor planets Juno and Vesta

	Half of long axis of orbit (relative to Earth's orbit)	Eccentricity	Period in orbit (years)	Mass (gm)	Radius (km)
Juno	2.670	0.256	4.37	2×10^{22}	110
Vesta	2.361	0.088	3.63	10^{23}	190

From this analysis, we see that the final score is 4 clear successes, 2 clear failures, 2 cases somewhat uncertain, and 2 no-contests. This is the kind of situation about which each of us can form our own opinion. It would hardly be reasonable for one worker to criticize another for holding a different opinion. For myself, I have always felt Bode's "law" to be a general reflection of a process whereby the planets accumulated out of a swarm of much smaller bodies which at one time orbited the Sun, bodies that started by being even smaller than the asteroids. If one thinks in terms of a gradual process of accumulation, with one body joining another, and with larger bodies eating up smaller ones, then it seems clear that two major bodies will never be formed having closely similar values of a. In a competition between bodies in similar orbits, one body will essentially swallow all the others. It is therefore to be expected that a system of "spheres of influence" must eventually be establishes, with the orbit of each large body well separated from the orbits of the other large bodies. Bode's law might then be an expression of these "spheres of influence."

§10.3. Some General Characteristics of the Solar System

Because the value of a for Mercury's orbit is more than 50 times greater than the radius of the Sun, and because the value of a for Pluto is about 100 times that for Mercury, it is not possible to draw a picture of the solar system to scale on any ordinary-sized piece of paper, unless the Sun is represented as a mere point, as in Figure 10.2. It is important not to get the erroneous idea from Figure 10.2 that the inner planets are packed close to the Sun. If one thinks of the Sun as having a radius of 1 meter, then the radius of the inner part of the solar system, including the planets Mercury, Venus, Earth, and Mars and the asteroids, is about 1 kilometer. The radius of the outer part of the solar system, including the planets Jupiter, Saturn, Uranus, Neptune, and Pluto, is about 10 kilometers.

Turning now to satellites, the newest data for the solar system are summarized in Table 10.7. Neither Mercury nor Venus appear in the table because no satellite moving around these innermost planets has ever been observed. It is immediately interesting to notice that the Moon belongs to a class of larger satellites, constituting Moon (Earth); Io, Europa, Ganymede, Callisto (Jupiter); Titan (Saturn); Triton (Neptune). The Earth-Moon system is also exceptional in that the Moon is much closer to the Earth in size than any other large satellite is to its primary planet, as can be seen from Table 10.8.

The fact that all the large satellites except the Moon belong to the outer group of planets (Jupiter, Saturn, Uranus, Neptune) suggests that the Moon may have originated in the same way as the other large satellites. It is a conceivable speculation that the Moon may once have been a satellite of one of the outer planets, and that, as a result of the combined gravitational influences of the Sun, a planet, and possibly other satellites, it became "perturbed" into an elongated orbit of the kind shown in Figure 10.4. In such a situation one of the inner planets could have captured the Moon, as we may suppose the

TABLE 10.7. *Satellites of the planets*[a]

Satellite	Average distance from planet (1,000 km)	Orbital period (sidereal days)	Direction of Motion in orbit (with respect to planet's rotation)	Tilt of orbit to planet's equator	Radius (km, where accurately known)	Mass (gm, where known)
EARTH						
Moon	384	27.3	same sense	$23\frac{1}{2}° \pm 5°$[b]	1,738	7.3×10^{25}
MARS						
Phobos	9	0.32	same	1°	13[c]	—
Deimos	23	1.26	same	2°	7[c]	—
JUPITER						
I Io	420	1.77	same	0°	1,830	7.3×10^{25}
II Europa	670	3.55	same	0°	1,460	4.8×10^{25}
III Ganymede	1,100	7.15	same	0°	2,550	1.5×10^{26}
IV Callisto	1,900	16.69	same	0°	2,360	9.5×10^{25}
V	180	0.50	same	0°	—	—
VI	11,500	251	same	28°	—	—
VII	11,700	260	same	26°	—	—
VIII	23,500	737	opposite sense	33°	—	—
IX	23,700	758	opposite	25°	—	—
X	11,700	253	same	28°	—	—
XI	22,600	692	opposite	16°	—	—
XII	21,200	631	opposite	33°	—	—
SATURN						
I Mimas	190	0.94	same	2°	—	4×10^{22}
II Enceladus	240	1.37	same	0°	—	7×10^{22}
III Tethys	300	1.89	same	1°	—	6.5×10^{23}
IV Dione	380	2.74	same	0°	—	1.0×10^{24}
V Rhea	530	4.52	same	0°	700	2.3×10^{24}
VI Titan	1,220	15.95	same	0°	2,440	1.4×10^{26}
VII Hyperion	1,480	21.28	same	0°	—	1.1×10^{23}
VIII Iapetus	3,560	79	same	15°	—	1.1×10^{24}
IX Phoebe	12,900	550	opposite	30°	—	—
X Janus	170	0.82	same	0°	—	—
URANUS[d]						
I Miranda	190	2.52	same	0°	—	1.2×10^{24}
II Ariel	270	4.14	same	0°	—	5×10^{23}
III Umbriel	440	8.71	same	0°	—	4×10^{24}
IV Titania	590	13.46	same	0°	—	2.6×10^{24}
V Oberon	130	1.41	same	—	—	1.1×10^{23}
NEPTUNE						
I Triton	350	5.88	opposite	20°	2,000	1.4×10^{26}
II Nereid	5,600	360	same	28°	—	3×10^{22}

[a]Unlisted planets have no known satellites. [b]Tilt varies from 18° to 29°. [c]Oblong shape; this "radius" equals one-half of the long dimension. [d]Note that the rotation of Uranus is retrograde.

TABLE 10.8.
Sizes of planets and large satellites

Planet	Radius of planet (unit, 1,000 km)	Satellite	Radius of satellite (unit, 1,000 km)
Earth	6.378	Moon	1.738
Jupiter	71.9	Ganymede	2.470
Saturn	60.4	Titan	2.5[a]
Neptune	23.5	Triton	2.0[a]

[a] Only two significant figures are given here because these bodies, being distant, appear very small in the telescope, and so their radii have not been determined more accurately.

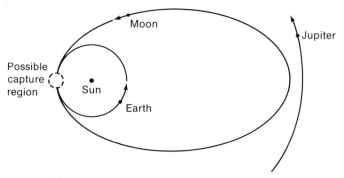

FIGURE 10.4.
The Moon may have formed together with the Earth, but it could conceivably have started its life as a satellite of the outer planets that subsequently became detached, took up an elongated orbit that came within the Earth's orbit, and was captured by the Earth.

Earth to have done. In this way the Moon may have started its life as a satellite of an outer planet, and ended as a satellite of the Earth. This speculation has the interesting consequence that lunar material would then be a sample from the outer solar system, and could have had a chemical history quite different from that of our more familiar terrestrial material.

A somewhat similar idea illustrated in Figure 10.5 has been suggested for the origin of Pluto. The orbit of Pluto departs more from a circular form than that of any other planet. If one traces out in detail the motions of Pluto and Neptune, it turns out that these two bodies occasionally approach each other quite closely. This could be explained if at one time Pluto had been a satellite of Neptune. Then Neptune would have possessed not one but two satellites of the large class, Pluto and Triton. This explanation would likewise require that Pluto, as a result of the combined gravitational influences of the Sun, Neptune, and Triton, would become detached into an elongated orbit about the Sun. But whereas for the Moon the elongated orbit would be required to take the Moon inward toward the Sun, for Pluto the elongation would have

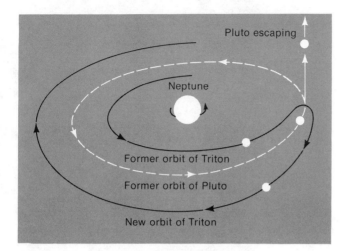

FIGURE 10.5.
Pluto may well be not a planet, but an escaped satellite of Neptune.
The diagram shows how the present satellite, Triton, may at some
stage have overtaken Pluto in its orbit around Neptune. Such an encounter could have speeded up Pluto enough to escape. It would also
have reversed the direction of Triton's motion.

taken Pluto farther away from the Sun. Both these cases are dynamically
possible, and depend on whether the body happens to gain or to lose "angular
momentum" during the complex motions which cause it to escape from the
parent planet.

For FIGURE 10.6,
see PLATE XI.
For FIGURES 10.11 to
10.14, see PLATES XII
to XV.
For FIGURE 10.17,
see PLATE XVI.

In Figures 10.6 to 10.18 we have a set of photographs of the nine planets
and of the Moon. That of Pluto tells us nothing of its surface characteristics.
Those of Mercury, Mars, and the Moon give clear views of a rocky surface,
which on Mars is accompanied by a white polar cap (or caps). Earth presents
a complex picture of cloud, sea, and land, whereas Venus, Jupiter, and Saturn
are wholly cloudbound. Uranus and Neptune are also cloudbound, although
this is determined from criteria other than direct photography.

Table 10.9 gives the measured reflectivities of the planets, of the four
asteroids Ceres, Pallas, Juno, and Vesta, of the Moon, and of the four large
satellites of Jupiter. The scale of measurement is such that a perfectly reflecting
sphere would have a value of 1. It is interesting to regroup the various objects
according to the nature of their surfaces, as in Table 10.10.

It is clear from the associations of Table 10.10 that cloudy surfaces have
comparatively high reflectivities, as also have white patches of frost on an
otherwise rocky surface. A wholly rocky surface, on the other hand, can have
exceedingly low reflectivity. The reflectivities for Mercury, Ceres, Pallas, the
Moon, and Callisto are indeed exceedingly low, about as black as coal. The
surface of the Moon is known to consist of a dusty layer of powdered rock.
It seems likely that the surfaces of Mercury, Ceres, Pallas, and Callisto are
similar.

FIGURE 10.7.
In view of the superb lunar photographs now available from the Apollo and Orbiter missions, we may tend to forget how very good the old pictures obtained with ground-based telescopes really were. The portion of the Moon shown here is about 1,200 kilometers across. (Courtesy of Hale Observatories.)

FIGURE 10.8.
Only from the space missions has it been possible to see the far side of the Moon. Note the large double-ring structure. There are several of these structures on the far side of the Moon, but none on 'our' side. The area shown here is about 1,000 kilometers across. (Courtesy of NASA.)

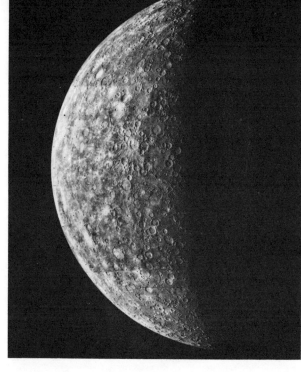

FIGURE 10.9.
A recent picture of the planet Mercury. Mercury is about 4,840 kilometers in diameter. (Courtesy of NASA.)

FIGURE 10.10.
Venus photographed in ultraviolet light from a distance of 720,000 kilometers, by Mariner 10. Winds produce a circulation of the planet in about 4 days. Cloud patterns are controlled by the wind. It has recently been suggested that the clouds may be droplets of sulfuric acid. The diameter of Venus is about 12,300 kilometers. (Courtesy of NASA.)

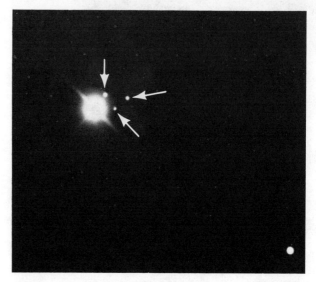

FIGURE 10.15.
Uranus and three satellites. (Courtesy of Lick Observatory.)

FIGURE 10.16.
Neptune and Triton. (Courtesy of Lick Observatory.)

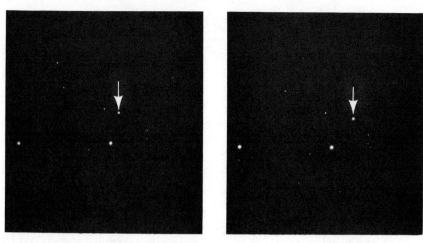

FIGURE 10.18.
The remote and little-known Pluto, well-named. (Courtesy of Hale Observatories.)

TABLE 10.9.
Brightness of planets and satellites

Planet or satellite	Reflectivity (albedo)	Magnitude[a]
Mercury	0.059	− 0.2
Venus	0.85	− 4.2
Earth	0.40	− 3.80
Mars	0.15	− 1.98
Jupiter	0.58	− 2.50
Saturn	0.57	+ 0.70
Uranus	0.80	+ 5.51
Neptune	0.71	+ 7.85
Pluto	0.15	+14.9
Ceres	0.028	+ 7.0
Pallas	0.053	+ 7.9
Juno	0.145	+ 8.7
Vesta	0.27	+ 6.2
Moon	0.068	−12.70
Io	0.73	+ 5.52
Europa	0.68	+ 5.67
Ganymede	0.34	+ 5.18
Callisto	0.13	+ 6.20

[a] Except for the Earth, magnitudes are as observed from the Earth; that for the Earth is as it would be observed from the Sun. Those for Mercury and Venus are at their greatest angular separation from the Sun; all others are for when the planet or satellite is in the opposite direction from the Sun.

TABLE 10.10.
Reflectivities of planets and satellites

Object	Reflectivity	Nature of surface
Venus	0.61	Wholly cloud
Jupiter	0.41	"
Saturn	0.42	"
Uranus	0.45	"
Neptune	0.54	"
Earth	0.34	Partial cloud cover
Mercury	0.060	Wholly rock, probably in powder form
Ceres	0.028	"
Pallas	0.053	"
Moon	0.070	"
Callisto	0.029	"
Mars	0.150	Rock with patches of white frost
Pluto	0.16	Probably like Mars
Juno	0.145	"
Vesta	0.27	"
Io	0.37	"
Europa	0.39	"
Gamymede	0.20	"

§10.4. Earth

Of all the planets we know most, of course, about the Earth. It has become clear during the past ten years that the rocky crustal features of the Earth change on a time-scale of about a hundred million years. The changes are brought about by movements like those of the treads on a caterpillar tractor. What is not known is whether the tractor-like movement is entirely shallow, or whether the circulation goes deep, as in Figure 10.20. We will return to this point in a moment. Several such movements are at work simultaneously, so that the Earth's surface constitutes a system of moving plates, as shown in Figure 10.19.

The rocks of the continents on which we live are of lower density than those of the ocean floors. The continents ride above the oceans, like lumps of cork in a liquid, because their lighter rock is floating in the heavier rock beneath them. The continents are moved around by the conveyer-belt action of the oceanic rocks, a phenomenon which was suspected for many years but which remained a subject of controversy so long as proof did not exist.

The simplest and most immediate argument for this movement of the continents comes from comparing the coastline of West Africa with that of eastern South America. When the first accurate maps became available in the eighteenth century, it was clear that these two coastlines are uncannily like each other, in the sense that they can be fitted together like two pieces of a jigsaw puzzle.

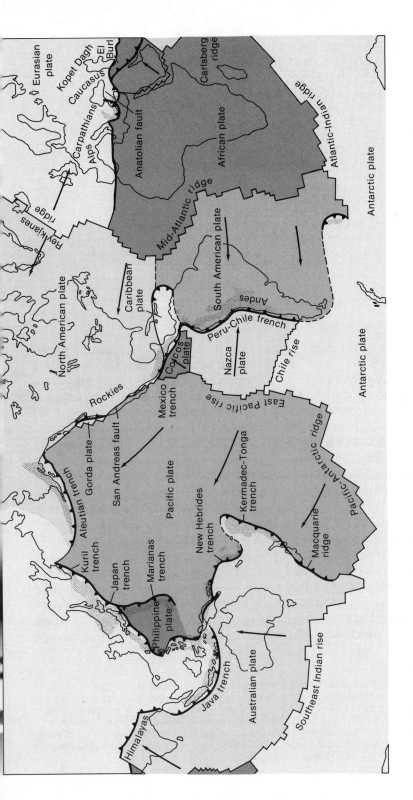

FIGURE 10.19.

The rocks beneath the oceans are known to move in a system of plates. Earth's lithosphere is broken into large, rigid plates, each moving as a distinct unit. The relative motions of the plates, assuming the African plate to be stationary, are shown by the arrows. Plate boundaries are outlined by earthquake belts. Plates separate along the axes of mid-ocean ridges, slide past each other along transform fault, and collide at subduction zones. It is not known, however, whether the tractor-like movement is shallow, affecting only a skin of the Earth, down to a depth of a few hundred kilometers. (After J. Dewey, "Plate Tectonics." Copyright © 1972 by Scientific American, Inc. All rights reserved.)

—————— Subduction zone

—————— Transform

—————— Ridge axis

- - - - - - Uncertain plate boundary

⟶ Direction of plate motion

Areas of deep-focus earthquakes

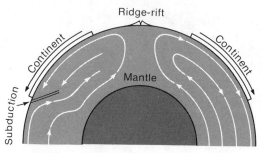

FIGURE 10.20.

We do not know whether the circulation goes into the
deep interior of the Earth. A simple model showing
how convection currents in the deep interior might be
the driving force of sea-floor spreading and continental
drift. Hot matter rises under the ocean ridge and flows
apart, carrying the plates along. (After E. Orowan,
"Origin of Ocean Ridges." Copyright © 1969 by Scientific American, Inc. All rights reserved.)

The shapes of the continents as we see them on a map depends, of course,
on where the continental rocks emerge out of the sea. Since the ocean water
level is scarcely very relevant to fitting continents together, it is better to seek
a fit at the edges of the continental shelves, the places where the continental
rocks give place to the oceanic rocks. These continental shelves are characterized
by a precipitous downward plunge to the ocean floor, often involving descents
of several miles. An attempt to fit South America and West Africa at the edges
of the continental shelves is shown in Figure 10.21. The fit turns out to be
better than for the shapes determined by sea-level, which is a significant
indication that the concept of moving continents, *continental drift* as it is usually
called, is correct. The implication is that Africa and South America were once
joined together. A rift developed between them, and they began to drift apart.
This is thought to have happened about two hundred million years ago.

The rock both on the west coast of Africa and on the east coast of South
America includes many varieties well-recognized by geologists, some of it
predating the time when Africa and South America are supposed to have
separated from each other; i.e., some of the rock is older than about two hundred
million years. If the two continents separated at that time, we would expect
such rocks to match each other, when we imagine the coastlines being brought

FIGURE 10.21.

The fit of South America to the west coast of Africa is much improved when the edges of
the continental shelves are used, instead of irrelevant sea-level shapes. Rocks of similar type
are then found to be associated together, as indicated by the dark areas. In a similar way, the
east coast of the United States can be fitted to Africa, while Europe meets both Africa and
the United States. This association together of the continents is believed to have existed about
200 million years ago. (After P. Hurley, "The Confirmation of Continental Drift." Copyright
© 1968 by Scientific American, Inc. All rights reserved.)

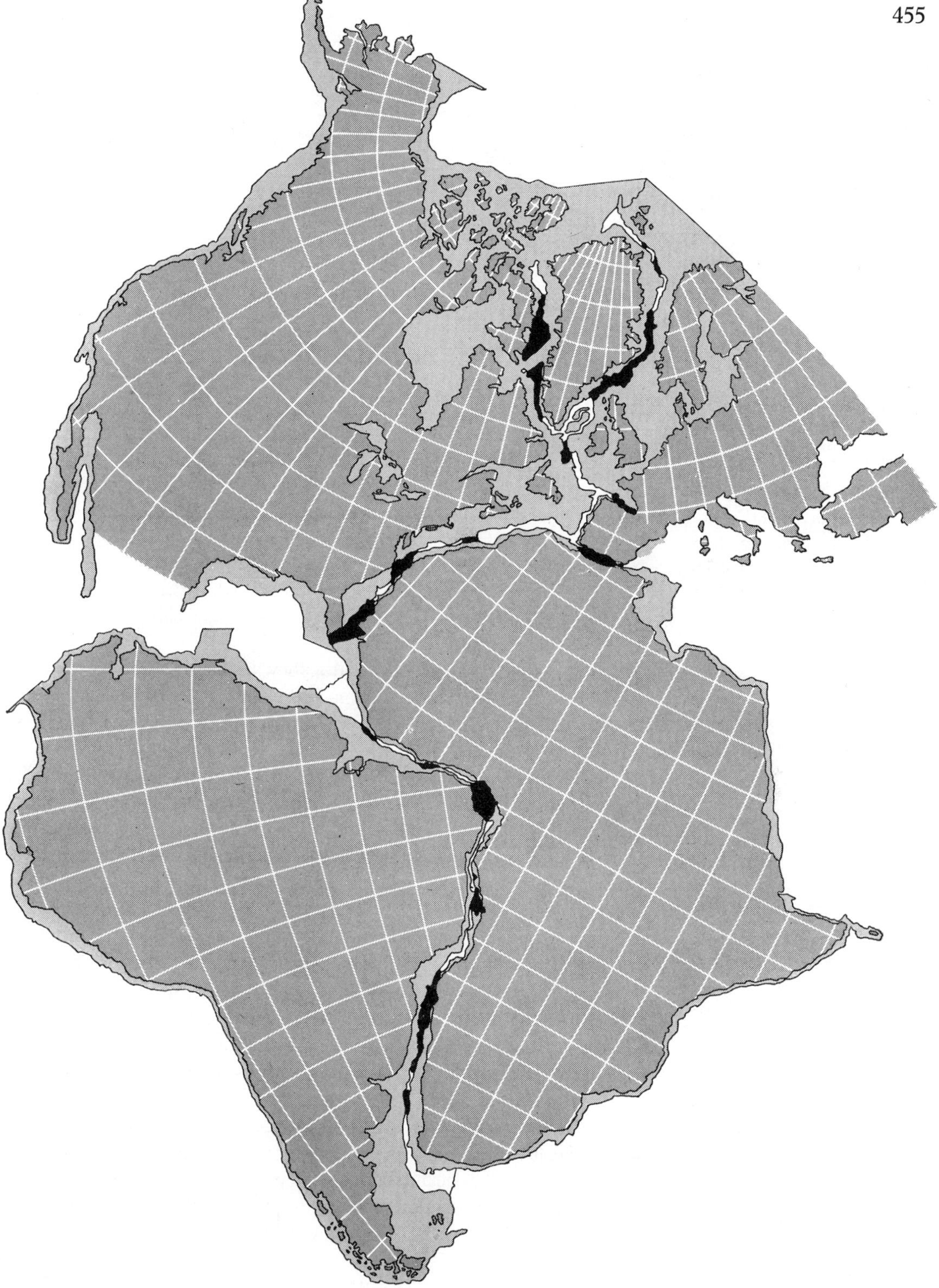

456

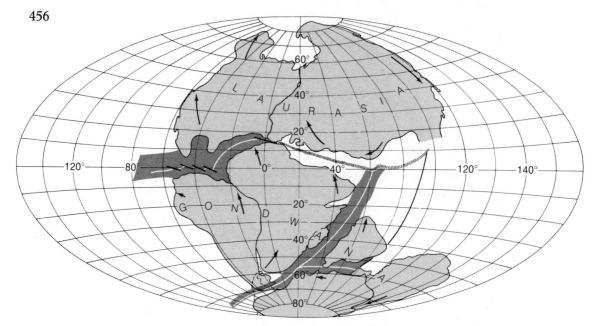

FIGURE 10.22.
Schematic fitting of the continents. (After R. Dietz and J. Holden, "The Breakup of Pangaea." Copyright © 1970 by Scientific American, Inc. All rights reserved.)

together. From Figure 10.21 we can see that they do. Such a test of continental drift is rigorous and specific, and leaves little doubt that Africa and South America really did separate from each other, however incredible the idea may seem.

Using similar criteria, geophysicists have come to the conclusion that the events which occurred some two hundred million years ago were on a still bigger and more cataclysmic scale than this, that indeed all the present continents have come from the breakup of two vast continents, to which they have given the names Laurasia and Gondwanaland. The way in which the present continents were then fitted together is shown in Figure 10.22.

Of course, the movements of the oceanic floors and of the continents are not like those of a manmade piece of machinery, on oiled bearings designed to make motion as frictionless as possible. Enormous frictional forces are brought into play in all these terrestrial movements, and it was for long a mystery to most geophysicists how the Earth managed to generate such forces. A driving engine is clearly needed. Nowadays there is a fairly general agreement about what the power supply for such an engine must be, namely, heating within the body of the Earth due to the radioactive decay of the elements uranium and thorium. Heat released in these decays remains trapped *in situ*, unless it can be transported to the surface by some form of motion. The idea is that the heat accumulates *in situ* until a softening of the inner rocks occurs, at which stage motion begins and develops into what we recognize at the surface as a system of plate movements. The forces necessary to split a continent apart comes from this motion. It turns out that such an operation places no strain on the heat engine within the Earth, which generates ample energy to power all the movements we see at the surface.

The main topic of present-day controversy concerns the depth at which the motions are generated, whether the plate movement is shallow or deep, as in Figure 10.20. The difficulty with shallow motion is to understand how the plates can develop horizontal motions of several thousand kilometers from an engine with a depth of only a few hundred kilometers. The difficulty with deep motion is that the rocks of the Earth would be expected to become more viscous, and therefore harder to move, the deeper one goes; so the question is whether the driving engine could be powerful enough for this motion. The controversy here is still unresolved.

The motions of the inner conveyor-belt system produce many interesting effects at the surface. Sometimes we may expect two continents to be pressed against each other, causing powerful compression and buckling of the continental blocks along their edge of contact. This is occurring now between Africa and Asia, with the edge of disturbance stretching from the Alps, through Yugoslavia, Turkey, and Iran, to the Himalayas. Or an oceanic plate may press against a continental block, again causing buckling and uplifting, as we see on the west coast of America, with the edge of contact stretching from Alaska all the way to southern Chile. Or two oceanic plates may have an edge between them. If the plates are both in downward motion at their edge of contact, as in Figure 10.23, a trench is formed, such as are found along the great deeps of the Pacific Ocean. But if the plates are both in upward motion at their edges of contact, as in Figure 10.24, a ridge is formed. The mid-Atlantic ridge is an example of such an upwelling of the plates. The maps in Figures 10.25 and 10.26 show these features.

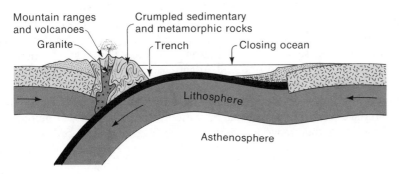

FIGURE 10.23.
Plates that are downward-moving at their edge of contact produce the great deeps of the ocean. (After R. Dietz, "Geosynclines, Mountains, and Continent-Building." Copyright © 1972 by Scientific American, Inc. All rights reserved.)

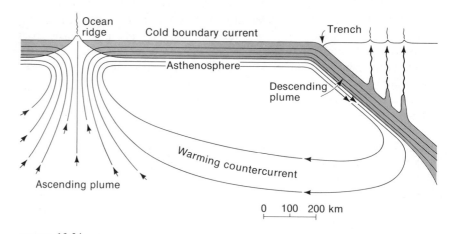

FIGURE 10.24.
Plates that are upward-moving at their edge of contact produce long ridges which rise up from the ocean bed. The vertical and horizontal scales are equal. (From F. Press and R. Siever, *Earth*. W. H. Freeman and Company. Copyright © 1974.)

The effect of friction along all these edges, whether between continent and continent, plate and plate, or plate and continent, is to produce heat. This causes rocks below the surface to be melted. Such fluid rock seeks to rise to the surface, and does so in the phenomenon of the *volcano*. Volcanoes and earthquakes are found along all these edges.

In plate movement of the kind shown in Figure 10.24, the upwelling material along the interface is also hot, simply because the rock has risen from the regions where it has been heated by the radioactive decays of uranium and thorium. The rock contains small particles which become magnetizable as they cool on reaching the surface, and are then magnetized by the Earth's own magnetic field. Although during a human lifetime the Earth's field does not change much, major changes occur on time-scales much shorter than the 100-million-year scale of the plate movements. Every few hundred thousand years, the magnetic field actually reverses its sign. If one were to imagine this happening now, the needle of an ordinary magnetic compass would swing around from pointing north to pointing south. What this means is that, as the rocks welling up at an interface such as that of the mid-Atlantic ridge become cool, they become magnetized in the direction that the Earth's magnetic field happens to be pointing at the time of their cooling. This causes the rocks that emerge from the interface to have different magnetization patterns, which will be retained by the rocks as they move horizontally away from the interface between the plates. Since a two-way motion is involved here—for example, if the interface is oriented north-south, there is motion from the interface both to the east

FIGURE 10.25.
The Pacific Ocean floor. (Courtesy of National Geographic Society. © 1968.)

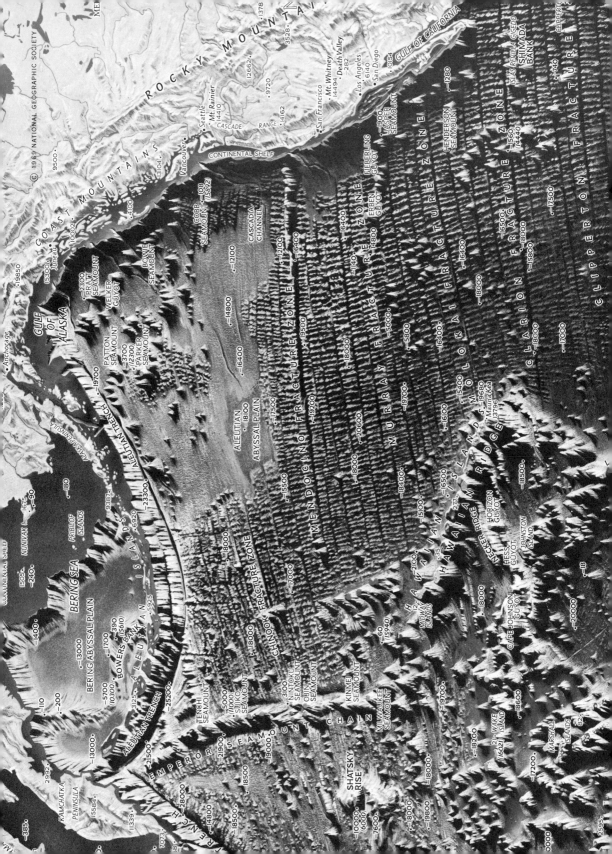

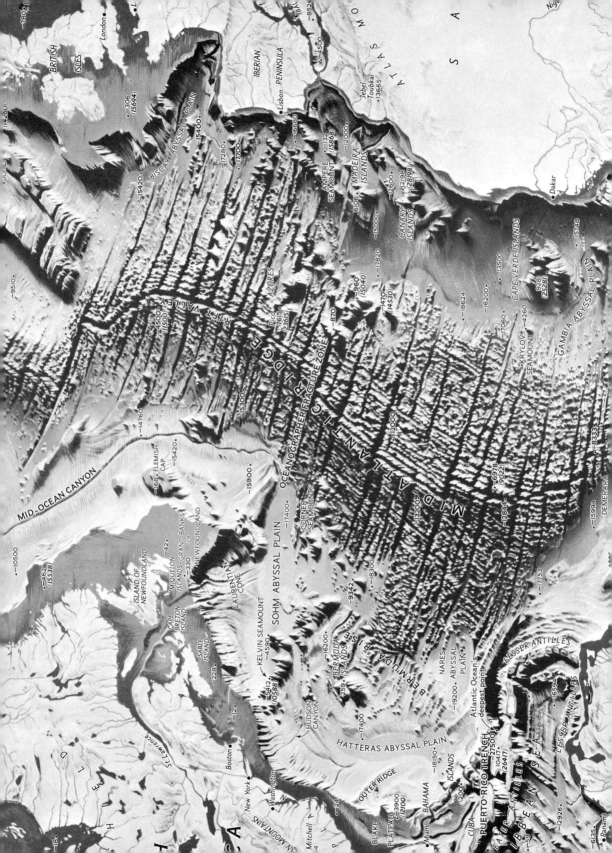

and to the west—we expect the magnetic variability to form matching patterns on the two sides of the interface. This provides a crucial test of plate motion. Results for the mid-Atlantic ridge are shown in Figure 10.27. The matching of details of the magnetic variations is so very good that plate motion must now be regarded as well-established.

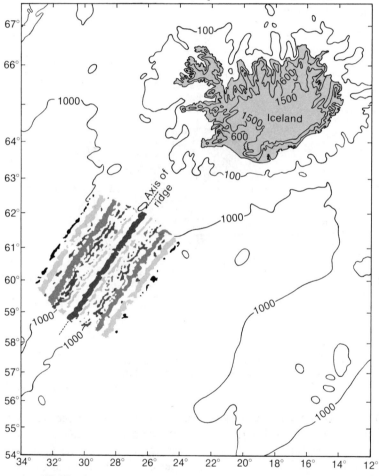

FIGURE 10.27.
The proof that the floors of the oceans are indeed moving: molten rock emerging from the Earth's interior becomes magnetized as it cools. The rock spreads from the ridge both westward and eastward, forming two patterns of magnetized strips on the ocean floor, which correlate with one another in remarkable detail. (After E. Orowan, "Origin of Ocean Ridges." Copyright © 1969 by Scientific American, Inc. All rights reserved.)

The Earth's magnetic field is thought to be maintained by electric currents flowing in a central metallic liquid core which extends to a radius of about 3400 kilometers, rather more than a half of the total radius of 6370 kilometers. Outside the core lies the mantle of rock, the surface of which we have discussed above in some detail. The core is thought by most astronomers and geophysicists to consist largely of molten iron, too hot—perhaps about 5,000°C—to be magnetic in the sense of an ordinary magnet, but not too hot to maintain a system of electric currents. The currents are believed to be generated spontaneously by movements of the liquid iron, with the Earth's magnetic field then arising from the pattern of the electric currents. The reverses occurring in the magnetic field are taken to correspond to reverses in the movement of the iron. The situation is illustrated in Figure 10.28.

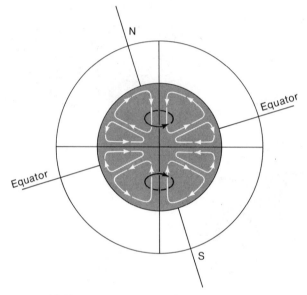

FIGURE 10.28.
This kind of circulation, in meridian planes in the Earth's core, could maintain the magnetic field of the Earth.

Although the iron core has a radius amounting to some 55 per cent of the total Earth radius, and although iron is considerably more dense than rock, the volume of the core is so much less than the volume of the mantle that the total mass of the iron is only some 30 per cent of the mass of the whole Earth. Mixed with the iron there are likely to be smaller quantities of the other "iron group" metals—titanium, chromium, vanadium, manganese, cobalt, and nickel. These same metals are also found in smaller quantities near the Earth's surface, where their mining forms the basis of our modern industrial society.

Turning back to Table 10.1, we can see that, in the broad characteristics of mass and radius, Venus is very similar to the Earth. We expect the internal structure of Venus also to be similar to that of the Earth, and hence that Venus will have similar movements of its surface rocks. It is not possible to check this directly with an optical telescope, since Venus is cloudbound. But information can be obtained by radio methods, or by means of an altimeter carried on a fly-by space vehicle, or by the radar technique mentioned at the end of Chapter 9. A map of the solid surface of Venus obtained by radar is given in Figure 10.29; it shows that Venus possesses high and low ground, as our prediction expects it to.

For FIGURE 10.29,
see PLATE XVII.

Although the internal properties of Venus are likely to be very similar to those of the Earth, the atmospheric properties are quite different. We have already noted that Venus is totally cloudbound, whereas Earth is only partially cloud-covered. The clouds are also of a different kind. Although the chemical nature of the clouds of Venus is still uncertain (a recent suggestion being droplets of sulphuric acid, H_2SO_4), it is clear that they are not water clouds like those of our own atmosphere. Indeed, it is clear that the essential difference between the Earth and Venus is that we have great quantities of water, whereas Venus has very little. Carbon dioxide, which may have been present in quantity in our atmosphere at one time, has now become locked away in the rocks, mostly as calcium carbonate, $CaCO_3$, which forms the main constituent of limestone. On the other hand, carbon dioxide has remained in the atmosphere of Venus, where it produces a pressure about a hundred times greater than that of our own atmosphere. This high-pressure atmosphere produces complex effects which force up the temperature at ground level. Thus on Venus the ground-level temperature is about $600°K$, much too hot for life. The problem of why the Earth has so much water and Venus so little will arise again in the next chapter.

§10.6. The Moon

Whereas Venus is a frustrating object to the terrestrial observer, the Moon is just the opposite. The surface of the Moon is magnificently accessible to an observer equipped with even a small telescope. For this reason, the Moon has been observed more assiduously than any other celestial object. The most obvious features are the large, dark, more or less circular basins, known as the maria, the general pock-marked, crater-strewn landscape, and the absence of apparent movement. Yet movement is certainly present in some degree on the Moon. Figure 10.30 is a picture taken by a space vehicle in orbit around the Moon. A stone, or some object, has plainly rolled downhill, leaving a series of marks rather like steps in a snowslope. And what shall we make of the sinuous meandering channels to be seen in Figure 10.31? How can these be explained

FIGURE 10.30.
An object at left center has rolled downhill on the Moon, leaving a sequence of marks like prints in a snowfield. (Courtesy of NASA.)

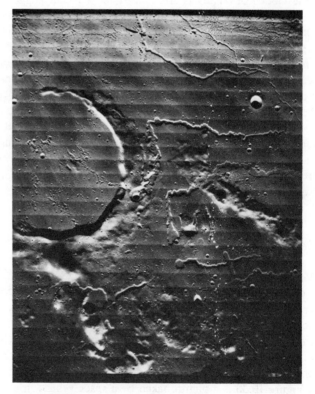

FIGURE 10.31.
The sinuous meandering channels on the Moon suggest the presence at one time of fluid. (Courtesy of NASA.)

except as caused by the flow of some fluid? Water? There is no water to be found (unequivocally!) on the Moon at the present time, but was this always so? Or could there be frozen water locked away below the visible surface? These questions remain unanswered.

By now it is widely agreed that many, if not all, of the craters were formed by objects that struck the Moon from outside. But there is still sharp controversy concerning the origin of the dark maria. Many believe them to have arisen from episodes of intense volcanic activity, in which molten lava, released from the Moon's interior, spread out over the maria basins. Others think the maria basins were also formed by the impact of missiles from outside, and that the Moon has always been a mass of cold rock, without volcanoes, and without plate movements like those of the Earth. They think the way we see the Moon now is essentially the way it was at the time of its formation.

Of these two opposed beliefs, I happen to favor the latter myself, believing that those who favor the lava-flow theory can maintain their position only by ignoring several cogent facts. The Moon always presents very nearly the same aspect to the Earth. There is a part of the Moon's surface, somewhat less than half of it, that cannot be seen at all by a terrestrial observer. This side of the Moon can now be "seen," however, with the aid of a vehicle sent out from Earth and arranged to orbit the entire Moon. The two halves, "our" half and the "other" half, are reconstructed in Figure 10.32. It is immediately apparent that there are no large dark maria on the far side of the Moon. How is this to be explained in terms of a volcanic theory? How does hot molten rock inside

FIGURE 10.32.
The two halves of the Moon. The half we see from the Earth is on the left. There are no dark maria on the far side of the Moon. (Courtesy of Prof. T. Gold, Cornell University.)

the Moon "know" whether it will be facing the Earth when it eventually breaks through to the surface? Faced with this awkward question, those who favor the lava-flow theory seem to me to perform a kind of mental sidestep, arguing something like this: "True, we can't explain this very curious fact, but neither can anybody else. In any case, the situation is so peculiar that it is quite inexplicable—which means we can ignore it."

Let us take a look at the altimeter records obtained by Apollo 15, and shown in Figure 10.33. These make it clear that the difference of elevation between

FIGURE 10.33.
The far side of the Moon is also much rougher than our side. The jagged peaks and troughs of the altimeter record for the far side correspond to the crater-strewn landscape visible on the far side. (Courtesy of NASA.)

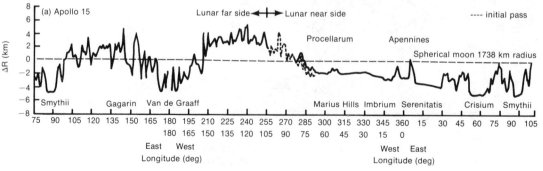

high ground and low ground is about twice as great on the far side of the Moon as it is on the near side. This difference is only explicable, it seems to me, if the low basins on the near side have been submerged by some kind of "filler" material, by a sea of material out of which only the high ground rises. Although this explanation might at first be thought to admit lava as a possible filler material, how do we then explain another important feature of the altimeter records, namely, that the high ground is much more spiky on the far side than it is on the near side? The filler material must be able to smooth the high ground as well as fill up the hollows in the low ground. A fine rock powder, able to compact itself into a friable rock under pressure, would seem to me a much better candidate for the filler than solid lava. Powder could fill out irregularities in the high ground, whereas it is hard to see how lava could achieve this at all.

Thinking in terms of powder instead of lava, how are we to understand the difference between the two sides of the Moon? In terms of transport of the powder. We require the powder to be able to move much more freely on the front side than on the back side. This could be connected with subtle electric effects which are known to occur when streams of high-speed particles from the Sun impinge on the Earth's magnetic field. Such effects immediately bring the Earth into the problem, opening up the possibility of a connection with the Earth.

It would not be profitable for us to pursue details of how the transport of powder might be connected with such electromagnetic effects. Instead, I would prefer to describe another phenomenon, apparently just as curious, and connected with a satellite. Jupiter emits very intense bursts of low-frequency radio waves. The timing of these bursts has been found to be correlated with the position in its orbit of the satellite Io. Here again we have a most subtle electrical situation involving a satellite, a situation whose reality is undoubted. Although the details of how Io is related to the electrical bursts of Jupiter may well be different from those of how the "Moon powder" is related to the Earth, the clear-cut facts of the Io-Jupiter situation suggest that certain remarkable processes may exist, and may be relevant to the Moon-Earth situation as well.

The essential point here is that quite subtle processes, processes which do not immediately hit the eye, *could* be responsible for transporting fine grains of rock with masses no greater than a small fraction of a milligram. But no subtle process can suffice to transport many trillions of tons of molten lava, not just horizontally but uphill. Only a cataclysmic process could do this, and it does not seem likely that any cataclysmic process inside the Moon would trouble itself about which side of the Moon happens to be facing the Earth.

Figure 10.34 shows the remarkable case of Mare Orientale. This multiple-ringed structure lies close to the limb of the Moon. Its genesis in terms of the impact theory from outside is immediately clear. The incoming object, provided it doesn't have too large a speed, produces a pool of molten rock

FIGURE 10.34.
The multiple-ringed structure of Mare Orientale. These are prob-
ably frozen waves from a circular pool of once-molten rock.
(Courtesy of NASA.)

in the region of impact. This pool is dynamically agitated into a circular wave
pattern centered on the point of impact. Molten rock continues to rise and
fall in a wave pattern, just as the ripples continue on a circular pool of water
after a stone has been dropped at the center. Eventually the rocks cool to the
point where viscosity stops the motion, however, and at this point the wave
pattern becomes frozen. It is such a fossilized wave pattern that we see in Mare
Orientalis.

Four similar structures are found on the far side of the Moon and none on
the near side, which is the opposite way round from the dark maria. Why?
Because on the near side such structures have been covered up by the filler
material. To me, it seems evident that, if the filler material could be removed
from the dark maria, quite a number of these double-ring structures would
also be revealed on the near side of the Moon.

A final thought about the Moon. Suppose Mare Orientalis had happened
to be exactly at the center of the full moon. What effect would such an 'eye'
gazing down balefully on the Earth have had on the beliefs of primitive peoples?
And how far would those beliefs have affected us today?

The surface of Mars has many features similar to the Moon, large basins and extensive cratering. On the other hand, Mars also has features which, before the voyage of the spacecraft Mariner 9, had not been expected. Perhaps the most remarkable discovery of this flight past the planet was the vast canyon in what is called the Coprates region, shown in Figure 10.35. This canyon has an over-all length of more than 2,500 km and an average width of about 150 km, although in one place it widens to as much as 250 km. A curious feature of this system is that no place can be seen where material taken from the excavation of the canyon has been deposited. There are no obvious piles of debris. Perhaps the least radical resolution of this strange situation is that the material which once filled the canyon has been transported away to other parts

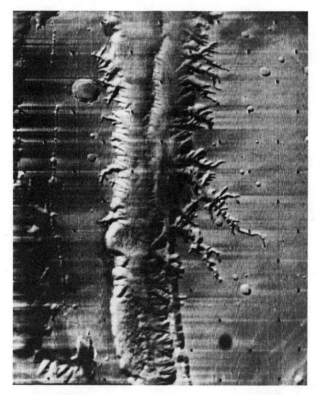

FIGURE 10.35.
A portion of the Coprates canyon of Mars. The canyon has an over-all length of more than 2,500 km and an average width of about 150 km. There appears to be evidence of fluid erosion, showing in the many channels that enter the canyon from the side. (Courtesy of NASA.)

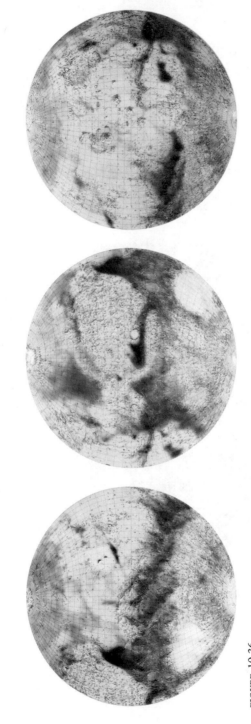

FIGURE 10.36.
A panoramic view of the surface of Mars, built from a mosaic of pictures taken by Mariner 9 in early 1972. The diameter of Mars is about 6,760 kilometers. (Courtesy of Lowell Observatory.)

of the planet by winds. The material would, of course, need to be in the form of small grains and would be deposited as a system of layers in other places.

A picture of the general form of the Martian landscape is shown in Figure 10.36. Wind appears to be the principal agent of erosion and change on Mars, at any rate at this time. The atmosphere of Mars consists largely of carbon dioxide, but water vapor is also present in a smaller proportion. Winds in the atmosphere are, of course, powered by the heat of sunlight. At first sight it seems strange that the weaker intensity of sunlight on Mars should be able to generate windstorms that are apparently more devastating in their effects than are terrestrial storms. The difference comes from the much lower density of the Martian atmosphere, which requires higher windspeeds for a given energy content within the atmosphere. There appear to be two linked sequences of events which lead to the sudden onset of violent storms.

Sequence 1
   Wind picks up dust.
   Dust absorbs sunlight, becoming heated.
   Heat transferred from dust to atmosphere.
   Windspeed increases because of increased heat content of atmosphere.

Sequence 2
   Wind picks up dust.
   Dust acquires high speed of wind.
   Fast-moving dust particles impinge on Martian surface, lifting up more dust.

Because of these self-amplifying effects, dust storms that obscure a large fraction of the area of Mars can be generated with startling suddenness. Such a storm was in fact in progress at the time that Mariner 9 first approached the planet.

These storms are responsible for many of the detailed surface markings, for patterns of dark and light streaks, and for the dune structure shown in Figure 10.37. The craters of Mars show significantly more erosion than do those of the Moon, as can be seen from Figure 10.38. This is almost surely due to the scouring action of the dust-laden Martian winds.

The average temperature on Mars is about 100°F lower than the average temperature on Earth. At midday on the equator of Mars, water-ice would reach its melting point, but during the Martian night the temperature drops below −120°F. Extensive glaciers of water-ice could exist on Mars. Because of their permanence, they would become covered by dust deposits carried by the winds, and so would not be immediately visible. The situation is different for carbon dioxide. Dry ice evaporates at a lower temperature than water-ice, and this temperature lies in the range of Martian variability. In the warmer spots, solid carbon dioxide will evaporate into vapor, and in the colder spots the vapor will condense into a frost of dry ice. This circumstance explains the changes in the Martian polar caps, which grow extensively during the winter

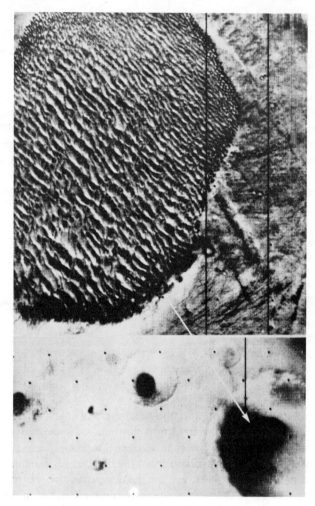

FIGURE 10.37.
A dune structure produced by the Martian winds.
(Courtesy of NASA.)

and evaporate during the Martian summer. It is thought that under a comparatively thin covering of carbon-dioxide frost there may lie extensive deposits of water-ice, although water-ice could persist, and may be present, in the nonpolar regions of Mars.

There are features of the surface topography of Mars which seem clearly to have been formed by extensive flows of liquid, as can be seen in Figure 10.35. Such features suggest the occurrence of flash floods rather than the steady flow of a river. The fluid responsible for these features is thought to have been water, and some controversy has occurred over the origin of the water. One

FIGURE 10.38.
The crater-strewn landscape of Mars, showing the effects of erosion due to winds. This flat map covers most of the surface of the planet; compare with Figure 10.36. (Courtesy of Lowell Observatory.)

possibility is that liquid water may be released from time to time from the interior of Mars, where the temperature is high enough for the water to be in liquid form. Another possibility is that the Martian climate may have long-period cycles, with episodes considerably warmer than at present. In such warmer periods, the glaciers, now frozen, would melt, and liquid water would flow on the surface of the planet. Whether the desert would bloom is another matter, for it is now thought rather unlikely that there is life on Mars.

It is curious that, whereas lunar observers have searched assiduously but without success for signs of recent volcanic activity on the Moon, several very large volcanoes have been found on Mars. Nix Olympica, shown in Figure 10.39, is an enormous, gently rising mound some 600 km in diameter, with a steep-walled central caldera.* The height of the center above the general level of the Martian surface has been estimated at 100,000 feet. The slopes of the rising cone appear to be smooth, being largely free from cratering. This general absence of cratering is interpreted to indicate that Nix Olympica is a quite recent feature. An age of two hundred millions of years has been suggested.

It has seemed natural to geologists to interpret these volcanoes in terms of terrestrial experience, although the Martian volcanoes are many times the scale of the largest terrestrial volcano. Thus the cones of the several volcanoes to be seen in Figure 10.40 are thought to consist of the usual volcanic materials, lava and ash possibly. This could be the correct explanation, although simple hydrostatic considerations would then require these Martian volcanoes to have remarkably deep roots, because the density of fluid rock is not much less than the density of solid rock, and it takes a great depth to give a sufficient head of pressure to force fluid rock to a height as great as 100,000 feet (30 kilome-

---

*An international committee recently changed (needlessly, I think) the name of Nix Olympica to Olympus Mons. It may be too much to hope that the original name will persist.

FIGURE 10.39.
The volcano of Nix Olympica, about 600 kilometers in diameter.
It has been estimated that this central part of the volcano may
rise as high as 30 kilometers above the general level of the Martian surface, which raises an interesting problem about what may
have created this "volcano." (Courtesy of NASA.)

ters). This aspect of the problem would be alleviated if the liquid that formed
these volcanoes had a density much less than that of rock. Ordinary water is
an obvious candidate for such a liquid. Could these structures really be vast
domes of ice, covered smoothly by the deposits laid down by many dust storms?
The question is an interesting one.

§10.8. Mercury

Mariner 10 recently reached Mercury, revealing a surface little different from
that of the Moon, as was shown in Figure 10.9. Since it seems likely that the
surface features of Mercury have been little changed since the planet formed,
the similarity with the Moon also indicates that the surface features of the Moon
have remained essentially unchanged during the past history of our satellite.

Somewhat to the surprise of astronomers, Mercury has turned out to have
a weak magnetic field, probably associated with the considerable quantities of

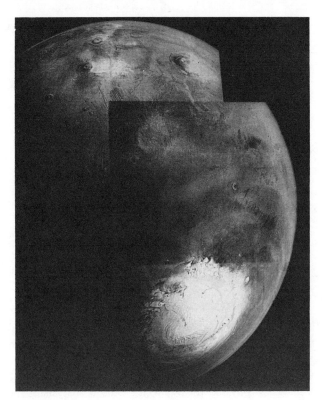

FIGURE 10.40.
A general picture of Mars, showing the southern polar cap, and
also, toward the top, several of the gigantic Martian volcanoes.
(Courtesy of NASA.)

iron which the planet is believed to contain. Mercury also seems to have a
thin atmosphere consisting of the noble gases, again a rather unexpected result.

Astronomers already were much surprised when radar measurements first
revealed that Mercury's rotation period is about 59 days, not 88 days as optical
measurements had suggested it to be.* Previously it was thought that Mercury
rotated on its axis in exactly the same length of time that it took to make a
revolution around the Sun, 88 days, and that Mercury therefore always turned
the same face toward the Sun, as the Moon always turns the same face toward
the Earth. The perpetually sunlit half of Mercury would then have attained
an extremely high temperature, while the perpetually dark side would have
had an exceedingly low temperature, there being no extensive atmosphere to
mediate between the two hemispheres. With a period of 59 days, the situation
is not as extreme. Nevertheless, the measured temperature reaches about 700°K

---

*The rotation period is exactly $\frac{2}{3}$ of the orbital period, and there is a reason why this should
be so. As a project, discover this reason.

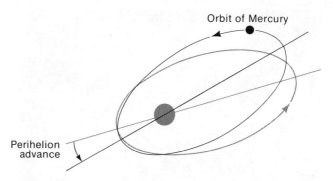

FIGURE 10.41.
The orbit of Mercury, with the eccentricity much exaggerated. In
the nineteenth century, the French astronomer U. J. J. Le Verrier
found the orbit to be turning around by an amount which could
not be wholly accounted for in terms of the gravitational forces
exerted on Mercury by the other planets. (From J. Brandt and
S. Maran, *New Horizons in Astronomy*. W. H. Freeman and
Company, Copyright © 1972.)

on the sunlit side, whereas on the dark side the temperature of the surface layers
of the ground fall below the lowest temperatures experienced on Mars.

The orbit of Mercury, with the eccentricity much exaggerated, is shown in
Figure 10.41. The elliptical shape of the orbit has played an important role
in the development of physical ideas. From a careful analysis of the observed
motion of Mercury, Le Verrier found that the major axis of the ellipse, which
turns relative to a fixed direction determined by the distant stars, was turning
slightly faster than it was expected to, as calculated from the gravitational forces
exerted on Mercury by the Sun and all the other planets. There was an excess
turning of about 43 arc seconds per century. What was the reason for this
discrepancy?

We saw in §10.1 how Le Verrier and Adams had used discrepancies between
the expected motion of Uranus and its observed motion to predict the existence of a new planet, Neptune. The success of this prediction suggested
to Le Verrier that he should try the same argument again. Perhaps there was
another planet to be discovered? It soon appeared that the hypothetical planet
would have to be small and to lie still nearer the Sun than Mercury. It would
therefore be difficult to observe. The best hope of detection would be when
the planet lay directly between the Earth and the Sun, for then it would appear
as a dark, circular, rapidly moving spot projected against the bright disk of
the Sun. Le Verrier calculated when such transits of his planet should occur,
and observatories throughout Europe prepared themselves to compete for the
honor of being the first to report the detection of Le Verrier's new planet. In
fact, none of them ever did, although many amateur astronomers claimed to
have done so. These claims were carefully investigated and were found one
by one to be defective. Gradually, astronomers came to realize that Le Verrier's

new planet, Vulcan, as he had already named it, did not exist. (I learned this story when I was young and drew a moral from it: that it is unwise to attempt to repeat one's successes!)

Another explanation of the excess rate of turning of the long axis of Mercury's orbit was well-known to astronomers in the nineteenth century. We have seen that the formula used by Newton for the gravitational force between two particles, a distance a apart with masses m_1, m_2, had the form

$$\frac{Gm_1m_2}{a^2}.$$

If this formula were not quite right, if the correct form were

$$\frac{Gm_1m_2}{a^2} + \frac{A}{a^4},$$

with A some still-unknown constant, then there would indeed be an excess rate of turning of the orbit of Mercury. But this seemed an unpleasantly arbitrary step to take, since it did nothing but explain the one detail of the motion of Mercury, and did so at the expense of destroying the simplicity of Newton's form for the force law.

Yet this seemingly unattractive explanation has turned out to be correct. Einstein was led to his theory of gravitation, which he called the *general theory of relativity,* by abstract physical considerations, rather than by a need to solve some explicit problem. When he had arrived at a detailed mathematical structure in which to express his ideas, it became necessary for him to compare his new theory in detail with that of Newton. Although the new theory had many facets to it which did not appear at all in the Newtonian theory, in those problems where Newton had been notably successful it turned out that the old calculations remained correct, except in one detail. The old simple Gm_1m_2/a^2 had to have an extra bit added to it, of precisely the above form, namely, A/a^4. Furthermore, the new theory actually gave an unequivocal way to calculate A, and so could be applied to the problem of the orbit of Mercury. It turned out to give just the required excess turning rate of 43 arc seconds per century. Notice that A was not adjusted to give this value. It emerged uniquely out of the theory. So what had seemed in the nineteenth century to be only an arbitrary adjustment to Newton's theory became a triumph of twentieth-century science.

§10.9. The Four Major Planets

We turn now to the four large outer planets, Jupiter, Saturn, Uranus, and Neptune. Some of their properties are set forth in Table 10.11.

All these planets are clearly much larger than any of the inner planets. When we come in the next chapter to discuss the problem of the origin of planets,

TABLE 10.11.
The major planets

Planet	Mean radius (Earth as unit)	Mass (Earth as unit)	Equatorial rotation period	Inclination of equator to orbit
Jupiter	10.97	318.00	$9^h\ 50^m$	3° 7′
Saturn	9.18	95.22	$10^h\ 14^m$	26° 45′
Uranus	3.69	14.55	$10^h\ 49^m$	97° 59′
Neptune	3.90	17.23	$15^h\ 40^m$	28° 50′

we will evidently need to explain why these outer planets are all so large. If one were to take nine planets, some large, some small, and to distribute their motions about the Sun at random, only rarely would it happen that all the large ones fell on the outside of the system. On a random basis we would expect big planets and small ones to be intermixed with each other.

In spite of their greater sizes, these planets rotate rapidly, in periods which range from 9 hours 50 minutes for Jupiter to 15 hours 40 minutes for Neptune. None of the inner planets rotate as quickly as this. It will be noticed that the axis of rotation of Jupiter is only slightly tilted to the plane of its orbit, but that the tilts are greater for the others. Indeed, the situation for Uranus is most peculiar. Imagine a planet rotating to begin with about an axis perpendicular to the plane of its orbit, and let the rotation be in the same sense as that of the orbital motion. This initial state of affairs is illustrated in Figure 10.42. Now apply a tilt to the axis of rotation until the situation becomes the one that is actually observed. The angles of tilt that would be needed for the various planets are given in the last column of Table 10.1. The value of 97° 57′ for Uranus means that this planet is "lying on its side," as illustrated in Figure 10.43. If the Earth's axis of rotation were oriented in such a way, our experience of the seasons of the year would be distinctly odd. There would be unbroken daylight near midsummer and unbroken night near midwinter, not just for the polar caps but for every place on the Earth. Why Uranus should be oriented like this is a problem related to the work of the next chapter.

The gas methane, with the chemical formula CH_4, is found in the atmospheres of all the major planets. The hydrocarbons which play such a big role here on the Earth can be formed by adding methane molecules together in a suitable way. It is ironic to reflect that, while we attach such great financial and political importance to the comparatively small quantity of oil found on Earth, there should be vast quantities of "natural gas" in the atmospheres of the major planets. As well as methane, ammonia in gaseous form is found, most of it being in Jupiter, a little in Saturn, and none in Uranus and Neptune, where it is very likely frozen out in a solid form. Hydrogen and helium must also be present in large quantities in the atmospheres of all the major planets.

Although the major planets are mainly composed of materials we usually think of as being gaseous, or easily made gaseous—hydrogen, helium, methane, ammonia, water—their actual atmospheres are surprisingly thin. For Jupiter

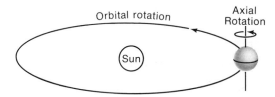

FIGURE 10.42.
To illustrate the meaning of the inclination value given for
Uranus in Table 10.12, start with Uranus rotating about an
axis perpendicular to the plane of its orbit, the sense of the
rotation and the sense of the orbital motion being the same.

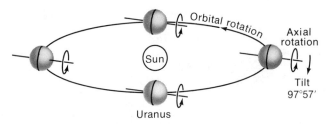

Uranus

FIGURE 10.43.
Then apply a tilt of 97° 57', leading to a situation in which Uranus
rotates essentially "on its side."

the depth of the atmosphere is on the order of five hundred miles. The high
pressure in the deep interior keeps these normally volatile materials, including
most of the hydrogen and helium, in liquid or solid form.

The solar constant of 1.39 kilowatts per square meter gives the flux at the
Earth of radiation from the Sun. Because Jupiter is considerably farther from
the Sun than the Earth, the solar flux at Jupiter is much less, about 0.05
kilowatts per square meter. Moreover, Jupiter reflects some 58 per cent of the
incident sunlight, so that only about 0.02 kilowatts per square meter is absorbed
by the planet. However, since Jupiter has a radius rather more than 10 times
that of the Earth, the interception area for Jupiter is more than 100 times the
terrestrial area, giving a total energy absorption for Jupiter of about 3×10^{14}
kilowatts, about three times the solar power absorbed by the Earth.

The natural expectation would be to find that Jupiter reradiates this same
power, but as infrared radiation, not as sunlight. The argument for such a
balance between input and output is that Jupiter's atmosphere, taken as a whole,
cannot continually either accumulate or lose energy. A balance must be reached
between input and output, just as it is in the Earth's atmosphere. Observation
by modern infrared astronomy has shown, however, that the energy output
from Jupiter's atmosphere is some two to three times greater than the solar
input. The inference from this surprising discovery is that the atmosphere must

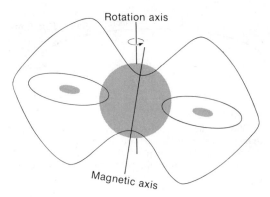

FIGURE 10.44.
A schematic diagram, showing the relation of the magnetic field of Jupiter to the axial rotation of the planet. Radioemission from Jupiter comes strongly from the shaded areas. The visible planet of Figure 10.12 is represented by the circle. (From J. Brandt and S. Maran, *New Horizons in Astronomy*. W. H. Freeman and Company. Copyright © 1972.)

be receiving energy from somewhere other than the Sun, and that the power delivered by this other source must be about 3×10^{14} kilowatts. The only plausible "other source" is the interior of the planet itself. In short, the power emerging from the interior of Jupiter must be on the order of 3×10^{14} kilowatts.

Heat escapes also from the hot interior of the Earth, but in an amount much less than the supply of solar energy, an amount of order 3×10^{10} kilowatts, only 1 part in 10,000 of the power from the interior of Jupiter. Yet it is precisely this heat source within the Earth that generates the plate motion that builds mountains, and that creates volcanoes and earthquakes. If the much smaller heat source in the Earth is capable of producing such a rich variety of phenomena, how much more such activity might we not expect on Jupiter? Some indication of this activity is to be gained from the remarkable belts and markings to be seen in the atmosphere of Jupiter. Of the markings, the best known is the Red Spot, shown in Figure 10.12. And the magnetic field, illustrated in Figure 10.44, with an intensity about 10 times the field of the Earth (and with a total energy 100,000 times greater), is another aspect of this activity. We saw earlier in this chapter that the terrestrial magnetic field is thought to arise from the movement of metal within the deep interior, and the magnetic field of Jupiter probably arises in the same way. However, the metal in Jupiter is not iron, but hydrogen! A glance at Table 7.1 shows that the element hydrogen is similar to the elements sodium, potassium, rubidium, cesium, and francium. Since these other members of the first column of Mendeleev's Table are all metals, it is to be expected that hydrogen will also become a metal when it is compressed to a solid or liquid form.

What is the source of the large internal energy of Jupiter? It is not primaeval heat, and not radioactive heat, for both these are inadequate. The source is apparently the slow shrinkage of the whole planet, by a fraction of a centimeter per year, the energy being supplied by the gravitational forces, which heat the interior just as in a condensing star. The cause of the shrinkage seems to be a slow dissolving of helium within the hydrogen. Conditions within Jupiter are remote indeed from terrestrial experience.

Saturn is very likely quite similar to Jupiter in its structure, although it has a smaller supply of internal heat. The belts observed at the surface of Saturn are less marked and less variable than those of Jupiter. There is also much less of red and brown colors, as can be seen from Figure 10.14. Instead, the equatorial regions of Saturn appear yellow and the polar regions green. Probably the colors of both planets arise from the condensation of small liquid particles, the particles being different for the two planets because the atmosphere of Saturn is colder than that of Jupiter. Indeed, markings like the Red Spot could simply be a temperature phenomenon, a place in Jupiter's atmosphere of abnormally high temperature due to an exceptionally large supply of heat from below.

Undoubtedly to the eye the most striking feature of Saturn is its magnificent system of rings, three flat concentric rings lying in the plane of the planet's equator. Galileo glimpsed the rings indistinctly in 1610. The division between the two main outer rings was first observed by Cassini toward the end of the seventeenth century. It was not until the middle of the nineteenth century that Bond observed the faint innermost ring. The rings are referred to nowadays as A, B, C, with A the outermost and C the innermost. Ring A is partly opaque and ring B almost wholly opaque to light, except near their edges, but stars can readily be seen through ring C. The rings are believed to consist of macroscopic-sized particles, probably of dry ice and water-ice. Their general thickness may be only a few kilometers, and their total material content may be no larger in its mass than the terrestrial oceans.

Radio observations have given a surprising result. Attempts to detect radio waves originating from the Sun and reflected by the rings, in the manner of Figure 10.45, have shown that these signals are very weak. Yet radio waves from the Earth, detected as illustrated in Figure 10.46, are reflected back to Earth comparatively strongly. This means that the rings have what is called "strong backscatter"; that is, the bodies that constitute the ring must have the property, shown in Figure 10.47, of always giving strong reflection in the direction from which the radio waves have come, but not in other directions. Water-ice has this property, provided it is in spheres that are about a meter in diameter. If rings A and B were entirely composed of such spheres, there would need to be so many of them that collisions between them would tend to grind the chunks into smaller pieces. This difficulty might be avoided if the rings are thinner than the old estimate of a few kilometers mentioned above, since the

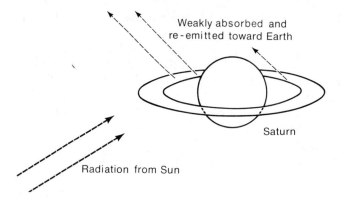

FIGURE 10.45.
Weak absorption of solar radiation by the rings of Saturn.

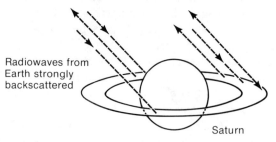

FIGURE 10.46.
Yet radiowaves from the Earth are strongly reflected by the rings.

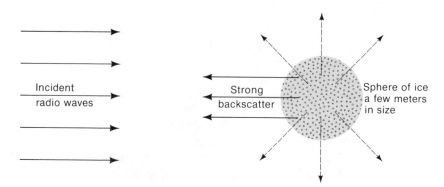

FIGURE 10.47.
Saturn's rings have what is known as strong "backscatter." Reflected waves are returned strongly in directions close to that from which the waves have come. Spheres of ice a meter or two in diameter have the required property.

thinner the rings the more gentle the collisions. Even so, we must expect the surfaces on the icy chunks to be considerably roughened by constant rubbing, and indeed there is radio evidence to show that this may be the case.

Uranus and Neptune appear in a telescope as small greenish disks, more like Saturn in color than Jupiter. Although in appearance Uranus and Neptune are undistinguished and seemingly uninteresting, in certain ways these two planets are the most important of the solar system, indeed, so important that we will devote much of the next chapter to them. So, leaving them aside for the moment, we end this chapter by taking a brief look at a very different class of object.

§10.10. Comets

Perhaps no astronomical object captures popular attention more strongly than the comets, and after seeing two brilliant ones myself, both in the year 1957, Comets Arend-Roland and Mřkos, shown in Figures 10.48 and 10.49, I can

FIGURE 10.48.
The comet Arend-Roland, April 29, 1957. (Courtesy of Hale Observatories.)

FIGURE 10.49.
The comet Mřkos, August 22, 1957. (Courtesy of Hale
Observatories.)

understand why. At its most brilliant, such a comet covers a substantial area
of the sky, in an arc stretching from the horizon to the zenith, appearing like
a luminous vertical sword with the point raised high above one's head (unlike
the sword of Damocles, which was pointed downward). In the Middle Ages
comets were regarded as harbingers of disaster, perhaps even of the end of
the world itself.

Yet comets are quite insubstantial compared to other objects we see in the
sky. The mass of a typical comet is thought to be no more than 10^{-8} of that
of the Moon, about the same as a sphere of solid material with a radius of
10 km. Of course, a chunk of material of this size could indeed cause a serious
disaster if it were in direct collision with the Earth. Unlike the Moon and the
planets, comets move in orbits that would in fact permit them on very rare
occasions to hit the Earth. The chance of a collision happening during any
one passage of a comet is on the average about one part in a billion, although
the chance would become higher if the comet were divided into several pieces.
About one new comet is observed per year, and there are likely to be several

others that escape detection. Taking all these factors into account, we can estimate that a substantial chunk of cometary material probably hits the Earth about once every 10,000,000 years.

Cometary orbits are mostly of a highly elongated form. A typical comet begins its journey a great distance away from the Sun, far out beyond the planet Pluto. It may take thousands, or even hundreds of thousands, of years to fall toward the inner part of the solar system. If the comet comes inside the Earth's orbit, collision may occur, with the low probability estimated in the previous paragraph. Should collision occur in the head-on circumstance of Figure 10.50, the relative speed between the Earth and the comet could be as high as about 70 km per second. The energy released in such a collision, even for a comparatively "small" piece of solid material, say, a sphere of 1 km radius, would blow the Earth's atmosphere out into space over an area of tens of thousands of square kilometers. The curious small button-shaped pieces of glass known as *tektites*, which are found strewn over certain localized areas of the Earth's surface, may possibly be bits of fine-scale debris from such collisions.

The collision of cometary material with the Earth is not to be confused with the collision of *meteorites*. Meteorites are pieces of solid material, usually a few inches or feet in diameter, which come from the region between Mars and Jupiter, the region of the asteroids (see the discussion in §10.2). For the most part, bodies in this region do not follow orbits that cross the path of the Earth, but occasionally they are disturbed by Jupiter, Mars, and each other into new paths that have a small chance of intersecting the Earth. Should such a perturbed body hit the Earth, the speed of collision would usually be of order 15 km per second, less than in the situation of Figure 10.50, because the meteorite

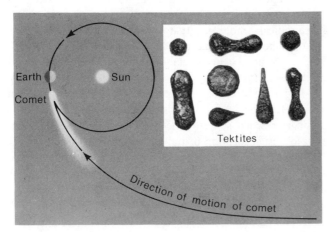

FIGURE 10.50.
A comet in head-on collision with the Earth. It may be that tektites are produced in violent collisions of this type, which can occur at speeds up to about 70 kilometers per second. (Photograph courtesy of Griffith Observatory.)

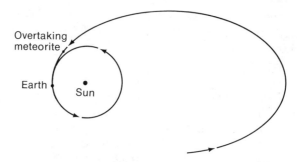

Overtaking
meteorite

Earth

Sun

FIGURE 10.51.
An overtaking collision of a meteorite with the Earth. Such collisions are much less violent than the head-on kind.

and the Earth follow paths that go around the Sun *in the same sense,* as in Figure 10.51. The exceptionally high collisional speed in Figure 10.50 results from the fact that in this configuration the Earth and the comet go around the Sun *in opposite senses.*

The mass of a comet was described above in terms of a sphere of solid material with a radius of about 10 km. It may be wondered how such a sphere of material, small compared to the radius of the Earth's orbit, could show itself as an object spread across a substantial area of the sky. This question is answered by the volatility of the initially solid compact material. As a comet comes toward the Sun, a fraction of the material evaporates into a gaseous form, a process which usually begins when the comet moves in toward the orbit of Mars. The process becomes stronger still as the Sun is approached more closely, particularly for the occasional comet which goes inside the orbit of Mercury.

The gases evaporating from the nucleus of a comet contain high concentrations of free radicals, molecules that are so chemically active they are hard to prepare in the laboratory, molecules like OH, CH, NH, NH_2. In addition, there are C_2, C_3, CN, and Na in detectable amounts. These gases form a "coma" surrounding the solid nucleus, a coma which may grow to a radius of order 100,000 km—large, but still not large enough to spread over much of the sky. The nucleus and the coma form what is called the *head* of the comet.

The next stage is when some of the gaseous atoms lose an electron: they become "ionized" by absorption of radiation from the Sun, a process studied in Chapter 4. Ionization being denoted by a plus sign, molecules of CO^+, OH^+, CO_2^+, N_2^+, are all formed. Such molecules have a positive electric charge, and they interact with the streams of particles which come from the Sun, particularly from solar flares, as we saw in Chapter 2 (see Figure 2.16). The ionized cometary molecules tend to become locked to these solar streams, in the manner of Figure 10.52, because of the magnetic field carried by the solar particles. When this happens, the comet is said to develop a "tail." The tail of a comet represents a streaming away of gas, peeled off, as it were, by jets of particles

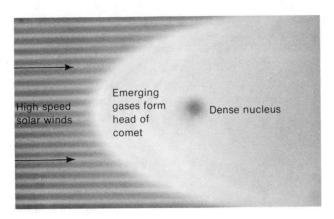

FIGURE 10.52.
The emergence of gases from the nucleus of a comet leads to the formation of the "head." Then pressure from a wind of particles from the Sun, possibly applied by a magnetic field, forces ionized cometary molecules into the "tail" of the comet.

from the Sun. The force producing a cometary tail is thus directed outward, away from the Sun, which is why tails point outwards, as in Figure 10.53.

The gas moving outward in a cometary tail is not recovered. Indeed, it is just this streaming away of the tail that produces the vast size of a comet. Usually the length of visible tail—visible because absorbed sunlight is reradiated by the molecules in the tail—is about a tenth of the radius of the Earth's orbit. However, for spectacular comets the visible tail can be as large as the Earth's

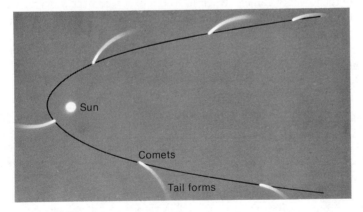

FIGURE 10.53.
The "tail" forms as a comet approaches the Sun. Throughout the passage of the comet, the tail always points *away* from the Sun.

orbit itself. It is the development of the tail which gives so impressive a scale to spectacular comets.

If the solid material of a comet were influenced only by the gravitational field of the Sun, the solid particles and chunks of material would follow highly elliptical orbits, taking them back to the extreme outer regions of the solar system from where they came. However, such a highly elliptical orbit can be markedly changed by perturbations due to the gravitational influences of the planets, particularly Jupiter. For this reason, comets do not usually return to their starting positions. The orbit of a comet can be altered decisively in two quite different ways, depending on the configuration of the planets in relation to the comet. The orbit may become such that the comet leaves the solar system entirely, and the orbit is then said to have become *hyperbolic*. The other important possibility leads to the situation of Figure 10.54. Here the comet does not retreat to the great distance from which it came, although it may still retreat to the distances of the outer planets. Instead of taking hundreds of thousands of years to go around the Sun, the comet now completes its orbit in only a few tens of years. The period for Halley's comet, shown in Figure 10.54, is 76.2 years.

Newton's theory of the planetary motions permitted highly elliptical orbits as well as nearly circular ones. The nearly circular motions of the planets had to be explained either in terms of the way the planets were set moving in the first place, or in terms of mutual influences between the planets that may have operated gradually to make their orbits more circular. Today we believe the nearly circular orbits of the planets are to be explained in the first of these two ways, as we will see in Chapter 11. Yet in Newton's time it was not clear which of the two was correct, and one could ask if there could be other, nonplanetary bodies which followed the highly elliptical orbits permitted by the theory. Comets were possibilities. Edmund Halley in 1705 published orbits of more than twenty comets which had been observed over previous centuries and for which he thought the records reliable. He found three of the orbits to be similar, those for comets observed in the years 1531, 1607, and 1682. The even spacing of about 76 years between these appearances suggested that the same body had appeared each time. Adding a further 76 years, Halley therefore predicted that the comet would return again about the year 1758. Naturally when 1758 arrived the comet was avidly sought, and was in fact observed, on Christmas night, by George Palitzsch. The last return of Halley's comet was in 1910. The next will be in 1986.

Once a comet has been perturbed as much as Halley's comet has been, it must experience further orbital disturbances, particularly from the four major planets. The gradual effect of these perturbations is to round off the orbit more and more. Halley's comet will eventually move in an orbit whose greatest distance from the Sun is much less than at present; it will no longer move out to the region of the major planets, the orbit having a scale more comparable to its distance of nearest approach to the Sun.

The graveyard of the comets lies in the general region between Mars and

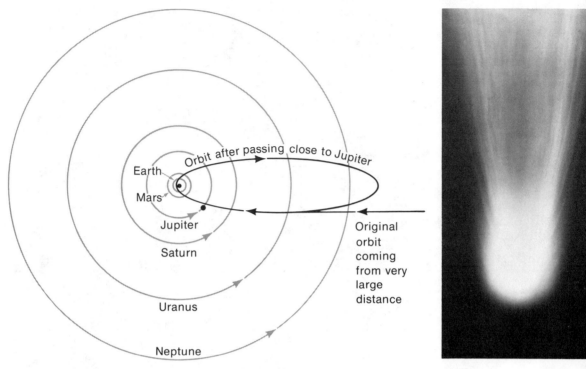

FIGURE 10.54.
In (a), a comet in an orbit with very long period becomes perturbed by the gravitational influence of Jupiter into an orbit of much shorter period. This may well have happened to Halley's comet (b), which now has a period of 76.2 years.

Jupiter. There are now about twenty well-known comets in this region, moving around the Sun in periods of about five years, with the orbits shown in Figure 10.55. When this stage is reached, most or all of the original volatile material has been evaporated and lost. These *short-period comets*, as they are called, have therefore lost the ability to grow spectacular tails. A comet, as it approaches the last phase of its life, consists of a swarm of refractory particles, which probably range in size from tiny sub-pinhead grains up to considerable chunks of material. Originally all these refractory particles may have been frozen into the volatile ices, like the dirty ice at the snout of a terrestrial glacier. Such comets then suffer the further indignity that their solid particles, particularly the tiny ones, are pulled more and more away from any nucleus that may remain, so that the small particles come to be spread in a kind of tube, as in Figure 10.56, around the whole of the orbit. This effect arises from the slightly different gravitational force which the Sun exerts on different bits of the comet.

Sometimes the orbit of a comet comes moderately close to the path of the Earth. Although the chance of collision with the Earth is small so long as the

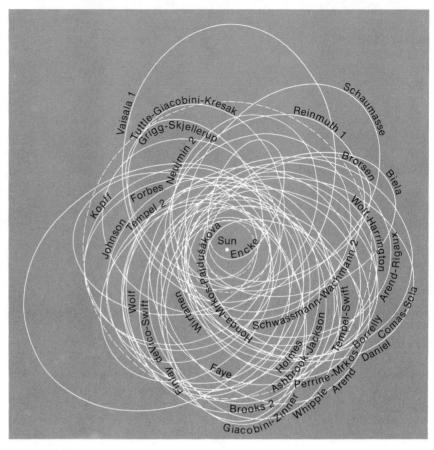

FIGURE 10.55.
The orbits of short-period comets lie mostly in the region between Mars and Jupiter. This is
the graveyard of the comets. (From N. B. Richter, *The Nature of Comets*, Methuen, 1963, by
permission of Barth Verlag, Leipzig, Germany.)

comet remains a compact object, collisions become more and more probable
as the comet is gradually pulled apart by the gravitational action of the Sun.
When the situation of Figure 10.56 is reached, with a swarm of small particles
strewn around the whole orbit, such particles will strike the terrestrial atmosphere whenever the Earth comes close to the path of the comet. Being tiny
and moving at high speed, the particles are almost immediately evaporated
on entering our atmosphere. The heat generated by the motion produces a trail
of hot gas often visible at night from ground level. Such momentary streaks
of light across the sky are called *meteors*.

The Earth crosses about a dozen cometary streams, as illustrated in Figure
10.57. It does so at various parts of its orbit, i.e., at various times in the year.

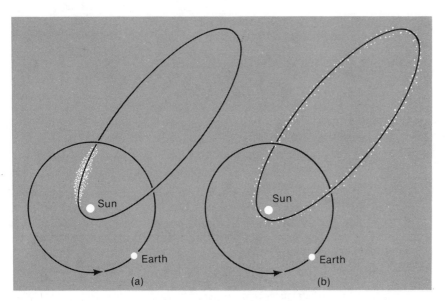

FIGURE 10.56.
A comet breaks up into particles that gradually spread around the whole orbit, as in the change from (a) to (b).

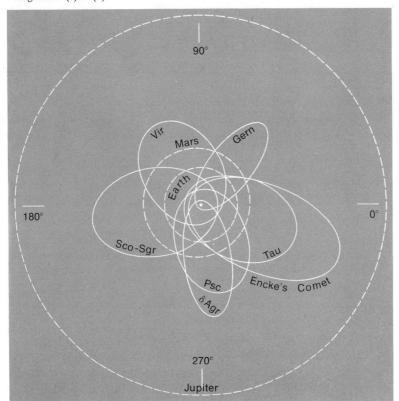

FIGURE 10.57.
Meteor streams in orbits close to the plane of the Earth's motion around the Sun. These streams have resulted from the gradual breakup of comets which once moved in similar orbits, as Encke's comet is here shown to do. (From N. B. Richter, *The Nature of Comets*, Methuen, 1963, by permission of Barth Verlag, Leipzig, Germany.)

The best months for observing meteors are August and November.

The comets are believed to be material left over from the primaeval process in which the Sun and the planets were formed. When we see the momentary flash of a meteor in the sky, we might remember that the particle responsible for it was an old particle, a survivor from the episode in which the Earth itself was born.

General Problems and Questions

1. State the Titius-Bode law and assess its value.

2. What is an asteroid?

3. How were the planets Neptune and Pluto discovered?

4. The solar system has six other satellites comparable in size and mass to the Moon. What are the names of these six, and to what planets do they belong?

5. Do the satellites of a planet obey Kepler's third law?

6. Verify your answer to the previous problem for the cases of the 4 large satellites of Jupiter.

7. How does gravity at the surface of the Moon compare with that at the surface of the Earth?

8. Which planet has the largest surface gravity?

9. How do the planets and satellites vary in their reflectivities (albedos)? What inferences can be drawn from these variations?

10. Using positions at opposition, which outer planet exerts the strongest gravitational force on the Earth? Which is the second strongest, and which the third?

11. Compare the forces of Problem 10 with that exerted on the Earth by Venus when at inferior conjunction (Venus between Earth and Sun).

12. Obtain a general estimate for the gravitational force exerted on the Earth by the whole of our galaxy. In what direction does this galactic force act? Compare your estimate with the gravitational effect of a nearby star taken at a distance of 10^{19} cm.

13. Why does the gravitational force of the galaxy, although much smaller in magnitude than that of Jupiter, nevertheless have a much greater long-term effect on the motion of the Earth than Jupiter does?

14. What effect does the galactic force have on the motion of the whole solar system?

15. Discuss the process of plate movements, especially in terms of how they affect the surface features of the Earth. Discuss the evidence for such movements, and state the general time-scale during which they have a major influence. Speculate on what the Earth might be like if such movements did not occur.

16. Following up the answer to Problem 15, what is thought to be the basic energy source for volcanoes, earthquakes, and the uplifting of massive mountain ranges?

17. To what extent is Venus similar to the Earth, and in what respects is Venus different?

18. Suggest an explanation for the altimeter record of Apollo 15 other than that proposed in the text.

19. In what respects are the surface features of Mars different from those of the Moon?

20. Compare the surface features of Mercury with those of the Moon.

21. How did a major physical discovery emerge from the observational determination of the rate at which the long axis of the orbit of Mercury turns?

22. Draw up a table giving the main differences between the group made up of the four inner planets and the group made up of Jupiter, Saturn, Uranus, and Neptune.

23. In what way is Uranus very peculiar?

24. What gases are found in the atmospheres of Jupiter, Saturn, Uranus, and Neptune?

25. Which two planets are known to have rather strong magnetic fields?

26. From measurements of the infrared radiation emitted by Jupiter, what inferences have been made about the internal structure of Jupiter?

27. What are the rings of Saturn thought to consist of? What peculiarity has recently been discovered in how radiowaves are scattered by the rings?

28. What are: (a) comets, (b) meteors, (c) tektites, (d) meteorites?

29. What molecules are found to be present in cometary gases?

30. Describe the history of Halley's comet, and discuss what the long-term future of this comet is likely to be.

Chapter 11:
The Origin of the Planets

Nowadays most astronomers believe the planets were formed by the same general process that led to the condensation of the Sun. It is then a seemingly logical program to use our knowledge of star formation (Chapter 6) to seek a theory of the origin of the planets. We could attempt to work from the primordial conditions that appear to be operative in an object like that shown in Figure 11.1, for example. However, this program would have the disadvantage, explained in Chapter 6, that obscuration by dust in such objects as Figure 11.1, and quite generally in dense interstellar nebulae where star formation is taking place, interferes with observation. Rather does it seem preferable therefore to use the known properties of the planets to infer, by a kind of detective story, how they were formed. This is the line of attack to be followed in this chapter. We will find that, using only simple facts about the planets, we can obtain step by step a theory which succeeds in explaining most of the outstanding features of the solar system.

§11.1. The Mass, Density, and Composition of the Planets

The average density of a planet is obtained by dividing the mass by the volume. Writing m for the mass and r for the radius, the average density is therefore given by $m / (4\pi r^3 / 3)$. Putting $m = 5.977 \times 10^{27}$ grams, $r = 6.371 \times 10^8$ cm,

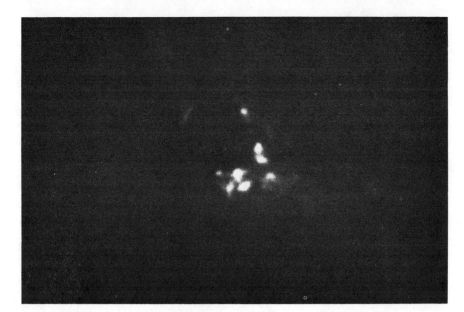

FIGURE 11.1.
Star formation is believed to be taking place within this object, known as Herbig-Haro No. 2.
(Courtesy of Lick Observatory.)

for the Earth gives an average terrestrial density of 5.517 grams per cm³. Notice that not every cubic centimeter within the Earth need have a mass of 5.517 grams. The ordinary rocks of our everyday world have densities between about 2.5 and 3.2 grams per cm³. Common metals like iron have densities of about 8 grams per cm³. Evidently, then, a mixture of rock and metal can be chosen to give an average density of 5.517 grams per cm³, the value for the Earth.

It might be thought that we have arrived here at a simple proof of a mixed composition for the Earth—part rock, part metal—as the Earth was stated to be in Chapter 10. But the argument is not in itself sufficient to establish this conclusion, because any material when squeezed under high pressure takes on a higher density than it has under ordinary laboratory conditions, where the pressure is quite small. All the densities given in the previous paragraph refer to low-pressure conditions, not to conditions inside the Earth. Hence to establish that the Earth is a mixture of rock and metal, it is necessary to show that rock alone, when squeezed under the pressures which exist inside the Earth, cannot yield as high an average density as 5.517 grams per cm³. Calculations have been made that lead to this conclusion. Although doubts have been expressed of their validity, most geophysicists believe these calculations to be correct, and it is widely accepted therefore that the Earth possesses a central core of iron and other common metals surrounded by an extensive rocky mantle—as we assumed in Chapter 10.

Using the data of Table 10.11, we can easily find the average densities of the major planets. Indeed all we need do is multiply the average density of

TABLE 11.1.
Densities of the major planets

Planet	Mass (Earth as unit)	Density (grams per cm³)	Main constituents
Jupiter	318.00	1.33	Hydrogen and helium
Saturn	95.22	0.69	Hydrogen and helium
Uranus	14.55	1.6	Water, carbon dioxide, ammonia
Neptune	17.23	1.6	Water, carbon dioxide, ammonia

the Earth, 5.517 grams per cm^3, by the mass values of Table 10.11, and then divide by the cubes of the radii. The reason for this simplicity of procedure is that the data of Table 10.11 were given with the mass and radius of the Earth as units. The density values are shown in Table 11.1.

The column of Table 11.1 headed "Main constituents" refers to the materials that contribute most to the masses of the planets. There could also be long lists of minor constituents, but minor constituents will not be our chief concern in this chapter. Our first objective is to follow the reasoning which leads to the materials given in the rightmost column of Table 11.1.

From Table 7.1 it will be recalled that main-sequence stars like the Sun are very largely composed of hydrogen and helium. The amounts by mass of the next most important constituents of the Sun are shown in Table 11.2.

The CNO elements together with Ne make up about 1.6 per cent of the Sun's mass, and the heavier elements make up about 0.4 per cent; about 2 per cent of the mass of the Sun is neither hydrogen nor helium. The amount of helium is usually estimated to be about 23 per cent (although there are reasonable arguments to suggest a somewhat lower value, about 20 per cent). This

TABLE 11.2.
Main constituents of the Sun (excepting hydrogen and helium)

Element	Symbol	Fraction by mass[a]	
Carbon	C	~0.0040	
Nitrogen	N	~0.0015	0.014
Oxygen	O	~0.0085	
Neon	Ne	~0.0020	
Magnesium	Mg	~0.0007	
Aluminum	Al	~0.0007	
Silicon	Si	~0.0007	0.0035
Sulfur	S	~0.0004	
Iron	Fe	~0.0010	

[a]The symbol ~ is used here to remind the reader that, although these numbers are believed to be substantially correct, they are not claimed to be exact. The numbers represent the fractions of the Sun's mass that are made up by these constituents.

gives 75 per cent hydrogen, 23 per cent helium, 1.6 per cent CNONe, and 0.4 per cent for the rest.

If we assume the planets were derived from solar material, we must relate these abundances to the main constituents of Table 11.1, and to the main constituents of the inner planets, Mercury, Venus, Earth, and Mars. The latter are mainly composed of iron, and of Mg Al Si in combination with oxygen. The remarkable fact emerges that the inner planets are made from material that constitutes only about 0.4 per cent of the mass of the solar material. Somehow this 0.4 per cent *became segregated from the rest in the process that led to the formation of the inner planets.*

Consider next the planets Jupiter and Saturn. Of all the planets, only these two possess hydrogen and helium in much the same large concentrations as in the solar material. Under terrestrial conditions the densities of hydrogen and helium are far lower than the densities of Jupiter and Saturn, but at the very high pressures within these planets the hydrogen and helium are highly compressed. Indeed, calculations have shown that the expected densities of the high-pressure states of these elements accord very well with a density of 1.33 grams per cm^3 for Jupiter and 0.69 grams per cm^3 for Saturn, the higher density of Jupiter being simply due to its larger mass, which compresses the hydrogen and helium more than the lower pressure within Saturn.

The combined mass of Jupiter and Saturn is equivalent to about 413 Earth masses. The total content of Mg Al Si S Fe in 413 Earth masses of solar material would be only some 1.4 Earth masses, whereas the combined mass of the four inner planets amounts to about 2 Earth masses. This point emphasizes the remarkable degree of segregation of the inner planets. Separating out *all* the metal and *all* the common rock-forming elements from a mass as large as Jupiter and Saturn combined would still be scarcely sufficient to provide for the inner planets.

The planets Uranus and Neptune have been left until last, because the facts about them lead to a far-reaching conclusion. Pressures within Uranus and Neptune are significantly lower than within Jupiter and Saturn, too low to compress hydrogen and helium to densities anywhere near the observed values of about 1.6. Nor can these planets be composed of the same kind of material as the inner planets, since such materials would give densities much higher than 1.6. What, then, are the main constituents of Uranus and Neptune? In attempting to answer this question, we can naturally turn to the CNO elements and to the molecules which these elements form in combination with hydrogen and with each other, H_2O, NH_3, CH_4, CN, CO, CO_2, and N_2 being the most important. The commonest of these would be H_2O, NH_3, and CH_4, and under the pressures within Uranus and Neptune 'ices' formed from these materials would indeed take on densities near the known value of 1.6 grams per cm^3.

Accepting that water, methane, and ammonia are the principal constituents of Uranus and Neptune, and remembering that these substances form about 1.4 per cent of solar material, we see that the amount of solar material from which the H_2O, CH_4, and NH_3 must have been segregated had to have been some 70 times the sum of the masses of Uranus and Neptune, i.e., 70×31.78

Earth masses, or about 2,200 Earth masses. We are therefore led to this remarkable conclusion: the original mass of the material from which all the planets in the solar system were formed must have been substantially greater than the present-day total mass of the planets, which is only about 450 Earth masses. Originally, the amount of planetary material must have been at least 2,200 Earth masses, the main fraction of which must subsequently have been lost in some way. The lost material consisted of some 1,800 Earth masses of hydrogen and helium. How did this loss occur?

The answer is rather obvious: through the evaporation of hydrogen and helium from the region of Uranus and Neptune. These regions are on the outskirts of the solar system, where the gravitational pull of the Sun is least able to restrain light gases like hydrogen and helium from streaming away into space. The gravitational pull of the Sun in the region of Jupiter and Saturn is stronger, and is still stronger in the inner region of Mars, Earth, Venus, and Mercury. The situation is illustrated in Figure 11.2. In this figure, the planetary material in the outer regions is taken to be gaseous, the condensation of H_2O, CH_4, and NH_3 into Uranus and Neptune occurring at a later stage. This point is important, because the hydrogen and helium must have escaped *before* the

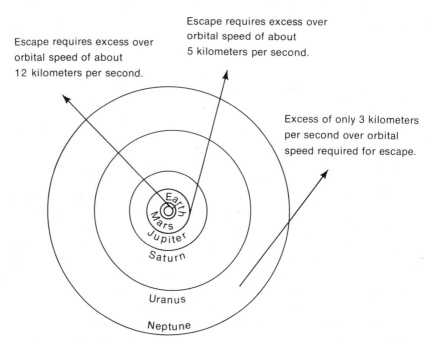

FIGURE 11.2.
The gravitational pull of the Sun is stronger in the inner part of the solar system than it is in the region of Uranus and Neptune. The light gases, hydrogen and helium, can escape most readily therefore from the periphery of the solar system.

formation of Uranus and Neptune, otherwise these planets would simply have picked up the hydrogen and helium through their own gravitational pull, and would then have become monster planets like Jupiter but of still larger mass, about 1,000 Earth masses each if the hydrogen and helium were shared equally.

Heat is required to give atoms of hydrogen and helium sufficient speeds to escape from the solar system, and the required heat must have come from the primordial Sun itself. The present-day output of solar radiation could hardly drive off hydrogen, still less the heavier atoms of helium, from the periphery of the solar system. It is true that heating would be stronger if the gas were nearer the Sun, but then, as we saw in the preceding paragraph, the gravitational holding forces would also be stronger. The apparent dilemma here is resolved, however, by the circumstance that the primaeval Sun was much brighter than the present-day Sun. This is a general property of the mode of formation of solar-type stars, as we have already studied in Chapter 6 (see Figure 6.24). Instead of the general temperature in the vicinity of the Earth being some 10 to 20 degrees Centigrade (283 to 293°K), as it is now, the temperatures in the region of the inner planets lay between about 1,000 and 1,300 degrees Centigrade. Everywhere in the vicinity of the inner planets, conditions were like the interior of a blast furnace. The temperature of a gas cloud in the outer parts of the solar system would similarly be significantly higher than it is now, and would indeed be sufficient for hydrogen and helium to evaporate in the manner we require.

§11.2. The Primordial Sun

To proceed further toward our goal of understanding how the planets came into being, imagine the following experiment. Add energy to the Sun, causing it to expand, and continue doing so until the Sun becomes so big that it contains all the planets. After becoming immersed in the body of the Sun, the planets would become heated and would evaporate into gas, and the planetary gas would become well-mixed with the general material of the Sun. Now remove energy, so that the bloated Sun shrinks gradually back to its normal size. Will there be any important difference between this hypothetical Sun and the actual Sun? Yes, the hypothetical Sun would be found to be spinning around with an equatorial speed of about 100 km per second, much faster than the presently observed speed of about 2 km per second. It is remarkable that swallowing up the planets would make such a difference in the spin of the Sun. The implication to be drawn from this imagined experiment is that the actual Sun spins very slowly.

Still more remarkable, if we were by some magic able to pull back the 1,800 Earth masses of hydrogen and helium that escaped from the solar system, and were to make a similar experiment, letting the Sun absorb not only the present planets but the other 1,800 Earth masses as well, the resulting spin would be faster still. The equatorial rotation speed of the hypothetical Sun would become

FIGURE 11.3.
The phenomenon of spin-up. The ice skater increases his spin by drawing arms and legs toward the axis of rotation. (From A. Baez, *The New College Physics*. W. H. Freeman and Company. Copyright © 1967.)

about 1,000 km per second, and this would be too much for stability. The gravitational forces holding the Sun together would not be strong enough to restrain the rotary forces that would come into play (the mv^2/a forces discussed in Chapter 9), and our hypothetical object would simply burst apart. This further imagined experiment shows us that, if the whole of the 2,200 Earth masses of planetary material were originally a part of the primordial Sun, then spin-up effects, of the kind which we studied in Chapter 6, must have occurred at some stage during condensation. Our next step must be to reconsider these spin-up effects.*

Any object possessing what is known as *angular momentum* spins up as it shrinks. The process is familiar from experience in everyday life. The ice skater wishing to set up a fast spin starts the rotation with arms and legs as widely spaced as possible, and then closes up the whole body toward the axis of spin, as illustrated in Figure 11.3. What happened in our imagined experiment was that by swallowing up the planets, and still more by taking up the reconstituted 1,800 Earth masses of hydrogen and helium, the Sun acquired angular momentum. This caused a spin-up in the manner of the ice skater as the Sun condensed back to its original radius.

---

*The reader may care to refer back to what was said in Chapter 6 concerning such effects (Figures 6.14 and 6.15), although the following discussion is self-contained.

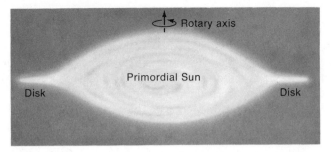

FIGURE 11.4.
A stage during the condensation of the primordial Sun when spin-up
became severe enough for material at the equator to separate away
into a disk.

Spin-up causes a large body to develop polar flattening. A distinct polar
flattening can be seen, for example, in Figure 10.12. This flattening of Jupiter
is caused by the short period of rotation of the planet, 9 hours 50 minutes.
The primordial Sun, containing the planetary material, would have become
more and more markedly flattened as spin-up proceeded. A stage would eventually be reached, with the polar axis about half the equatorial axis, when the
rotary forces at the solar equator would cause material to separate away into
a disk, leading to the situation illustrated in Figure 11.4. Calculation shows
that this stage would have been reached when the equatorial radius was about
30 times the present radius, about 20 million kilometers, which is about half
of the present-day radius of the orbit of Mercury.* It is interesting that the
configuration of Figure 11.4 occurred at a stage of condensation when the scale
was comparable with the inner dimension of the planetary system.

§11.3. An Historical Interlude

It is of interest to follow the developments after the stage of Figure 11.4
according to an older line of argument, since this will bring out the difficulties
that faced theories of the origin of the planets which were current in the first
half of this century. In doing so, we entirely omit the magnetic spin-down
process discussed in Chapter 6; we will return to magnetic spin-down once
it has become clear that no satisfactory alternative is available.

With the formation of the disk of Fig. 11.4, the rotary forces on the main
condensation are temporarily relieved. But as shrinkage continues the main
condensation spins up again, and the rotary forces reassert themselves. Once
again the main condensation is required to lose material into the surrounding

---

*The reader seeking to check this calculation should take into account the refinement men-
tioned in the footnote at the end of Appendix IV.1.

disk, which therefore continues to grow as the inner body contracts. An intermediate stage of the process is illustrated in Figure 11.5, and a late stage in Figure 11.6, when the disk has become both extensive and massive. Calculation shows that about 1/3 of the whole initial supply of solar material would thus come to reside in the disk. In making this calculation, we are supposing that the disk first forms when the radius of the main condensation is half the radius of the present orbit of Mercury. The sequence of Figures 11.4, 11.5, and 11.6 therefore leads to a disk of much greater mass than is needed for the planets. Even allowing generously for the eventual evaporation of hydrogen and helium, the mass of material needed to form the planets is only some 1 per cent of the solar mass, 30 times less than the result of such a calculation. Moreover, the disk would lie within the orbit of Mercury, instead of being outside—as we observe the planets to be. Obviously, then, we have arrived at a situation having little resemblance to the conditions required to produce the planets of the solar system. Such a disk as we have described would eventually form itself into a *star*, separated from the primary star by a distance

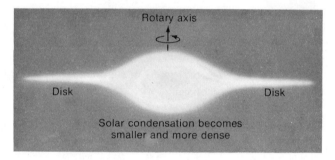

FIGURE 11.5.
Without magnetic spin-down, the central condensation continues to lose material into the disk.

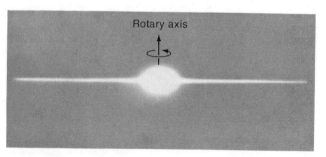

FIGURE 11.6.
In the end, the mass of material in the disk becomes of stellar, not planetary, order.

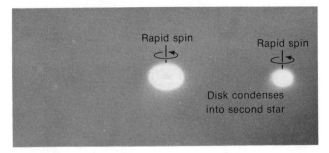

FIGURE 11.7.
Condensation of the disk into a compact body then gives a binary
system, not a planetary system.

only a little greater than the sum of the diameters of the two stars, which would
proceed to revolve around each other in a period of a few days. Such systems
are in fact known, and are called *contact binaries*. With Figure 11.6 evolving
into Figure 11.7, we thus have a reasonable theory of the origin of contact
binaries, but not of the origin of planets.

The theory we have just been discussing is essentially the one proposed
during the nineteenth century by Laplace, and the difficulties we have encountered are just those which led many astronomers in the first half of this century
to follow entirely different lines of attack. Perhaps a wandering star at one
time came close to the Sun, drawing out from the Sun by its gravitational pull
a thin bridge of material which subsequently condensed into planets? Perhaps
the Sun was at one time a member of a binary, and perhaps a wandering third
star happened to come along from outside, and drew the Sun's companion off,
leaving behind a bridge of material which condensed into the planets? Perhaps
the Sun's companion exploded into a supernova, leaving behind only a trace
of material which condensed into the planets? These were the sort of idea that
was tried. They were known as *catastrophic theories,* and they all would have
required our system of planets to be an exceedingly rare phenomenon, and
hence for life as we know it to be rare. Perhaps we were the only living creatures
in the whole of our galaxy? This concept seemed attractive to theologians, and
to many scientists as well.

§11.4. Magnetic Spin-Down

Figure 11.4 shows the beginning of disk formation, and at this stage there is
not a great deal of mass in it. Suppose the amount to be what we require for
the origin of the planets, not more than 1 per cent of the solar mass. The
question to be answered is how to *prevent* further supplies of material from
being added to the disk.

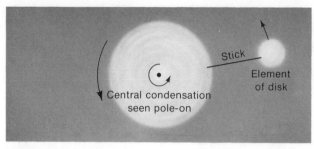

FIGURE 11.8.
Imagine a stick to connect the central condensation to an element of
material of the disk.

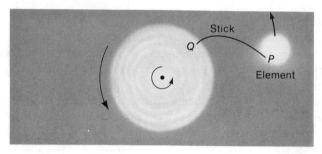

FIGURE 11.9.
Since the rotation period of the disk is longer than that of the central
condensation, the stick becomes bent. If there is whip in the stick,
the effect is to brake down the rotation of the central condensation.

Suppose the disk to become somewhat separated from the main condensation,
and imagine a stick, oriented initially straight out along a radius, that connects
the main condensation to an element of the disk, as in Figure 11.8. The time
required for one revolution of the outer point of attachment P will be different
from that needed for the inner point Q. By Kepler's third law, these two times
of revolution, T_P and T_Q, say, will be related to the distances a_P, a_Q, of P
and Q from the center by the equation $(T_P/T_Q)^2 = (a_P/a_Q)^3$. Since a_P is greater
than a_Q, it follows that T_P is a longer time interval than T_Q. Hence the stick
of Figure 11.8 will become bent as shown in Figure 11.9. If the stick has whip
in it, the effect will be to pull the disk forward in its rotation and to exert
a corresponding drag backward on the main condensation. The second of these
effects produces a braking action on the main condensation, whereas the first
effect causes the element of the disk at P to spiral slowly outward, as in Figure
11.10. If this were to happen for all the material of the disk, not just for the
element at P, the total effect would be to cause the whole disk to move slowly
outward in the manner of Figure 11.11. As the disk moved outward, the rotation
of the main condensation would become more and more checked. Should this

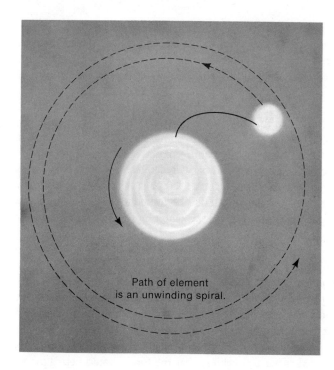

FIGURE 11.10.
And to cause the element of the disk to spiral outward.

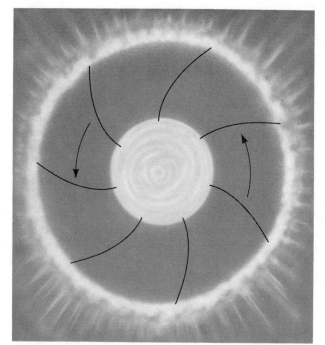

FIGURE 11.11.
The effect of many such sticks is to cause the whole disk to move steadily outward.

braking action be strong enough, the continued spin-up of the shrinking inner condensation would be prevented, with the consequence that further material would *not* be added to the disk. The mass in the disk would remain suitable for the formation of the planetary system, and our problem would thus be solved.

Figure 11.11 shows the difficulty which prevented this idea from being followed by the astronomers of fifty years ago. It shows a gap opening up between the solar condensation and the expanding ring of gas. How can an actual physical process operate in the manner of the stick in Figure 11.9, it being required to cross a gap? No ordinary dynamical process, no process depending solely on the motion of the material, can do so. We need a new idea at just this stage of the argument. The development of such an idea was a turning point in our understanding of the whole problem of the origin of planets.

It will be recalled from Chapter 9 how Kepler believed the planets were kept in their orbits by a force which pushed them in the direction of their motion. This was essentially the idea of Figure 11.9, but whereas Kepler believed such a push to be necessary simply to maintain a closed elliptic orbit for a planet, we know now that the force of Figure 11.9 produces motion in a slowly unwinding spiral. However, Kepler's idea was close to the argument we are following in the present problem. Moreover, the further part of Kepler's idea gives the clue to what we are seeking, that the force should be magnetic. The stick of Figure 11.9 is to be a magnetic field which connects the primordial Sun to the outer, expanding ring of gas. A magnetic field can cross a gap, as we well know from the use of magnets in everyday life (see Figure 11.12). This is the distinction between the older ideas and the modern point of view.

For our present purposes, we do not need to specify the precise configuration of the magnetic field. *Any* configuration extending between the central condensation and the disk will supply the whip action of Figure 11.9. The magnetic connection of condensation and disk need not even be maintained at all times. The connection could be sporadic, occurring in bursts, provided only that the net effect was to spin down the central condensation fast enough to compensate for any spin-up caused by the continuing contraction of the main solar condensation. For the situation of Figure 11.4, with the dimension comparable to the radius of the orbit of Mercury, the calculated contraction time is a few thousand years; the magnetic field must slow down the spin of the solar condensation faster than this. The consequent requirement for the strength of the magnetic field is that it have an intensity comparable to the magnetic field in present-day sunspots. In short, we require the whole surface of the primaeval solar condensation to be like a huge sunspot, like the situation shown in Figure 11.13. We can express this condition by saying that the whole surface would need to be extremely active. Because of the rapid rotation, and because of the disturbance caused by the formation of the disk itself, this is an entirely reasonable condition. Indeed, newly formed stars are actually observed to be highly disturbed at their surfaces, and it is thought that magnetic fields play a crucial role in their intense activity.

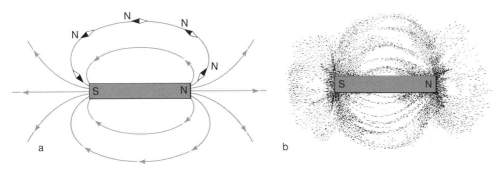

FIGURE 11.12.
The field around a magnet has an effect both on a compass needle (a) and on iron filings (b), showing that magnetic fields can cross a gap. (Diagram a is from G. Christiansen and P. Garrett, *Structure and Change*. W. H. Freeman and Company. Copyright © 1960. Diagram b is from A. Baez, *The New College Physics*. W. H. Freeman and Company. Copyright © 1967.)

FIGURE 11.13.
During condensation the primordial Sun is required to be active like a sunspot, but over the whole of its surface. (Courtesy of Hale Observatories.)

§11.5. Segregation by Condensation

In our discussion in Chapter 6 of the problems of star formation, we saw that a newly forming star of solar type approaches the main sequence as in Figure 6.24. Except in the phase just before it enters the main sequence, when nuclear reactions are beginning to supply energy, the star follows a downward track in the luminosity-color diagram. This downward track is one that corresponds to a nearly constant surface temperature of about 4,000°K. This would also be the temperature of the material of the disk when the disk was first formed, as in Figure 11.4. At such a temperature, the material of the disk would be entirely gaseous. However, as the disk moves outward in the manner of Figure 11.11, the temperature falls more and more below 4,000°K. Given a sufficient decline of temperature, small solid and liquid particles will condense within the planetary gases, thereby segregating materials that have high boiling points, like Fe, MgO, SiO_2, CaO, Al_2O_3, from the more volatile materials, which still remain in gaseous form.

It is of interest to work out values for the temperature within the disk at those stages in its expansion where the distance from the primordial Sun was comparable with the radii of the orbits of the present-day planets. A selection of temperature values are given in Table 11.3, worked out on the basis of a radius of 20 million kilometers and a surface temperature of 4,000 °K for the primordial Sun.

TABLE 11.3.
Temperature of gas cloud during formation of the planets

Planet	Radius of orbit (10 million km unit)	Temperature of material in outward-moving disk (°K)
Venus	10.82	1,600
Earth	14.96	
Jupiter	77.83	500
Saturn	142.8	
Uranus	287.2	300
Neptune	449.8	

This calculation bears out the point, mentioned already, that at the stage where the disk material had expanded to a scale comparable with the radii of the orbits of Venus and Earth, the temperature was about like that of a blast furnace. Only highly refractory substances like Fe, MgO, SiO_2, CaO, Al_2O_3, could condense out of the gas at this stage. If the condensations of these substances became reasonably large, say, a few meters in diameter, they would be left behind as the main gaseous materials of the disk, particularly the hydrogen and helium, continued to expand in the manner of Figure 11.11. Consequently, the readily condensable materials, provided they formed into bodies of appreciable size, would be left in orbits occupying the inner part of the solar system. The magnetic forces impelling the outward motion of the gas do not act effectively on compact bodies, and hence there would be no systematic outward spiraling motion for any considerable chunks of iron or of rocky material that were left behind after the gas had gone.

Not all the refractory material need remain on the inside of the solar system, however. Refractory materials in the form of fine smoke particles would be swept along by the gas. Only the larger pieces would remain on the inside of the solar system. Hence we have a sorting process both with respect to condensation properties and with respect to particle size. It is not yet clear whether sorting with respect to size could produce some degree of chemical fractionation among the refractory materials themselves; probably it would. The situation for these materials, which eventually formed the inner planets, is shown in Figures 11.14 and 11.15.

There is a strong argument to show that a process of this kind, involving an essentially complete separation of gas from solid, must have operated in the process of formation of the Earth. Light gases, particularly hydrogen and helium, can be evaporated thermally; so we cannot in principle interpret the deficiency of light gases on the Earth as certain proof of a complete gas-solid separation. The light gases might somehow have been evaporated thermally from heavier gases—although attempts to build an evaporation model for the

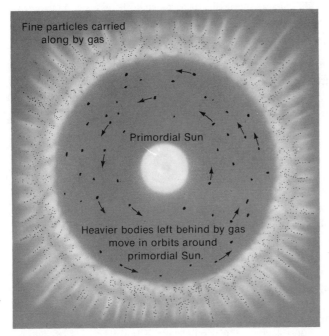

FIGURE 11.14.
Fine particles of condensed material would be swept along by the outward-moving gases, but larger pieces of rock and metal would be left behind.

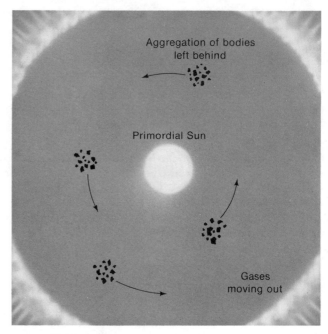

FIGURE 11.15.
With the gases gone, the condensation of the pieces of rock and
metal eventually led to the formation of the inner planets.

inner planets have always run into serious technical difficulties. No such idea
can be used for the heavy noble gases, krypton and xenon, however. Since the
amount of these gases on the Earth is found to be exceedingly small, we can
therefore be certain that an essentially complete gas-solid separation must have
taken place.

It seems, then, that the solid materials of the Earth must have become
essentially wholly separated from gas, which is just what the continued outward
movement of the disk of gas in Figure 11.11 would lead to. Such a complete
separation as is required to explain the nearly total absence of krypton and
xenon implies a corresponding absence of any constituents that would be
gaseous at $1,600°K$, including water, carbon dioxide, and nitrogen, which are
of course found on the present-day Earth in comparatively small but appreciable
amounts. Where then did these volatile substances come from? We shall return
to this important question in the next chapter, where we will be concerned
with life, for which the volatiles are particularly important, and with the
occurrence of life in the universe.

Another important aspect of Table 11.3 is immediately apparent. The temperature of $300°K$ ($\equiv 27°C$) is below the boiling point of water, and is thus
of the order at which water will condense out of the planetary gases. This is
in the region of Uranus and Neptune. Since water is probably the main

constituent of the latter planets, this is a further satisfactory result. At the low vapour pressure within the planetary gas, an actual temperature of 300°K would be somewhat high for the condensation of water, and certainly for carbon dioxide and ammonia. However, this is not an embarrassment, since the temperature values of Table 11.3 have been calculated for a fixed radius of the solar condensation, namely, for the situation of Figure 11.4, when the disk of planetary material first separated from the central condensation. As the disk of material expands, the main central condensation shrinks, and by the stage at which the disk reaches the region of Uranus and Neptune, the central condensation could well have shrunk in size by the amount necessary to lower the temperature estimate of 300°K, say, to 200°K. At the latter temperature, water will certainly condense, and CO_2 and NH_3 as well.

§11.6. General Conclusions

Let us return now to our main argument, and summarize the conclusions we have reached.

1. As the primordial solar nebula condensed to smaller and smaller size, there was a spin-up that increased the rotary forces until a disk of material quit the main central condensation. The mass content of the disk was about 1 per cent of the whole nebula.

2. A magnetic coupling was set up between the main condensation and the disk. This coupling produced a stabilizing spin-down of the main condensation, at the same time causing the disk to develop into an expanding ring of gas.

3. Refractory materials, iron and rock-forming substances, condensed as solids and perhaps liquids at an early stage in the expansion of the ring. Larger condensed bodies remained behind as the gas continued to expand. By eventually joining together, these bodies formed themselves into the inner group of planets.

4. The spin-down of the primordial Sun became essentially complete at the stage where the ring of gas came to occupy the outer parts of the solar system, particularly when the gas reached the region of Uranus and Neptune.

5. Hydrogen and helium evaporated away at the farthest distances reached by the gas, leaving water, carbon dioxide, and ammonia to form the main constituents of Uranus and Neptune. These planets did not form, however, until after the evaporation of hydrogen and helium was essentially completed.

6. The continued decrease in the radiative output of the condensing primordial Sun eventually caused the temperature in the region of Jupiter and Saturn to fall sufficiently that water-ice or dry ice could condense from gas in this region. Unlike the situation for Uranus and Neptune, there was still hydrogen and helium present in the region around Jupiter and Saturn. As these latter planets grew, their icy cores developed appreciable gravitational fields, which enabled them to acquire the hydrogen and helium in which the cores were embedded.

Although this picture does not deal in subtleties or fine detail, it at least succeeds in bringing together the important features of our planetary system. It explains an important major fact which other theories have not succeeded in grappling with, namely, the fact that the Sun rotates slowly. It explains the division of the planets into four inner ones of small mass, surrounded by four outer ones of much larger mass, and it connects the actual planetary masses with the main materials of which they are constructed.

Many questions remain: How did an initial swarm of small bodies aggregate themselves into a planet? How long did this process take? What about the satellites of the planets? What about Pluto? Some further progress toward answering these questions is made in the Appendixes to this section. Rather than concern ourselves here with these details, we shall pass on now, and in the next chapter, to consider some much broader problems.

§11.7. The Number of Other Planetary Systems

By working directly from the situation we now find in the solar system, we have largely been able to avoid speculative assumptions about the initial state of the primordial solar nebula. We have been led inevitably to a situation in which the nebula was required to spin up, reaching the critical point of disk formation at the stage where it had condensed to a scale somewhat less than the orbit of the innermost planet, Mercury. Now we ask if such a spin-up was a reasonable property for a newly forming star to have had.

This question involves a consideration of the general problem of star formation. It is easily demonstrated that such a spin-up property is to be expected. Indeed, in Chapter 6 we already encountered spin-up even at the earliest stages of star formation. Only if there was magnetic braking, similar in principle to that used here for the planets, could stars be formed at all. It seems, then, as though a two-stage situation is involved, a first spin-up occurring before the primordial solar nebula had even formed. Following a resolution of this first rotational crisis, along the general lines of Chapter 6, the solar nebula entered a contracting phase which developed into a second spin-up situation. This second spin-up was the one which led to the origin of the planets.

At all events, we can be confident that some form of spin-up must take place during the condensation of every star, and some way of resolving the rotary crisis must have been found in all cases. Three ways in which resolution might take place can be distinguished.

1. The stellar condensation grows a disk that is steadily driven through magnetic coupling to greater and greater distances, a planetary system being eventually formed.

2. The stellar condensation grows a disk to which it is coupled magnetically, as in 1, but the coupling is so strong that, instead of promoting a smooth outward expansion, it causes the disk, as soon as it begins to form, to be immediately blown catastrophically at high speed away from the main con-

densation. The spin-up crisis in this case is only resolved very temporarily, and the process must be repeated many times.

3. If there is *no* effective magnetic coupling—the opposite situation from 2—the stellar condensation continues to grow a disk until the central star is finally formed. In this case the disk is sufficiently massive to form a second star, and a contact binary probably ensues.

For a star following process 2, spin-up is always occurring, with the condensation proceeding always at the very edge of rotational stability. The condensation is perpetually growing a new small disk, which is then almost immediately shattered. When condensation to a star is completed, there is still a near-crisis situation, with a continuing tendency for the star to splutter off material. A high rate of spin is retained. Process 1, on the other hand, leads to stars with slow spin, a distinction which shows itself in actual observations of stars. Binaries apart, stars with masses comparable to the Sun all spin slowly, as can be seen from Figure 11.16, and presumably therefore have experienced process 1. Stars of large mass, on the other hand, say, 5 ⊙ or more, spin rapidly

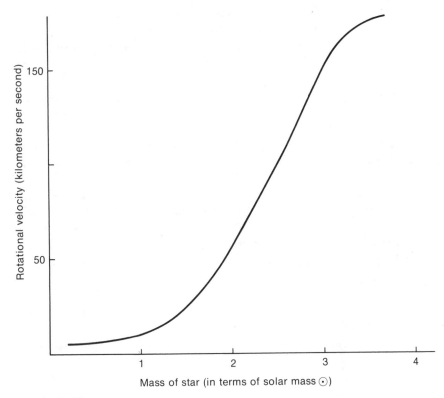

FIGURE 11.16.
Rotation versus mass for main-sequence stars. Single stars with masses of solar order are observed to spin slowly, as the Sun does. Such stars probably all have planetary systems. Stars of larger mass spin much more rapidly and may well *not* have planetary systems.

and have presumably experienced process 2. Sporadic disk formation is indeed observed in many massive, rapidly spinning stars.

Stars of large mass are rare compared to stars with masses of solar order, and contact binaries are also comparatively rare. Most stars are slowly spinning like the Sun, and have presumably experienced process 1. There are about 10^{11} of them in our galaxy. Taking all galaxies together, there are of the order of 10^{20} slowly spinning stars of small mass, presumably with planetary systems. We may conclude therefore that the number of potential sites for the origin of life is utterly vast: 100,000,000,000,000,000,000. We will consider in the next chapter the extent to which life may have occupied these sites.

General Problems and Questions

1. From their densities, what can be inferred about the chemical compositions of the planets?

2. In what respects are the chemical compositions of the planets (a) similar to, (b) different from, the composition of the Sun?

3. If it is supposed that the planets were derived from the primordial solar condensation, how are the composition differences of Problem 2 to be explained?

4. What argument supports the view that initially the Earth had little in the way of an atmosphere, and of volatile materials in general? Which two of the noble gases are particularly relevant to this argument?

5. If the planets were derived from a primordial solar condensation, what argument leads to the conclusion that the primordial condensation must have experienced spin-up to a rotationally unstable state?

6. What is the distinction between a rotationally unstable condensation evolving into a binary star system and such a condensation evolving into a single star surrounded by a planetary system?

7. Explain how the number of planetary systems in the galaxy can be estimated.

8. How was it possible for the highly refractory materials of which the four inner planets are mainly composed to have become segregated from the volatile components of the planetary gases?

9. Which substances formed solid condensations in the region of Uranus and Neptune?

10. How would the solar system have been different if hydrogen and helium had not evaporated from its periphery?

Chapter 12:
Life in the Universe

At the end of Chapter 11, we estimated there might well be as many as 10^{11} planetary systems in our galaxy. Among so large a number, we could be reasonably sure that many will be like the solar system, even if our study of the origin of the solar system had not made it clear that much the same processes will operate for all stars of solar type. That small planets like the Earth lie on the inside of the solar system was seen to be not an accident, but a circumstance that we can expect to be common.

This circumstance thus disposes of one of the main difficulties which scientists used to have in estimating the likelihood of life existing elsewhere in the universe. When the structure and origin of the solar system were less well-understood than they are today, astronomers thought that conditions suited to life might result only by chance, with the chance very slight, and they especially thought so in the days when catastrophic theories of the origin of our system mainly occupied their attention. The distance of the Earth from the Sun was a frequent topic of discussion. What was the chance, it was asked, of a planet like the Earth being at just the right distance from the parent star for temperature conditions to be appropriate for life? We now see that the chance is high—not low, as people used to think—for several reasons. First, since several small planets will probably form on the inside of the system, there is a quite large chance that one of them will be at the right distance. If the

Sun were hotter, then Mars would be more appropriate for life than the Earth. If the Sun were less luminous, then Venus would be more appropriate. Second, during an extended period of time, the Sun changes its luminosity to a considerable degree. Because of this change, a planet where temperature conditions were unsuitable for life at one epoch could become suitable at another epoch. We see therefore that the "zero order" condition for the occurrence of life on planets is likely to be satisfied in a large fraction of our 10^{11} cases. By a "zero order" condition, I simply mean the availability of a planet of appropriate size, like the Earth, where temperature conditions are within the range in which terrestrial life can exist.

Our biological concepts have undergone a similar evolution. Before our current understanding of the chemical basis of life was achieved, it seemed as if the origin of life on the Earth might have depended on a chain of accidents. Suppose the probability of a suitable outcome at each link in such a chain were $1/10$. Then with ten links the probability of an outcome favorable for the emergence of life would be as low as 10^{-10}. The advantage of having 10^{11} possible sites for life then becomes largely lost. Our situation on the Earth would become nearly unique. But this argument was actually based on ignorance. As more has become known about the basic chemical structure of life, it has become more and more clear that very little of what happened here on the Earth happened by chance. A similar basic chemical structure can be expected to have occurred elsewhere. It is, of course, always possible that living structures, very different from any we can yet conceive of, may exist elsewhere—but we need not worry about such speculations for our purposes here, and can base our discussion on what is actually known.

§12.1. The Chemical Basis of Terrestrial Life

The critical information content of living organisms is carried by a structure made up from only six molecules: a sugar (S), a phosphate (P), and four bases (A and T, G and C), the chemical formulas for which will be given in a moment. Imagine a set of railway tracks built from S and P, as in Figure 12.1, with ties joining pairs of S molecules. The permissible forms of tie are built from the bases in the manner shown in Figure 12.1, the possible pairings of the bases being A and T together, and G and C together. Imagine Figure 12.1 to be given a rotation of 180°. Except for a switch in the order of the base pairs, the situation is unchanged, as can be seen from Figure 12.2. The combinations T and A, and C and G, are also permissible, so that the ties of Figure 12.2 are valid. *Invalid* forms of tie are shown in Figure 12.3.

In an actual railway track, the ties must all be of the same length; the same is true here. The base pairings A and T, G and C, lead to ties of the same length. An attempt to pair A with C or G, or to pair T with G or C, would give ties of the wrong length. So it is simply the physical dimensions of the pairings that decide which is valid and which invalid.

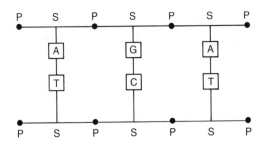

FIGURE 12.1.
Imagine two parallel tracks built from a sugar molecule (S) and a phosphate molecule (P), with the ties made up from one or other of the base pairs (A,T), (G,C).

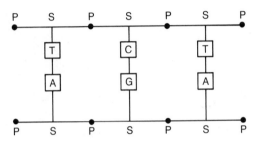

FIGURE 12.2.
After 180° rotation, the (A,T) pair is changed to (T,A), the (G,C) pair to (C,G). These are also valid combinations.

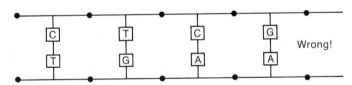

FIGURE 12.3.
Invalid pairs are shown here.

Imagine Figure 12.1 to be extended into a very long chain, with each tie in the chain being one of the four permitted pairings (A, T), (T, A), (G, C), (C, G). The particular pattern of ties along the track contains all the information required for the development of a plant or animal. In practice, the track is twisted into a double helix, the form of which is shown in Figure 12.4, and this helix is itself coiled up in such a way that it occupies astonishingly little space.

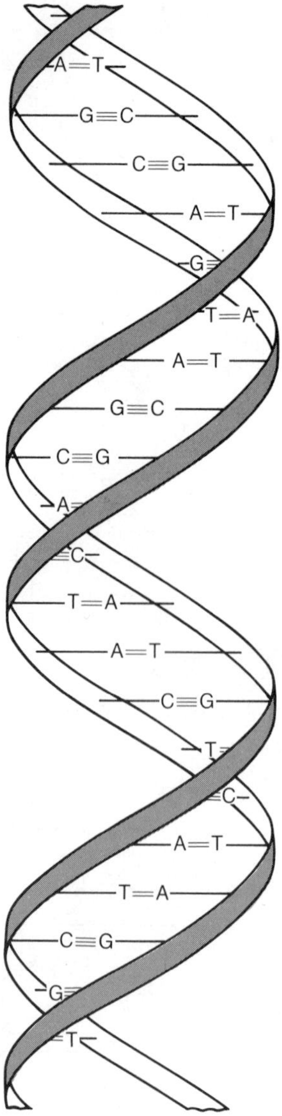

FIGURE 12.4.
The parallel tracks of Figure 12.1 are coiled now into a double helix, which can itself be coiled so as to occupy remarkably little space. (Based on J. D. Watson and F. M. C. Crick, *Nature*, 171, 1953, 964.)

It would be a long and intensely tedious walk along the railway track from New York to Los Angeles. Taking the ties to be a little less than a yard apart (just too small to match one's stride), there are about six million of them. In a single human cell our "railway track" has something like six *thousand* million ties, and the whole structure is packed away tidily within not much more than a hundredth part of a millimeter—about the dimension at which light begins perceptibly to turn corners (see Figure 5.26). Of course, a major reason for the smallness of the biological cell is the smallness of the six basic molecules themselves, the scale of a tie in the chain being only about 3×10^{-8} cm. Yet 6×10^9 such ties would still stretch for 2 meters, if they were laid out in a straight line like an actual railway. The packing efficiency in the biological cell is increased not only by twisting and coiling, but also by breaking the track itself into pieces; in the human cell there are 46 such pieces, known as *chromosomes*.

We saw in Chapter 7 that atoms joined together into molecules form hybrid electron shells. The extra electron from Na supplies the electron needed to complete the shell of C1, thereby giving the firmly bound molecule of NaC1. We shall find, when we take a detailed look at the chemical structure of Figure 12.1, that the same is true of the way the S and P molecules are joined along the two tracks, and of the joining of the ties to the tracks. Thus all the junctions in each half of Figure 12.1, shown separately in Figure 12.5, are firmly "bonded." However, the halves are not strongly bonded in the same way; so we can redraw the ties of Figure 12.1 as in Figure 12.6, with the dotted connections thought of as weak links. Although any one such link is easily destroyed, it is not easy to pull the two halves of Figure 12.1 bodily apart by a simple *mechanical force*, just because there are so many links; a single strand may be weak, but a rope woven from many strands can be exceedingly strong. Yet if the dotted connections of Figure 12.6 can be annulled *one by one by chemical means*, the two halves will easily come apart, as they are shown to be in Figure 12.5.

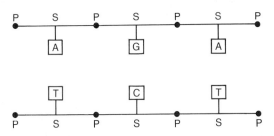

FIGURE 12.5.
The ties of Figure 12.1 have here been broken, in such a way that the base pairs are separated.

FIGURE 12.6.
The separation of Figure 12.5 occurs because the A to T, and the G to C, connections are weak. This is indicated by the dotted bonds between the base pairs.

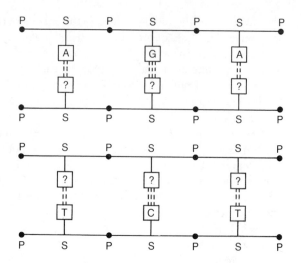

FIGURE 12.7.
Sometimes the two parts of Figure 12.5 become completely separated, and the biological cell attempts to recover the original double structure by growing two new tracks. Consistent with the rules for valid and invalid pairs, how are bases to be attached?

Suppose the two halves of Figure 12.5 to become wholly separated, so that they can never be fitted together again. Suppose that two new S and P chains are grown, as in Figure 12.7, and suppose further that the chemical situation is now changed to one in which the double tracks of Figure 12.7 are impelled to grow new systems of ties. *Because of the bases already attached,* there is only one way in which new ties can be grown consistently with the rules that determine valid base pairs, namely, in the manner shown in Figure 12.8. But the two halves of Figure 12.8 have now each become the same as in Figure 12.1. So the basic genetic information of the cell has thus been *duplicated.* This

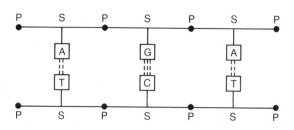

FIGURE 12.8.
The solution to this problem takes the unique form shown here. The remarkable point emerges that we now have two copies of the original structure.

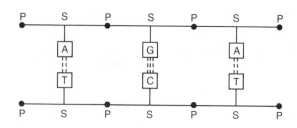

is the principle which enables biological cells to replicate themselves, a principle which turns on the critical rule that A goes only with T, and G goes only with C.

All the information necessary for the construction of a plant or animal is contained in the double-tracked structure we have been describing. Given one copy of this basic information, more copies can be obtained, when the process described in the preceding paragraph is carried out in a living cell. But should all copies be destroyed, nothing can be done to recover the plant or the creature, which thus becomes *extinct*, for we have no way of artificially putting together as many as 6×10^9 ties *in the correct order*, even if we knew what the order would be.

TOPICS FOR DISCUSSION:

1. According to modern biology, the difference between you and a dog, or a cow, a rat, a bird, or a grain of wheat, lies in the number and order of the ties along the two S-P tracks. An (A, T) tie in a dog is just like an (A, T) tie in you, and the S-P tracks have the same structure. How far do you consider this knowledge to be consistent with the biblical concept (*Genesis* 1:26–29) that plants and animals were provided by God for Man's use, i.e., exploitation? Do you find the biological knowledge to have moral implications?

2. The fact that the links within the (A, T) and (G, C) pairs are weak means that individual ties are often broken. Granted that biological cells possess an efficient repair process, ties can nevertheless become inoperable, or can even be changed from one kind of tie to another. Such changes, known as *mutations*, usually have only a minor effect on the plant or animal in question. Sometimes, however, the changes can be drastic and tragic—as we know well from children born with serious defects. In light of such tragic cases, how desirable do you think it would be for biologists to be able to change artificially the distribution of genetic information on the S-P chains? How do you balance the advantages and disadvantages of an artificial control of genetic information?

3. Because the basic genetic structure lacks long-term stability—for the reason given in 2—there is always the possibility that genetic information may be changed. Since natural variations are essentially random, and since random insertions into complex information are much more likely to garble the content than to improve it, the outcome of such variations will almost always be deleterious. Yet, on the "monkey and typewriter" principle, improvements of information content will occasionally occur. Natural selection is the process whereby these occasional improvements are retained by a plant or animal, whereas the frequent deleterious variations are rejected. Satisfy yourself that you understand the general idea of natural selection.

4. Natural selection works by accepting the few and rejecting the many. If you are a socialist, how does natural selection fit your political and social concepts? Do you consider that natural selection should be stopped? If it could be, what would happen in the long term?

5. Most people of a religious pursuasion are worried by the fact that there is so much pain and suffering in the world—the dilemma of St. Augustine. This biological discussion makes clear that pain, suffering, and extinction are absolutely guaranteed, are built into the whole system of genetic variability and of natural selection. If you are a religious person, why do you think God adopted this system? Was God a prisoner of His own laws, as were the gods of Nordic mythology? Was God not able to invent a system of laws that would have avoided the pain and suffering? Notice that one cannot pass off these issues in the manner of theological apologists, by attributing evil to the Devil, unless the Devil is to be credited with inventing the laws of physics, chemistry, and biology—and if he is, what is the status of God?

6. The immediate effect of an encounter with the ideas of biology is that distressing questions are raised—no matter what one's political philosophy may be. These questions are just as distressing for the Marxist-Leninist as for his sworn enemy, the "bourgeois capitalist." In the U.S.S.R., the study of biology has for the past thirty years been almost totally emasculated. In the U.S.A., on the other hand, biology is an exceedingly active and productive science. If you are a Marxist-Leninist, do you think it has been an advantage to the U.S.S.R. to attempt to deny the nature of the world in this way? Is it an advantage to rewrite history the way you would like it to have been, rather than the way it was?

---

§12.2. DNA, RNA, and Proteins.

We come now to the question of how the genetic information, stored in the manner described above, is used in the biological cell. From a logical point of view, the method is similar to the way plans are used in the construction of a large building. We would not permit everybody engaged in such a construction to have access to a single unique set of plans. Copies of the main plan are made, often copies limited to only a portion of the whole structure. Similarly, musicians in an orchestra do not play from the composer's full score. Parts are extracted from the full score, and the musician plays from the part for his own instrument. Just as we would mark a copy to distinguish it from the master plan, so biological copies are constructed to be distinguishable. A slightly different sugar molecule, say, S′, is used to form S′-P tracks, and a different base, say, U, is used instead of T. Otherwise the logic of the structure of the copy is the same as that of the original.

The function of the copies is to build substances known as *proteins*. These are long molecules built from *amino acids,* of which 20 play key roles. The makeup of a protein is $\boxed{A}\bullet\boxed{A}\bullet\boxed{A}\bullet$ . . . , with each of the boxes marked A referring to one or another of the 20 amino acids. The junctions marked • are obtained by removing a molecule of water in the following way. All amino acids are of the form shown in Figure 12.9, where R is one of many possible side-chains. Two amino acids with side-chains R_1 and R_2 are adjacent to each

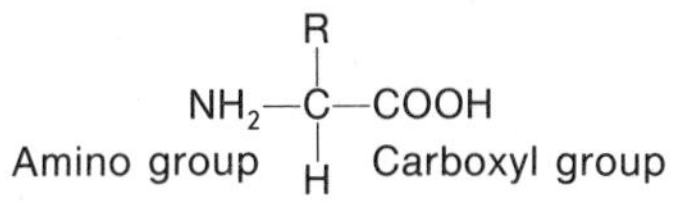

$$NH_2-\underset{\underset{H}{|}}{\overset{\overset{R}{|}}{C}}-COOH$$

Amino group   Carboxyl group

FIGURE 12.9.
The schematic structure of an
amino acid.

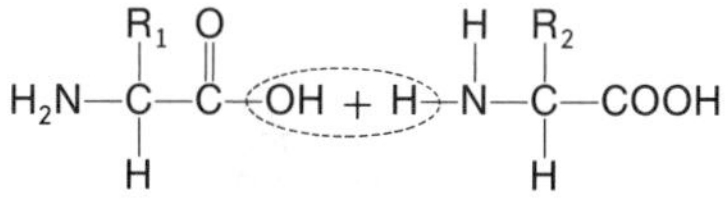

$$H_2N-\underset{\underset{H}{|}}{\overset{\overset{R_1}{|}}{C}}-\overset{\overset{O}{\|}}{C}-OH + H-N-\underset{\underset{H}{|}}{\overset{\overset{R_2}{|}}{C}}-COOH$$

FIGURE 12.10.
To link two amino acids, a molecule
of water is eliminated.

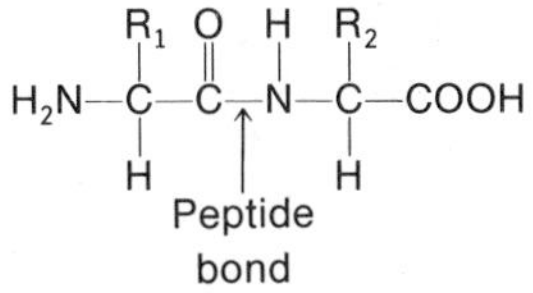

$$H_2N-\underset{\underset{H}{|}}{\overset{\overset{R_1}{|}}{C}}-\overset{\overset{O}{\|}}{C}-\overset{\overset{H}{|}}{N}-\underset{\underset{H}{|}}{\overset{\overset{R_2}{|}}{C}}-COOH$$

Peptide
bond

FIGURE 12.11.
Linked amino acids.

other in Figure 12.10. A molecule of water, H_2O, is eliminated to give the
structure of Figure 12.11. This is done by pulling OH from one amino acid
and H from the other. When many of the 20 important amino acids are linked
together in this way, we have the protein situation illustrated in Figure 12.12,
in which the group of atoms shown in Figure 12.13 alternates with that of
Figure 12.14—the latter being known as a *peptide linkage*. We shall find below
that the strong bonds of attachment between S and P in Figure 12.1, and also
between S and the bases A, T, C, G, are obtained similarly, by the elimination
of a molecule of water.

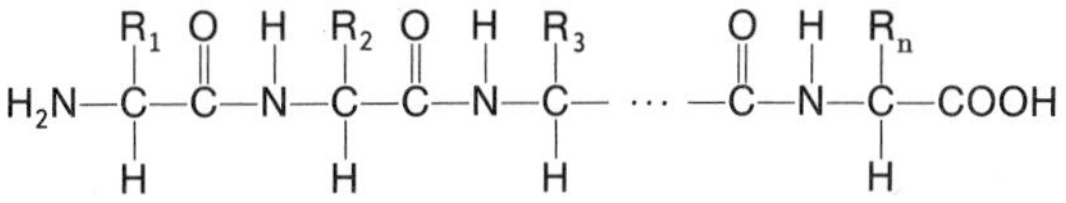

$$H_2N-\overset{\overset{R_1}{|}}{\underset{\underset{H}{|}}{C}}-\overset{\overset{O}{\|}}{C}-\overset{\overset{H}{|}}{N}-\overset{\overset{R_2}{|}}{\underset{\underset{H}{|}}{C}}-\overset{\overset{O}{\|}}{C}-\overset{\overset{H}{|}}{N}-\overset{\overset{R_3}{|}}{\underset{\underset{H}{|}}{C}}-\cdots-\overset{\overset{O}{\|}}{C}-\overset{\overset{H}{|}}{N}-\overset{\overset{R_n}{|}}{\underset{\underset{H}{|}}{C}}-COOH$$

FIGURE 12.12.
Many linked amino acids form a protein.

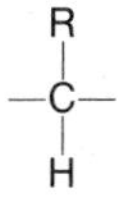

$$-\underset{\underset{H}{|}}{\overset{\overset{R}{|}}{C}}-$$

FIGURE 12.13.
Thus a protein consists of a long chain in which the group
shown here alternates with that of Figure 12.14.

$$-\overset{\overset{O}{\|}}{C}-\overset{\overset{H}{|}}{N}-$$

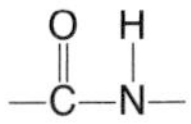

FIGURE 12.14.
This group of atoms is referred to as a peptide linkage.

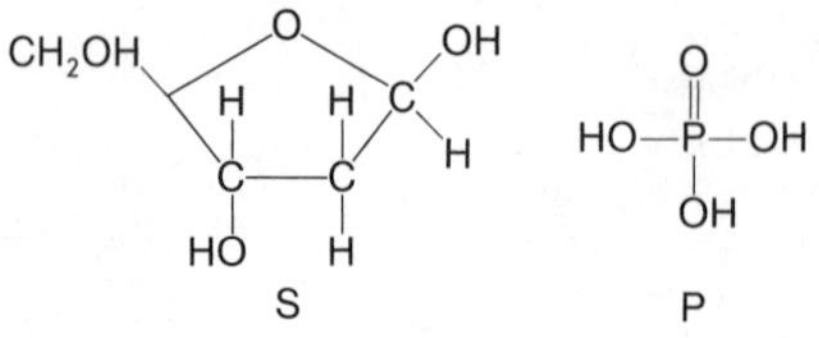
CH₂OH
O
OH
H
H
C
H
C
C
HO
H
S
O
HO—P—OH
OH
P

FIGURE 12.15.
The chemical structure of the sugar S and the phosphate P.

At this point it will be useful to specify the chemical structures of our six basic molecules. The molecules S and P are shown in Figure 12.15. A linkage of S and P is fashioned by eliminating a molecule of H_2O formed from one of the H atoms of P and one of the OH combinations of S. The forms of A, T, C, G, are shown in Figure 12.16. Linkages to S are again fashioned by eliminating a molecule of H_2O formed from an H atom of A, T, C, or G, and the OH group of S. All these linkages are illustrated in Figure 12.17. An abbreviated notation is adopted for simplicity in the latter figure. An unnamed junction point is a C atom if four lines radiate from it, is a CH group if three lines radiate from it, and is a CH_2 group if two lines radiate from it. In the attachment of any one of the bases A, T, C, or G, to S, notice that a carbon atom of the sugar is always attached to a nitrogen atom of the base.

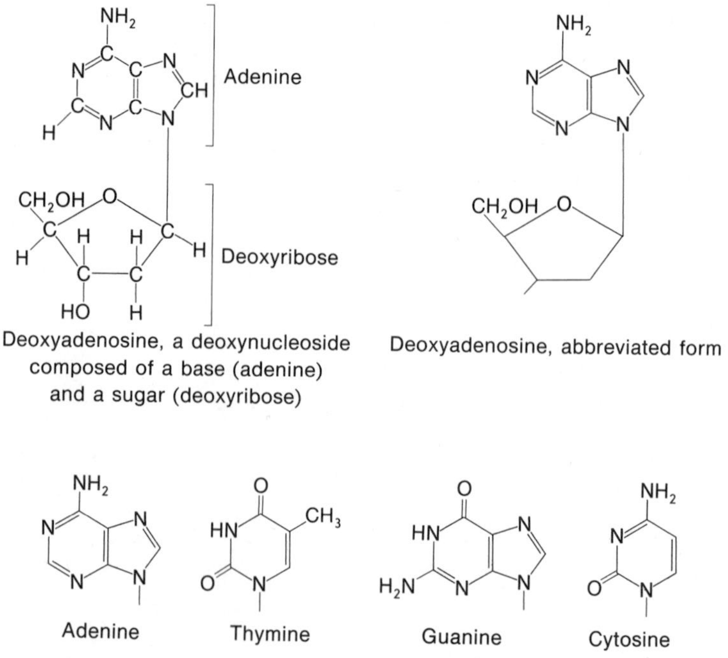
Adenine
Deoxyribose
Deoxyadenosine, a deoxynucleoside composed of a base (adenine) and a sugar (deoxyribose)
Deoxyadenosine, abbreviated form
Adenine
Thymine
Guanine
Cytosine

FIGURE 12.16.
The chemical structure of the bases A, T, G, C.

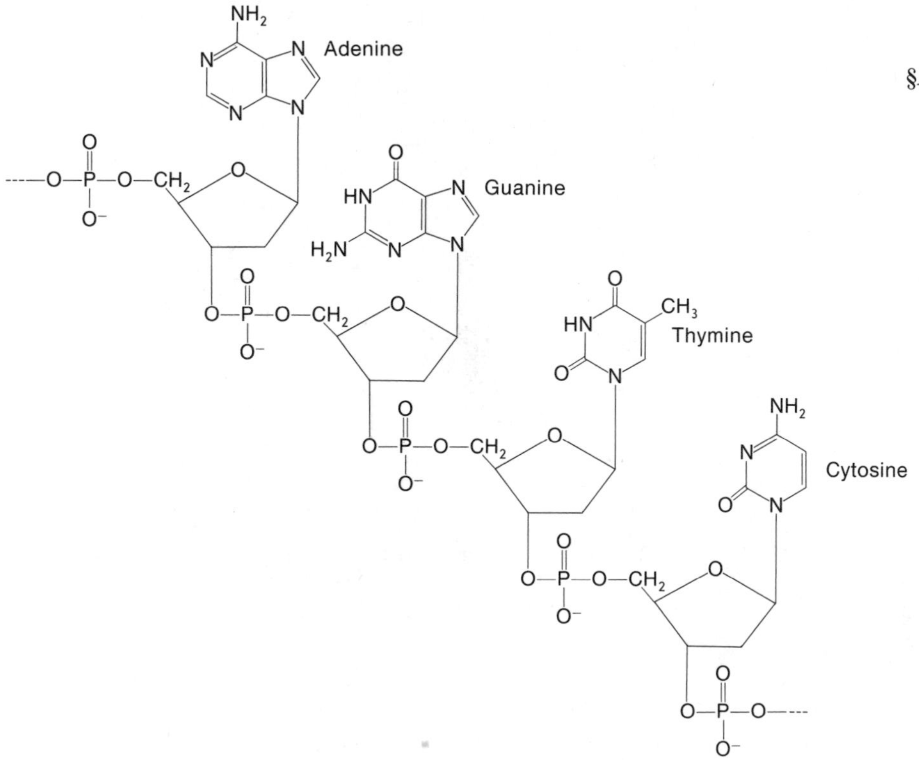

FIGURE 12.17.
The details of the ties of the four bases A, G, T, and C to the sugar-phosphate chain.

The following short diversion is necessary to understand a detail of Figure 12.17. We wrote the molecule formed by a sodium atom and a chlorine atom as NaCl, the "spare" electron of the sodium serving to complete a hybrid shell in the chlorine. By a hybrid shell we meant that in the quantum-mechanical sense the electron paths discussed in Chapter 4 do not entirely quit the sodium atom; the paths of high probability go near both atoms and thereby provide the "bond" that binds the molecule. This situation refers to the two atoms taken *by themselves*. When molecules of NaCl are dissolved in water, a different situation arises. The molecules of water themselves have electrical properties that appreciably disturb the paths of the binding electron in the NaCl, in such a way that the electron comes essentially free of the sodium atom. The electron is still held by the chlorine, however, which therefore becomes negatively charged, since the chlorine now has one electron more than the 17 protons in the chlorine nucleus. This property is denoted by writing Cl^-. And the sodium atom, having lost an electron, has become positively charged, and is therefore written as Na^+. So when NaCl is dissolved in water, the atoms separate

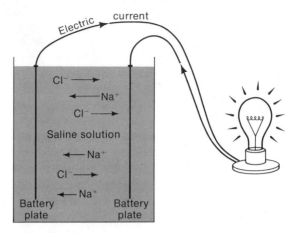

FIGURE 12.18.
When common salt, NaCl, is dissolved in water, the sodium
and chlorine atoms separate, with the chlorine carrying away
the extra electron of the sodium, the atoms then being denoted by Na⁺ and Cl⁻. These charged *ions* permit an electric
current to be passed through the solution. This is (in essence)
the reason why an electric current is able to flow in a car bat-
tery.

into Na⁺ and Cl⁻. It is hard to pass an electric current through pure distilled
water, but an electric current passes easily through a saline solution, the current
being carried by a systematic motion of Na⁺ relative to Cl⁻, as indicated
schematically in Figure 12.18.

The detail of Figure 12.17 requiring explanation is the meaning of O⁻. If
the P molecule were taken by itself, we should write OH. But in solution this
group separates into H⁺ and O⁻. The O⁻ remains attached to the phosphorus
atom, as shown in Figure 12.17, but the H⁺ is free to move away, and is
therefore not shown in Figure 12.17.

The technical name of the S molecule of Figure 12.15 is *deoxyribose,* and the
respective names of the A, T, C, and G bases are *adenine, thymine, cytosine,*
and *guamine.* Notice that the OH groups of Figure 12.15 occur at different
places in the deoxyribose molecule, and notice in Figure 12.17 how two of these
differently placed OH groups are systematically connected along the chain. In
the usual biochemical notation, the group from CH₂OH would be denoted by 5,
say OH(5), and the other one by OH(3). Reading Figure 12.17 from upper left
to lower right, we could denote the sequence of S and P molecules by the

$$
\begin{array}{ccccccc}
5 & 3 & 5 & 3 & 5 & 3 & 5 & 3 \\
\end{array}
$$

P • S • P • S • P • S • P • S • P,

$\longrightarrow$

sequence being read from left to right.

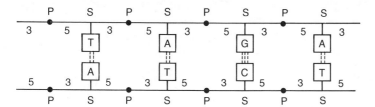

FIGURE 12.19.
The opposite orientation of the two sugar-phosphate chains of DNA.

The sense in which we elect to read such a chain is a matter of convention, but when there are two SP chains, as in Figure 12.1, we can ask whether they have the same sense or opposite senses—this question is not a matter of convention only. The answer, established by biochemical research, is that the two chains have opposite senses, as in Figure 12.19. Notice in this figure that the way the 3 and 5 connections between the S and P molecules are arranged is unchanged by a rotation through 180°. Subject to this arrangement of the S and P linkages, and with the chemical designations given in Figures 12.15 to 12.17, the basic genetic structure described above is known as deoxyribosenucleic acid (DNA).

The fitting together of A and T, C and G, to form weakly bonded pairs is shown in Figure 12.20. There are two weak connections for A and T, and three such connections for C and G. In all cases the atoms connected by dots cannot form hybrid shells by sharing electrons directly between themselves. Hence all these connections are weak, with consequences that have already been discussed.

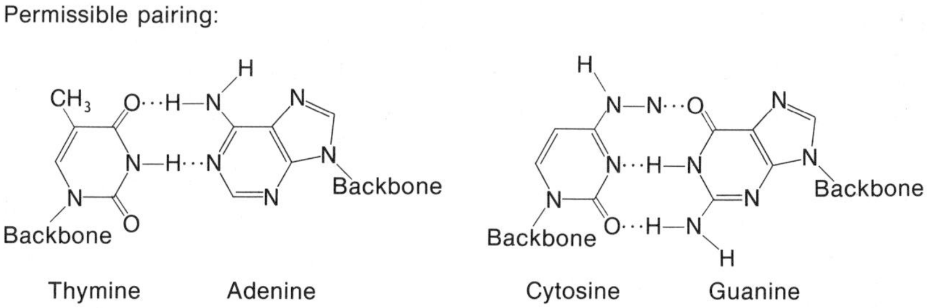

Permissible pairing:

| Thymine | Adenine | | Cytosine | Guanine |

FIGURE 12.20.
Details of the permissible base-pair connections, showing the weak (dotted) linkages.

We remarked above that, when copies are made of the basic genetic structure, the sugar S is replaced by S′, and T is replaced by U. The chemical forms for S′ and U are shown in Figure 12.21. A comparison with Figure 12.15 shows that S′ differs from S in having an OH group where S has an H atom. The

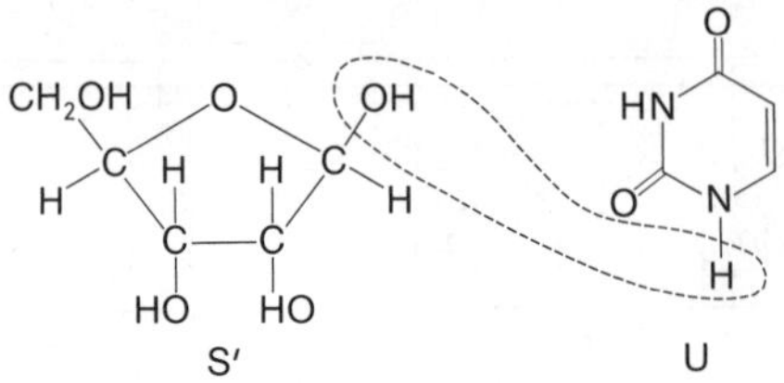

FIGURE 12.21.
In RNA, the sugar ribose (S′) takes the place of deoxyribose
(S), and uracil (U) takes the place of thymine (T).

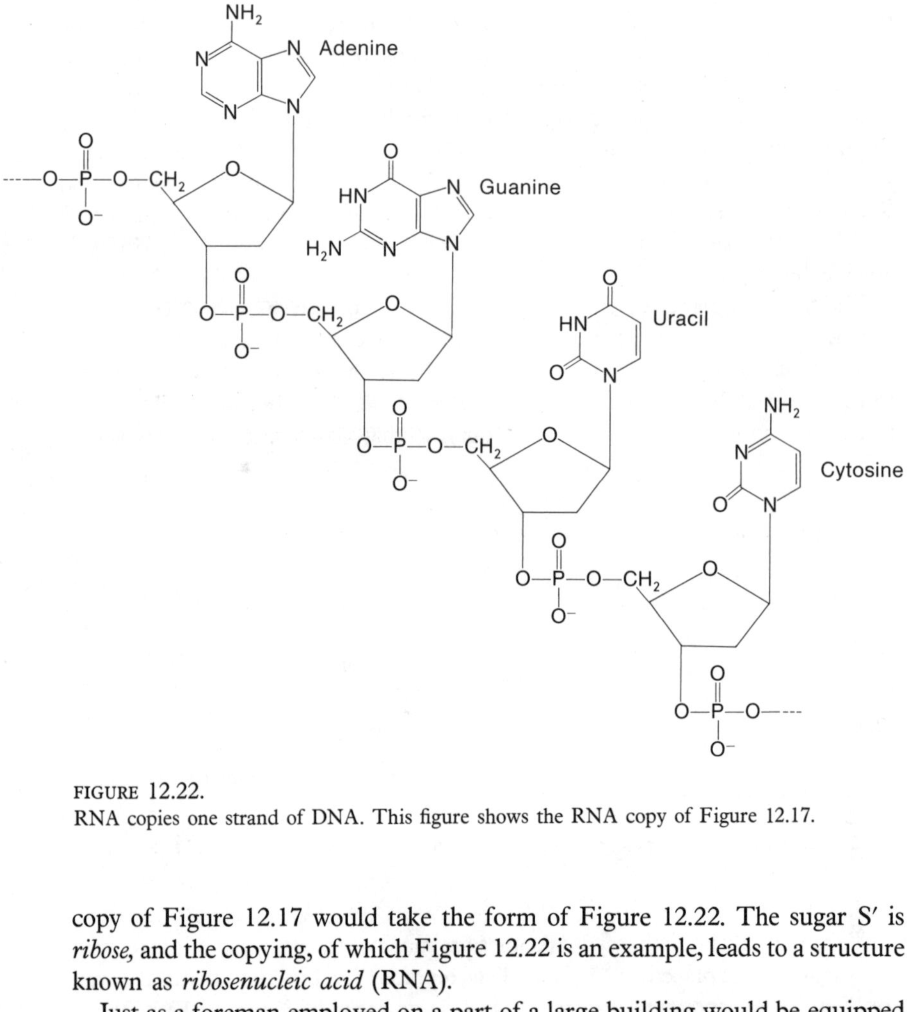

FIGURE 12.22.
RNA copies one strand of DNA. This figure shows the RNA copy of Figure 12.17.

copy of Figure 12.17 would take the form of Figure 12.22. The sugar S′ is
ribose, and the copying, of which Figure 12.22 is an example, leads to a structure
known as *ribosenucleic acid* (RNA).

Just as a foreman employed on a part of a large building would be equipped
with abbreviated plans sufficient for his section, and just as he would issue

still more abbreviated plans to his individual workmen, who would then proceed to the practical job of measuring and constructing, so it is in the biological cell. It is sufficient in RNA to copy just one of the two strands of DNA, Figure 12.22 being the copy of the single DNA strand shown in Figure 12.17. We already saw above that one strand of DNA implies the other, as in Figure 12.7. Consequently there is no loss of information in this single-strand copying, provided the RNA always copies the *same* strand. We shall assume this to be so, although there are still technical biochemical problems in understanding how the RNA "knows" it is always copying the same strand. We have the situation of Figure 12.23, with S′ (ribose) replacing S (deoxyribose) and with the attached bases being A, C, or G according to the DNA attachments, and with U in RNA appearing in place of T in DNA.

Think now in terms of the analogy of a reel of magnetic tape. Tapes are used in the operation of digital computers for the storing of records, many records being usually written on a single tape. The situation biologically is the same. The total length of DNA, if all chromosomes are included, is sufficient in the higher animals for the storing of about a million records, each record being what in classical biology was known as a *gene*. The information on a roll of magnetic tape is stored in the form of magnetized areas. In RNA the information is stored by the attached bases, which are to be read in units of three-at-a-time, as in Figure 12.24. And just as magnetic tape is read in a particular sense—the tape has a "beginning" and an "end"—so RNA is read in a particular sense, the sense defined by the $5 \longrightarrow 3$ attachment of S′ and P, as shown by the arrow of Figure 12.24. Furthermore, just as a specific marker

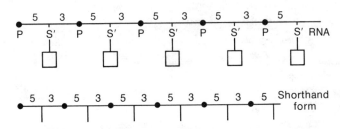

FIGURE 12.23.
Schematic form of RNA.

FIGURE 12.24.
Dividing RNA into consecutive groups of three bases. Such records are used to determine the structure of a protein, in accordance with the amino-acid designation given in Table 12.2.

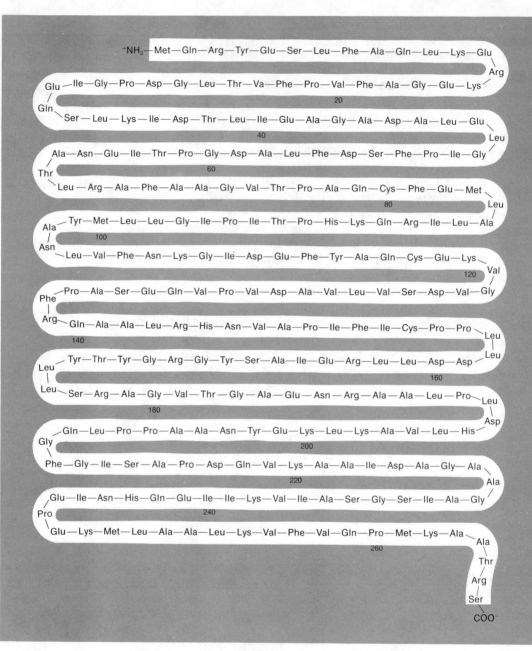

FIGURE 12.25.
An example of the amino-acid sequence of a protein.

is used on magnetic tape to indicate the beginning of a record, so the group of bases AUG defines the beginning of a biological record. And just as another magnetic marker is used to indicate the end of a record, so any one of the base triplets UAA, UAG, or UGA defines the end of the biological record. Thus we now have the situation of Figure 12.24, with the scope of the record determined by a beginning, then by a number of triplets, of which there are typically about a thousand, and by an end.

How is the information contained in such a record actually used? In the construction of a protein, whose general form we have examined already in Figure 12.12. Each triplet in the RNA message defines one of the R side-chains of Figure 12.12, the first triplet after AUG defines R_1, the second defines R_2, and so on, until the end of the record is reached. The synthesis of the specific protein determined by the genetic record is then complete. The total genetic information, sufficient for about a million records, is thus sufficient for the construction of about a million proteins.

As we have already stated, there are 20 biologically important amino acids distinguished by the R side-chains of Figure 12.12. Their names are given in Table 12.1, and their relationships to the base triplets of RNA are given in Table 12.2. An example of a protein for which the order of the amino acids has been determined is shown in Figure 12.25. The designations of Table 12.2 are often referred to as the *genetic code*.

TABLE 12.1.
The amino acids

Alanine (Ala)	Leucine (Leu)
Arginine (Arg)	Lysine (Lys)
Asparagine (Asn)	Methionine (Met)
Aspartate (Asp)	Phenylalanine (Phe)
Cysteine (Cys)	Proline (Pro)
Glutamate (Glu)	Serine (Ser)
Glutamine (Gln)	Threonine (Thr)
Glycine (Gly)	Tryptophan (Trp)
Histidine (His)	Tyrosine (Tyr)
Isoleucine (Ile)	Valine (Val)

Proteins so constructed serve like the worker bees of a hive. Because they can take part in a vast multiplicity of chemical reactions, they are the operating agents of life. They do the job of constructing the cell, of providing it with an outer protective wall, of building new DNA and new RNA, of controlling chemical balances within the cell, and ultimately of enabling the cell to display its most remarkable property—of being able to divide into two, with the daughter cells having the same structure and properties as the parent.

TABLE 12.2.
The genetic code

		SECOND BASE				
		U	C	A	G	
FIRST BASE	U	UUU Phe UUC Phe UUA Leu UUG Leu	UCU Ser UCC Ser UCA Ser UCG Ser	UAU Tyr UAC Tyr UAA *** UAG ***	UGU Cys UGC Cys UGA *** UGG Trp	U C A G
	C	CUU Leu CUC Leu CUA Leu CUG Leu	CCU Pro CCC Pro CCA Pro CCG Pro	CAU His CAC His CAA Gln CAG Gln	CGU Arg CGC Arg CGA Arg CGG Arg	U C A G
	A	AUU Ile AUC Ile AUA Ile AUG Met	ACU Thr ACC Thr ACA Thr ACG Thr	AAU Asn AAC Asn AAA Lys AAG Lys	AGU Ser AGC Ser AGA Arg AGG Arg	U C A G
	G	GUU Val GUC Val GUA Val GUG Val	GGU Ala GCC Ala GCA Ala GCG Ala	GAU Asp GAC Asp GAA Glu GAG Glu	GGU Gly GGC Gly GGA Gly GGG Gly	U C A G

THIRD BASE

*** = termination
The codons read in the 5'→3' direction.
From F. Crick, "The Genetic Code III." Copyright © 1966 by Scientific American, Inc. All rights reserved.

§12.3. What is Life?

The ability of a cell to replicate itself is a necessary property of life. And in organisms made up of many cells, the familiar plants and animals, the ability to replicate the large-scale organization of many cells is also implied by the word "life." But replication alone is not usually considered by biologists to be sufficient to define "life." A further condition is that replication should not depend on the cell's importing highly complex substances, particularly proteins. The virus is able to replicate itself when supplied with suitable complex substances, but not otherwise. A virus is not usually taken to be "alive." It is interesting to consider how far this further condition should be pushed. In popular conversation on dietary matters, we often hear it said that such-and-such a daily intake of proteins is necessary for our well-being. So on first thought our situation might seem similar to that of the virus. The proteins that we eat are not used directly, however. The human digestive system dissociates them into their constituent amino acids, which are later reassembled into the particular proteins required by the human organism. So we eat proteins as a source of amino acids, not immediately for the proteins themselves. And we eat carbohydrates in order to supply the S molecules, and phosphorus in vegetables to supply the P molecules. Given sugars, certain of the amino acids, and phosphorus, the human and other animals can replicate themselves. This criterion is usually judged to define the meaning of "life." Although widely used in biology, it is not a satisfactory criterion, however, being obviously anthropomorphic in its conception, i.e., designed so that humans are judged to be "alive." Indeed, if this criterion is taken strictly, humans actually fail to qualify for being alive, since there are just a few necessary substances which we must take in our diet and which we cannot build from simpler sources. These few substances we meet frequently in everyday life; they are the familiar *vitamins*. Presumably our remote ancestors, perhaps hundreds of thousands of years ago, were actually able to build the vitamins directly, but at some stage the human DNA developed an error, and the ability has now been lost. We ourselves, and our less remote ancestors, survive by acquiring these few required vitamins indirectly, by eating plants, or by eating other animals that do not suffer from this same genetic error.

Suppose it were possible for a plant to think. Surely a plant would set its definition at a chemical level suited to its own abilities. It would argue that in order to be "alive," it is essential to need only very simple inorganic materials: water, carbon dioxide, nitrogen, and various trace minerals. Creatures like ourselves, requiring anything as complex as sugars and amino acids, would be dismissed in much the way we disregard the virus. We would be looked on as an incorrigible pest, like the virus, existing only by ingesting complex chemical substances already synthesized by that truly living organism, the plant. This is not just a piece of whimsical flim-flam. It shows that our usual definition of "life" cannot stand up under serious analysis.

TOPIC FOR DISCUSSION:
Do you consider it possible to give precise meaning to the categories of "alive" and "not alive"? Can more progress be achieved by considering life as a whole? Is there an "inanimate world" and a "living world" that can be rigorously distinguished?

---

§12.4. The Basic Materials of Life

How is it that plants are able to produce the whole chemical range of life —amino acids, proteins, nucleic acids—from much simpler inorganic materials? Because plants make use of sunlight through the help of the substance chlorophyll, with the chemical structure shown in Figure 12.26. Although the molecule of chlorophyll is certainly complex compared to inorganic molecules like $NaCl$, H_2O, CO_2, it is far less so than DNA, RNA, or a protein like that in Figure 12.25.

Chlorophyll uses the energy of sunlight to convert H_2O and CO_2 into the sugar glucose. Energy is stored in the glucose molecule and is available for release in many ways. Sugar can be burnt, releasing the energy as heat and

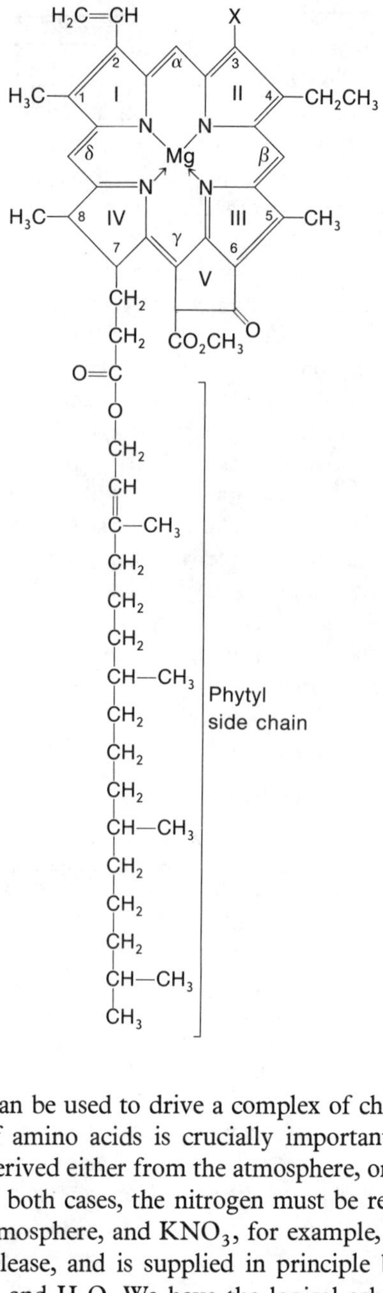

FIGURE 12.26.
The structure of chlorophyll.

light. Or, more subtly, the energy can be used to drive a complex of chemical reactions. Of these the building of amino acids is crucially important. The nitrogen present in amino acids is derived either from the atmosphere, or more usually from nitrates in the soil. In both cases, the nitrogen must be released from stable molecules, N_2 in the atmosphere, and KNO_3, for example, in the soil. Energy is required for this release, and is supplied in principle by the conversion of glucose back to CO_2 and H_2O. We have the logical scheme:

Sunlight + CO_2 + H_2O $\longrightarrow$ Glucose;

Glucose + nonenergetic substances

$\longrightarrow$ Energetic molecules + CO_2 + H_2.

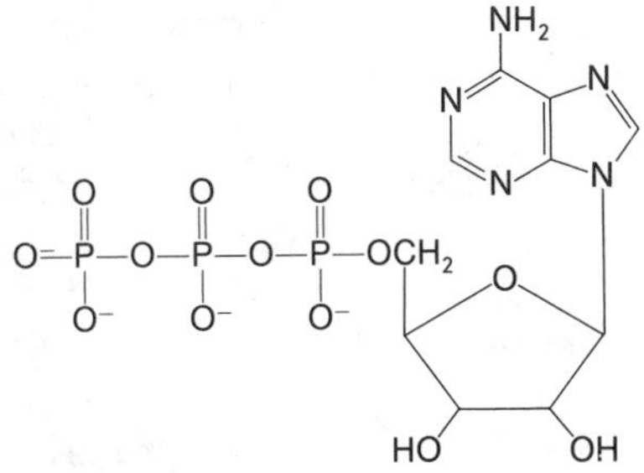

FIGURE 12.27.
The structure of ATP.

The energetic molecule ATP, built in this way, with the structure shown in Figure 12.27, plays a crucial role in a wide variety of biochemical processes. We spoke above of DNA being copied into RNA, and of the RNA being used in the construction of proteins. Energy is needed for all this, and ATP is the principle agent that supplies the energy.

The ability of plants to produce glucose is therefore important, not only as a source of sugar—needed to form the SP chain of DNA—but also as the energy supply for all life. What would happen if plants were to lose the ability to produce chlorophyll, as we humans have lost the ability to produce the vitamins? Unless we could somehow manage to carry through all the processes of synthesis which plants carry out—the production of sugars and amino acids—in factories, economically and in huge quantity, life on the Earth would come to an end.

TOPIC FOR DISCUSSION:
During the war in Viet Nam, large areas were deliberately defoliated, with the aim of denying food supplies to the enemy. From a military point of view, this tactic appeared no different from the blockades and sieges which have always been carried out in time of war. To some scientists, however, the policy of deliberately destroying the capacity of plants and trees to make use of sunlight in the synthesis of the basic materials of life had a worse connotation than the starving of human beings would have had. They saw the defoliation as being aimed against life itself. Do you consider this view to have been soundly based? What do you understand by the concept of "blasphemy"? To what extent can this concept be applied to an act of deliberate defoliation?

Although from a biochemical point of view plants and trees supply a vital link between simple inorganic materials and the much more complex molecules necessary for animal life, from an astronomical point of view the simple inorganic materials are not so "simple." There is a critical problem in understanding how materials like H_2O, CO_2, and N_2 came to be present on the Earth.

We saw in Chapter 11 that early conditions on the Earth resembled the interior of a blast furnace, and that volatile substances like water could not have been present then. We also saw that the very low terrestrial concentration of the heavy noble gases krypton and xenon strongly supports this picture. From where, then, did the Earth acquire its volatile materials? These materials include not only water, carbon dioxide, and nitrogen, but other volatiles formed by atoms of H, C, O, N, such as methane (CH_4), ammonia (NH_3), and hydrocyanic acid (HCN), which are thought to have played an important role in the origin of life. It also includes some heavier substances, such as compounds of mercury and lead.

If the Earth was initially much too hot to have retained these volatile substances, then obviously they must have all been acquired later. The question is how and from where? Perhaps small quantities of material continued to stream out of the solar condensation. Once the Earth was formed enough to have an appreciable gravitational field, it could have acquired the volatiles and have held them even at a high temperature, provided adequate quantities were available from such a streaming motion. The suggested situation is illustrated in Figure 12.28.

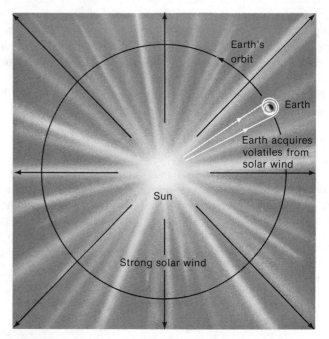

FIGURE 12.28.
One way in which the Earth could have acquired its volatile materials would have been by gravitational accretion from an intense solar wind. This suggestion suffers, however, from the difficulty that it would give too much krypton and xenon in relation to nitrogen and oxygen.

At first sight, the comparatively small quantities of volatiles required by the Earth appears to make this suggestion attractive. The amount of water in the terrestrial oceans is only about .025 per cent of the total mass of the Earth. The amount of CO_2 locked away in the $CaCO_3$ of limestone rocks is perhaps ten times less than the amount of water (about .0025 per cent), and the amount of N_2 is about three hundred times less (.000075 per cent). Compared to the bulk of the Earth, the required amount of volatiles is therefore quite small.

Although H_2O, CO_2, and N_2 might have been acquired by the Earth in this way, the CH_4, NH_3, and HCN required to promote the origin of life are not so readily obtained. The latter substances would not form in hot gases from the Sun. To obtain them, H_2O, CO_2, and N_2 must be mixed with an excess of H_2 under conditions of high density and low temperature. These conditions occurred in the outer planets, but it is hard to see how they could have occurred on the Earth.

A seemingly fatal objection to the idea of "volatiles from the Sun" comes from considering once again the very low concentration of krypton (Kr) and xenon (Xe) in the terrestrial atmosphere. Because most of the oxygen in gaseous material streaming from the Sun would join with H_2 to form H_2O, the terrestrial ratios Kr/H_2O, Xe/H_2O, should in such a picture be comparable with the solar values of Kr/O, Xe/O. Yet the terrestrial values of Kr/H_2O, Xe/H_2O, are in fact about a thousand times less than the solar values of Kr/O, Xe/O, a very large discrepancy that is inexplicable unless segregation between H_2O and Kr, Xe, occurred in the planetary material. For segregation to occur, it is necessary for H_2O to condense as ice, leaving Kr, Xe gaseous. This situation happened in the region of the outer planets, but it is hard to see how it could have happened in the region of the Earth. It seems more profitable, then, to seek an origin for the volatile materials in the region of the outer planets, rather than directly from the Sun.

TABLE 12.3.
Aggregation of Uranus and Neptune

Stage	Mass of bodies (gm)	Number of bodies	Time-scale (years)
1	2×10^{21}	10^8	300
2	5×10^{25}	4,000	10^6
3	2×10^{27}	100	10^7
4	10^{29}	2	3×10^8

Details of the way in which many small bodies formed the planets Uranus and Neptune are considered in Appendix IV.2, a similar problem for the inner planets also being discussed. The outcome is found to be the condensation picture set out in Table 12.3. The four stages of the table are intended to be in general a series of separate steps, although some overlap between one stage and another is likely to occur. For example, some bodies from stage (1) would survive into stage (3), and would then experience a randomization of motions caused by the gravitational force of the quite large bodies formed at this stage. The effect on the smaller bodies would be to cause them to thicken the thin disk formed in the beginning, and also to spread the disk, both inward and outward. Thus the small bodies remaining at stage (3) in the region of Uranus and Neptune must have developed into a kind of spray which engulfed the whole solar system. The outermost part of this spray would consist of smaller bodies going far away from the Sun, perhaps considerably beyond the distances of Uranus and Neptune. Such an extended distribution of bodies would tend to survive the collisional sweeping of stage (4), and so could persist to the present day. It is plausible to associate such bodies with the *comets*. The situation is illustrated in Figure 12.29.

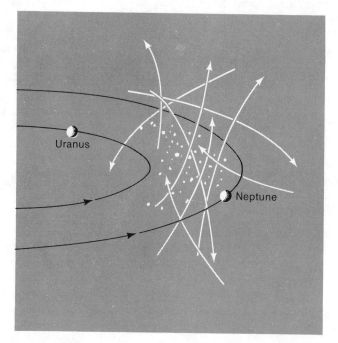

FIGURE 12.29.
Comets may well have been formed as a spray of icy bodies, developing in the region of Uranus and Neptune, which came to engulf the whole solar system.

The nature of the comets was discussed in Chapter 11, where we saw that molecules of CN, CH, NH, OH, NH_2, CO, CO_2, and N_2 are present in the "head" and the "tail" of comets. The highly active free radicals, CN, CH, NH, OH, and NH_2, can be regarded as fragments of HCN, NH_3, and H_2O that

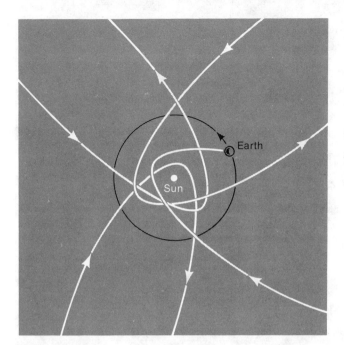

FIGURE 12.30.
Such icy bodies may also have been perturbed into orbits bringing
them to the inner part of the solar system, where they would have
collided from time to time with the newly formed inner planets.

have been detached by the action of sunlight. It seems, then, that comets consist
of just the material needed to provide for life on the Earth.

Bodies similar to the comets would also be perturbed into orbits bringing
them to the inner part of the solar system, as in Figure 12.30. Such bodies,
composed of H_2O, CO_2, CH_4, and NH_3, would cross the paths of the inner
planets and would collide with them from time to time. The mass of a single
object of stage (2) of Table 12.3 is of order 10^{25} gm, enough to supply all the
terrestrial volatile substances. Hence a single collision of a stage (2) object would
be adequate to meet the requirements of life on the Earth. Alternatively, there
could be many collisions of stage (1) objects; about 1,000 would be needed
to supply the required materials.

The popular fascination with comets was noted in Chapter 11. Although
this popular reaction may have little foundation in strictly astronomical terms,
it could very well have a sound emotional basis, for a comet-like object from
the region of Uranus and Neptune could have been responsible for the origin
of life on the Earth. It is remarkable that the difference between the Earth
being a barren sterile object, like the Moon in Figure 12.31, and being the
splendidly complex and active object of Figure 12.32, should turn on such a
comparatively small quantity of volatile materials.

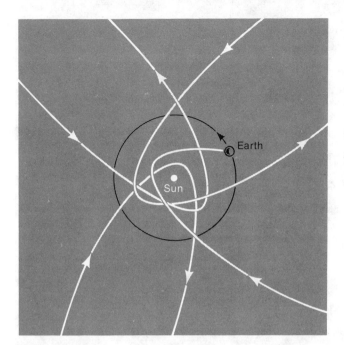For FIGURE 12.32,
see PLATE XVIII.

FIGURE 12.31.
The sterile surface of the Moon. (Courtesy of NASA, Apollo 13.)

§12.5. The Origin of Life

Given the Earth in its orbit around the Sun, and given such volatiles as H_2O, N_2, NH_3, CO_2, and HCN, how are we to proceed to an understanding of the genesis of life itself?

The chemical details preceding the origin of life as we understand it were undoubtedly complex. They are usually supposed to have occurred in an early atmosphere that was quite hot, so that all the volatiles were present as vapor and as clouds. Sunlight impinging on the vapor and clouds, as in Figure 12.33, is thought to have had effects similar to that of chlorophyll in our present-day plants, namely, to have induced the synthesis of amino acids and sugars, and of other molecules of a comparable complexity. These all fell to the ground in rain. Mostly they went into dilute solutions in the oceans, but some rain fell into more concentrated pools and lagoons. In these more concentrated pools the building of long-chain molecules is thought to have occurred, perhaps at the surfaces of particles of clay. Over long periods of time many chemical

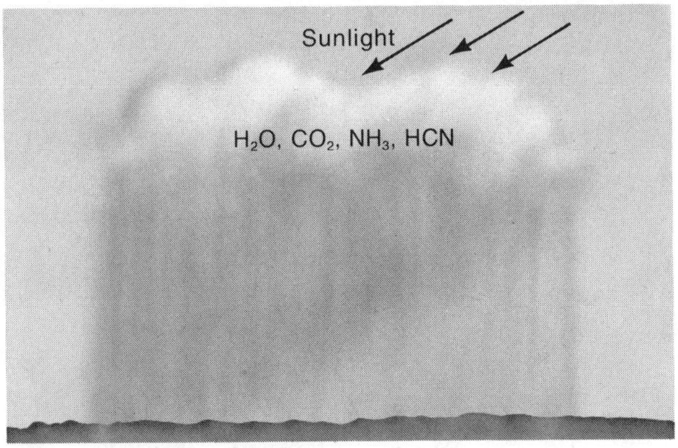

FIGURE 12.33.
Sunlight impinging on vapor containing water, nitrogen, ammonia, carbon dioxide, and hydrogen cyanide produced a continuous supply of amino acids and sugars which fell to the ground as rain.

combinations were "tried," until at last the first self-replicating system arose. This system became the basis of life.

Such in brief outline is the usual view. A rival theory takes the early temperature of the Earth to be low. Instead of the life-forming volatiles being in the atmosphere, they are taken to be in the ocean, and in more concentrated land-locked pools. For a low-enough temperature, water would freeze out as ice, causing dissolved CO_2, NH_3, and HCN to become still more concentrated in the isolated pools. Moreover, at low temperature there would be little water vapor in the atmosphere, and much of the ultraviolet component of sunlight, at present absorbed by atmospheric water vapor, would penetrate readily to ground level, where its effect on such concentrated solutions might be to produce amino acids and sugars, with an outcome similar to that of the first theory.

Astronomical considerations suggest that the Sun, after its early high-luminosity phase, should have been less luminous than it is now. The past history of the Sun can be represented by a track in the luminosity-color diagram, as shown in Figure 12.34. This past history, calculated according to normal astronomical theory, would favor the low-temperature theory of the origin of life. Yet laboratory studies of the building of long molecules shows that high temperatures are to be preferred, in order to establish linkages of the kind we discussed above, with a molecule of H_2O being pulled away, OH being taken from one half of the link and H from the other half. Such linkages do seem less conveniently established at low temperatures.

Table 12.4 sets out the general aspects of the past history of the Earth. When displayed in the form of the "clock" of Figure 12.35, most of the entries of Table 12.4 are seen to be crowded into the last small sector of the figure. During

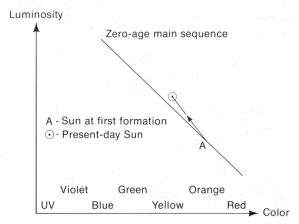

FIGURE 12.34.
The usual theory of the past history of the Sun. At the time of its formation, the Sun lay on the zero-age main sequence. As time passed, the Sun then evolved with steadily increasing luminosity, implying that the Earth should now be significantly warmer than it was several billion years ago.

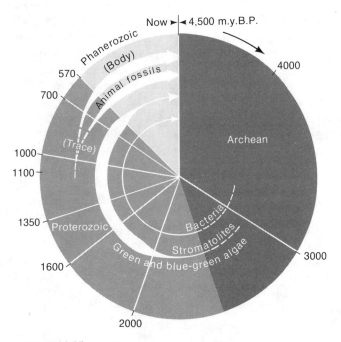

FIGURE 12.35.
But biological evidence suggests the reverse of this. The early forms of life were high-temperature forms (compare with Table 12.5).

TABLE 12.4.
Events in history[a]

Event	Approximate time	
	(Years ago)	(Seconds ago)
Explosion of the cosmic fireball	13 billion	4×10^{17}
Oldest stars in our galaxy formed	10 billion	3×10^{17}
Formation of the sun	5 billion	1.5×10^{17}
Formation of the earth with present mass	4.5 billion	1.4×10^{17}
Primary broth arises as oceans begin to form	4.2 billion	1.3×10^{17}
Formation of the oldest known rocks on earth	3.5 billion	1.1×10^{17}
Formation of exhaled atmosphere completed	3.5 billion	1.1×10^{17}
Oldest fossils formed—bacteria and blue-green algae	3.4 billion	1.1×10^{17}
Formation of oceans (at approximately their present volume) completed	3.0 billion	1.0×10^{17}
Plants began oxygen production	2.0 billion	6×10^{16}
Atmosphere formed as now known	1.0 billion	3×10^{16}
Production of abundant fossil record	600 billion	1.8×10^{16}
Sun began last revolution about galactic center	200 million	6×10^{15}
Dinosaurs were dominant life form	100 million	3×10^{15}
Rocky Mountains formed	80 million	2.5×10^{15}
Brightest stars on the main sequence began nuclear reactions	3 million	9×10^{13}
Man developed to Advanced Australopithecus	1 million	3×10^{13}
Early *homo sapiens* appeared	250 thousand	8×10^{12}
Modern man appears (Cro-magnon)	35 thousand	1.1×10^{12}
North celestial pole in same direction as now	26 thousand	8×10^{11}
Man's historical record began	5 thousand	1.5×10^{11}
Pluto began last revolution around the sun	248	8×10^{9}
Sun at the same point in sunspot cycle	22	7×10^{8}
Earth at same point in orbit around the sun	1	3×10^{7}
	TIME (DAYS AGO)	
Moon at same phase	29.5	2.6×10^{6}
Earth at same rotation position with respect to the sun	1	8.64×10^{4}
	TIME (MINUTES AGO)	
Sunlight arriving at earth now left sun	8	5×10^{2}
Moonlight arriving at earth now left moon	—	1.3
While the light traveled from moon to earth, the central star of the Crab nebula spun around 39 times.	—	—

[a] From *New Horizons in Astronomy*, by J. C. Brandt and S. P. Maran. W. H. Freeman and Co. Copyright © 1972.

TABLE 12.5.
*Approximate upper thermal limits for different groups
of organisms (after T. D. Brock)*

Organism	Upper temperature limit (degrees Centigrade)
Animals, including protozoa	45–51
Eucaryotic microorganisms (certain fungi and the alga cyanidium caldarium)	56–60
Photosynthetic procaryotes (blue-green algae)	73–75
Nonphotosynthetic procaryotes (bacteria)	above 90

most of the history of terrestrial life, only simple one-celled animals and plants have existed. It has only been in the last 20 per cent or so of the Earth's history that more complex forms of life have emerged.

Consider now the life forms of Figure 12.35 in relation to Table 12.5, which gives the highest temperatures at which the various life forms are found today in hot springs of the Yellowstone National Park. It is seen that the sequence of development of life on the Earth—bacteria, blue-green algae, fungi, protozoa—corresponds exactly to the descending temperature sequence of Table 12.5. In spite of the results of orthodox astronomical theory, these data undoubtedly support a high-temperature theory. Indeed, to prevent the later life forms from appearing too soon, there must have been temperatures of about 75°C on the Earth some three billion years ago. This temperature seems impossible given the orthodox view of the past behavior of the Sun.

There is, however, a very unorthodox astronomical theory which would actually require a temperature of some 75°C for the terrestrial oceans of three billion years ago. This theory has profound implications, not just for a study of the solar system, but for our understanding of the whole universe. In this theory, the gravitational constant G in Newton's formula, Gm_1m_2/a^2 in the notation of Chapter 9, would decrease slowly with time, and all the stars and all the galaxies would have been brighter in the past than they are now. I personally believe this theory would explain a great many facts, although some astronomers would not agree with this opinion. The main difficulty with the theory, again in my own view, lies in its relation to the basic physical laws. We shall return to this issue later in the Appendixes to Section VI.

Quite apart from the relation of Figure 12.35 to the data of Table 12.5, there is another good reason for thinking that living cells were faced during much of the Earth's history with rather high temperatures. To survive at high temperature, a cell must be better insulated against heat than it needs to be at lower temperatures. At high temperature the cell is thus subject to constraints that do not exist at lower temperatures, and that must affect the ability of one cell to join with another. Unless there were such a constraint, it is hard to understand why, for so much of the past, life was limited to single-celled organisms. Of course, there could be a constraint other than temperature, but the data of Table 12.5 certainly give a strong indication that temperature was indeed the factor which impeded for so long the emergence of complex life forms on the Earth. If so, there are likely to be profound implications here for astronomy.

§12.6. Intelligence

Our emotional attitude to "life" is not really a chemical matter at all. Although the difference between a well-loved person being alive and being dead may turn on certain subtle chemical processes, this is not at all the way we feel about it. There is something else involved. Most people who would never dream of strangling a dog do not hesitate to swat a mosquito. Yet the chemistry of the mosquito is basically the same as that of the dog. Rather, we distinguish between higher and lower animals according to the complexities of the nervous systems with which the animals are endowed. A nervous system is basically electrical in its operation, so that an animal is made up of a chemical system plus an electrical one,

$$\text{Animal} \equiv \text{Chemical replication} + \text{Electrical system}.$$

The more the electrical part dominates the summation, the "higher" the animal, and the more the electrical system happens to match our own system, the better we regard the animal. For humans, the more similar the other person's brain system is to our own, the better liked, or the better loved, the person is. Similarity, or otherwise, distinguishes "us" from "them."

At a certain level of electronic complexity, a level set just a little below our own capacity, we rather arbitrarily introduce the concept of intelligence. We acknowledge a spark of intelligence in a dog, but the behavior of a cat strikes us as "independent" rather than intelligent. Essentially as a matter of definition, any creature with an electrical system more complex than our own would be endowed with high "intelligence."

We have already discussed the need of animals for an input of amino acids and sugars, substances which plants can synthesize for themselves. Animals must acquire such substances either by eating plants or by eating other animals. Basically, all animals are scroungers, living on the stored chemical energy of others. It was precisely to assist in the process of scrounging that the electrical systems possessed by animals developed. Since the better the electrical system the better the scrounger, biological evolution has operated to increase steadily for many millions of years the complexity of animal electronics. Since we distinguish the level of an animal by the complexity of its electronics, it follows that the higher the animal the greater the scrounger, with Man sitting at the top of the pyramid.

The electrical system in Man has indeed become so subtle that our scrounging for energy has now extended well beyond the eating of plants and of other animals. We scrounge extensively today on nonliving materials. The discovery of fire made use of the decay products of trees, wood, as an energy source. The burning of coal and oil were further steps along the same path, and in the modern nuclear power plant we have achieved the use of entirely nonorganic materials as an energy source. This access to nonanimal sources of energy has developed with increasing rapidity, to a point in our modern society where we can clearly see that either some more restrained pattern of behavior must be applied in future years or the evolution of our species will end itself in a catastrophic social explosion. It is in these obviously critical circumstances that we have begun wondering how things may have fared with other creatures living on planets moving around other stars.

§12.7. Interstellar Communication

The topic of interstellar communication raises many problems, some technical, some of general interest. Let it be said immediately that the only feasible mode of communication between creatures living on planets moving around different stars would seem to be by radio. This matter has been looked into with care, and the optimum radio frequency has been found to be of the order of 3×10^9 oscillations per second, corresponding to a radio wavelength of about 10 centimeters. A vast array containing 900 individual radiotelescopes, each with a diameter of 100 meters—similar to the largest fully steerable radiotelescope yet constructed, at Bonn, Germany—has been proposed. Such an array would give full expression to our present-day ability to construct an instrument capable of achieving interstellar communication. The proposal, known as *Project Cyclops,* is illustrated in Figure 12.36. It has been estimated that such a multiple radiotelescope would be capable of transmitting information to, or receiving information from, any planet situated anywhere in our galaxy, and that the cost of building it would be comparable with the cost of the lunar landings of the *Apollo* program. It is worth some emphasis that actual travel, in the sense of *Apollo,* to distant stars *cannot* be achieved by any foreseeable extension of present-day technology, and that even if travel were physically possible, it would take very much longer than an interchange of messages through a system like *Project Cyclops.* The basic difficulty is that no known process could endow a space ship with a speed comparable to the speed of a radio message. Nor does it seem consistent with what we know about physics to think that such a speed could ever be achieved by a spaceship. People with a strong compulsion to believe in "deep space travel" usually take refuge in the argument that not everything in physics is yet known. Remarkable discoveries surely remain to be made. Perhaps a quite unexpected discovery, analogous to Becquerel's unexpected discovery of radioactivity, will show one day how space travel should properly be achieved? Against this seemingly persuasive argument, one

FIGURE 12.36.
A schematic drawing of the massed radiotelescopes of *Project Cyclops*. (Courtesy of NASA.)

should note that, although remarkable discoveries are very likely to be made, such discoveries cannot contradict what we already know to be true, otherwise the Universe would contain contradictions. It is the nature of remarkable discoveries that they open up entirely unexpected fields, not that they overturn already existing fields of knowledge.*

---

*At a much lower level of argument, it has been suggested, on the basis of equivocal archaeological evidence, that some millenia ago creatures of superior intellect actually landed on the Earth. If this were so, it is a pity the creatures never chose to leave *un*equivocal evidence of their visit(s). They could have been done so by leaving a monument constructed from any high-technology metal, for example, titanium. It is important to distinguish clearly between what is possible and what is not possible. *Project Cyclops* is possible. Visits from outer space are only dreams of a never-never land.

The difficulties of physical travel seem at first sight to be a decline in romantic possibility, a loss of richness in the scheme of things. A little thought soon shows, however, that precisely the opposite is true. If physical travel from one planetary system to another were feasible, then the first creatures to become technologically capable of space travel would be likely to spread themselves everywhere through the galaxy. It would be only too likely that the galaxy would thus come to have only one form of intelligent creature. This, indeed, would be a loss of richness. With space travel not possible, however, creatures in one planetary system cannot interfere with the physical development of creatures in other systems. Many possibilities, with great potential richness, are then permitted.

Are other intelligent creatures likely to exist? We have seen there is little, if anything, in our own planetary system that appears to be due to distant chance. To be sure, if we knew there to be only *one* other planetary system in our galaxy, the odds would be against its containing a planet like the Earth, at an appropriate distance from an appropriate central star, with a similar rotation speed, with a similar tilt of the axis of rotation, thereby giving similar seasons, with a similar distribution of H_2O, CO_2, and other volatiles at the surface, and so on. But the chance of a favorable situation would not be all that small, perhaps one in ten or one in a hundred, but not much less than that. And since we have not just one other planetary system, but on the order of 10^{11} of them, it is clear there must be many millions of other places where the physical conditions for life are just as suitable as they are here in the solar system.

Given similar physical conditions in another system, what would be the chance of a similar chemical evolution leading to the origin of life? Here again it seems that the chance would not be all that small. Amino acids and sugars can readily be synthesized by ultraviolet light; so the genesis of these basic substances can hardly be in doubt. Linkage into long-chain molecules seems more uncertain, but there is such a well-founded chemical logic in the structures of large molecules like DNA, RNA, ATP, and the proteins that this logic must surely have been frequently repeated, perhaps again with a probability of one in ten, or one in a hundred, but not much smaller than that. This still yields more than a million other planetary systems on which life may be taken to have arisen. Looking out at the stars on a clear night, we would find the nearest other system in which life has arisen to be not more than a hundred light years away, very easily within range of the powerful array of radiotelescopes planned for *Project Cyclops*.

Would life always tend to become "intelligent" in the sense discussed in §12.6? From our consideration of the nature of an "animal," it seems clear that the development of an electrical system would very likely occur for all animals, everywhere. Because of the need to search for food, "eyes" would be a normal development. Animals with eyes are then likely to prey on each other, with biological evolution forcing the further development of "weapon systems": claws, teeth, and, above all, a brain. The logical sequence leading to the emergence of a thinking brain appears inevitable; so we can expect it to have developed quite generally.

We come now to what appears to be the most uncertain question of all. Given a suitable planet, given the origin of life, given the emergence of intelligence to a level at least equal to ourselves, for how long *on the average* can we expect such an intelligence to persist? Denote this average lifetime of an intelligent creature by L. We can hardly expect that L will be at all comparable to the total age, say, T, of our galaxy. With L less than T, the number of planetary systems with intelligent creatures existing *at the same time,* and therefore able to communicate with each other, must be less than our total of a million favorable cases by the ratio L/T. Only a fraction L/T of the million cases will overlap with each other. Hence it follows that if L/T were as small as 10^{-6}, there would be no effective overlap at all, and interstellar communication would not be possible. There would have been intelligent "chaps" in the past, there will be more of them in the future, but today, effectively, there would only be ourselves.

But is L/T likely to be as small as 10^{-6}? Since T is about 10^{10} years, such a small fraction would imply that the average lifetime of an intelligent creature is only 10,000 years. Is such a short period to be expected? This is the critical question toward which we have been working throughout this chapter. We end by considering possible answers to it.

The thought that our capacity to execute a project of the technological quality of *Project Cyclops* might last for only ten thousand years seems at first sight to be a pessimistic assessment of the future of the human species. But in view of the facts, is it not rather an optimistic assessment? When one contemplates the huge human populations that have grown with startling suddenness within only a century, when one contemplates the excessive modern pressure on natural resources, it is hard to summon much confidence in a future extending more than a few decades. Devastating crises, one feels, must overtake the human species within less than a hundred years. We are living today, not on the verge of social disaster, but within the disaster itself. This is exactly what the news media report to us every day.

We have seen that the phenomenon of "intelligence" is an outcome of aggressive competition. Intelligence and aggressiveness are coupled together inevitably by their biological association. An intelligent animal anywhere in the galaxy must necessarily be an aggressive animal, and must necessarily become faced at some stage by the same kind of social situation as that which now confronts the human species. Inevitably, then, "intelligence" contains within itself the seeds of its own destruction. Can any solution be found to this inherent difficulty?

We can approach this final question by considering what conditions would be needed here on the Earth to enable our species to continue to maintain itself at a high technological level for more, or much more, than ten thousand years. A far smaller population would be needed, pressing only gently, if at all, on the resources of the Earth. It is hard to see our strident, competitive present-day society evolving smoothly in a more or less trouble-free way to the needed lower population level, or to see the persistently quarrelsome present-day human temperament changing voluntarily. To achieve such a change, psychologically as well as physically, an extensive selection of the human gene pool would probably be necessary. Some few individuals probably exist today with the necessary qualities, and it is from the progeny of these few individuals that the population of the future would have to come; the remainder of humanity, bearing the characteristics of our aggressive past, would have to become as extinct as the dinosaurs. In the violent future which lies ahead of us, these things may come to pass; yet I regard it as more probable that they will not. Inevitably, it seems the human species must then relapse back to its primitive condition. It seems that our moment of "intelligence," in a technological sense, will be exceedingly brief, that our ability to build and maintain *Project Cyclops* will not last for more than a century or two, perhaps for not more than the next few decades.

I see the uncertainties which now lie close ahead for the human species as being an inevitable obstacle in the way of the emergence of any long-term intelligence. I see it as an obstacle every bit as formidable as the early physical problem of obtaining an appropriate planet moving around an appropriate star, and every bit as crucial as the origin of life itself. I suspect that many creatures may reach our present stage of development, but that only a few could go any further. Perhaps the chance of successfully surmounting the obstacle is as high as one in a hundred. Suppose each successful creature then has a life span of a hundred million years. On these reasonably favorable assumptions, the number of long-term intelligent species at present alive in our galaxy would not exceed about a hundred. It is among these fortunate ones that I would expect interstellar communication to be now taking place. The nearest of them would be unlikely to be less than 3,000 light years away from us.

It is to be observed that, for a species with a long-term future of a hundred million years ahead of it, a necessary interval of a few thousand years between the transmission of a message and the reception of a response to it would not seem a serious impediment. There would be ample time for many messages to be interchanged. For us, however, it is unlikely that much popular or political support will be forthcoming for *Project Cyclops* once it is understood that perhaps many centuries would be needed to obtain a positive result from it. Only if results could be promised in the short term would I expect such a project to receive public support. This I take to be clear evidence of the ephemeral nature of our modern society. We have no faith in tomorrow. A species with real confidence in its future would not hesitate to give expression to such a magnificent concept.

TOPICS FOR DISCUSSION:
1. In what respect is the chemical part of the equivalence Animal ≡ Chemical replication system + Electrical system necessary for intelligence? Could a purely electronic system be intelligent?

2. At what stage in the development of animals do you consider the phenomenon of "consciousness" to have arisen?*

3. If you reject the somewhat gloomy assessment given above of the future prospects of the human species, what prognosis do you offer yourself?

General Problems and Questions

1. What arguments can be given in favor of the view that physical conditions similar to those in the solar system would occur in a considerable fraction of other planetary systems?

2. Without necessarily giving the detailed chemical form of the basic molecules, describe the structure of the double helix.

3. What is a chromosome? Describe the principle whereby a chromosome is copied.

4. What are: (a) amino acids, (b) proteins, (c) RNA, (d) genes?

5. How is the code of protein synthesis used by RNA?

6. Why does an electric current pass freely through seawater, but not through distilled water?

7. Discuss the characteristics we associate with "life."

8. Why is sunlight crucially important to the existence of life on the Earth?

9. Life depends chemically on certain rather volatile substances, very different from the highly refractory materials of which the Earth is mainly constituted. In what way, or ways, could the Earth have acquired these volatile materials?

10. Which of your suggestions in answer to Problem 9 might be expected to apply in other planetary systems?

---

*I do not think any problem has worried me so persistently as the issues which relate to this question. If one disregards oneself, and considers only the consciousness of others, then the problem is not unduly troublesome. As with the concepts of "alive" and "not alive," there are probably no hard and fast categories. It is reasonable to say that simple animals are not "conscious," that the higher animals are "conscious," and that a continuous range exists between these extremes. But one's own subjective consciousness seems to me to present difficulties of a higher order. The problem is so severe that some philosophers have sought to avoid it by claiming subjective experience to be an illusion, and therefore outside the range of reasoned discussion. This device is not satisfactory, however. The physicist uses subjective experience to sharpen his description of the world. Using quantum mechanics, he calculates the *probabilities* of events occurring, but he uses subjective experience to decide whether the events actually occur or not, thereby changing the probabilities to unity or to zero as the case may be. Without this constant sharpening of his world description, the physicist would be faced by an increasingly blurred picture of the world. It seems as if the world makes explicit decisions of which we are aware consciously, but which we can never calculate from the laws of physics. I find myself best satisfied with the idea that subjective consciousness is associated in some way with these explicit decisions, being thus a part of the world which the physical laws do not claim to describe.

11. Given that on the primitive Earth organic molecules containing some tens of atoms were synthesized by the effect of sunlight, how do you think the first living creature might have arisen from the association of these organic molecules? Would a system of random trial fittings of one molecule to another be sufficient to generate a complex self-replicating system?

12. Which kind of life forms were the first to arise on the Earth? What special property characterizes these early forms?

13. Granted that animals are partly chemical in operation and partly electrical, the development of intelligence is seen to be contingent on an increasing complexity of the electrical part. Extrapolating this trend, might intelligence without the chemical part be possible?

14. Suppose humanity were eventually replaced on the Earth by a purely electronic intelligence capable of deep thought and of sensitivity. Would the extinction of humanity then be a matter of regret? If your answer is affirmative, why?

15. What is *Project Cyclops*, and what is its aim?

16. If in past ages high intelligences landed by spaceship on the Earth, what motivated them never to leave a piece of advanced technology behind them when they left—e.g., an obelisk of titanium metal?

17. Assuming our present civilization is *not* going to collapse, what crucial technological changes will be essential (a) 50 years from now, (b) 500 years hence, (c) 10,000 years in the future?

18. Assuming our present civilization *is* going to collapse, would you consider it desirable to hang on as long as possible, until all natural resources are exhausted, or would early collapse, leaving some resources unconsumed, be preferable? How far do you think that the wishes of the individual in questions like this may not be in the best interest of the species as a whole? How far do you think it the social function of religion to persuade us to bow to communal needs in circumstances where such needs are in conflict with our individual interests?

Appendixes to Section IV

Appendix IV.1. The Formation of Jupiter and Saturn

The flares which occur at the surface of the present-day Sun (see Figure 2.16) cause streams of particles to be ejected from the Sun at speeds of about 1,000 km per second. These streams carry magnetic fields with them, in the manner illustrated in Figure IV.1. In our discussion of the origin of the planets, we considered in Figure 11.9 a schematic model in which radially mounted sticks had a mechanical effect similar to the effect of the magnetic "lobes" carried by the particle streams of Figure IV.1 It is reasonable to suppose that such a process was much more intense at the time of formation of the planets than it is today, and that it led to a general magnetic connection of the form of Figure IV.2 between the primordial Sun and the outer disk of planetary material.

It is curious that, whereas the magnetic spiral structure of Figure IV.2 trails behind the direction of rotation about the primordial Sun, the trajectories of particles in the outward-moving streams spiral the opposite way, as in Figure IV.3. The impact of such high-speed particles on the disk of planetary material, shown also in Figure IV.2, adds angular momentum to the disk, causing the planetary material to move outwards in the manner discussed in Chapter 11.

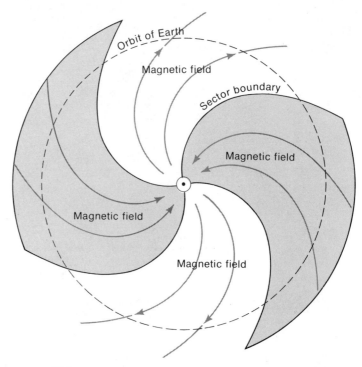

FIGURE IV.1.
Streams of particles from the Sun carry a magnetic field with them. Each
individual stream forms a separate sector. The rotation of the Sun causes
the field to assume a spiral structure. Can you think of an explanation of
why this should be so? In which sense would you expect the Sun to be rotating in this diagram, and in which sense does the Earth go round its orbit?
(After J. Brandt and S. Maran, *New Horizons in Astronomy*. W. H. Freeman
and Company. Copyright © 1972.)

PROBLEM:
Consider again the mechanical stick model of Chapter 11. Satisfy yourself that
when the sticks develop "whip," in the configuration of Figure 11.9, the material
at the ends of the sticks actually moves forward (in the sense of the rotation)
with respect to the material of the outer heavy ring. This forward motion, like
that of Figure IV.3, communicates rotation to the outer ring.

The forces driving the disk of planetary material outward, with the material
following slowly expanding spiral orbits (see Figure 11.10), are imposed at
the inner edge of the disk. Although it is possible that a magnetic field, coiled
like a wound-up clockspring, exists within the disk, causing a torque to be

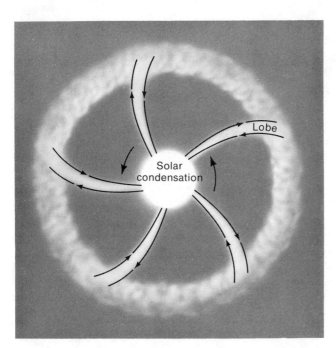

FIGURE IV.2.
A multiple magnetic lobe (sector) connection of the primordial Sun to the disk of planetary material.

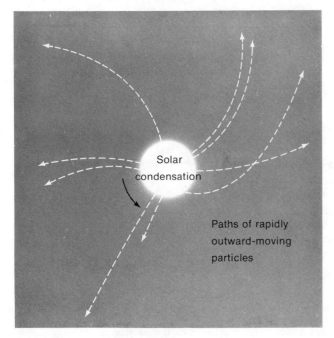

FIGURE IV.3.
The trajectories of the outward-moving particles pick up a rotation about the primordial Sun in an opposite sense to the magnetic lobe structure of Figure IV.2. If the solar condensation were not itself rotating, the particle motions shown here would twist the field in a sense opposite to that of Figure IV.2, as we might intuitively expect. It is the rapidity of rotation of the primordial Sun that leads to the apparently contradictory situation of Figure IV.2.

transmitted from the inside to the outside, it seems more likely that the gas on the inside simply pushes the gas on the outside, which moves away from the primordial Sun simply because it is being pushed. To generate sufficient pressure for this, the gas must remain reasonably dense, which means that it will expand as a rather well-defined ring, instead of as an extensive flat disk. This creates a problem for the formation of Jupiter and Saturn. Since we require the gas to move out to the region of Uranus and Saturn, where hydrogen and helium are evaporated, it follows that the gas will pass beyond Jupiter and Saturn. Where, then, did the material of Jupiter and Saturn come from?

Let us approach this question by considering the evaporation of hydrogen and helium in a little more detail. In Chapter 11 a value of 300°K was given for the order of the temperature of the planetary gases at the stage where they reached the region of Uranus and Neptune. This is the value one obtains for a primordial Sun of radius 2×10^{12} cm and of surface temperature 4,000°K, provided it is assumed that the planetary gases absorb the radiation from the Sun, in the manner of Figure IV.4. In such a situation an energy balance is achieved, with the planetary material reradiating, at infrared frequencies, the energy which it receives from the Sun. The estimate of 300°K applies to the planetary material with this balance established.

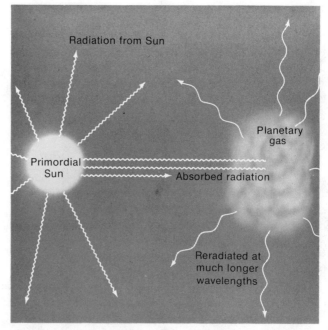

FIGURE IV.4.
Planetary material absorbs the incident radiation from the Sun. An energy balance is then set up, with the material re-emitting the absorbed sunlight, but doing so at much lower radiation frequencies.

Now gases, particularly hydrogen and helium, are not efficient absorbers of the red light that would be emitted from a primordial Sun with a surface temperature of 4,000°K. It is true that water vapor and carbon dioxide are better able to absorb such light, but even they could not absorb enough to establish the balance we have just assumed to exist. However, the planetary material probably contained several Earth masses of refractory material, Fe, MgO, SiO_2, CaO, in the form of fine smoke particles; in Chapter 11 we saw that only fairly large chunks of the refractory material would fall out of the gas and remain in the region of the inner planets, small particles being carried along by the gas. Such particles are efficient absorbers of light, and several Earth masses of them distributed throughout the planetary material would certainly make it quite opaque to radiation from the primordial Sun.

Even if refractory particles were not present, a similar situation would arise from crystals of water-ice. Since the gas alone is an inefficient absorber, its temperature would simply fall to the point where the condensation of normally volatile materials would begin. Of the common volatiles, this would happen first for water-ice, which would condense until sufficient radiation was absorbed to prevent the temperature from falling any further. This would happen for a temperature between 200°K and 250°K.

In either case, the heating which is to lead to the evaporation of hydrogen and helium comes from the absorption by small solid particles of radiation emitted by the primordial Sun. Now, for a temperature of between 200°K and 300°K, a gas composed largely of hydrogen and helium would attain a thickness of about 3×10^{13} cm in a direction perpendicular to the plane of its orbit about the center. It would thus have a thickness equal to about 1/10 of its distance from the center, as in Figure IV.5. Solid particles in such a distribution of gas must always tend to fall to the plane of the orbital motion; they will fall like raindrops down on to this plane, and if permitted to do so would form themselves into a thin ring, as in Figure IV.6. Hence the gases near the plane will be heated by these solid particles more than the gases farther away, and a convective motion will result, in the manner of Figure IV.7. Rising columns

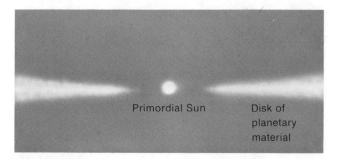

Primordial Sun   Disk of planetary material

FIGURE IV.5.
The distribution of the planetary gas is a disk increasing in thickness with distance from the primordial Sun.

558

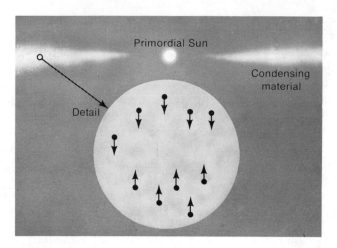

FIGURE IV.6.
Condensing solid material falls like rain toward the central plane of
the distribution.

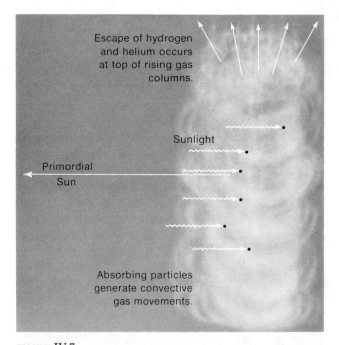

FIGURE IV.7.
Condensed solid material is highly efficient in absorbing sunlight. As
the condensed material falls in the manner of Figure IV.6, heating by
the primordial Sun generates circulating currents. Hydrogen and helium both escape in the region of Uranus and Neptune from the top-
most part of columns of gas that are thus rising from the central
plane.

of gas tend to carry the solid particles upwards (i.e., perpendicular to the plane), but gravitation tends to make them fall back toward the orbital plane. It is from the very tops of such rising columns of gas that the evaporation of hydrogen and helium takes place, as is also shown in Figure IV.7.

Only when the density of gas becomes very small can an individual atom escape freely, since in a high-density gas an atom is impeded by collisions with other atoms. This is why escape must take place from the tops of the columns of Figure IV.7. Now in order to escape entirely from the solar system, a particle in the region of Uranus and Neptune must have a speed of about 8.5 km per second. In a gas at 300°K, a particle is very unlikely to acquire such a speed from the heat motion alone; a temperature of several thousand degrees would be needed. The evaporation of hydrogen and helium could not therefore have been of a simple thermal kind. Nor could the convective motions have made much of a contribution. It must be remembered, however, that the gas already possesses a speed of about 6 km per second, due to its orbital motion about the primordial Sun. Thus only an additional 2.5 km per second is required, *provided the additional motion has the same direction as the orbital motion.* Although hydrogen molecules in a gas at 300°K have an *average* thermal motion somewhat less than this, enough of them have thermal motions as high as 2.5 km per second for such a mode of escape to be quite efficient.

This means that escaping atoms all possessed an exceptional rotational motion about the primordial Sun of about 8.5 km per second, instead of the 6 km per second that we would calculate from Kepler's third law. Their escape decreased the average angular momentum of the material remaining within the disk, causing the ring of gas to *decrease* its radius. If we suppose that expansion away from the primordial Sun, due to the process in Figure IV.2, precedes the evaporation of the gas, then there will be a subsequent contraction of the ring of gas caused by the evaporation itself. Although these two oppositely directed effects on the radius of the ring of gas may not have been entirely separated in time, it is reasonable to suppose that evaporation persisted for longer, and hence that the final effect was to decrease the radius of the ring of gas.

Denote by m_i the mass of the ring of gas at the stage where the process of Figure IV.2 ceased, and let r_i be the radius at this stage. Write m_f for the final mass of gas after all evaporation ceased, as it must eventually, since higher and higher speeds are needed for molecules to escape from the solar system as the radius of the ring of gas decreases. Let r_f be the radius when evaporation ceased. Then it is not hard to show the approximate validity of the proportionality $r_f/r_i = m_f/m_i$. Let us see what this relation implies for the formation of Jupiter and Saturn.

We will suppose that the final mass m_f went to form Jupiter and Saturn. Hence we require m_f to be about 400 Earth masses, mostly contributed by Jupiter. In Chapter 11 we saw that, in order to explain the composition and densities of Uranus and Neptune, m_i had to be about 2,200 Earth masses. Hence we arrive at a value of about 1/5 for m_f/m_i, giving $r_f = r_i/5$. This relation provides an excellent prediction for the radius of the orbit of Jupiter. In the region of Uranus and Neptune we have r_i about 25 times the radius of the Earth's orbit, so that r_f comes out at about 5 times the radius of the Earth's orbit. Comparison with Table 10.4 shows that this result is very close indeed to the actual radius of the orbit of Jupiter. It is too small, by a factor of about 2, to match the radius of Saturn's orbit, but this is not a serious worry because the simple relation $r_f/r_i = m_f/m_i$ applies only to the bulk of the material—and the bulk of the material is actually resident in Jupiter, not in Saturn.

Would there be solid or liquid particles within the gas at this stage? Some fine grains of refractory material may well have remained all along within the gas. In the original total of some 2,200 Earth masses of planetary gas which separated from the primordial Sun, we might expect six to seven Earth masses in the form of refractories, about one-third of which fell out as fair-sized chunks to form the four inner planets. The remaining four to five Earth masses would be carried out, as a fine smoke, to the region of Uranus and Neptune, and most of it probably remained there. Perhaps one or two Earth masses of refractories went to form Jupiter and Saturn.

Of the volatiles of high concentration, water condenses most readily. Most of the water, making up about half of all the volatiles, must have condensed in the region of Uranus and Neptune.* If the solar luminosity remained high enough during the escape of the light gases, the less easily condensed volatiles, particularly CO_2 and H_2S, may have remained in gaseous form. If so, the eventual condensation of CO_2 and H_2S, occurring as the solar luminosity declined (see Figure 6.24), would have taken place in the region of Jupiter and Saturn. It therefore seems possible that the formation of the latter two planets began with the aggregation of bodies of dry-ice and frozen H_2S, with hydrogen, helium, and other, less-abundant gases, such as N_2, CO, and the noble gases, being added by gravitational accretion.

Appendix IV.2. The Aggregation of Planets

The picture we have developed of the origin of planets requires their aggregation to take place from solid bodies, chunks of material initially a few meters in size. The composition of the bodies for the various groups of planets is set out in Table IV.1. We now ask how this aggregation from chunks into planets took place. How long did it require? These questions will be considered first for a swarm of water-ice chunks in the region of Uranus and Neptune.

Each small chunk of material, *if it were alone,* would pursue an orbit like that of a planet, a nearly circular ellipse with the Sun at one of the foci. But the chunks of material are not alone. To make up 2×10^{29} grams, the present combined mass of Uranus and Neptune, needs about 10^{23} chunks each 1 meter

---

*If Uranus and Neptune thus contain only half the volatiles, the mass of the original planetary material must have been about 4,500 Earth masses. This circumstance has been taken into account in the calculation described on page 501 (see the footnote on that page).

TABLE IV.1.
Major constituents of the planets

Planet	Composition
Mercury Venus Earth Mars	Fe, MgO, SiO_2, Al_2O_3, CaO
Jupiter Saturn	CO_2, H_2S (?)
Uranus Neptune	H_2O

in size. Too much individuality in the motion of a particular body leads to collisions with other bodies. Collisions persist so long as there is any appreciable inconformity in the motions of the various bodies. Eventually the many chunks take up very nearly circular orbits all in the same plane; a thin disk is formed, in a time-scale of only a few orbital revolutions, say, about a thousand years.

Although the gravitational force between a meter-sized chunk of water-ice and its nearest neighbors is very weak, the *difference* between the solar gravitational forces acting on two neighboring chunks is also small. So the possibility must be considered that, once orderly motion has become established, local gravitational effects are able to produce small-scale aggregations in the manner illustrated in Figure IV.8. Taking 2×10^{29} grams for the total mass of all the chunks, and taking the chunks to move in orbits with a scale 25 times that of the Earth's orbit, it can be calculated that the radii of the aggregation zones of Figure IV.8 must be about 3×10^{10} cm, i.e., about 50 times the radius of the present-day Earth. The total quantity of material contained in such a zone is about 10^{21} grams, and the time-scale for the formation by gravitation of a swarm of bodies of this mass is only one or two orbital periods. After local gravitational forces have pulled the regions of Figure IV.8 together, compact icy bodies with a diameter of about 100 km are formed. Such bodies, in size, mass, and composition, are rather similar to the nuclei of large *comets*.

The picture so far is that the initial modest-sized chunks settle quickly into a flat disk, and that this disk then forms into 100 million much larger bodies of cometary scale. The time required for this first stage is short, not much more than a thousand years. What happens next?

Consider the orbit of a particular comet-sized object. Around very nearly the same orbit, there will be many similar objects, as shown schematically in Figure IV.9. The orbits of such objects will never be rigorously the same, however, nor will the orbital periods be rigorously the same. One such object will slowly catch up on another. As they pass each other, gravitation can once again cause the objects around the orbit to come together and so to join into a single larger body, as indicated in Figure IV.10. To specify numbers, the circumference of the circle of Figure IV.9, when divided into bits each

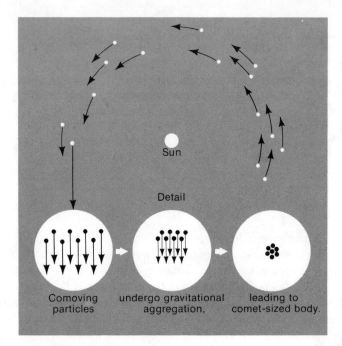

FIGURE IV.8.

Once any initial random motions have become damped out within a distribution of meter-sized chunks of water-ice, local gravitational effects can produce the first step in an aggregation process. Comet-sized bodies are eventually formed in this way.

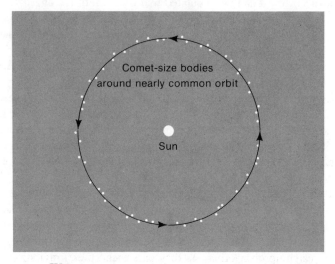

FIGURE IV.9.

Many comet-sized bodies have very nearly the same orbit. From time to time, such bodies come close to each other.

FIGURE IV.10.
Then gravitation produces a further measure of aggregation.
Ultimately bodies about the size of Mars are formed.

with a size equal to that of the zones of Figure IV.8, has about 5×10^4 bits, which is the number of objects that can be considered to possess essentially the same orbit in the sense of Figure IV.9. The aggregation of all such bodies, in the manner of Figure IV.10, thus yields much larger objects, each with a mass of about $(10^{21}) \times (5 \times 10^4) = 5 \times 10^{25}$ grams. Such larger objects would have radii of about 2,000 km, and there would be about 4,000 of them. They would be objects similar in size and mass to the larger *satellites* of the planets. The time-scale for the aggregation process of Figure IV.10 to be completed around the whole orbit of Figure IV.8 can also be calculated, and is about 10^4 orbital revolutions, about 10^6 years, much longer than the time-scale for the formation of the comet-sized objects.

Starting with meter-sized chunks, we have now described two stages of aggregation, the gravitationally controlled local zones of Figure IV.8, and the addition around the orbit of Figure IV.9. These led to objects like the comets and like the satellites, respectively. Suppose we try repeating the same sequence, but replace the meter-sized chunks with which we started by the satellite-sized bodies reached in the second of the above stages. That is to say, we use Figures IV.8 and IV.9 again, but now start with some 4,000 objects each with a mass of about 5×10^{25} grams. This leads to more aggregation, after which we may repeat the sequence yet again, leading to still more aggregation. At first sight, it might seem as if, by continuing in this way, we could eventually arrive at a total aggregation into just one or two bodies, which we could finally identify with Uranus and Neptune. A detailed investigation shows that this is not the case, however. The repetitive process we have just considered is found to *converge*. That is to say, however many times the sequence of Figures IV.8 and IV.9 are repeated, there is a degree of aggregation beyond which this sequence cannot go. There is a limiting mass of about 2×10^{27} grams, which still leaves about 100 bodies in the region of Uranus and Neptune. A further development is therefore required in order that these bodies, now of the scale of a small planet like Mars, shall join together to form into Uranus and Neptune.

Imagine, then, a hundred bodies, each with mass rather greater than that of Mars, moving around the Sun in a hundred equally spaced circular orbits, as in Figure IV.11. Manifestly such a regularly spaced sequence of orbits cannot persist. Whenever two bodies in adjacent orbits pass each other, as in Figure IV.12, perturbations will be caused by their mutual gravitational force. This mutual influence is not strong enough for there to be much chance that the bodies will approach each other and join together, but it is strong enough to destroy the precise ordering of their orbits. The orbits become slightly non-circular, and cross over each other, as in Figure IV.13.

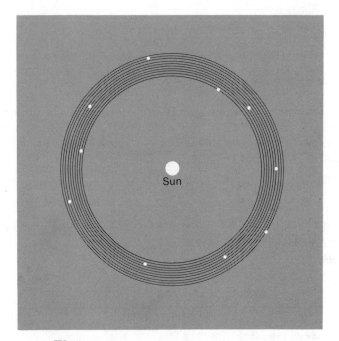

FIGURE IV.11.
This leads to a situation with about 100 bodies of the size of small planets moving in orbits lying in the general region where the orbits of Uranus and Neptune are now found.

FIGURE IV.12.
Neighbors pass each other as before, but gravitation now promotes disorder in the orbits rather than aggregation.

FIGURE IV.13.
The encounter of Figure IV.12 can lead to neighboring orbits crossing each other.

FIGURE IV.14.
A third neighboring orbit then gives a further measure of disorder.

The body in the next orbit will also approach those in the elliptic orbits of Figure IV.13, and may cause these orbits to become still more elliptic, as in Figure IV.14. The possibility exists therefore that disorder will slowly spread throughout the orbits of Figure IV.11. Since each body of Figure IV.11 passes its neighbors within about a hundred revolutions, about 10,000 years, such total disorder must arise in this way within less than a million years.

Suppose enough disorder has been set up that an appreciable fraction of the initially regular orbits of Figure IV.11 cross each other. Such a situation implies that the mutual gravitational forces between the bodies have generated variations of speed in the motions of the bodies that are about 10 per cent of their initial values; i.e., 10 per cent fluctuations in the speeds of the bodies in the configuration of Figure IV.11 will produce the situation of Figure IV.15. In this extensively overlapping situation, bodies may meet and join together, but it is important to notice that such further aggregations now occur *randomly*, whereas the aggregations of Figures IV.8 and IV.9 were systematic. The probability of random encounters is not large, especially since the same variations of motion which produce the overlap effect of Figure IV.15 also cause the orbits to depart from their initial plane. Thus 10 per cent variations cause the planes of the orbits to become inclined at angles up to about 5°, as in Figure IV.16. The *volume* in which the bodies move is thereby much increased.

FIGURE IV.15.
Eventually a highly disordered situation results. This is actually the distribution of orbits of short-period comets (Figure 10.55), but with the cometary names removed we gain a clear idea of what a disordered situation can mean. We have now to think of bodies of the order of the size of Mars moving in these orbits. Collisions occur between them from time to time, giving a slow evolution into fewer and fewer bodies, until at last only one or two large planets are left, and the process of planetary formation is effectively complete.

Because the final stage of aggregation must thus depend on random collisions of one body with another, the time-scale is now much increased. Calculation shows that about 300 million years are required for this final stage.

Bringing together the whole of the above discussion, we have the conclusions set out in Table IV.2.

The last line of the table has been "idealized" to a final aggregation of just two bodies, agreeing with what we know to have actually happened. There is no way, in a random calculation of the kind discussed above, to decide exactly what the final number of bodies must be. This number will depend on the details of the complicated distribution of orbits illustrated in Figures IV.15

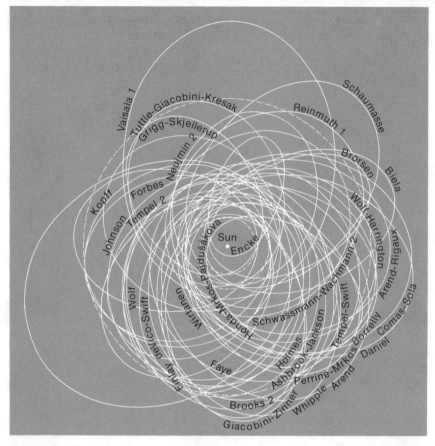

FIGURE IV.16.
The orbits of the comets shown in Figure IV.15 are not all in the same plane. If the same situation was true for the orbits of bodies in the early solar system, this would slow down the final process of random collision, so that this final stage occupies much more time than the preceding stages of aggregation.

and IV.16. The details might well have been such that three planets were finally formed instead of two. Indeed, the fact that the axis of rotation of Uranus lies nearly in the plane of its orbital motion (see Figure 10.43) strongly suggests that the last stage in the formation of Uranus consisted of a collision between two bodies of comparable sizes. If this collision had not occurred, the solar system might well have had three planets in the region of Uranus and Neptune. It will be hard for any theory to decide the precise details of the last stages in the formation of the planets, since these can be much affected by freak events.

A discussion of the aggregation of the inner planets can be given in the same sequence of steps as for Uranus and Neptune. However, the time-scales for the corresponding stages are shorter in proportion to the orbital periods, i.e., by a factor of about 100. Thus the total time for the formation of the inner

TABLE IV.2.
Aggregation of Uranus and Neptune

Stage	Illustrated in Figure	Mass of bodies (grams)	Number of bodies	Time-scale (years)
1	IV.8	2×10^{21}	10^8	10^3
2	IV.9	5×10^{25}	4,000	10^6
3	IV.8 and IV.9 repeated	2×10^{27}	100	10^7
4	IV.15 and IV.16	10^{29}	2	3×10^8

TABLE IV.3.
Aggregation of the inner planets

Stage	Illustrated in Figure	Mass of bodies (grams)	Number of bodies	Time-scale (years)
1	IV.8	2×10^{17}	5×10^{10}	10
2	IV.9	10^{23}	10^5	10^4
3	IV.8 and IV.9 repeated	5×10^{25}	200	10^5
4	IV.15 and IV.16	5×10^{27} (Earth, Venus)	2	3×10^6
		5×10^{26} (Mars, Mercury)	2	

planets was no longer than about three million years. Details of the various stages are given in Table IV.3. The orders of the bodies formed are *cometary* in stage 1, *asteroidal* in stage 2, *satellital* in stage 3, and *planetary* in stage 4. Once again the final details cannot be calculated, since freak events may have been involved.

The situation for Jupiter and Saturn follows similar lines, with time-scales intermediate between those of Tables IV.2 and IV.3. However, a fifth stage must be added for Jupiter and Saturn, a gravitational accretion of hydrogen and helium, and of other, less abundant gases, such as the noble gases. This would be comparatively rapid.

Appendix IV.3. Other Planetary Systems

The picture we have developed for the origin of the planets depends in an important way on the thermal evaporation of hydrogen and helium from the outskirts of the solar system. If there had been no such evaporation, Jupiter and Saturn would not have been formed. Instead, Uranus and Neptune would have been extremely massive planets, containing essentially the whole of the initial planetary material, some five to ten times greater than the combined

TABLE IV.4.
Nearby stars that seem to have planets

Star	Distance (light years)	Mass of star (Sun = 1)	Mass of companion (Jupiter = 1)	Orbital period (years)
Cin 2347	27	0.33	20	24
70 Oph	17	0.89	10 } 12 }	17 } 10 }
Lae 21185	8	0.35	10	8
Kruger 60A	13	0.27	9	16
61 Cyg A	11	0.58	8	5

TABLE IV.5.
Barnard's Star

Planet	Mass (Jupiter = 1)	Orbital period (years)
B_1	1.1	26
B_2	0.8	12
B_1	1.26	24.8
B_2	0.63	12.5
B_3	0.89	6.1

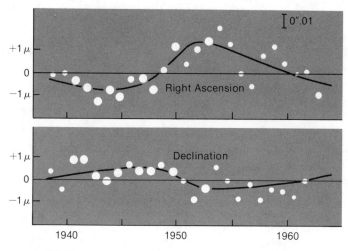

FIGURE IV.17.
A star with a massive planet would experience a very slight periodic wobble in its position on the sky, which would show as slight variations of right ascension and declination. Observations of Barnard's star (see Figure 7.15) are shown in this figure, and they do seem to indicate such a wobble, during a period of about 24 years. (From P. Van de Kamp, in *Sky and Telescope*, 26, 1963, 8.) *Project:* Examine Van de Kamp's article and discover for yourself the meaning of the quantity μ in terms of which the variations of right ascension and declination were measured.

masses of Jupiter and Saturn. Now, the evaporation of hydrogen and helium must have been contingent on the details of our particular solar system. Had the solar condensation been sufficiently less luminous, or had the scale of the solar system been sufficiently less, evaporation would not have taken place, and large monster planets would then have been formed.

These considerations are of interest in relation to planetary systems other than our own. We can expect the situation in some systems to have favored a similar evaporation of hydrogen and helium, but not in others. Thus planets like Jupiter and Saturn would sometimes arise, but in other systems much more massive planets would be formed, in addition to small inner planets composed of refractory materials, like Venus and the Earth, which we expect to be formed in all systems.

From delicate observation of certain nearby stars, slight wobbles of position in the sky have been detected. The wobbles are similar in character to, but much smaller in amplitude than, those observed for stars that are components of wide binary systems. Companion stars are not observed in these systems, but invisible companion planets may perhaps be present—even large planets would not be directly visible at the distances of even nearby stars—and the wobbles have been analyzed from this point of view. Results are shown in Table IV.4. The planetary masses given in Table IV.4 range from about 2,500 to about 6,000 Earth masses, being thus of the same general order as the mass inferred in Chapter 11 for the total mass of planetary material initially present in the solar system. The examples of Table IV.4 would result from situations in which the thermal evaporation of hydrogen and helium did not take place. A system from which hydrogen and helium evaporated, and which is thus similar to the solar system, is set out in Table IV.5.

Barnard's star, with a mass about 15 per cent of that of the Sun, is about six light years away. The two distinct sets of values in Table IV.5 refer to two different ways of analyzing the observations shown in Figure IV.17, the first set of values being from van de Kamp and the second from Suffolk and Black. The two ways of analyzing the data agree reasonably well for the planets designated B_1 and B_2, but Suffolk and Black think that a third planet of shorter orbital period is also present. The difference is a measure of the uncertainties. From the observations it seems clear that one or more planets comparable to Jupiter must be present in this system.

It will be noted that all the stars in these two tables have masses less than that of the Sun; in fact, most nearby stars have masses less than the Sun does.

Appendix IV.4. The Age of the Solar System

It is remarkable that the age of the solar system, 4.6×10^9 years, can be determined with quite sharp precision. The method described here involves the two elements, lead (Pb) and uranium (U). Lead has four stable isotopes, ^{204}Pb, ^{206}Pb, ^{207}Pb, ^{208}Pb, of which the first three play a role in the age-determination method. Two isotopes of uranium, ^{235}U and ^{238}U, are found in naturally occurring materials. Both uranium isotopes are unstable, and emit α particles (helium nuclei). The characteristic decay time for ^{235}U (i.e., the time in which half the ^{235}U decays) is 7.1×10^8 years, that for ^{238}U is 4.5×10^9 years. Subsequent decays of the product nuclei are comparatively rapid. Each such decay is accompanied by the emission of either an α particle or a β-ray, as we saw in Chapter 7. The decay series of ^{238}U, ending at ^{206}Pb, was set out in Table 7.2. The decay series of ^{235}U, ending at ^{207}Pb, is given in Table IV.6.

Suppose we consider a sample of solid material formed an interval of time t ago, and suppose the history of the sample to be such that neither uranium nor lead has been added to it or taken away from it. The amount of ^{204}Pb must be the same now as it was when the sample was formed, because there are no decays leading to ^{204}Pb. But the amounts of ^{206}Pb, ^{207}Pb, will be greater now than they were originally, because of the radioactive decays starting at ^{238}U, ^{235}U. The increases in ^{206}Pb, ^{207}Pb, depend on two factors: on the concentration of uranium in the material in question; and on the length of time t which has been available for the decay processes to occur. In order to be able to deduce t from the measured ratios ^{206}Pb/^{204}Pb, ^{207}Pb/^{204}Pb, one must be able to separate these two factors. This is done by a method that turns on the following point. Consider two samples of the same age, one with twice the uranium concentration of the other. Then the amounts of ^{206}Pb, ^{207}Pb, produced from the uranium will be twice as great in the sample of higher concentration. That is to say, the first of the two factors, the uranium concentration, affects ^{206}Pb, ^{207}Pb, in the same way. On the other hand, if we consider two samples that originally had the same uranium concentration, one of which is twice the age of the other, then the two isotopes ^{206}Pb, ^{207}Pb, are not affected in the same way, since the characteristic decay times for ^{238}U, ^{235}U, are not the same. Although the first factor affects the two lead isotopes in the same way, the second factor affects them differently. After a fairly straightforward

TABLE IV.6.
The principal decay series of ^{235}U

Parent nucleus	Characteristic decay time	Ray emitted	Daughter nucleus
^{235}U	7.1×10^8 years	α	^{231}Th
^{231}Th	25.5 hours	β	^{231}Pa
^{231}Pa	3.2×10^4 years	α	^{227}Ac
^{227}Ac	21.8 years	β	^{227}Th
^{227}Th	18.5 days	α	^{223}Ra
^{223}Ra	11.4 days	α	^{219}Rn
^{219}Rn	3.96 seconds	α	^{215}Po
^{215}Po	1.78×10^{-3} seconds	α	^{211}Pb
^{211}Pb	36.1 minutes	β	^{211}Bi
^{211}Bi	2.15 minutes	α	^{207}Tl
^{207}Tl	4.79 minutes	β	^{207}Pb (stable)

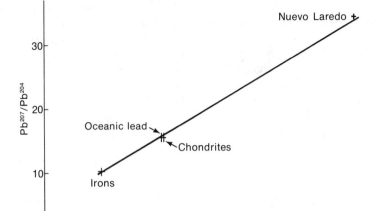

FIGURE IV.18.
The plot of $^{207}Pb/^{204}Pb$ versus $^{206}Pb/^{204}Pb$. (Courtesy of C. Patterson, California Institute of Technology.)

bit of mathematics, this circumstance can be shown to result in the following situation.

Plot the ratios $^{206}Pb/^{204}Pb$, $^{207}Pb/^{204}Pb$, in a diagram of the form of Figure IV.18. For samples of the same age, but with different relative concentrations of lead and uranium, and satisfying the "closed" condition that neither uranium nor lead has been added or taken away from them, the resulting points lie on a straight line. Moreover, the age t of the samples can be determined from the slope of this straight line. The calculation of t involves the ratio $^{235}U/^{238}U$, and the characteristic decay times of ^{235}U, ^{238}U, but these quantities are all measured and therefore are available in the calculation.

The practical problem in determining the age of the solar system is to obtain suitable samples of material dating from the very early history of our system. The meteorites are believed to have formed essentially contemporaneously with the inner planets, which we saw above to have condensed very quickly, within only a few million years. Furthermore, meteorites are also believed to satisfy the "closed" condition that neither lead nor uranium has been added to or taken away from them. A point for the iron meteorites and one for the particular meteorite Nuevo Laredo are shown in Figure IV.18. These two points, being well-separated, serve to define accurately the slope of their straight-line connection. This turns out to yield an age close to 4.6×10^9 years, which is thus the age of the meteorites, and hence of the inner planets also.

Samples of terrestrial material satisfying the closed condition that neither uranium nor lead has been added to or taken away from them during the whole history of the Earth are quite unobtainable. All the individual samples to be obtained at the Earth's surface have been subject to changes in their lead and uranium concentrations. However, material from the ocean bottoms is believed to be a well-mixed sample of all terrestrial rocks taken together, and so to have the same relation of uranium to lead that a sample obeying the closed condition would have had. The values of $^{206}Pb/^{204}Pb$, $^{207}Pb/^{204}Pb$, for such oceanic material determine a point in Figure IV.18 that is seen to lie very close to the line determined by the meteorites. This is taken to verify the age of 4.6×10^9 years for the Earth.

PROBLEM:

Rocks recovered from the Moon have ages mostly between about 2×10^9 years and 4×10^9 years. Lunar geologists believe these rocks date from epochs when great quantities of lava emerged from the lunar interior. It is also believed that the dust found at the lunar surface has been formed by a powdering of such rocks. Yet the age of the dust turns out to be 4.6×10^9 years. Consider the implications of this result.

Appendix IV.5. The Age of the Galaxy

The present-day measured value of the ratio $^{235}U/^{238}U$ is about 0.00725. Because the isotopes of uranium are not appreciably fractionated by chemical processes, this value is obtained from essentially all samples of naturally occurring material. Since ^{235}U has been decaying faster than ^{238}U during the whole history of the solar system, there was more ^{235}U in relation to ^{238}U when the planets were formed than there is now. In fact, if there has been no *production* of uranium, by the process of rapid (r) neutron addition discussed in Appendix III.3, then the ratio $^{235}U/^{238}U$ behaves with respect to time (measured backward) in the manner of Figure IV.19. In this figure, time is plotted directly, the ratio $^{235}U/^{238}U$ logarithmically. The straight-line behavior depends on the condition of zero uranium production being satisfied. This line has been continued backward in time to an epoch some 6.5×10^9 years ago, when $^{235}U/^{238}U = 1.42$. This last value is believed to be the isotope ratio at the time of formation of the uranium itself, being calculated from the details of the process of rapid neutron addition. Hence, if all the uranium were produced in a single galaxy-wide event, that event must have occurred some 6.5×10^9 years ago. If we assume that this event was the formation of the galaxy, then our galaxy would be about 6.5×10^9 years old.

An independent check on this idea can be made by considering the ratio $^{232}Th/^{238}U$. Nuclei of ^{232}Th also decay by emitting α particles, with a characteristic time of 1.4×10^{10} years, which is longer than the decay time

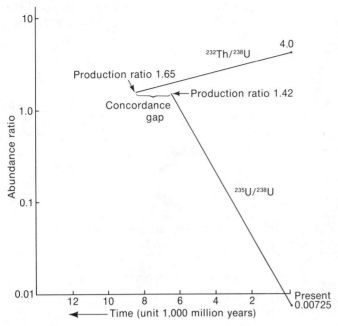

FIGURE IV.19.
The hypothesis that uranium and thorium were produced in a single galaxy-wide event leads to a disagreement between the value of 6.5 times 10^9 years given by the two uranium isotopes, and the value of 8.5 times 10^9 years given by the ratio $^{232}Th/^{238}U$, for the time which has elapsed since the assumed galaxy-wide event.

$(4.5 \times 10^9$ years) of ^{238}U. Hence the ratio $^{232}Th/^{238}U$ was less in the past than it is today. Under the condition of zero production, this ratio follows the straight line also shown in Figure IV.19, where the present-day value has been taken as 4.0. The line has been drawn backward to a time about 8.5×10^9 years ago when the value was 1.65, this being the initial ratio determined by the details of the process of rapid neutron addition. According to Figure IV.19, the production of uranium and thorium—if it occurred in a single, galaxy-wide event—must have taken place 8.5×10^9 years ago, which is not concordant with the 6.5×10^9 years given by the two uranium isotopes. Thus the implication of Figure IV.19 is that uranium and thorium production did not happen all at once, but has been happening throughout the history of the galaxy.

Some scientists have sought to avoid this conclusion by arguing that the present-day value of $^{232}Th/^{238}U$ should be set rather lower than 4.0. If one takes 3.3 instead of 4.0 for the present-day value of $^{232}Th/^{238}U$, the two straight lines of Figure IV.19 would concur at the epoch some 6.5×10^9 years ago. If uranium and thorium have been subject to differential chemical fractionation, the measured value of about 4.0 in terrestrial materials could be spurious—so runs the argument. This doubt has largely been settled, however, by the recent measurements of ^{232}Th/^{238}U in samples of lunar material, which are thought to have been less affected by chemical segregation than terrestrial samples may have been. The measured value of ^{232}Th/^{238}U in lunar samples is about 4.2, and the difference from 4.0 *increases* the discrepancy discussed above.

The problem we face is to obtain a plausible model in which uranium and thorium are produced continuously and in such a way that the two estimates of the age of the galaxy, one from ^{235}U/^{238}U, the other from ^{232}Th/^{238}U, are concordant. Now, any continuous process, whatever form it takes, must *increase* the estimate of 8.5×10^9 years given by the straight-line behavior of ^{232}Th/^{238}U in Figure IV.19. *Hence the galaxy must be older than 8.5×10^9 years.*

The recent model discussed by W. A. Fowler leads to the situation shown in Figure IV.20. The straight lines used during the age of the solar system

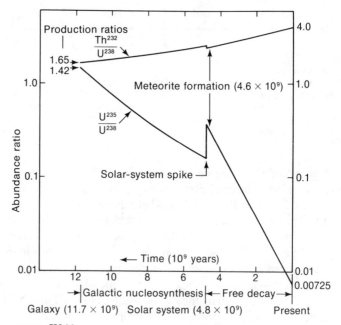

FIGURE IV.20.

The hypothesis that uranium and thorium have been produced continuously during the lifetime of the galaxy leads to a concordance in which the age of the galaxy is determined to be about 12 times 10^9 years. There is evidence of uranium and thorium production in an event which immediately preceded the formation of the solar system. Although such a possibility was included in the making of this diagram, it does not appreciably affect the result for the age of the galaxy. (Courtesy of W. A. Fowler, California Institute of Technology.)

correspond to a condition of zero uranium and thorium production within the solar system. The remarkable features of Figure IV.20 are the sudden breaks in the curves, occurring at a time immediately preceding the formation of the solar system. These breaks imply a sudden infusion of newly produced uranium and thorium into solar-system material at a time very close to its formation. According to the ideas of Chapter 6, all stars are born in clusters. It is thought that a massive star belonging to the cluster in which the solar system was formed became a supernova, and caused a sudden production of uranium and thorium by the process of rapid neutron addition. There are technical details about the meteorites which seem to require such a remarkable local event to have taken place, and this is why its effects have been included in Figure IV.20.

According to Fowler's model, the age estimates given by $^{235}U/^{238}U$, $^{232}Th/^{238}U$, are concordant at about 11.7×10^9 years ago. If the first production of uranium and thorium by massive stars occurred shortly after the galaxy was formed, then the galaxy can be only a little older than this, say, 12×10^9 years, which is perhaps the best estimate for the age of our galaxy that our present knowledge will allow us to make.

References for Section IV

For general reading, see J. C. Brandt and S. P. Maran, *New Horizons in Astronomy* (San Francisco, W. H. Freeman and Co., 1972).

Concerning the Moon, see *The Moon*, volume 7 (1973). Concerning the argument for compacted powder and for its transport, see T. Gold, "Conjectures about the Moon," *ibid.*, p. 293.

That the clouds of Venus might be sulfuric acid was suggested independently by G. Sill and A. T. Young, "Are the Clouds of Venus Sulfuric Acid?" *Icarus,* 18 (1973), 564.

The surface features of Mars are discussed by J. F. McCauley, M. H. Carr, J. A. Cutts, W. K. Hartmann, Harold Masursky, D. J. Milton, R. P. Sharp, and D. E. Wilhelms, "Preliminary Mariner 9 Report on the Geology of Mars," *Icarus,* 17 (1972), 289. See also many other papers in Volumes 17 and 18 of *Icarus.*

The rotation period of the planet Mercury has been found to be exactly 2/3 of its orbital period. The reason for this was explained by G. Colombo, "Rotational Period of the Planet Mercury," *Nature,* 208 (1965), 575. See also G. Colombo and I. I. Shapiro, *Special Report 188R*, Smithsonian Astrophysical Observatory.

Concerning the outer planets in general, see R. L. Newburn, Jr., and S. Gulkis, "A Survey of Outer Planets and their Satellites" *Space Science Reviews,* 14 (1972–73), 179.

The cause of the energy emerging from the interior of Jupiter has been discussed
by E. E. Salpeter, "On Convection and Gravitational Layering in Jupiter
and in Stars of Low Mass," *Astrophysics J.*, vol. 181 (1973), letter page 83,
and by R. Smoluchowski, "Dynamics of the Jovian Interior," *Astrophysics
J.*, vol. 185 (1973), letter page 85.

For the reflection of radio waves by the rings of Saturn, see R. M. Goldstein
and G. A. Morris, "Radar Observations of the Rings of Saturn," *Icarus,*
20 (1973), 260.

The classic statement of the structure of DNA is F. H. C. Crick and J. D. Watson,
"Molecular Structure of Nucleic Acids," *Nature,* 171 (1953), 737.

For a discussion of the genetic code, see F. H. C. Crick, "The Genetic Code
III," *Scientific American*, October 1966, p. 55, and *Symposia on Quantitative
Biology*, Volume XXXI (1966), Cold Spring Harbor Laboratory.

Evidence concerning the temperature behavior of organisms in the Yellowstone
Hot Springs is from T. D. Brock, "Life at High Temperatures," *Science,*
158 (1967), 1012.

Section V:
Radioastronomy

James Clerk Maxwell, 1831–1879. Photo courtesy of the Cavendish Laboratory, University of Cambridge.

Chapter 13:
Radioastronomy

During the last two decades, radioastronomy has grown rapidly in importance. Remarkable objects, previously unknown, have been discovered by its techniques—for example, pulsars and quasars, which are considered in this section. I have deferred the discussion of radioastronomy until now, because it will form a useful prelude to the final section, Section VI, which deals with *cosmology*, the structure of the universe taken as a whole.

§13.1. Recapitulation of the Basic Properties of Radiation

In Chapter 4 we saw that radiation is given many names—γ-rays and x-rays, ultraviolet and visible light, infrared and radiowaves—according to the frequency ν of oscillation of the radiation. We saw that a particle with electric charge, moving slowly compared to the speed of light, in the oscillatory manner of Figure 13.1, can induce a similar oscillatory motion in other charged particles. We referred to the phenomenon of Figure 13.1 as a radiative interaction. We distinguished in Chapter 4 between the source particle and the detector particle according to the time-sense of their relationship: the oscillations of the detector particle occur at a later time than those of the source particle.

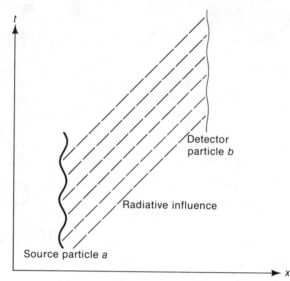

FIGURE 13.1.

A source particle sets a detector particle in oscillation, the swing of the latter being small compared to that of the source particle.

We also discussed an idealized procedure whereby an observer uses the detector particle in the following way. He sets up a "receiver" designed to recognize the oscillation of the detector particle whenever ν happens to be close to a frequency ω, a known quantity, variable by the observer. The observer then varies ω until a positive response is obtained in the receiver. The frequency ω is then the same as the frequency ν. In this way the frequency of oscillation of the source particle can be found, even though the source particle may be situated millions of light years away from us.

An ordinary radio receiver can be used in exactly this way. Suppose we wish to discover the oscillation frequency ν of some distant transmitting station. We simply turn the dial of our receiver until the clearest signal response is obtained, and then read off ω from the dial.

This method of measuring ν is the same in all the many forms of equipment used by physicists and astronomers for the detection of radiation. Although the logic of detection is always the same, the practical means by which the detection is carried out depends on the value of ν itself, for reasons that are worth noticing.

It is never possible to work in practice with single particles for source and for detector, as in Figure 13.1. The effect of a single distant source particle would be much too weak to be detected. Many source particles are needed, all oscillating with essentially the same frequency ν, and it is their combined interaction with the detector that we then observe.

Of course, the astronomer has no control over the number of source particles—he confines his observations perforce to situations where there are enough source particles that detection is possible. But the astronomer does have control over his detecting equipment. He can seek to improve the arrangement of Figure 13.2 by using the focusing method of Figure 13.3, where the small oscillations of many detector particles are used to produce a much larger oscillation of a second-stage set of detector particles. In chapter 5, in our discussion of telescopes, we considered the practical aspects of the general

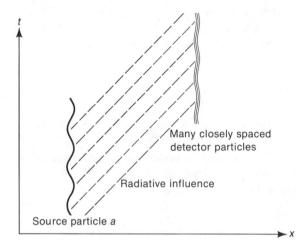

FIGURE 13.2.
In practice many detector particles are used.

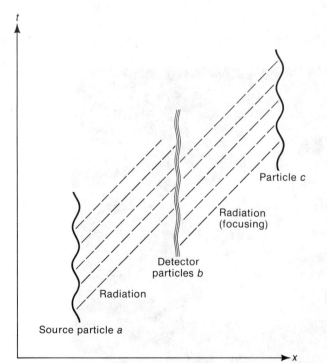

FIGURE 13.3.
The small amplitude of oscillation of many detector particles *b* are used to make particle *c* oscillate with a larger amplitude. Should *c* be *prevented* from moving unless the radiation frequency ν is close to some assigned frequency ω, which can be determined by the observer, we have the principle of the radio receiver.

principle as applied when ν is of the order of 10^{15} oscillations per second—that is, for visible light. When ν is of order 10^{18} oscillations per second, the practical details whereby physicists carry out the concept behind Figure 13.3 are very different from those for visible light. Again the details are different when ν is of order 10^9 oscillations per second. Although the principle remains the same, the different practical means for applying the principle determine the name given to the radiation—x-rays at $\nu = 10^{18}$ oscillations per second, radio waves at $\nu = 10^9$ oscillations per second. The techniques used for radio were discussed in Appendix II.2. In this and the next chapter we will be concerned with the results obtained by the application of these radio techniques.

§13.2. The 21-cm Radiation of Atomic Hydrogen

Hydrogen atoms emit radio waves with an intensity adequate to be detected readily with our telescopes. The frequency of the emission is 1.42×10^9 oscillations per second, centrally located in the combined radio and microwave bands used in astronomy, which extend from about 10^7 to 10^{11} oscillations per second. The corresponding wavelength λ, determined from $\lambda = c/\nu$, is close to 21 cm.

FIGURE 13.4.
Radiowaves from the galaxy M51 have been used to build a picture in which brightness is an indication of the intensity of the radiowaves received from the galaxy. (From D. S. Mathewson, P. C. van der Kruit, and W. N. Brouw, *Astronomy and Astrophysics*, Vol. 17, 1972.)

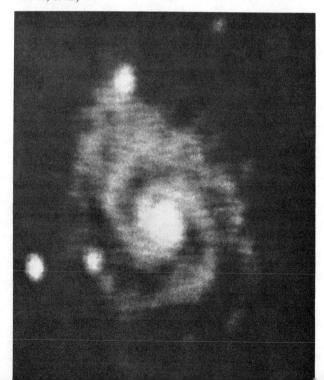

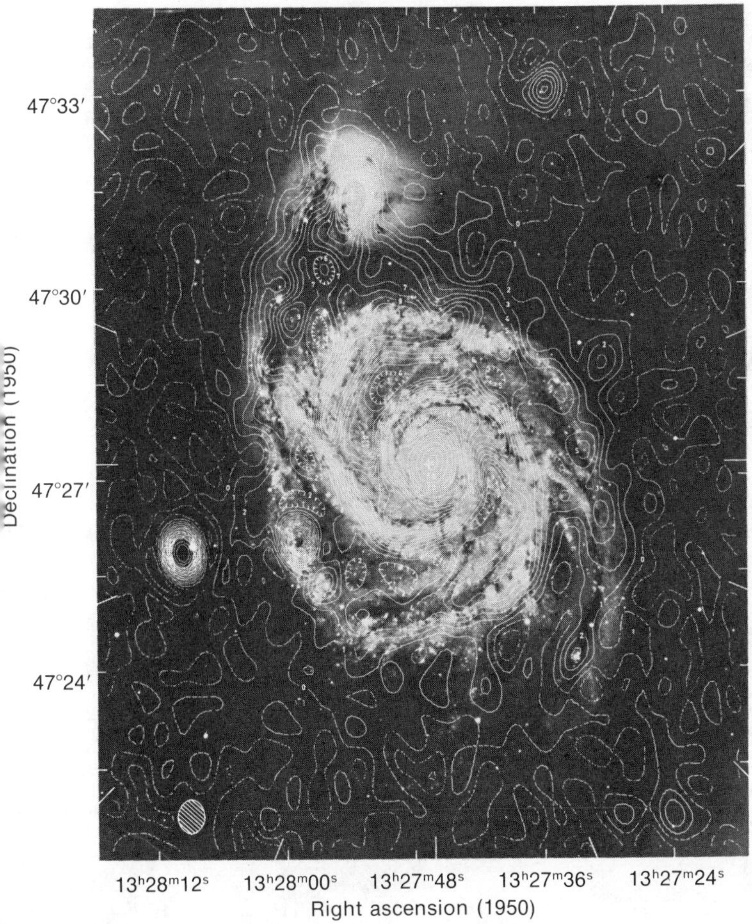

FIGURE 13.5.
Here the radio intensity is presented as a contour map, with the optical picture of M51 superimposed. The galaxy is about 100,000 light years in diameter. (From D. S. Mathewson, P. C. van der Kruit, and W. N. Brouw, *Astronomy and Astrophysics,* Vol. 17, 1972.)

Observation at this frequency permits us to estimate how much hydrogen gas in atomic form there is in the interstellar spaces within our galaxy. With modern large radiotelescopes, similar measurements can be made for other nearby galaxies. The results range from about 1 per cent up to about 10 per cent of the total mass of a galaxy. The larger proportions of atomic hydrogen tend to be found among the smaller galaxies, those of comparatively low total mass, say 10^{10} $\odot$, although not all galaxies of small mass possess much gas. The Magellanic Clouds (Figures 8.7 and 8.8) are small galaxies that possess considerable quantities of interstellar gas. The results of these 21-cm radio observations for our own galaxy have also been seen already, in Figure 3.7. Results obtained recently in Holland for the galaxy M51 (NGC 5194) are shown in Figures 13.4 and 13.5. The optical picture of M51 is shown in Figure 13.6. Notice that the radio picture has two bright spots which do not appear on the optical picture.

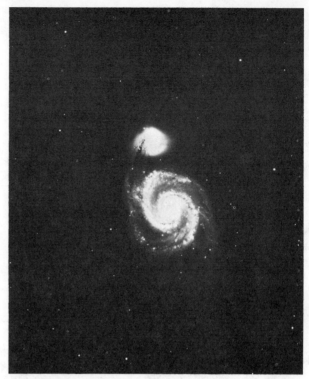

FIGURE 13.6.
The optical picture of M51.
(Courtesy of Kitt Peak National Observatory.)

§13.3. Molecules

It used to be thought that most of the interstellar hydrogen existed as individual atoms, and was thereby detectable by the 21-cm radio observations. Molecules of hydrogen, H_2, do not emit this particular radiation, and so cannot be detected by this radio technique, whereas monatomic hydrogen can be. Yet it is now known from other (non-radio) data that as much interstellar hydrogen may exist in molecular form as in atomic form. Molecular hydrogen exists in compact dense clouds, with perhaps as many as a million particles per cm^3, whereas atomic hydrogen is much more uniformly spread, at a low density of only one or two atoms per cm^3. It is even possible that much more H_2 than H may exist, but so far it has not been possible to determine with any certainty whether this is so. It is possible therefore that considerably more interstellar gas may exist in our galaxy than has hitherto been supposed. If it does, most of the gas would be in dense clouds, of the kind which we considered in Chapter 6 in discussing problems of star formation.

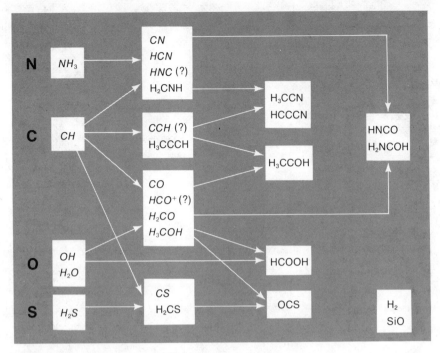

FIGURE 13.7.
The 26 compounds that have been identified in interstellar gases. The arrows indicate chemical similarities, not reactions. The formulas set in italic type indicate molecules that have been detected outside the galactic center. (From D. Buhl, *Sky and Telescope*, 45, no. 3, 1973, 156.)

Contained within the same dense clouds as the molecular hydrogen are many other kinds of molecules. These other molecules have recently been detected by the methods of radioastronomy, mostly at high frequencies, ν about 10^{11} oscillations per second—i.e., $\lambda = c/\nu$ only a few millimeters in wavelength. Work at such very short radio wavelengths is often referred to as "millimeter-wave astronomy," a special name being used because the technical method used to apply the general principle of Figure 13.3 contains features that are different from those of the older methods.

The molecules that have been detected at the time of writing are given in Figure 13.7. They are all detected by means of the specific frequencies they emit, in accordance with the method of Figure 13.3, each type of molecule having different characteristic frequencies.

It will be seen that Figure 13.7 contains some molecules, HCN, for example, that were considered in Chapter 12 to be important for the origin of life. Here, then, we have evidence that life-forming molecules exist everywhere in the galaxy. The quantities present in just one cloud are comparable with the mass of a whole cluster of stars—enormous compared to the masses of mere planets.

FIGURE 13.8.
The Lagoon Nebula, about 40 light years in diameter, in Sagittarius (M8, NGC 6523). (Courtesy of Hale Observatories.)

In view of the wide-spread occurrence of such quantities of these molecules, in clouds like those shown in Figures 6.1 and 6.3, or in Figure 13.8, we can have some confidence that life really does exist everywhere throughout our galaxy.

§13.4. Radio Waves from Hot Gas Clouds

Source particles do not always have the straightforward oscillatory behavior of Figure 13.3. Imagine a gas composed of hydrogen atoms that are hot enough for the electrons to have become free of the protons, $H \longrightarrow p + e$. It is a

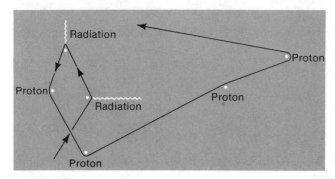

FIGURE 13.9.
Electrons in a hot gas have paths of irregular form.

general rule of thermal motions that less massive particles move faster than more massive particles; the free electrons in hot hydrogen gas, being of much smaller mass than the protons, consequently move much faster. We may imagine the protons to form a slowly changing matrix through which the electrons move at comparatively high speeds. The paths of the electrons, treated classically, are not straight lines, since the electrons are intermittently colliding with the protons, and with each other. It follows that, instead of the regular oscillations of Figure 13.3, the electrons have irregular paths of the forms illustrated in Figure 13.9. In such collisions, radiation is sometimes emitted, and this is also shown schematically in Figure 13.9.

There is a remarkable relation between Figures 13.3 and 13.9 that is far from being obvious and which illustrates in an excellent way the great power of advanced mathematics. It had been proved already a hundred and fifty years ago by the French mathematician Jean B. J. Fourier (1768–1830) that the complex path of Figure 13.9 is a mixture of simple paths with the form of Figure 13.3. The explicit problem solved by Fourier was to determine the mixture of oscillations of the form of Figure 13.3 that *when added together* give precisely the same motion as Figure 13.9. The reason why this mathematical problem is so important for physics is that an electron with the complex path of Figure 13.9 has the same radiative effect as the sum of the constituent oscillations of the form of Figure 13.3 into which it can be separated by the method of Fourier. Since we know how to cope with the motion of Figure 13.3, we therefore can cope with that of Figure 13.9.*

---

*Notice that it is the physical behavior of radiation which makes Fourier's mathematical problem useful. An electric particle in complex motion behaves like the summation of simple oscillatory motions. A similar method *cannot* be used, however, for the gravitational effect of a particle with complex motion. Thus in Einstein's theory of gravitation the effect of a particle with the path of Figure 13.9 is *not* the same as the sum of the oscillatory motions into which it can be resolved by the method of Fourier. Thus the radiative and the gravitational effects of a moving particle behave quite differently; gravitation is said to be *nonlinear*, whereas radiative effects are *linear*.

The method of Fourier therefore enables us to deal with the radiative effects of the free electrons moving in a hot gas with complex paths of the form of Figure 13.9. Instead of a radiative oscillation at just one frequency, we now have radiation throughout a whole range of frequencies. This difference is expressed by saying that instead of "line radiation" we now have a *continuum* of radiation. When radiation from a hot gas is observed with the aid of a tunable receiver, we no longer have a unique yes-or-no result with respect to the receiver frequency ω. Previously we had a positive response in the receiver only when ω was close to the source frequency ν. Now that there is no unique source frequency, a response is observed in the receiver throughout a wide range of values of ω.

This is not to say that all values of ω give the same response in the receiver. Some values of ω may well give a stronger signal than other values. Indeed, as ω is varied, the receiver response will usually vary. By studying these variations, one can in fact make useful inferences about the properties of the source particles themselves. For electrons in a hot gas cloud, it is possible to make inferences concerning the temperature and density of the gas. It is in this way, by determining the response variations of a tunable radio receiver, that a temperature of about 10,000°K is inferred for the gas cloud shown in Figure 13.8. The density of electrons in such clouds is often found to be of order 10^3 per cm^3. Since the electrons belong mostly to hydrogen atoms, this value also gives the density of hydrogen atoms. It will be noted that 10^3 per cm^3 is intermediate between the one or two neutral hydrogen atoms per cm^3 that are found throughout large volumes of the galaxy by using the line frequency at 1.42×10^9 oscillations per second, and the high densities of about a million per cm^3 for H_2 in the molecular clouds. It will be apparent that, by using different techniques, one can obtain information about quite different astronomical objects and situations. Every significant variation in the method of observation gives a different view of the universe.

§13.5. The Radiative Effects of Fast-Moving Particles

The discussion so far has been contingent on the assumption that the source particles move at speeds much less than that of light. It will now be of interest to remove this restriction. We do so for a source particle having the simple oscillation of Figure 13.10. From what we have discussed so far, we might expect detector particles to move with the same oscillation frequency as that of the source particle, as in Figure 13.2. Instead a detector particle moves in the much more complex way illustrated schematically in Figure 13.11. As with radiation from the free electrons of a hot gas, this complicated behavior can be analyzed by the method of Fourier; it can be represented as a summation of many simple oscillations with many frequencies. Although the source particles may themselves be oscillating with a well-defined frequency, say, ν, the

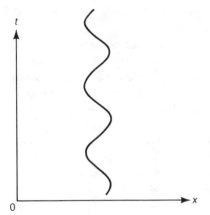

FIGURE 13.10.
A source particle in regular oscillation but
moving at a speed close to that of light.

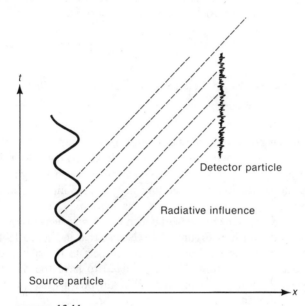

Detector particle

Radiative influence

Source particle

FIGURE 13.11.
The motion of a detector particle is complex, its oscillations
being much more rapid than those of the source particle in
Figure 13.10.

detector particles have oscillations in the sense of Fourier that cover a wide
range of frequencies. As with the radiation from a hot gas, it is not possible
therefore to use a tunable receiver to detect a unique oscillation frequency of
the detector particles. The strength of the receiver response will change slowly
as we vary the tunable frequency ω. In other words, we will have the impression

of a continuum of radiation, as we do for a hot gas, even though the oscillation frequency ν of the source particles is well-defined.

At first sight we might expect to obtain the strongest response in the receiver when ω is reasonably close to ν—and so we do when the speed of the source particle is only moderately close to the speed of light. On the other hand, when the speed of the source particle becomes very close to that of light, the strongest response occurs in the receiver at values of ω that are *very much larger than* ν. It is possible for ν to be a low frequency, say, 1,000 oscillations per second, and yet for the strongest receiver response to occur at ω of order 10^{15} oscillations per second—i.e., at the frequencies of visible light. A situation of this kind is believed to occur in the Crab Nebula, shown in color in Figure 13.12. Even though ν is estimated to be about 1,000 oscillations per second, the white light from the inner part of the Crab is generated in this way (but not the outer red light, which is "line radiation" from neutral hydrogen atoms). Indeed, response signals are observed from the Crab even at frequencies in the x-ray and γ-ray ranges, 10^{18} to 10^{21} oscillations per second.

For FIGURE 13.12, see PLATE XIX.

PROBLEM:

What is the wavelength for a frequency of 3×10^{20} oscillations per second?

Particles moving at speeds close to that of light are not easily deflected from their paths by collisions with other particles. So at first sight we might expect the high-speed electrons in the Crab Nebula to move in straight paths. This would mean no oscillation and therefore no radiative interaction. So what is it that causes these electrons to undergo oscillation? The answer is that the Crab Nebula contains a magnetic field which forces the electrons to move along helical paths of the form of Figure 13.13. Notice that in Figure 13.13 we have returned to a plot of the electron path with respect to the spatial dimensions x,y,z, and that the magnetic field is considered to have the direction of the z dimension. If we now follow the method of Chapter 4 whereby the behavior with respect to the time t is also specified in three separate diagrams, one of x versus t, one of y versus t, and the third one of z versus t, we have the situation of Figure 13.14. Evidently both the x and y dimensions have the required oscillatory behavior, both with the same oscillatory frequency. Collapsing the x,y, dimensions into a single dimension, we thus arrive at the picture of Figure 13.10. The behavior with respect to the z dimension, corresponding simply to uniform motion, has no effect on the radiative problem and does not therefore need to be considered in any detail.

Hence we have answered the basic and crucial question of the source of the white-light radiation from the Crab Nebula, and also of the source of γ-rays and x-rays from the Crab. It is the effect of a magnetic field within this

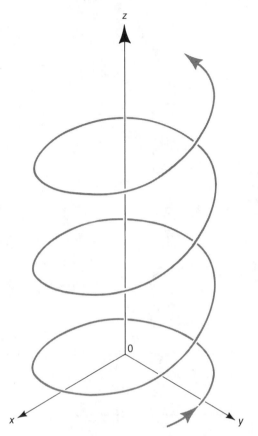

FIGURE 13.13.
Electrons in the Crab Nebula, moving at speeds
close to that of light, follow helical paths, because
of the presence of a magnetic field.

remarkable object that leads to the electron oscillations which then generate
these radiative interactions. From detailed calculations it is found that a magnetic field with an intensity of about one-thousandth of the Earth's magnetic
field will suffice to produce the required electron oscillations. At first sight this
might seem a rather weak field, but it must be remembered that the magnetic
field of the Crab is maintained throughout a region with a diameter of the
order of 10 light years, which is enormous compared with the scale of the
Earth's magnetic field, or with the Sun's magnetic field, or indeed with the
scale of the magnetic field of any ordinary star.

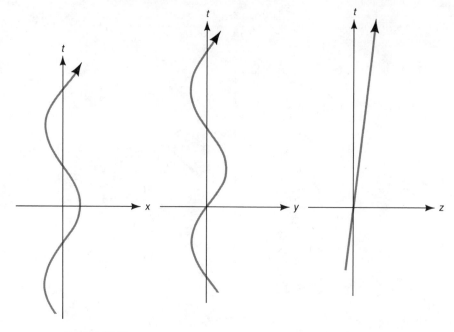

FIGURE 13.14.
The (x,t), (y,t), (z,t) diagrams for a magnetic field aligned in the z direction (see Figures 4.3 and 4.4).

§13.6. The Crab Nebula: An Historical Diversion

The Crab Nebula is believed to be the remains of a supernova which occurred nearly a thousand years ago. The outer material, the red filaments of Figure 13.12, are observed today to be in rapid outward expansion. When traced backward in time, this outward expansion suggests that an explosion from some centrally located object occurred about a thousand years ago. Such an argument does not, of course, give a precise date for the explosion of the supernova. Yet a precise date is available from the records of Chinese astronomers, who observed the light which accompanied the explosion. The supernova was seen on July 4 in the year A.D. 1054 as a "guest star visible by day like Venus." It remained as bright as Venus until July 27. Thereafter it faded gradually, to become eventually invisible to the naked eye by April 17, 1056, almost two years after the outburst. Moreover, the day-by-day variations estimated by the Chinese agree very well with the behavior of supernovae observed today (see Figures 7.26 and III.19). The position in the sky given by the Chinese astronomers for their "guest star" agrees, within expected uncertainties, with the present-day position of the Crab Nebula.

It is curious that no similar accounts have been found in European or Arabic records. An apparition in the sky at least as bright as Venus can scarcely have been missed. It remained exceedingly bright for about a month, and it is hard to believe that European, North African, and Arabian skies could have been cloud-covered for as long as this. Rather must one suppose that the supernova was seen by millions of people. Most would be illiterate and therefore unable to record it. What the specialized class of chroniclers chose to record seems to have turned on their beliefs. In China, it was thought that terrestrial events could be foretold from occurrences in the sky, and therefore peculiarities in the sky were avidly searched for and noted. In Europe, on the other hand, the monkish chroniclers believed the heavens to be the personal handiwork of "God," and therefore "perfect" and not subject to change. Any record to the contrary would have been heretical, and would have provoked an angry response from theologians and philosophers, just as did Galileo's discovery of spots on the Sun, more than five centuries later.

TOPIC FOR DISCUSSION:
The preceding remarks raise the uncomfortable question of how far even today the observations of astronomers depend on their beliefs. Most would vehemently deny any such connection, and in a social or religious sense this is probably true. But is it true in a scientific sense? Do we tend to pay more attention to those observations that seem to confirm our scientific beliefs than to those which seem to contradict them?

Two pictographs shown in Figure 13.15 have been reported by William C. Miller. They were discovered at different sites during exploration of the cave dwellings of the Pueblo culture of the North American Indians, a culture which spanned the year A.D. 1054. Miller believes the two pictographs refer to a moment when the Moon in its monthly motion across the sky approached close to the "guest star" of the Chinese. Another interpretation might be that the object close to one of the horns of the crescent is the planet Venus, but as Miller points out such a close approach of Venus to the Moon occurs every two or three years. The fact that only two among very many pictographs have ever been discovered with an undoubted astronomical connotation, namely, the two reproduced in Figure 13.15, suggests that the Pueblo Indians were not usually interested in astronomical events, and that only a quite uncommon circumstance would have impelled them to make these remarkable visual records. What else but the supernova of A.D. 1054? But did the Moon really approach so close to the supernova, and did it do so for an observer in Arizona, but not for an observer in China—otherwise the Chinese astronomers would almost surely have recorded the event? Miller found the answers to both questions to be affirmative, thereby providing strong confirmation that these two pictographs really do refer to the supernova of 1054. The chroniclers of Europe were too prejudiced to record the event, but not so the Indians of North

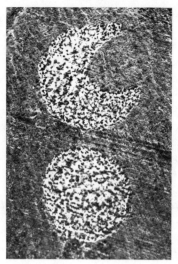

FIGURE 13.15.
Two Pueblo Indian pictographs. That on the left is from Navajo Canyon, that on the right from White Mesa. (Courtesy of William C. Miller.)

America. Unable to write, they nevertheless managed to leave a record behind them.

The reader will note that the two pictographs of Figure 13.15 are different, being reflections of each other in a vertical axis. Miller notes that the artists chipping out the separate pictographs would have had their backs turned to the sky, and would thus have experienced a left-right ambiguity. Does one mentally invert the left-right sense or not?

PROBLEM:
Imagine yourself looking over your shoulder at the horns of the Moon. Would you draw the Moon with the actual left-right sense, or would you draw it as if you turned round full face toward the sky?

---

§13.7. A Personal Recollection

The ideas discussed above concerning the nature of the Crab Nebula were formulated in the decade from 1950 to 1960. Although astronomers became convinced of their correctness, a big mystery remained. Radiative interactions cause source particles to lose energy, an effect illustrated in Figure 13.16. Both

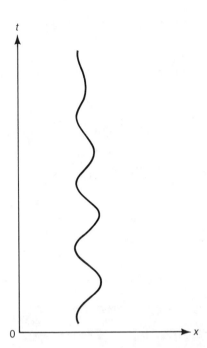

FIGURE 13.16.
A radiating particle loses energy. For an electron moving in a magnetic field, both the amplitude and the frequency of oscillation become less. Only the (x,t) diagram is shown
here (cf. Figure 13.14). Actually, many more
oscillations than are shown here would be
needed for the particle to lose appreciable energy.

the amplitude and the frequency of oscillation become less. This energy loss
is particularly marked for electrons moving at speeds close to that of light,
as the electrons in the Crab Nebula are thought to do. It could be calculated
that indeed the electrons responsible for the white light of Figure 13.12 would
lose most of their energy in only a few years; yet the Crab Nebula was about
900 years old. High-speed electrons produced in the explosion observed by
the Chinese in A.D. 1054 would long ago have lost their energy, and hence
their ability to radiate visible light. So where did the electrons responsible for
Figure 13.12 come from, if not from the actual explosion of the supernova?

   The late Walter Baade, who up to the time of his death was responsible for
much of the optical work on the Crab Nebula, had identified a peculiar yellow
star near its center which he believed might have been responsible for the
explosion of 1054. He had also detected moving ripples in the nebulosity itself.
These and other matters connected with the Crab were topics of discussion
at a conference of astronomers which met in Brussels in the summer of 1958.
In an outside-the-conference-room talk with Jan Oort from Leiden, Holland,
we wondered if Baade's star might be showing fast light variations, and we
speculated that such variations might be connected with the continuing source
of the high-speed electrons which we knew were needed to supply the white
light of the Crab. We were sufficiently impressed by the idea to refer it to
Baade, asking if it could be checked observationally. Baade immediately wanted
to know how fast the light variations might be, and I replied that they might

occur in a few seconds, basing my estimate on the concept of a pulsating white-dwarf star. Baade seriously considered the possibility of taking a series of photographs at about one-second intervals, which was the only feasible method at that time. In any event, however, he made no attempt in this direction, so far as I am aware, presumably because the photographic method proved too insensitive. It was only when more refined electronic devices replaced the photographic plate that the problem could be successfully tackled. It then turned out that variability is in fact present, as can be seen in Figure 13.17, but not in the one or two seconds that I had estimated. The oscillation frequency was found to be about 30 per second, i.e., each oscillation being performed in only 0.033 seconds. This much shorter oscillation period implies that the object responsible for the light variations must be much smaller than a white dwarf star. The object is nowadays thought to be a *neutron star,* the

FIGURE 13.17.
The oscillating star-like object in the Crab Nebula (NP 0532). The whole of the Crab Nebula was shown in Figure 4.12; find NP 5032 in that figure. (Courtesy of S. P. Maran, Kitt Peak National Observatory.)

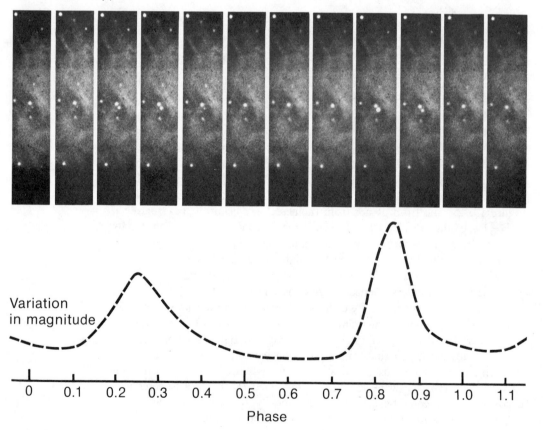

properties of which we discussed in Chapter 7. It belongs to a class of objects known as *pulsars*, which were first discovered in 1967 by S. J. Bell, A. Hewish, and their collaborators in Cambridge, England.

The point of my anecdote is to emphasize yet again the importance of the technical resources available to the astronomer. Having the right kind of idea is not in itself sufficient. The idea must be supported by instruments that are sophisticated enough to enable the idea to be applied and developed. Otherwise the idea remains sterile, as it did following the 1958 conference. It was only some ten years later, when quite new methods were available, that a further step could be made. By then Baade was dead, and the investigation had passed to a younger generation.

§13.8. Pulsars

A pulsar, as its name implies, emits a regular sequence of pulses. Although pulses of light are detected from the Crab, the pulses are best discovered by radio observations. Two hundred or so of these remarkable objects have now been found. The spacing between two pulses is known as the pulsar *period*. Typically, pulsar periods lie in the range from .25 second to about 2 seconds, although much shorter periods can occur, only 0.033 seconds for the Crab. The spacing between two pulses is usually much longer than the duration of an individual pulse, as can be seen from the example shown in Figure 13.18.

The pulses are thought to arise from rotation, the radiation being emitted in the fashion of a lighthouse beam, as is shown schematically in Figure 13.19,

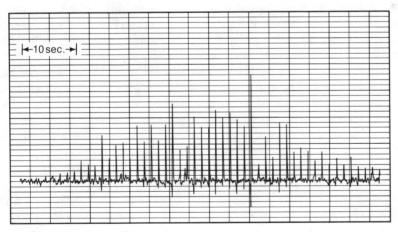

FIGURE 13.18.
A pulsar record. (Courtesy, Commonwealth Scientific and Industrial Research Organization.)

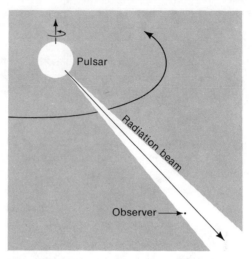

FIGURE 13.19.
Pulsars are thought to behave like the rotation beam
of a lighthouse.

with the radiation coming from a particular spot on the surface of the object,
or from a particular place in an atmosphere around it. Since pulsars generate
high-speed electrons which escape outward, and which in the Crab light up
the whole Nebula, it is reasonable to suppose that high-speed particles are also
present in the immediate surroundings of the objects themselves. The radiative
emission giving rise to the lighthouse beam of Figure 13.19 is thought to be
connected with such high-speed particles, and with their behavior in strong
magnetic fields. The details of the emission are not straightforward, however.

It is clear that any object going through a complete rotation in only 0.033
seconds, like the Crab pulsar, must be small in size compared to ordinary stars
like the Sun. If spun even at a rate of one rotation per 1,000 seconds, the Sun
would immediately fly apart under the influence of powerful rotary forces. Even
much more compact stars, white dwarfs, could not be spun as fast as the Crab
pulsar. The only known stable objects which can be spun as fast as this are
the neutron stars, which in Chapter 7 we saw to be only a few kilometers in
radius, much smaller than the Earth.

Neutrons in bulk can be stable, provided there are neither too many of them
nor too few. About as many as are in the Sun, 10^{57}, can exist in a stable
configuration, but if there were appreciably more, they would shrink into what
is called a *black hole*. (The properties of black holes are discussed in Appendix
VI.3.) Stable neutron stars could also exist with appreciably less than 10^{57}
neutrons. However, most of the observed pulsars are thought to have just about
this number. What one might ask is so special about it? What is magic about
10^{57}?

Here is something odd to think about. In Section I we considered the universe on the scale of the most distant observed galaxies, explicitly, on a scale of 10^{28} centimeters. This is some 10^{22} times greater than the radius of a typical neutron star. The total number of particles contained in such a grand-scale picture of the universe is known to be about 10^{79}. The ratio of this to our magic 10^{57} is also 10^{22}. Hence we have a remarkable equality

$$\frac{\text{Scale of universe}}{\text{Scale of neutron star}} = \frac{\text{Number of particles in universe}}{\text{Number of neutrons in star}}.$$

It is this equality that specifies the magic 10^{57}.

Let us end this chapter by another strange thought. The mass of the neutron is slightly greater than the mass of the proton plus the mass of the electron. This is why a neutron by itself changes in about 1,000 seconds into a proton, electron, and antineutrino, the reaction $n \longrightarrow p + e + \bar{\nu}_e$. If, on the other hand, the sum of the proton and electron masses had exceeded the neutron mass, we should have had $p + e \longrightarrow n + \nu_e$. Consider what would then happen to the Sun and to other stars, and to the planets. Their constituent protons and electrons would quickly decay into neutrons. The neutrinos produced in these decays would escape out into space, since neutrinos have very little interaction with other particles. The resulting bodies would then quickly collapse into neutron stars or into black holes. The universe, although it would still consist of particles which the physicist considers to be similar to the particles of our familiar every day world, would *look* fantastically different. Furthermore, radiative reactions, which play such an important role in our world, would become of quite minor significance, since neutrons and neutrinos have no electric charge.

All this is quite straightforward. The curious aspect arises from the fact that the mass of the neutron is only slightly greater than the sum of the proton and electron masses, by only about one part in a thousand. So one can feel that things might quite easily have been the other way around, in the sense of the preceding paragraph. Is this a matter of chance? Already in Appendix III.9 we had a somewhat similar situation. We saw how the production of the important chemical elements, in the appropriate proportions to be suitable for biological requirements, was contingent on certain very detailed and apparently coincidental properties in the nuclei of these elements. We find then that the world as we know it depends in critical respects on what appear at first sight to be favorable accidents. But could these circumstances not be accidents at all? If the value of $m_n - (m_p + m_e)$, where m_n, m_p, m_e, stand for the masses of the neutron, proton, and electron, were variable from one part of the universe to another, we could argue that of necessity we must find ourselves in a part where m_n exceeds $m_p + m_e$; for otherwise the local portion of the world would be so critically different that creatures like ourselves could never live in it.

This idea would make the universe a much vaster and more intricate structure than we usually suppose it to be. What should be wrong with that, one might ask? Nothing, in principle. In practice, however, we would have to cope with problems that are far more sophisticated than the ones we are accustomed to, but this is not a satisfactory logical reason for rejecting such an idea.

I would myself take a fairly firm bet that this kind of notion represents a valid glimpse of the nature of things. Such a universe would be vastly richer in its properties than the universe conceived of in present-day physics. It would have not just one chemistry, but an infinite variety of chemistries, with all that this might imply for the varieties of life forms. Although it is hard to predict the direction in which radical progress from our present knowledge may turn out to lie, it is safe, I believe, to say that science becomes richer in its content the more it progresses. The universe as it will be understood in the future will be richer in its properties than our present-day concept of the universe. Since we have in this speculation an example of an almost incomparable increase of richness, there is, I think, a chance that such a speculation may turn out to be correct.

General Problems and Questions

1. Give the frequency ranges and the corresponding wavelength ranges used to describe the following forms of radiation: γ-rays, x-rays, ultraviolet, visible light, infrared, microwaves, radiowaves. How does it come about that these different names are used to describe essentially the same radiative phenomenon?

2. Describe the procedure whereby a radio receiver is used to determine the intensity at a particular oscillation frequency of waves from a cosmic radio source.

3. How is 21-cm radiation from atomic hydrogen generated?

4. What information concerning the distribution of atomic hydrogen in other galaxies has been discovered by means of the 21-cm radiation?

5. Discuss the many molecules which have been discovered in the gas clouds of our galaxy. What form of technique was used in this work?

6. In what important respect does the radiation from a hot gas cloud differ from that emitted by molecules? Discuss in general terms the mathematical method used to analyze the radiation from a hot gas cloud.

7. Discuss the difference between the emission of radiation by slow-moving particles and by particles moving with speeds close to that of light. Give an example where the latter kind of situation is important.

8. What are pulsars?

9. If the sum of the masses of the proton and electron were (by some magic) to become larger than the mass of the neutron, what would happen to the stars?

Chapter 14:
The Most Powerful
Radiosources

§14.1. Radiogalaxies

Events in galaxies sometimes cause explosions of exceedingly great violence. Such explosions are believed to occur in the very central regions of the galaxies. Particles with speeds approaching that of light are shot out, usually in two oppositely directed streams, as shown schematically in Figure 14.1. Should such a stream impinge on a cloud of gas that extends throughout, or even beyond, the galaxy, the high-speed motion outward of the particles will be checked by the magnetic field that always exists in such a cloud. An equally rapid motion of the particles around the direction of the magnetic field is then set up, as in Figure 14.2. Such a circling motion causes electrons to emit radiowaves by a process similar to that which takes place in the Crab Nebula. This process is often referred to as *synchrotron* radiation, since this form of radiation was first detected in the laboratory in a machine called a synchrotron. The effect is to produce two patches of radio emission, as illustrated in Figure 14.3. Radiogalaxies frequently show this double pattern of emission, examples being given in Figure 14.4.

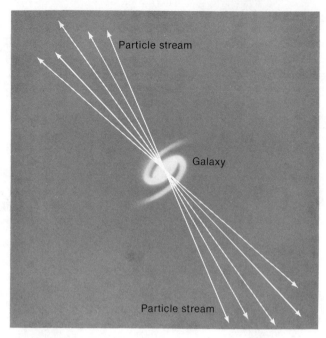

FIGURE 14.1.
In violent outbursts from galaxies, particles are shot out with speeds approaching that of light, usually in two oppositely directed streams.

FIGURE 14.2.
Should the streams in Figure 14.1 impinge on an external gas cloud, radiowaves are emitted in the presence of a magnetic field.

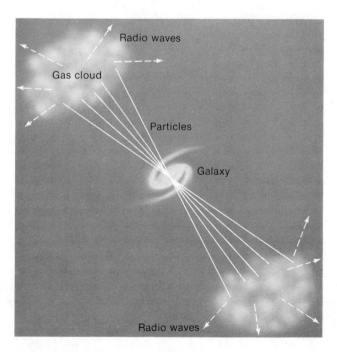

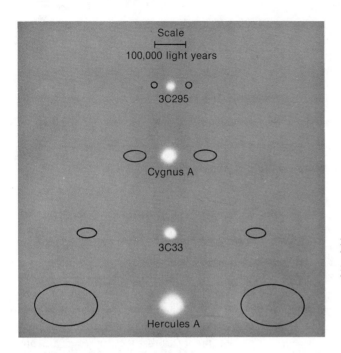

FIGURE 14.3.
The effect of the situation in Figure 14.2 is to produce two patches of radio emission. The names here refer to explicit radio sources.

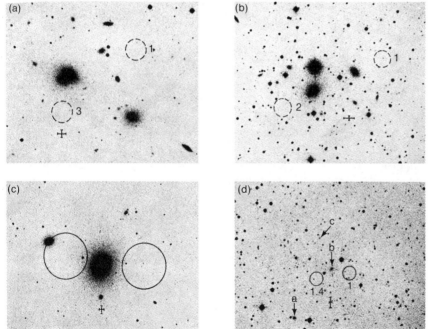

FIGURE 14.4.
Examples of radiosources with double patches of emission, for (a) 3C 40, (b) 3C 66, (c) 3C 270, (d) Hercules A. These are *negatives* of the usual photographs, with the galaxies now appearing dark against a light sky. (From P. Maltby, T. A. Mathews, and A. T. Moffet, *Astrophysical Journal*, 137, 1963, 156. University of Chicago Press.)

The outbursts from radiogalaxies are much more energetic than those of supernovae, even though a supernova can temporarily become as bright as a galaxy, as we saw in Figure 7.26. The explosive energy in a supernova is of the order of 10^{51} ergs, 10^{10} ergs being equal to 1 kilowatt second—i.e., the energy consumed by a 1 kilowatt electrical device in a time of 1 second. The energy of one supernova is equivalent to the consumption of some 100,000,000,000,000,000,000,000,000 billion barrels of oil, far beyond the dreams of an Arabian Shaikh, or even beyond those of the President of an oil company. Yet the outbursts of radiogalaxies release a still vaster amount of energy, ranging from about 10^{54} ergs on the low side to about 10^{60} ergs in the most extreme outbursts.

So great indeed is the energy that one may wonder where it can come from. In Chapter 7 we saw that leptons and baryons have antiparticles. To the electron e there is the positron e^+; to the muon μ there is μ^+; to the neutrinos ν_e, ν_μ there are the antineutrinos $\bar{\nu}_e$, $\bar{\nu}_\mu$. To the proton p there is the antiproton $\bar{p}$; to the neutron n the antineutron $\bar{n}$; and similarly for the other six baryons. Any particle-antiparticle pair can undergo annihilation. When astronomers speak of matter-antimatter annihilation, they usually have in mind the particle-antiparticle pairs (e, e^+), $(p, \bar{p})$, and $(n, \bar{n})$. The release of energy from such a matter-antimatter annihilation is more than a hundred times more efficient than the hydrogen-to-helium conversion processes that we studied in Chapter 7. Even so, it would require the total annihilation of a mass equal to a million times the mass of the Sun to yield as much energy as is released by the most violent explosions of radiogalaxies.

Some astronomers have speculated that a mass millions of times the Sun's mass may actually be annihilated in this way, but it has proved difficult to make such a theory work satisfactorily when details are considered. It is more usually supposed that a large mass in the central regions of a galaxy becomes highly compact, hundreds of millions of solar masses being confined in a region not much larger than the solar system. Because of the very strong gravitational fields, the speeds of motion of such compact masses would be expected to approach the speed of light. It is considered that sudden changes in these motions may lead to the explosive ejection of enormous streams of particles, streams that give rise to the radiogalaxies.

Further examples of radiogalaxies are shown in Figures 14.5 to 14.7. These galaxies are of course exceptions, perhaps one galaxy in a thousand being of the kind which astronomers term a "radiogalaxy." Evidence of explosion shows clearly in Figure 14.5, the galaxy M82. A still more immediate case is shown in Figure 14.8, where a remarkable jet emerges from the center of the galaxy M87, which has already been seen in a longer photographic exposure, in Figure 8.10. The light emitted by this jet comes from high-speed electrons moving in helices of the form of Figure 13.13 about the direction of a magnetic field. Thus the optical light from the jet of M87 is "synchrotron" radiation, not starlight. The galaxy M87 also emits radiowaves by the same synchrotron

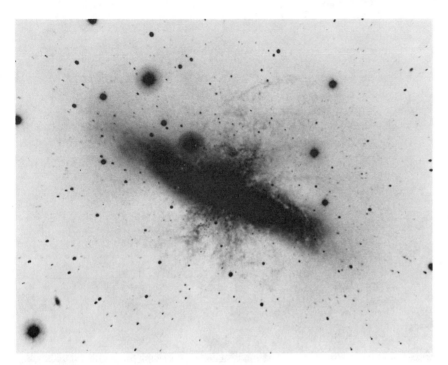

FIGURE 14.5.
The galaxy M82, where evidence of a recent explosion can be seen. (Courtesy of Hale Observatories.)

FIGURE 14.6.
The radio source Cygnus A, a very intense source. (Courtesy of Hale Observatories.)

FIGURE 14.7.
The comparatively nearby system of NGC 5128. This is a giant elliptical galaxy with a diameter of about 100,000 light years, out of which a vast cloud of gas and dust seems to have emerged. (Courtesy of Hale Observatories.)

FIGURE 14.8.
The remarkable jet of the galaxy M87 (NGC 4486), with a visible length of about 10,000 light years. Compare with Figure 8.10. (Courtesy of Hale Observatories.)

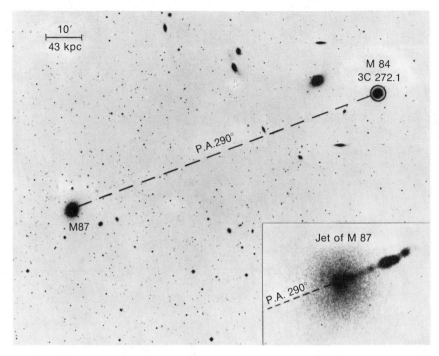

FIGURE 14.9.
The jet of M87 points almost directly toward the galaxy M84, which is also a radiosource. Note that 1 kpc equals 3,260 light years. (Courtesy of Prof. G. R. Burbidge, University of California.)

process. It is interesting and curious that the nearest radiogalaxy to M87 is the system of M84, shown in Figure 14.9, the direction joining M87 to M84 being closely the same as the direction of the jet. Is this mere happenstance, or does it imply a connection between M84 and M87? According to the usual ideas, the coincidence of the directions must be ascribed to chance, since most astronomers feel the implication that M84 has been shot out of M87, or that both galaxies have been shot out of some invisible intermediate object, to be too strange to merit serious attention.

Another curious case is shown in Figure 14.10. This is the system of NGC 7603, with a satellite galaxy apparently connected to the main galaxy by both a clearly marked arm and a fainter arch. A Doppler-shift technique (described in detail in Appendix II.8) has shown that, whereas the main galaxy is moving away from us at a speed of about 8,000 km per second, the satellite galaxy is moving away at a speed of about 16,000 km per second. Unless this apparent association is a remarkable fluke, the least radical explanation of these results is that the satellite has been shot out of the main galaxy at a speed of at least 8,000 km per second. The concept that one galaxy can disgorge another in this way is highly unorthodox, however, and is not accepted by most astronomers.

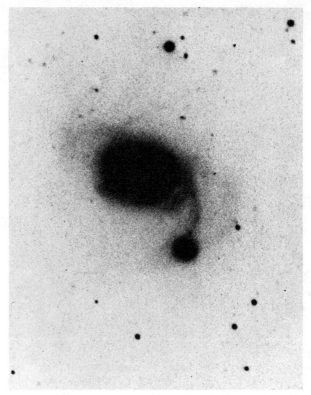

FIGURE 14.10.
The strange case of NGC 7603. The smaller galaxy appears to be
connected to the main galaxy not only by a clearly marked arm
but also by a fainter arch. Yet the Doppler shift for the smaller
galaxy (16,000 km per second) is twice that for the main galaxy
(8,000 km per second). (Courtesy of H. Arp, Hale Observatories.)

§14.2. Quasars

Imagine the nucleus of a radiogalaxy taken by itself. The radio properties of
such a nucleus would resemble those of a class of objects known as *quasars*.
However, the quasars are thought to be often much brighter optically than
galaxies.

The optical light of quasars displays a continuum of frequencies which is
thought to be produced by the synchrotron process, as in the jet of M87. There
are also line radiations from atoms; hydrogen, helium, carbon, nitrogen, oxygen,
neon, magnesium, silicon, sulfur, and iron have all been detected in quasars.
Usually, but by no means always, these line radiations are in emission lines.
Typical radiation distributions ("spectra") are shown in Figure 14.11. These
spectra are characteristic of a moderately diffuse gas, say, with a density of

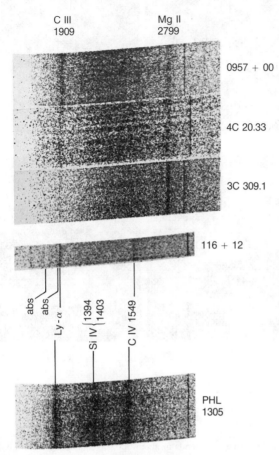

FIGURE 14.11.
The spectra in negative form of several quasi-stellar objects (quasars). Such
spectra are characteristic of a moderately diffuse gas. They show emission
lines of common atoms—hydrogen, carbon, magnesium, silicon. (Courtesy of
Dr. R. Lynds, Kitt Peak National Observatory.)

the order of 10^6 atoms per cm³. They are highly distinctive and can readily
be distinguished from the spectra of other kinds of astronomical object. It is
generally believed that, whereas the continuum light is generated in the compact
central region of a quasar, the line radiations come from a much more extended
cloud of gas with dimensions in the range from 10 to 100 light years.

When photographed directly against the sky in the ordinary way, quasars
appear like faint, rather blue stars, as can be seen from the examples shown
in Figure 14.12. After examining Figure 14.12, one may wonder how, given
all the ordinary stars that are scattered over the sky, quasars are ever discovered.
The first method uses radio observations. Radioastronomers can nowadays
determine with great accuracy the positions in the sky from which radio

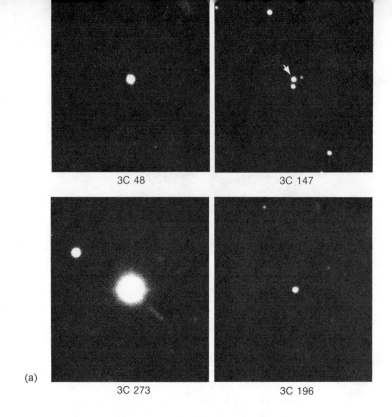

3C 48

3C 147

3C 273

3C 196

(a)

FIGURE 14.12.
(a) When photographed directly, the quasi-stellar objects (quasars) appear like ordinary stars. (b) A long exposure showing the jet of 3C 273. According to the usual interpretation of the redshift of quasi-stellar objects, the jet extends to about 300,000 light years from the center of the system. (Courtesy of Hale Observatories.)

(b)

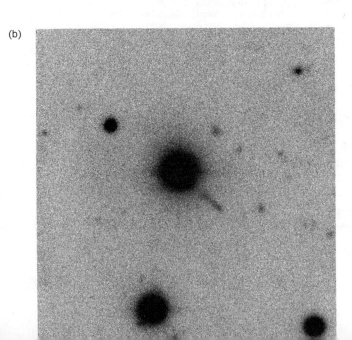

emissions are detected; such patches of emission are called "radio sources." Sometimes the patches of emission are large and extended, especially for radio radiation generated by rapidly moving electrons in our own galaxy. Other patches of emission on the sky are highly localized, especially the quasars. With their positions accurately known, it is a comparatively straightforward task to search optical photographs of the relevant parts of the sky. When an optical object is found to be coincident with a radio position, there is a presumption that the optical object is associated with the source of the radio emission. To decide whether this is so or not, the light from the object is then analyzed to discover its frequency distribution. With the "spectrum" thus available, it becomes immediately clear whether or not the object is a quasar. Should the spectrum be of the characteristic quasar type, the initial presumption of a connection with the radio emission is taken to be confirmed. The source is then said to be "identified."

A second method of discovering quasars is simply to search a restricted area of the sky for faint blue star-like objects, and then to obtain spectra of them. Those which are quasars are readily picked out from their spectra. The disadvantage of this second method is that only a very limited area of sky can be searched in this way within any reasonable amount of time. There is the advantage, on the other hand, that quasars can be discovered in this second way whether or not they happen to be sources of radiowaves. The surprise was to find that most quasars are *not* sources of radiowaves. Although the radio method first led to quasars being discovered, it has turned out that the property of radio emission is a comparative rarity, just as intense radio emission is a rarity for galaxies—only about one galaxy in a thousand is a radiogalaxy.

It will now be realized that the word "quasar" used in the above sense is a misnomer. "Quasar" is a contraction of "quasi-stellar radio source," which came into widespread usage at a time when it was thought that all such objects were radio sources. It is therefore quite meaningless to use "quasar" for objects that are not radio sources at all. For this reason many astronomers prefer the more general description "quasi-stellar object," which is applicable whether or not the object happens to be a source of radiowaves. This more logical name is usually contracted simply to QSO.

In mathematics one learns that precision of notation helps greatly toward understanding. Most students find Einstein's general relativity to be less difficult than they had anticipated, because of the splendid notation used to develop this theory. I recall Dirac telling me, more than thirty years ago, when he had just invented the 'bra-and-ket' notation for wavefunctions, of his intention to rewrite the whole of his famous book on quantum mechanics in terms of the new notation. At the time, I thought this an astonishing thing to do. Today I understand very well the reason for Dirac's decision. Indeed, I now suspect that the prime requirement for being a creative mathematician is an ability to invent compact notation with a meaning that is instantly clear to the eye. From here on, therefore, we will drop the imprecise term "quasar," and will use QSO instead.

How do we know that QSO's are appreciably brighter than galaxies? The first step in answering this question is to understand the so-called "redshift" phenomenon. (This was discussed in some detail in Appendix II.8, but for the benefit of the reader who may have elected to omit that material, the following less technical discussion is sufficient.) The line radiations in the light from a distant galaxy or a QSO have precisely defined frequencies which can be measured. As examples, we might have the Balmer lines of hydrogen, or the H and K lines of the singly ionized calcium atom. The ratios of the measured frequencies of the lines to each other are always the same in a single spectrum, whether from an extraterrestrial source or in the terrestrial laboratory. However, the absolute frequency of a line in a spectrum from a galaxy or QSO is always found to be less than the corresponding frequency measured in the laboratory. By analogy, suppose we think of the line radiations as corresponding to the notes of a piano. Although the "piano" which constitutes a distant galaxy or QSO is in perfect *relative* tune, the absolute pitch of such a "piano" is lower than that of a similar "piano" here on the Earth. A quantity z is used by astronomers to express this lowering of pitch, namely,

$$1 + z = \frac{\text{A pitch on Earth}}{\text{Pitch in distant object}}.$$

If we write ν_{lab} for the frequency of a certain line radiation as measured in the laboratory, and ν_{obj} for the frequency of the same line observed in the light from a distant object, we have $1 + z = \nu_{lab}/\nu_{obj}$.

PROBLEM:
Explain why this relation can also be written in the form $1 + z = \lambda_{obj}/\lambda_{lab}$, where $\lambda_{obj}, \lambda_{lab}$, are the corresponding wavelengths. Writing $\Delta\lambda$ for the difference $\lambda_{obj} - \lambda_{lab}$, satisfy yourself that $z = \Delta\lambda/\lambda_{lab}$. Since both $\Delta\lambda$ and λ_{lab} are measurable, z can be determined.

Now that we understand what z means, and how it can be measured, we can return to the problem of determining the relative brightness of QSO's and galaxies. For galaxies, the measured z values are found to increase systemetically with the distances d, the distances being determined by the method discussed in Chapter 8. This systematic behavior shows itself when a diagram like Figure 14.13 is constructed, with the values of z and d plotted logarithmically. The remarkable result emerges from the measured z and d values that galaxies which are structurally similar to each other (for example, the brightest galaxies in similar clusters) tend quite markedly to fall on the straight line of Figure 14.13.

Turning to the QSO's, *the assumption is made that the z and d values are related in the same way as for galaxies.* However, most QSO's turn out to have z values appreciably greater than those of the measured galaxies, so that an

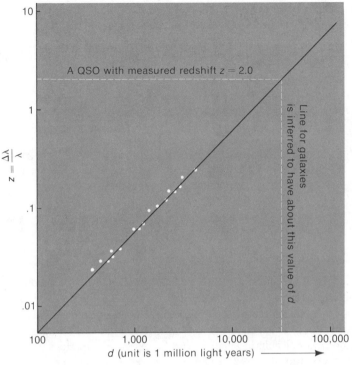

FIGURE 14.13.
Schematic illustration of the observed relation between the redshift z and the distance d, for galaxies that are structurally similar to each other. An extrapolation of this relation is used, together with the measured z values for quasi-stellar objects, to infer distance values for the latter.

immediate comparison with galaxies is not possible. To cope with this difficulty, the line of Figure 14.13 is extended until it reaches large enough z values, and the QSO distances are then read off from the line, as indicated schematically in Figure 14.13. With the d values for the QSO's determined in this way, it is possible to compare their intrinsic luminosities with those of the galaxies. As stated above, it is found by this method that most QSO's are brighter than the most luminous galaxies, some QSO's very much so, and some indeed at least 100 times more luminous than the brightest galaxies.

PROBLEM:
A QSO and a galaxy have the same apparent brightness. The measured z value for the QSO is considerably greater than that of the galaxy, however. Satisfy yourself that, provided the method of Figure 14.13 does indeed give the correct value of d for the QSO, then intrinsically the QSO must be much brighter than the galaxy.

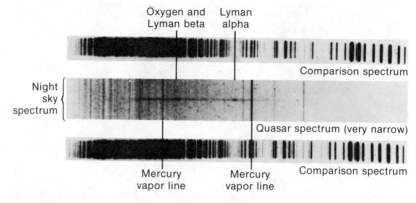

FIGURE 14.14.
The spectrum of a quasi-stellar object of very large redshift, OH 471 with z = 3.40. (Courtesy of R. F. Carswell and P. A. Strittmatter, Steward Observatory.)

The measured values of z for galaxies, whether ordinary galaxies or radio-galaxies, have rarely turned out to be greater than 0.3. On the other hand, values of z up to about 3.5 have been observed for quasars, an example being shown in Figure 14.14. It is reasonable to consider whether the straight-line relation of z and d, applicable at small z, can be extended to values as large as 3.5. Should the relationship of z to d become curved at large z? This is an important question, which we will consider in later chapters, but the answer to it does not significantly affect the issues we are discussing here.

It is more appropriate here to ask *why* should QSO's be appreciably brighter than galaxies? Start with an ordinary galaxy, and suppose that intense emission of radiation by a synchrotron process becomes operative in the compact central nucleus. Indeed, let the nucleus become so inordinately bright that when seen from a great distance the nucleus overwhelms the normal starlight of the galaxy. Seen from a distance only a central brilliant point of light can then be distinguished. Many astronomers believe this to be a QSO. In short, a QSO is a galaxy in which the nucleus has temporarily become exceedingly bright.

§14.3. Some Heterodox Considerations.

The picture developed above of the nature of QSO's would probably never have been questioned by anybody had it not been for the discovery that both the light and the radio emission from QSO's sometimes vary significantly on a time scale of a year or less. The observed variations are of such a nature that they must result from corresponding intrinsic variations in the QSO's themselves. The straightforward interpretation of this behavior requires the source of emission to be very compact indeed, but the radiation within the QSO's would then be so exceedingly intense that it would quench the high-speed electrons giving rise to the radiation itself—a case of killing the goose that lays the golden eggs. This difficulty could be avoided if the QSO's in question were not situated at the great distances obtained from the assumption of Figure 14.13—namely, that the z and d values of QSO's are related in essentially the same way as for galaxies—but were only at distances comparable to those of nearby galaxies.

As one of the people involved in this theoretical work, I was fully aware that such an argument does not disprove the usual picture of the nature of the QSO's. Difficulties of this kind can all too easily arise from a failure of one's own imagination. Yet the situation suggested that it would be wise to keep a weather eye open for any observational data which might suggest that the z values of QSO's are not related to their d values in the manner of Figure 14.13.

It was therefore of considerable personal interest that a radio source previously misidentified with the bright galaxy NGC 4651 turned out to be associated with a QSO situated some 3 arc minutes away from the galaxy itself, as can be seen in Figure 14.15. The probability of such an apparent association being due to chance was about one in ten, provided one formulated the problem in the following way. Given all the QSO's of the radio catalogue to which this one belonged, the 3C catalogue, and given all the galaxies in the New

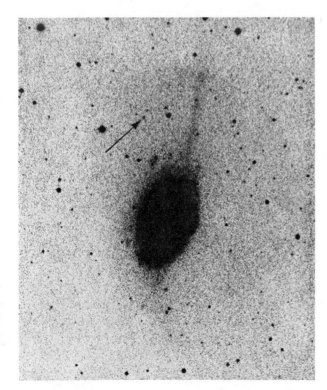

FIGURE 14.15.
The galaxy NGC 4651 with a nearby QSO of the 3C catalogue, 3C 275.1. (From A. R. Sandage, P. Veron, and J. Wyndham, *Astrophysical Journal,* 142, 1965, 1307. University of Chicago Press.)

General Catalogue (NGC), what was the probability of finding one of the QSO's within 3 arc minutes of one of the galaxies? The answer already given, about 1/10, was small enough to be interesting, but was not small enough to force one to believe that the association was not due to chance.

However, by 1970 three more cases had come to light. Together with NGC 4651 they are shown in Figure 14.16. In all four cases, the QSO's had 3C

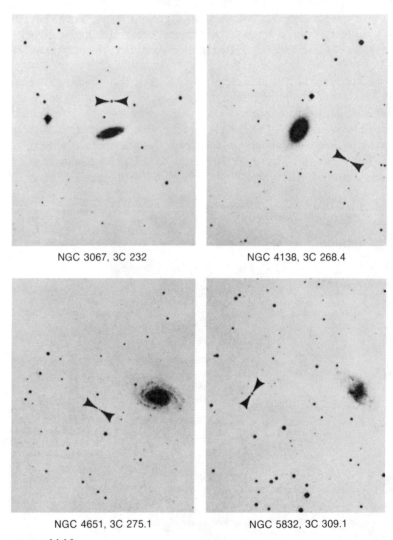

NGC 3067, 3C 232   NGC 4138, 3C 268.4

NGC 4651, 3C 275.1   NGC 5832, 3C 309.1

FIGURE 14.16.
Four cases in which QSO's of the 3C catalogue, having comparatively large redshifts, lie close in the sky to galaxies of the NGC catalogue. (From E. M. Burbidge, G. R. Burbidge, P. M. Solomon, and P. A. Strittmatter, *Astrophysical Journal*, 170, 1971, 233. University of Chicago Press.)

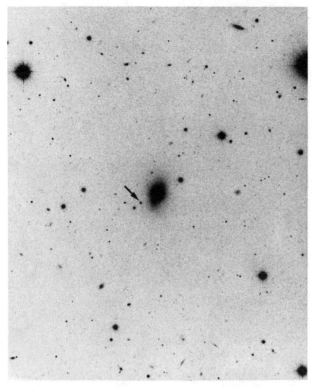

FIGURE 14.17.
A fifth case; here the QSO 3C 455 lies only 23 arc seconds from
the galaxy NGC 7413. (From H. C. Arp, E. M. Burbidge, C. D.
Mackay, and P. A. Strittmatter, *Astrophysical Journal*, 171, 1972,
L41. University of Chicago Press.)

designations and the galaxies belonged to the NGC catalogue. This permitted
the chance that all four apparent associations were accidental to be worked
out without undue difficulty or ambiguity. The chance that the associations
were accidental turned out to be less than one in 250, a result small enough
to demand serious consideration. The matter was under debate by astronomers
when a fifth case (shown in Figure 14.17) came along, again of a QSO from
the 3C catalogue and of a galaxy from the NGC catalogue. Taken by itself,
the chance that this fifth case is accidental is only one in 100; so the chance
that all five apparent associations are accidental is about one in 25,000. On
all normal ways of assessing probabilities, it became necessary at this stage
to say that the five associations of Figures 14.16 and 14.17 are *not* due to chance.

If one makes this assessment of the situation, one must argue that the
observed large z values of the QSO's do not arise in the same way as the z
values of galaxies. Working along these lines, one can appeal to the strong
gravitational fields which must in any case exist within the QSO's. Line
radiation from atoms situated deep inside a gravitational field experiences a

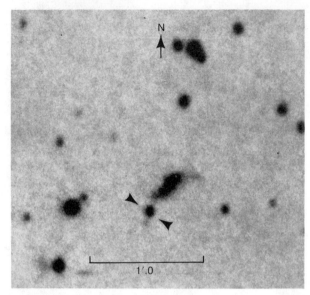

FIGURE 14.18.
The QSO with catalogue designation PKS 2020-370, only 11 arc
seconds away from a spiral galaxy in the southern sky. (Courtesy
of J. G. Bolton, Commonwealth Scientific and Industrial Research Organization.)

FIGURE 14.19.
The radiosource 1953-325, believed to be a QSO, lying 25 arc
seconds from the galaxy at the center of the field. (Courtesy of
J. G. Bolton, Commonwealth Scientific and Industrial Research
Organization.)

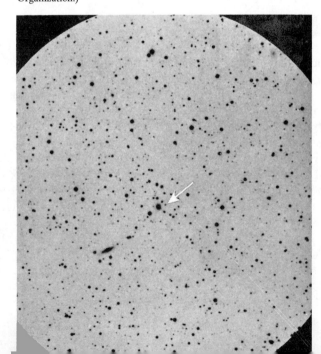

reduction of frequency as it emerges from an object. In principle, therefore, the observed frequency shifts of the QSO's, expressed by their z values, could be of gravitational origin, in which case it would be entirely wrong to attempt to infer d values in the manner of Figure 14.13. Astronomers have been reluctant to take this possibility seriously, however, largely because it has proved difficult to obtain a satisfactory understanding of the detailed emission of the line radiation. Although this reluctance may prove to be well-founded, it is in just such a situation that a failure of imagination may occur. We know comparatively little about the physical properties of strong gravitational fields, and here our ignorance falls at a sensitive spot.

Four further cases of QSO's lying near bright galaxies are shown in Figures 14.18 to 14.21. It is harder to calculate probabilities that these apparent associations are accidental than it was for the five cases of Figures 14.16 and 14.17 simply because the QSO's do not belong to a well-defined sample, and because the galaxies do not belong to the same catalogue. The two in Figures 14.18 and 14.19 come from the southern hemisphere. The two in Figures 14.20 and 14.21 differ from all the others, in that the QSO's are not radio sources. It is hard to know how to calculate a probability at all for Figure 14.21, since the system of NGC 3561 is very peculiar.

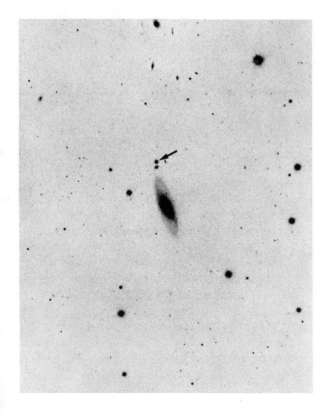

FIGURE 14.20.
A QSO that is not a radiosource, lying about 25 arc seconds from the galaxy IC 1746. (Courtesy of H. Arp, Hale Observatories.)

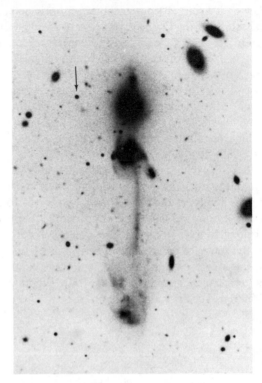

FIGURE 14.21.
A QSO that is not a radiosource, lying close to the remarkable system of NGC 3561. (Courtesy of H. Arp, Hale Observatories.)

§14.4. QSO Pairs

The two QSO's with catalogue designations 3C 286 and 3C 287 have z values of 0.849 and 1.055, respectively. On the distance interpretation of Figure 14.13, these two objects would be at significantly different distances and they would not be expected to be similar to each other, any more than any two QSO's are similar to each other. Yet in their optical and radio emissions, and particularly in the angular sizes of their patches of radio emission on the sky, these two objects are essentially identical twins. The close similarity of the angular sizes is especially noteworthy, because patches of radio emission as large as those of 3C 286 and 3C 287 are exceptional among the QSO's.

Two other seemingly connected physical pairs of QSO's, one with the designations Ton 155 and Ton 156, and the other denoted simply by 1548 + 115, have recently been investigated. Ton 155 and Ton 156 are about 35 arc seconds apart on the sky, and the two QSO's of 1548 + 115, shown in Figure 14.22, are only 4.8 arc seconds apart. The z values for these two

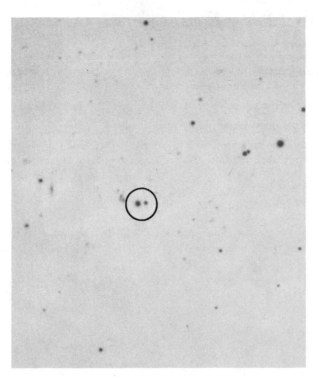

FIGURE 14.22.
The double QSO's of 1548 + 115 are separated by only 4.8 arc
seconds, but have very different redshift values. Notice the nearby
group of faint galaxies. This object is a triple radiosource. Negative photographs are much better for distinguishing objects as
faint as these. (Courtesy of H. Arp, Hale Observatories.)

TABLE 14.1.
Two pairs of quasi-stellar objects.

Name	Separation	Magnitude	Redshift (z)
Ton 155	35 arc seconds	16.9	1.703
Ton 156		16.4	0.549
1548 + 115			
Brighter component	4.8 arc seconds	about 17	0.4359
Fainter component		about 19	1.901

pairs are given in Table 14.1. Like those for 3C 286 and 3C 287, the z values
within each pair are quite different. Yet the chance that these two apparent
pairs are just accidental associations on the sky is small, again suggesting that
there are aspects of the QSO's that we do not yet understand.

General Problems and Questions

1. What are radiogalaxies?

2. What speculations have been put forward concerning the source of the energy released in the outbursts of radiogalaxies?

3. Describe the general properties of quasi-stellar objects (quasars).

4. How are quasi-stellar objects discovered?

5. Discuss the argument used to infer the distances of quasi-stellar objects, assuming their redshifts arise in the same manner as those of galaxies.

Appendix to Section V

Appendix V.1. The Counting of Radiosources

Just as the optical astronomer can measure the apparent magnitude of an object, the radioastronomer can measure the power from a radio source, by means of a radio receiver tuned at some chosen frequency. We are all familiar from our everyday experience with the general procedure to be used. We find that any practical receiver does not tune at just one exact frequency; it is always possible to hear a radio program even though the receiver is mistuned by a small amount. This small amount is known as the *bandwidth* of the receiver. What the radioastronomer does is to measure the power from a source using a receiver with a certain standard bandwidth.

In practice, the radioastronomer does not have a free choice of the frequency at which he works, simply because so much man-made radio emission is being generated all the time all over the Earth. There would be little chance of measuring the radio waves from a cosmic object if there were even very slight competition from a manmade source, since by everyday standards the power received from a cosmic source is very small indeed, just because astronomical distances are so very large. It has been estimated that all the radio power received by all the world's radiotelescopes operating for a decade would not raise the temperature of a spoonful of water by as much as a millionth of a degree. So what the radioastronomer must do is to work at some frequency which nobody else is using. Such special frequencies must be agreed upon by international arrangement. Indeed, radioastronomers have been allotted a few special frequencies, and it is at these special frequencies that all observations in radioastronomy must be made.

Using one of the frequencies allotted to him, the radioastronomer can measure the power received from a source. And he can survey the sky, or a portion of it, in order to count the number of sources that are more powerful than some assigned amount, say, S. That is, he counts the number of small patches of the sky that give power readings in his receiver that are greater than S. Denote this number by $N(S)$. How do we expect N to behave as S is varied in this experiment?

This question can readily be answered if we make the following simplifying assumptions:

1. All radiosources are intrinsically alike.

2. The radiosources are distributed uniformly in space.

3. The large-scale properties of spacetime are similar to its small-scale local properties.

Making a logarithmic plot of N versus S, we find that the expected behavior of N follows the simple straight line of Figure V.1. The number N increases as S becomes smaller, because smaller S implies increasing distance, and there are many more sources at large distances than there are closeby.

The results of the most recent survey, carried out at the National Radioastronomy Observatory at Greenbank, W. Virginia, are shown in Figure V.2. The observations agree very well with the expected straight line of Figure V.1, except at high and low values of S. Throughout a considerable intermediate range of S, the agreement is good. The departure of the observed counts from the strict straight line of Figure V.1 is significant and important at low values of S, but is not important at high values of S. The total number of sources affecting the high S end of Figure V.2 is about 400. The deviation from the expected straight line of Figure V.1 would be removed if about 40 more sources were added to this 400. Now the square root of 400, namely, 20, represents what is called a *standard deviation*, so about two standard deviations are involved in the deviation of the observations at high S from the expected straight line of Figure V.1. This is well within the range of a normal statistical fluctuation, and hence is of no far-reaching significance.

The tendency of the observed points to fall below the straight line of Figure V.1 at low values of S is not a statistical fluctuation, however. The number $N(S)$ of sources at low S is so large that fluctuations have little effect. Rather must the falloff of the observed points be ascribed at low S to a failure of

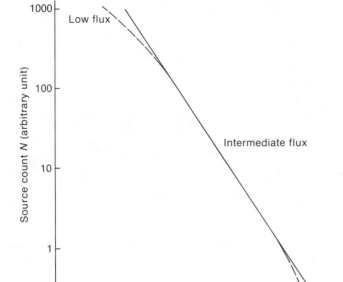

FIGURE V.1.
The straight-line behavior of the number N of radio-sources more powerful than some assigned S, expected on the assumptions of intrinsic similarity and of uniformity of distribution, and on the hypothesis that spacetime has large-scale properties similar to its small-scale local properties. Evidence for this kind of behavior is shown in Figure V.2.

one, or more, of the above three assumptions. It is usually supposed that, at the large distances implied by low values of S, it is (3) which fails. That is to say, the geometrical structure of spacetime, when taken on a large-enough scale, is different from the geometry of which we have local experience. This problem will form one of the main topics of the next section, where we will be concerned with the structure of the universe in the large. We end this section by discussing a strange and mysterious situation, one which springs on us in the very place where the observations of Figure V.2 seem to fit the expected line of Figure V.1, at the intermediate values of S.

To come to grips with this issue, we notice, first, that most of the sources contributing to $N(S)$ are radiogalaxies. The QSO's make about a 15 per cent contribution to the source counts, which is not sufficient to have a critical effect on the behavior of N with respect to S.

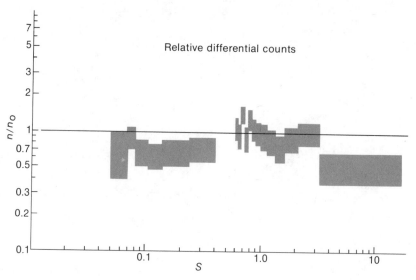

FIGURE V.2.

Radiosource counts in which the range of S has been divided into bins, in such a way that equal numbers of sources would be found in each bin *if* the distribution were of the form of the straight line of Figure V.1. The lefthand scale gives the ratio of the actual count n for each bin to a constant number n_0 which is arbitrarily chosen. The bin at the highest range of S has a deficit of sources. Otherwise there is no appreciable deviation from the straight-line situation of Figure V.1 until the lefthand row of bins is reached. The results for these left-hand bins show that $N(S)$ falls off at low S, as indicated by a dotted line in Figure V.1. (From K. Kellerman, M. Davis, and I. Pauliny Toth, *Astrophysical Journal*, 170, 1971, L1. University of Chicago Press.)

Now, for galaxies, we can be assured that measurements of z values lead to the d values given by using the line of Figure 14.13. (Notice that this line is established from pairs of z and d values derived from observations of galaxies. Then, with the line established, we no longer trouble ourselves to measure d values any more. Since measuring z is relatively easier than measuring d, what is done—with z known—is simply to read off d in the manner of Figure V.3.) If z values were available for all the radiogalaxies counted at intermediate S values, those for which the behavior of $N(S)$ agrees with the line of Figure V.1, we would thus know the d values for all these radiogalaxies. This would yield the general distance scale corresponding to the intermediate S values, and would enable us to assess assumption (3) in relation to this distance scale.

What has actually been done so far is to measure z values for a modest fraction of the radiogalaxies which contribute to N at the intermediate values of S. These measured z values lead to quite large d values, indeed, to d values so large that it is a surprise that assumption (3) would seem to hold good on so large a scale. Such a situation would go contrary to the considerations to

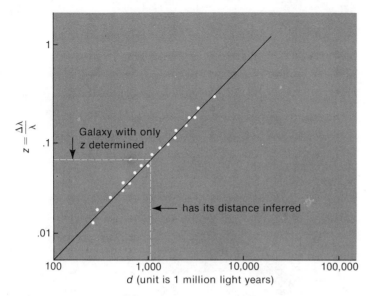

FIGURE V.3.
With the relation between z and d established by observing a number of galaxies, distance values for other galaxies are simply read off from the resulting curve, once an observed redshift is obtained.

be given in Chapter 16. It would involve a major reassessment of all our ideas on cosmology, a radical position which astronomers would be reluctant to accept unless the alternatives were first eliminated. The alternative most favored by astronomers comes from abandoning (2), the second of the three assumptions used in arriving at the straight line of Figure V.1. Then we could not arrive at this straight line (which actually fits the data at the intermediate S values) *unless* (3) *were also invalid*. The idea is to have *both* (2) and (3) invalid, but in such a way that the expected behavior of N with respect to S continues at the intermediate values of S to follow the straight line of Figure V.1—a fortuitous compensation for two errors in the previous argument!

The sense in which (2) is required to break down is that we must have a higher density of sources at large distances than at smaller distances. Does this mean that our own galaxy is situated at the center of some kind of hole in the distribution of radio sources? Yes, but not a spatial hole, a *time hole* of the kind illustrated in Figure V.4. As we look to greater distances, we look farther back in time. We require sources to have been more frequent in the past than today. In the manner of Figure V.4, we still preserve uniformity in space *at any given moment of time.*

This favored interpretation of the situation is often referred to as an *evolutionary universe*, because the distribution of radio sources, although uniform

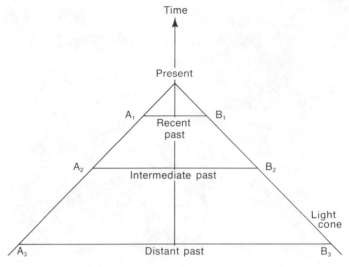

FIGURE V.4.
The requirement for a higher density of sources at larger distance rather than at smaller distance does not necessarily imply any lack of spatial uniformity. We can see more sources at large distances because we are then looking farther backward in time. Here the density of radiosources is the same at A_1 as at B_1, but is less than at A_2 and B_2 which in turn are less than at A_3 and B_3.

at any given moment, is considered to change with time. As a means of fitting the so-far available z value measurements for a fraction of the radiogalaxies to our ideas of the permitted range of local geometry, and to the observations of Figure V.2 at intermediate S values, such an evolutionary universe has one peculiar defect, namely, the fortuitous compensation mentioned above. At present, there is no plausible theory to explain this compensation. The usual view is that the compensation happens accidentally.

A less usual view is that, unlike the minority for which z values are available, most of the sources with intermediate S values are not very far away, and so their optical faintness arises because they happen to be intrinsically faint. Although this further possibility contradicts a mystique that has grown up in radioastronomy, namely, that all radiogalaxies are optically very bright and similar to each other, it satisfies the facts without requiring any coincidence. At present, it is the most straightforward way to resolve the problem. It amounts simply to taking the straight-line part of Figure V.2 at its face value, as showing us that most of the sources at intermediate S values are comparatively closeby, within the range of applicability of the local structure of spacetime. If this is so, the observations of Figure V.2 cannot be considered to have far-reaching consequences for the study of the large-scale properties of the universe.

General Problems and Questions

1. What is the $N(S)$ relationship obtained from the counting of radiosources?

2. If a logarithmic plot of the observed $N(S)$ relationship had turned out to be a certain straight line, what inferences would you have drawn?

3. What inferences can in fact be drawn from the observed $N(S)$ relationship?

4. What is an evolutionary universe?

References for Section V

For general reading, see *Frontiers in Astronomy: Readings from the Scientific American* (San Francisco: W. H. Freeman and Co., 1970).

For a discussion of "astrochemistry," see D. Buhl, "Molecules and Evolution in the Galaxy," *Sky and Telescope*, 45 (1973), 156, and P. M. Solomon, "Interstellar Molecules," *Physics Today*, 26, no. 3 (1973), 32.

Many of the early papers concerning quasi-stellar objects appeared in *Quasi-Stellar Sources and Gravitational Collapse*, edited by I. Robinson, A. Schild, and E. L. Schucking (University of Chicago, 1965).

The discovery of pulsars was reported by A. Hewish, S. J. Bell, J. D. H. Pilkington, P. F. Scott, and R. A. Collins, "Observation of a Rapidly Pulsating Radio Source," *Nature*, 217 (1968), 709.

The discovery that regular light variations come from the pulsar in the Crab Nebula was made by W. J. Locke, M. J. Disney, and D. J. Taylor, "Discovery of Optical Signals from Pulsar NP0532," *Nature*, 221 (1969), 525.

Section VI:
Cosmology

Georg Bernhard Riemann, 1826–1866.
Courtesy of Foto Schubert, Göttingen.

Chapter 15:
Universal Geometry
and Cosmology

In this section we will study the universe as a whole, which is the subject of the field of *cosmology*. We will begin by setting forth the main arguments, so that the reader will be able to see the general plan of this section; the reader will also find it helpful to refer back to Appendix II.4.*

We cannot be sure that the rather simple properties of space and time that we know about from local laboratory experiments will continue to hold true on the scale of the whole universe. However, if we admit the most general form of large-scale geometry that can be studied with the aid of advanced mathematics, trying to understand cosmology becomes an exceedingly difficult task. In cosmology, therefore, we seek a middle road between these two extremes. We admit that universal geometry may be more complex than local geometry, but we require it to satisfy certain sensibly chosen simplifications, which are generally referred to as "principles." This compromise avoids the rather horrifying complexity of the most general situation, but still leaves a considerable range of possibilities for how the universe may be structured. Since the ultimate objective of the cosmologist is to discover a unique geometry for

*In general, I have been able to keep the main text independent of the Appendixes so far, but here we have run into an exceptional situation. The reader may also find it advantageous to refer to Appendix V.1.

the universe, criteria are sought that can restrict this range of possibilities still further. One form of restriction comes from Einstein's theory of gravitation. The other comes from observation, from what is called "observational cosmology."

The preceding paragraph comprises the topics to be considered in this chapter. We will find that this method of attack has been only partially successful: although the range of possibilities has indeed been greatly restricted, the method has not yet succeeded in indicating a single, unique structure that the universe must have.

A common property of all the possible structures that remain after the criteria mentioned above have been applied is that the universe possessed an "origin." In Chapter 16 we will consider the implications of this result. Throughout Chapter 16 nothing differs in principle from the ideas held by most astronomers and physicists, although a new technique for presenting the argument is introduced. This technique permits questions to be asked that would otherwise be quite outside the range of an introductory text. To understand the nature of this new technique, we should first notice that in the usual discussions all the possible structures mentioned above depend on assuming that the masses of particles are fixed and immutable. This constancy is advantageous for discussing most physical problems. However, we can make a trade: we can give up the simplicity of having the particle masses immutable in order to achieve a simpler form of geometry. There is an *exact* mathematical equivalence,

Particle masses constant + complex universal geometry $\equiv$

Particle masses variable + simpler universal geometry,

provided the variability of the particle masses satisfies certain precisely defined mathematical rules. A question now arises: how simple can the universal geometry be made? It turns out, again by exact mathematics, that *the simpler universal geometry can be identical with local geometry,* the same geometry that was studied in Appendix II.4. It is this far-reaching geometrical simplification that will permit the discussion to be extended outside what can usually be covered in an introductory text.

All this we shall encounter in Chapter 16. At the end of Chapter 16, we will reach a situation concerning the "origin" of the universe which the reader may or may not regard as satisfactory. If the situation appears satisfactory, there is no need to proceed any further, but if the reader—like the author—feels that the situation appears incomplete, then two further short chapters must be considered. Chapter 17 discusses what is known as the "steady-state" model, a heterodox form of cosmology, considered by most astronomers to be contradicted by certain observed facts; in Chapter 17 we will consider how far this is true. Chapter 18 also differs from the usual ideas about cosmology, but along lines very different from those of the steady-state model. What this difference is we may leave over until Chapter 18 is reached.

Let us return now to the lefthand side of the above equivalence, and take the usual view that particle masses are fixed. It is of interest here to ask what is meant by "local." Explicitly, what is the scale beyond which local geometry must be abandoned in favor of a more complex geometry? The answer is that local geometry can be taken as remaining approximately valid to the distances of the most remote galaxies plotted in Figure II.59, reproduced here as Figure 15.1, or in the more recent diagrams given by Sandage, as shown in Figures 15.2 and 15.3. The largest observed z values for these galaxies are about 0.2, corresponding to a distance of about 10^{27} cm. This is the scale on which local geometry can be considered to remain approximately valid, with the galaxies possessing the Doppler expansion velocities discussed in Appendix II.8. Although this scale may seem exceedingly large compared with the distances we have been concerned with until now, from here on we must think about still greater distances, 10^{28} cm. or more. On this larger scale, we must broaden our geometrical ideas, and *we must no longer continue to think in terms of Doppler motions for the galaxies,* since the concept of Doppler motions belongs to local geometry. The reader must be on his guard about this last point. Once local geometry is abandoned, Doppler motion, in the sense discussed in Appendix II.8, must be abandoned also. Failure to appreciate this point will lead to serious conceptual difficulties.

§15.1. The "Principles" of Cosmology

Because of the enormous generality of the subject of cosmology, we make two very broad assumptions at the outset.

1. We assume that the local geometrical rules found by actual experiment (and discussed in Appendix II.4) would also have been found to apply locally if we had lived in any other place or at any other time. This assumption is known as the *principle of equivalence.**

2. We assume that, when observed on a large enough scale, the universe is both isotropic and homogeneous. This means there are no distinguishable spatial directions and no distinguishable spatial positions, except insofar as directions and positions can be based on a particular local distribution of galaxies—e.g., the Coma cluster of galaxies, or NGC 7603. An observer may know where he is from some local peculiarity, such as the remarkable structure of NGC 7603 (Figure 14.10), but he cannot know where he is by observing the large-scale distribution of galaxies. This assumption is known as the *cosmological principle.* Notice that the restrictions it imposes are *spatial,* not temporal. Since the large-scale distribution of galaxies may change with time, a long-lived observer could discover what epoch he is living in from the large-scale distribution which the galaxies happen to have at a particular moment.

---

*There are other, apparently quite different ways of stating the principle of equivalence, but such ways can be shown to be really the same as the above statement.

The assumptions set out above are not inevitable or unique in any way. Although nobody, so far as I am aware, has ever suggested a departure from (1), departures from (2) have indeed been proposed. We can, for instance, dispense with isotropy but retain spatial homogeneity. An observer would then be able to discern a directional structure in the universe, but would still be unable to discover variations from place to place, except in the local sense discussed above. This would imply that the large-scale features of the universe had a kind of crystalline structure. Such a possibility was considered some years ago by Kurt Gödel.

Another possibility, which has received widespread discussion, was proposed some 25 years ago by Hermann Bondi and Thomas Gold. Instead of making assumption (2) less restrictive, as Gödel did, they made it more restrictive, by denying the possibility of an observer determining the epoch at which he lives. According to Bondi and Gold, there are to be no changes of the large-scale structure of the universe with respect to *time*; the universe on a large scale is to be featureless both spatially and temporally. When formulated in this way, assumption (2) was referred to by Bondi and Gold as the *perfect cosmological principle*. It led to what became known as the *steady-state model* of the universe. This model turned out to be fertile in its consequences, and will be considered later, in Chapter 17. For the present, we shall continue with assumptions (1) and (2) in the forms stated above.

§15.2. Universal Time

Suppose an observer in each galaxy sets up a system of local time measurement, in the way discussed in Appendix II.4. In practice, whenever we set up a clock we always have in mind some reference point, often referred to as a zero point, relative to which time is measured, because we cannot measure a moment of time; we measure intervals of time. In modern astronomy the zero point is usually set at 0.0 h on Jan. 1, A.D. 1950. Although such a parochial way of setting a reference point is satisfactory enough for purely local purposes, and although observers in other galaxies could also have highly individual ways of setting their zero points, such a procedure would not be at all useful should the observers hope to correlate their experiences, and eventually to build a picture of the large-scale structure of the universe. Is there a procedure, then, for setting the zero points in a way that would be uniform for all such observers?

To deal with this question we consider that, although our system of local geometry may well fail for very distant galaxies, our local geometry remains valid for galaxies that are not too far away. For these, we have strong evidence, in Figures 15.1, 15.2, and 15.3, for the relation $v = Hd$, with v obtained from cz, the quantity z being given by $z = \Delta\lambda/\lambda_{\text{lab}}$ (see Chapter 14), and with d obtained in the manner discussed in Chapter 8. Hence by determining v and d values for a number of nearby galaxies, the Hubble constant H is obtained.

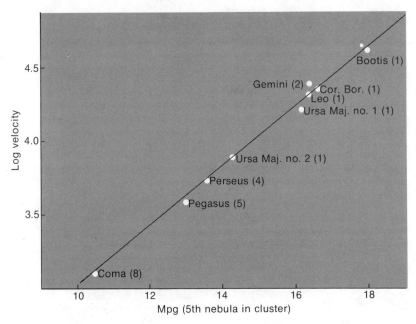

FIGURE 15.1.
The relation between apparent magnitude m and the logarithm of the Doppler velocity v, obtained by Hubble and Humason.

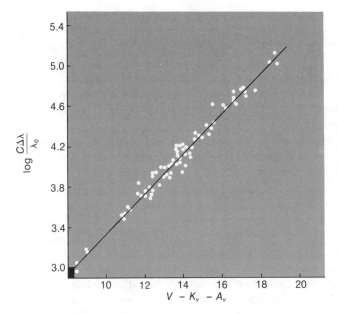

FIGURE 15.2.
Since $z = \Delta\lambda/\lambda_0$, the lefthand scale is $\log(cz)$. The bottom scale gives the apparent visual magnitude of the galaxies (the brightest galaxy in each of 84 clusters), after certain corrections represented by K_v, A_v, have been applied. (From A. R. Sandage, *Astrophysical Journal*, 178, 1972, 12. University of Chicago Press.)

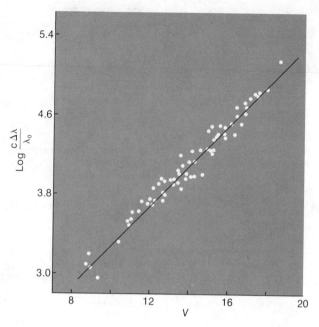

FIGURE 15.3.
A diagram similar to Figure 15.2 for radiogalaxies. For certain technical reasons, the corrections K_v, A_v, have not been applied in this case. (From A. R. Sandage, *Astrophysical Journal*, 178, 1972, 12. University of Chicago Press.)

According to the recent work of Sandage, described in Chapter 8, the reciprocal of H, $1/H$, is about 18 billion years. From $v = Hd$, it is seen that $1/H = d/v$, which is the length of time *going backward* that would be required to bring all the nearby galaxies together into a clump *if they have always been moving apart at their present speeds.* This interpretation of the meaning of $1/H$ is illustrated in Figure 15.4.

A similar determination of H carried out at a different epoch would in general yield a different value. The distance d of a galaxy increases with time. Hence H must change unless v increases in exactly the same way as d, which v does only in the steady-state model. In the other cosmological systems at present under consideration the value of H changes with the epoch. In the future it will be less than it is now. In the past it was greater than it is now. According to the "principles" of cosmology, these changes of the Hubble constant are to be the same for all observers in all galaxies. By using the variability of H, it is possible for all such observers to adjust the zero points of their clocks in a systematic way. We could imagine the observers fixing on a particular value of H through an interchange of messages. Every observer could agree, for example, to set the zero point of his clock at the moment when

$1/H = 18 \times 10^9$ years, although they would need to specify the "year," not by the Earth's motion around the Sun, but by a specified number of oscillations of the radiation emitted in a specified transition of a specified atom.

A correlated time system of this kind, illustrated in Figure 15.5, is known as a system of *universal time*. From here on we shall use the concept of time in this sense—or in an equivalent way.

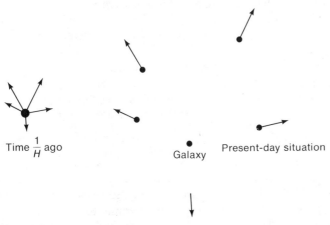

Time $\dfrac{1}{H}$ ago

Galaxy   Present-day situation

FIGURE 15.4.
The quantity $1/H$ is the length of time, going backward, that would be required to bring all the nearby galaxies together if they have always been moving apart at their present speeds.

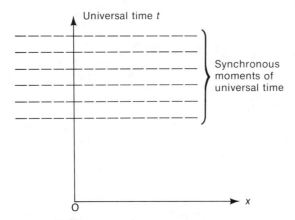

FIGURE 15.5.
A system of universal time can be determined provided each observer sets the zero point of his clock in an agreed upon way.

The first of the two principles in §15.1 restricts the geometry of the entire universe to a class of geometries first studied by Georg F. B. Riemann (1826–1866). Hence we require the universe to satisfy *Riemannian geometry*. However, this first principle, the equivalence principle, tells us nothing about how to select one of the infinitely many geometries of this kind. The equivalence principle is thus useful in directing our attention to an appropriate class of geometries, but it is unable to help us discover the appropriate member of this class. To proceed further, we must appeal to the second principle, the cosmological principle, which restricts the possible universal geometry still further, in the following way.

Suppose we consider the positions $G_1, G_2, \ldots, G_n$, at a certain moment of universal time, say, t, of a set of n galaxies, and suppose we join G_1 to G_2, G_2 to $G_3, \ldots, G_{n-1}$ to G_n, G_n to G_1, by straight lines to form a polygon. We can also do the same thing for the same galaxies at a different moment of time, say, t', thereby obtaining a second polygon. What the cosmological principle permits one to deduce (by a fairly sophisticated piece of mathematics) is that the second polygon must have the *same shape* as the first polygon. What we *cannot* deduce, however, is that the two polygons have the same scale; the scales can be different, as in Figure 15.6. The situation is that our procedure for setting up a system of universal time permits us to construct a polygon from any set of galaxies, and this polygon has a definite shape whatever the moment of universal time at which we choose to construct it.

As a special case of what has just been said, we can think of a set of three galaxies. The polygon then becomes a triangle. Can we apply the usual rules of Euclidean geometry to such a triangle? Particularly, could we have a right-angled triangle with Pythogoras' theorem applying to it? The cosmological

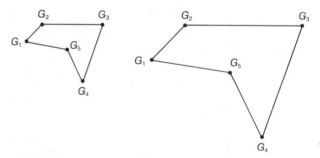

FIGURE 15.6.
Under the usual postulates of cosmology, the shapes of two polygons, obtained by joining the spatial positions of a set of galaxies at two different moments of universal time, must be the same. But the usual postulates of cosmology do not require the two polygons to have the same scale.

principle does not permit us to answer either "yes" or "no." What can be proved is that the geometry of such a triangle must be one of three types, which we may denote quite simply by the letters A, B, C. One of these, say, A, is indeed the geometry of Euclid, $A \equiv$ Euclid. The geometries B and C are similar to that of Euclid when the triangle is small, but are not at all the same as that of Euclid when the triangle is large.

From our starting principles, the principle of equivalence and the cosmological principle, we can go no farther than this. We have no means to distinguish between A, B, and C. Additional progress therefore depends on further precepts, or on some new physics, or on actual observations of a new kind, or on some combination of these possibilities. All such possibilities have in fact been tried, sometimes separately and sometimes interlinked with each other. The perfect cosmological principle introduced by Bondi and Gold is an example of an additional precept. Here we concern ourselves with the second and third possibilities, which we now discuss in turn.

§15.4. Einstein's Theory of Gravitation

By "new physics," we mean physics that goes outside the range of our local experience. Physics covering our local experience has already been incorporated within the principle of equivalence. It might therefore be thought that, since gravitation falls within local experience, nothing useful for understanding "new physics" can be learned from gravitation. But the local gravitational effects that have been studied are always weak, that is, cannot generate speeds of motion comparable with that of light for material bodies. Thus planets move with speeds that are only of order 0.01 per cent of that of light, and the stars in our galaxy move at speeds of order 0.1 per cent of that of light—still small. Nor can gravitation in our locality modify significantly the speeds of particles that are already moving close to the speed of light. By a strong gravitational field, we mean the converse: a strong field could generate, or significantly modify, speeds of motion close to that of light for material systems. Strong gravitation does not fall within our local experience, and it is from strong gravitation that something new is to be learned.

The sense in which gravitation acts is clear: it will slow down the expansion apart of the galaxies. Let us return to the polygon formed by a particular set of galaxies. We saw that a sequence of such polygons, constructed at different moments of universal time t, all have the same shape but may differ in scale. We can represent the scale by a quantity Q which varies as t changes. This variability we denote by $Q(t)$. We choose some moment as standard, say, t_0, and take the polygon at this moment as a reference scale with which to compare the polygons at other moments of time. Thus we arrange $Q(t)$ to be such that $Q(t_0) = 1$. (For example, we could take the epoch in which we are now living as our t_0.) Because the galaxies are separating from each other, $Q(t)$ will then be less than unity for t earlier than t_0, and greater than unity for t later than

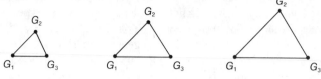

FIGURE 15.7.
Because the galaxies are separating apart from each other, the triangle formed by joining three galaxies was smaller in the past than it is now. In the future the triangle will be larger than it is now.

t_0. This is illustrated in Figure 15.7. The problem we have next to formulate is contained in the question: How will gravitation affect the behavior of Q? This question can be answered in general qualitative terms, in the following way. In Figure 15.8, we have a curve which we imagine to represent $Q(t)$. It will be seen that as t increases, the curve always turns in a clockwise sense. This is the qualitative meaning to be attached to the concept that gravitation slows the motions of the galaxies. If we drew a curve of the form shown in Figure 15.9 the opposite situation would arise, with the curve turning in an anticlockwise sense. Here would be an acceleration of the galaxies apart from each other. Since gravitation is an attractive force, the behavior of Figure 15.8 would seem correct.

Suppose we extend the range of t shown in Figure 15.8. With the curve of $Q(t)$ always turning in a clockwise sense, could we have the kind of situation shown in Figure 15.10, where $Q(t)$ passes through a maximum and then decreases with increasing t? Such a curve would imply that gravitation is strong

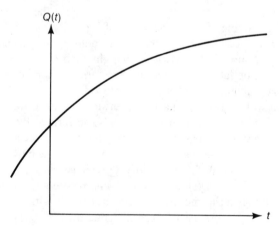

FIGURE 15.8.
Schematic form of the curve representing the behavior of the scale factor $Q(t)$ with respect to the time t, in which the curve always turns in a clockwise sense.

enough to put an eventual stop to the separation of the galaxies. This would happen at the time of the maximum of our curve, the time denoted by $t = T$ in Figure 15.10. Thereafter gravitation causes the galaxies to move toward each other, just the opposite situation from that which we actually observe at the present epoch.

This discussion shows both the strength and the weakness of qualitative arguments. A good qualitative argument serves to clarify concepts and to

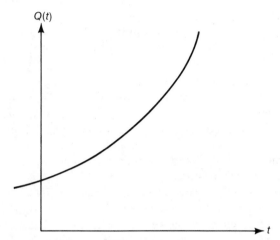

FIGURE 15.9.
Schematic form in which the curve of $Q(t)$ always turns in an anticlockwise sense.

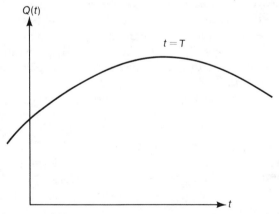

FIGURE 15.10.
With the curve always turning in a clockwise sense, $Q(t)$ could increase to a maximum and then start decreasing again.

formulate significant questions, but it mostly fails to answer the very questions which it raises. Qualitative considerations alone will not answer the question of the preceding paragraph. To proceed further we need a quantitative theory of gravitation.

From the study of local phenomena, two quantitative theories of gravitation have been formulated, the theories of Newton and Einstein. For over two hundred years Newton's theory, which we discussed in some detail in Chapter 9, was believed to represent all local effects with complete accuracy, but late in the last century and early in the present century, certain very small effects were found which could not be explained by Newton's theory. These were discussed in Chapter 10. The long axis of the orbit of the planet Mercury was found to possess a small rotation of about 43 arc seconds per century which could not be explained in terms of the Newtonian attraction on Mercury of the Sun and planets. This rotation turned out to be a natural consequence of Einstein's theory when this was formulated in 1915. Einstein's theory also predicted that light from distant stars would be appreciably deflected in passing the Sun, and this prediction was verified in 1919. Although this first verification was not too precise, many subsequent observations—especially in recent years—have shown that light and other forms of radiation behave very much as Einstein's theory requires them to do.

For these reasons alone, it would be natural to choose Einstein's theory rather than Newton's in seeking to answer our question concerning the behavior of $Q(t)$. There is a further reason, however, to which most scientists attach great weight. Einstein's theory possesses great intellectual strength. This is an essential requirement if we are to extend its range of application from local physics to the universe in the large—i.e., from weak gravitational effects to strong effects. Such an extension has of necessity to be an article of faith. It is essential that the subject matter of an article of faith be of great depth and subtlety. Einstein's theory conforms to this requirement. Nevertheless we could be making an error in this step. I used to think so myself, but for reasons to be discussed in Chapter 18 I now believe the case for extending Einstein's theory to the universe in the large to be very strong indeed. Since this is the general point of view—and it is the general view we are considering here—let us proceed on this basis.

Einstein's theory places further restrictions on the large-scale structure of the universe, but it still does not suffice to lead to a unique choice for the geometry of the universe. Particularly, it does not settle the form of geometry to be applied to our galaxy polygons, the forms we called *A, B, C* ($A \equiv$ Euclid). It does show, however, that if *A* is the appropriate form, then the curve of $Q(t)$ does *not* turn over in the manner of Figure 15.10. The galaxies continue forever expanding apart from each other. Indeed, in this case Einstein's theory determines an explicit result for $Q(t)$, shown in Figure 15.11. *By using the moment when $Q = 0$ as the zero point for fixing the clocks of all observers in all galaxies,* which is equivalent to the procedure suggested in §15.2, it can be shown that Q is proportional to $t^{2/3}$ [i.e., to the square of the cube root of t: $(t^{1/3})^2$].

FIGURE 15.11.
The behavior of $Q(t)$, according to Einstein's gravitational theory, for a large-scale geometry of type A. Here the unit for $Q(t)$ is chosen so that $Q = 1$ at $t = 1$.

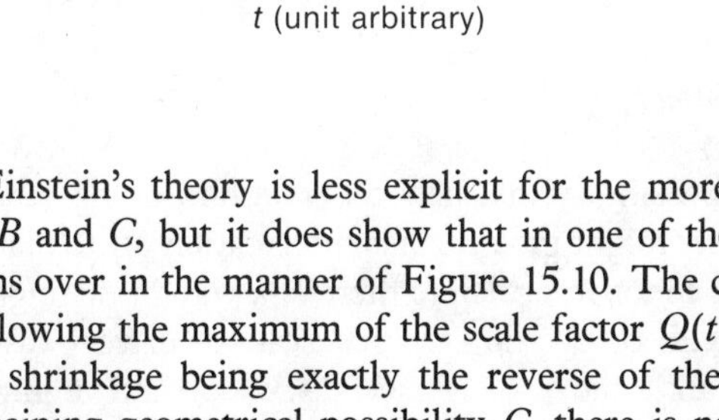

FIGURE 15.12.
The behavior of $Q(t)$, according to Einstein's gravitational theory, for the three forms of large-scale geometry: A, B, and C. Here again the unit for $Q(t)$ is to be chosen so that $Q = 1$ at $t = 1$, but this will require a different scale on the vertical axis for each of the three cases.

B

1   2

t (unit arbitrary)

Einstein's theory is less explicit for the more difficult geometries denoted by B and C, but it does show that in one of them, say, B, the curve of $Q(t)$ turns over in the manner of Figure 15.10. The curve has a symmetrical form. Following the maximum of the scale factor $Q(t)$, the galaxy polygons shrink, the shrinkage being exactly the reverse of their earlier expansion. For the remaining geometrical possibility C, there is no reversal of $Q(t)$. The three cases are shown together in Figure 15.12, it being supposed that the convention is adopted of setting the zero points of all clocks so that $t = 0$ corresponds to the moment when $Q = 0$, and that this is done irrespective of whether the geometry is A or B or C. Notice that, whereas the behavior of Q for geometry

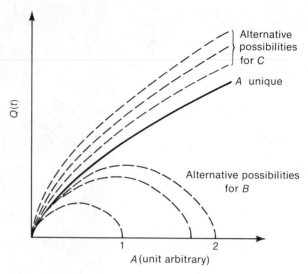

FIGURE 15.13.
The behavior of $Q(t)$ is not uniquely determined when the large scale geometry is either of type B or of type C. Here again unit is chosen to make $Q = 1$ when $t = 1$.

A is definitely known, many possibilities remain for geometries B and C. Although we can say that B and C have the general properties shown in Figure 15.12, Einstein's theory does not permit us to distinguish between the possibilities shown, for example, in Figure 15.13.

We shall restrict much of our discussion to case A, partly because this case has a unique behavior of the scale factor Q, partly because its Euclidean treatment of the galaxy polygons is simple, and partly for certain very general reasons which emerge in Chapter 18. This model for the universe is often referred to as the *Einstein–de Sitter model,* and we shall usually refer to it by this name.

§15.5. Observational Cosmology

We still have to consider the possibility that new observations, going outside our local region of the universe, might give further information on the large-scale structure of the universe. Since in principle there is a way to distinguish between the geometrical forms $A, B, C,$ using the observations of very distant galaxies, this is both an important topic of theoretical discussion and an important practical program. We end this chapter by considering the situation in some detail.

In Appendix II.8 we saw how to represent galaxies in a diagram with observed apparent magnitudes plotted horizontally and with the Doppler velocity v plotted logarithmically as ordinate. Similar diagrams devised by Allan Sandage were shown in Figures 15.2 and 15.3. It is undesirable now to plot Doppler velocities, however, because the Doppler interpretation of the observed redshift of the galaxies belongs to local geometry. We wish now to extend our discussion to much greater distances. All we need do to avoid using $v = cz$ is to plot the observed redshift z itself.

The nature of the redshift has been discussed previously, but it will be useful to rediscuss the matter. Characteristic line radiations from atoms, the H and K lines of singly ionized calcium atoms, for example, are observed for a particular galaxy. Denote the measured frequencies by v'_H and v'_K respectively. These frequencies are different from those observed for ionized calcium atoms in the terrestrial laboratories, say, v_H, v_K. However, the ratio v'_H/v'_K is the same as v_H/v_K. Define $1 + z = v_K/v'_K = v_H/v'_H$. Since v_K is larger than v'_K, and v_H is larger than v'_H, the "redshift" z is a positive number.

On the assumption that all the galaxies under investigation are intrinsically similar to one another, the Einstein-de Sitter model leads to the relation between z and apparent magnitude M shown in Figure 15.14.* In the next

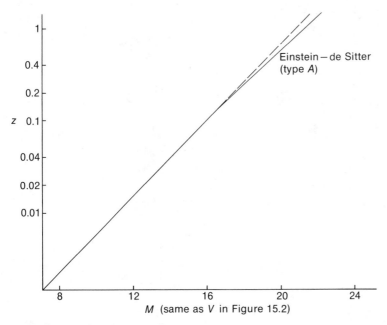

FIGURE 15.14.
The relation between z and M when the large-scale geometry is of type A (the Einstein-de Sitter model).

---

*In earlier chapters m was used to denote an apparent magnitude. Here we switch to M so that m can later be used to denote the mass of a particle.

chapter we will see how to obtain this particular relation. At small values of z the curve of Figure 15.14 is essentially the 45° line of Figure 15.2, but at large z the curve leans to the right of this line.

There are corresponding curves for geometries of classes B and C. Those of class C lean still farther to the right of the 45° line, between the curve of Figure 15.14, drawn again in Figure 15.15, and a second curve also drawn in Figure 15.15. Curves for geometries of class B take the form shown in Figure 15.16. At large z such curves lie markedly above that of the Einstein–de Sitter model. Thus the curve of the Einstein–de Sitter model separates those of class B from those of class C, as is shown in Figure 15.17.

The 45° part of Figure 15.17 applies within the limited range of local geometry. At larger distances there is in general a departure from this straight line, depending on the nature of the large-scale geometry. If indeed the galaxies observed at great distances were found to lie on the 45° line, as appears to be indicated by Figures 15.2 and 15.3, and if the galaxies were indeed intrinsically similar to each other, then the world geometry would be required to be

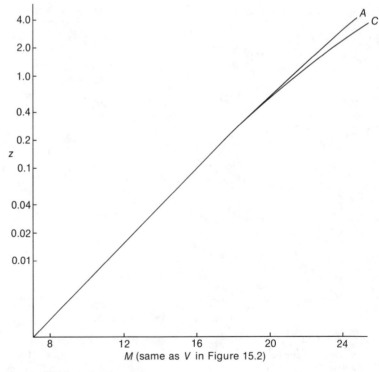

FIGURE 15.15.
When the large-scale geometry is of type C, the relation between z and M is a curve lying below that of Figure 15.14.

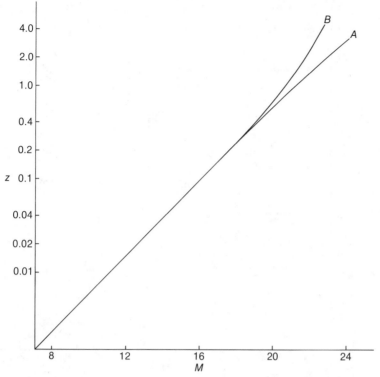

FIGURE 15.16.
When the large-scale geometry is of type B, the relation between z and M is a curve lying above that of Figure 15.14. Otherwise, the relation is not determined in detail.

of class *B*. Some astronomers do indeed believe Figures 15.2 and 15.3 show the large-scale geometry of the universe to be of class *B*, and hence that $Q(t)$ has the form of Figure 15.10. This is the basis of statements which are sometimes made, to the effect that the expansion of the system of galaxies will eventually cease and will be replaced by a contraction of the universe. However, many other astronomers feel that the difference between the 45° line and the curve of Figure 15.14 is so small, even for z as large as 1/3, that this conclusion cannot be considered reliable. Small intrinsic differences between the observed galaxies in Figures 15.2 and 15.3 would vitiate the conclusion.

If observations could be carried to values of z significantly larger than those of the galaxies in Figures 15.2 and 15.3, it would be less difficult to decide this matter, because the curve of Figure 15.14 for the Einstein–de Sitter model leans more and more away from the 45° line as z increases. If the 45°-line dependence of Figures 15.2 and 15.3 were found to be maintained up to $z = 1$, for example, the argument for regarding the universal geometry as being of

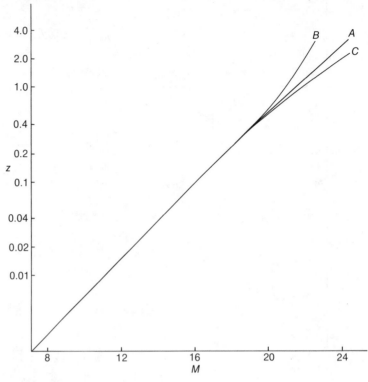

FIGURE 15.17.
The form of the z-M relation for the three types of world geometry.

class B would be strong. Such an extension to large z is technically difficult to achieve for galaxies, simply because galaxies at large z are very distant and are therefore exceedingly faint.

When the quasi-stellar objects (QSO's) were first discovered, it was thought that they would soon lead to a resolution of this issue, for the reason that many QSO's were found to have the required large z values, z greater than unity in many cases. It will be clear from Figure 15.18, which gives the z-M relationship for some 250 QSO's, why this hope has not been realized, however. This figure shows that the QSO's have a very large variation among themselves, and accordingly do not satisfy the essential condition of intrinsic similarity. In spite of this disappointment, it is likely that attempts will be made in future years to push the observations of Figures 15.2 and 15.3 to larger z values. Many new large telescopes are at present under construction throughout the world, and more observing time will become available, and new methods will become more sensitive than the older methods. The prize of determining the nature of the universal geometry is a rich one, and astronomers are not likely

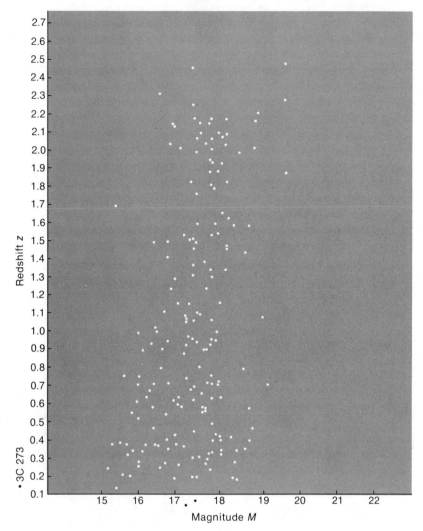

FIGURE 15.18.
The redshift-magnitude relation for some 250 quasi-stellar objects. Notice that
z is plotted directly, not logarithmically.

to be deterred for very long by technical difficulties from attempting to seize
it.

We have now covered all the topics which were planned for this chapter.
In the next chapter we will be concerned with an exceedingly peculiar result
contained within the above discussion. Figure 15.11 shows a moment at which
the scale factor Q was zero. This implies that the galaxy polygons considered

above were all shrunk down to nothing at all. This strange result applies irrespective of whether the world geometry is of class A, as in Figure 15.11, or is of class B or C. It applies inevitably so long as Einstein's gravitational theory is taken to apply to the universe in the large. This circumstance is often referred to by the graphic term of "big bang." We will have much to discuss in the remaining chapters about the meaning of this big bang.

General Problems and Questions

1. Instead of the assumption that local geometry can be extended to very large distances, certain "principles" are used in cosmological studies. What are these principles?

2. How does the "perfect" cosmological principle differ from those which are normally used?

3. Describe a procedure whereby observers in different galaxies can set up a system of universal time.

4. Join the spatial positions of three galaxies at a particular moment of universal time. What property does the resulting triangle have in common with a second triangle obtained by joining the same three galaxies at a different moment of universal time? In what respect may two such triangles differ?

5. Discuss the information concerning the triangles of Problem 4 given by Einstein's theory of gravitation.

6. In what important respect do astronomers seek to use observational procedures to overcome the limitations still present in cosmology even after the use of Einstein's theory?

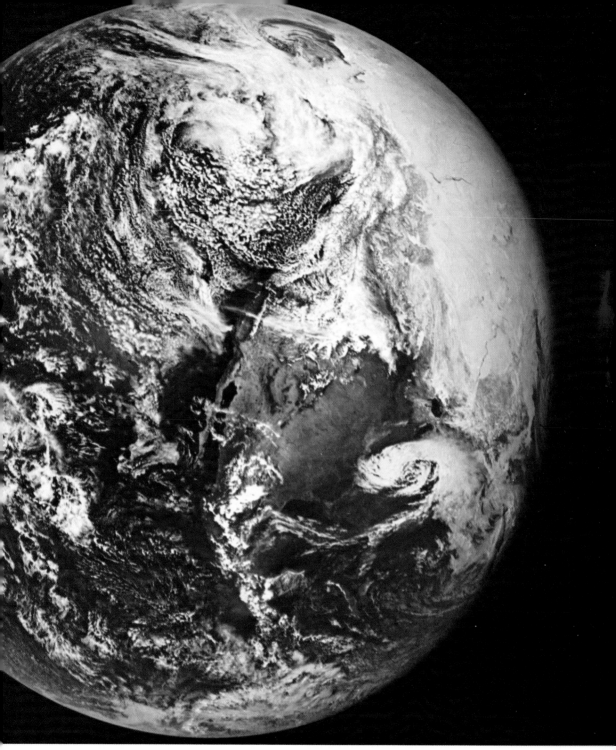

PLATE I (FIGURE 2.2).
Energy from the Sun maintains the biosphere of the
Earth. (Apollo 16, courtesy of NASA.)

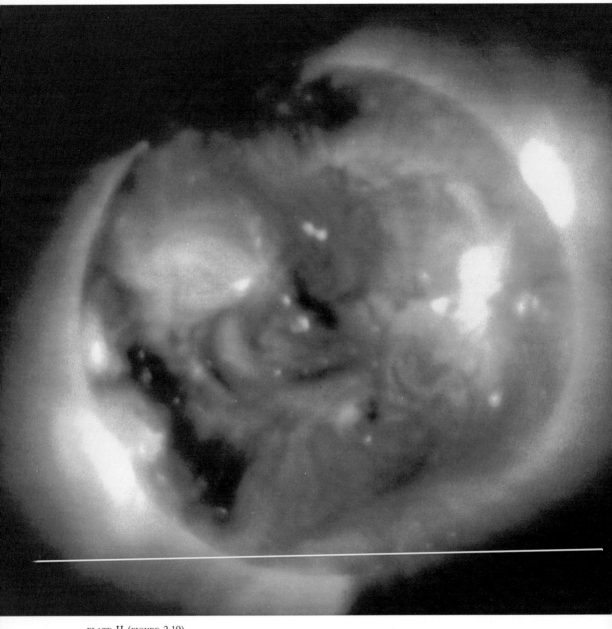

PLATE II (FIGURE 2.19).
X-rays emitted by the hot corona of the Sun. Notice how some spots are hotter than others. (Courtesy of American Science and Engineering, Inc.)

PLATE III (FIGURE 6.1).
The Orion Nebula, a cloud of gas in which stars are now
forming. (Courtesy of Hale Observatories.)

PLATE IV (FIGURE 6.3).
The Rosette Nebula, a more distant cloud, but still close to us within our
galaxy taken as a whole. (Courtesy of Hale Observatories.)

PLATE V (FIGURE 6.4).
The Horsehead Nebula, a *dark* nebula produced by myriads of fine dust
grains. (Courtesy of Hale Observatories.)

PLATE VI (FIGURE 6.29).
The galaxy M31 in color. The difference between the inner yellow and the outer
blue colors can be explained in terms of the Salpeter function. (Courtesy of Hale
Observatories.)

PLATE VII (FIGURE 6.31).
The colors seen in the galaxy
NGC 253 can be explained
like those seen in M31.
(Courtesy of Hale Observatories.)

PLATE VIII (FIGURE 6.33).
The bright stars of the Pleiades cluster are both luminous and blue, belonging to
the upper left of the main sequence. (Courtesy of Hale Observatories.)

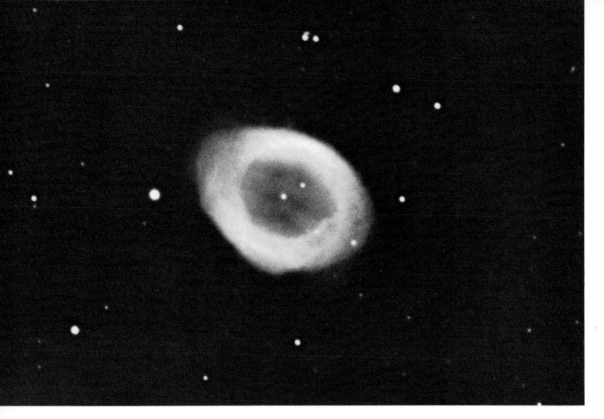

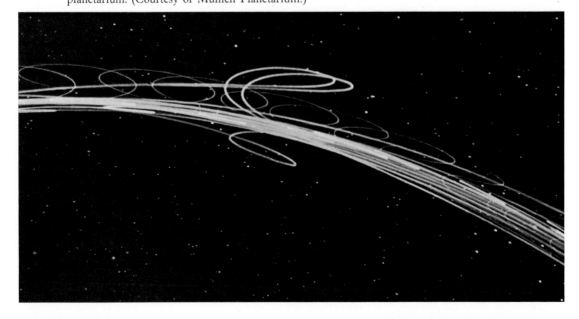

PLATE XI (FIGURE 10.6).
The full Moon, photographed on the Apollo 11 mission. This is the side of the
Moon seen from the Earth. The large dark patches, known as maria, have turned
out to belong entirely to this side of the Moon. (Courtesy of NASA.)

PLATE XII (FIGURE 10.12).
Jupiter with the satellite Io in transit. Photograph from Pioneer 10, taken at
11:02 P.M., Pacific Standard Time, December 1, 1973, at a distance of about 2.5
million kilometers from the planet. The Red Spot is easily seen, near the fainter
part of the limb. (Courtesy of NASA.)

PLATE XIII (FIGURE 10.13).
Another view of Jupiter, showing convection cells near the north pole, which cannot be seen at all from the Earth. These cells rise like thunderstorms do on Earth. (Pioneer 11, courtesy of NASA.)

PLATE XIV (FIGURE 10.11).
Ground-based photographs of Mars, showing in the right-hand picture the effects of a dust storm. (Courtesy of Lowell Observatory.)

PLATE XV (FIGURE 10.14).
The planet Saturn as seen in early 1973. The prominent dark gap in the ring system is Cassini's division which separates the outer Ring A from Ring B. (Courtesy of New Mexico State University.)

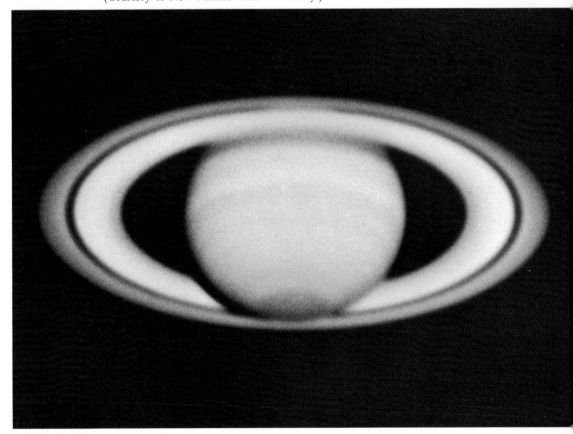

PLATE XVI (FIGURE 10.17).
Earth-rise over the Moon, from the Apollo 8 mission. When the Earth is seen
full from the Moon, it is some 25 times brighter than the full Moon seen from
the Earth. This would indeed be something for poets to wax lyrical about.
(Courtesy of NASA.)

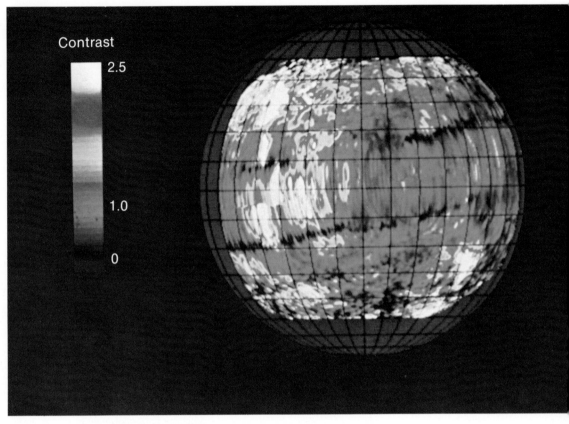

PLATE XVII (FIGURE 10.29).
Radiowaves penetrate through the clouds of Venus, permitting a map of the surface topography to be obtained. High ground (indicated by yellow) and low ground (indicated by blue) show a topography not unlike that of the Earth, except that the surface of Venus has large craters like those on the Moon. Similar craters formed long ago on the Earth would by now have disappeared through erosion. (Courtesy of Dr. D. B. Campbell, Northeast Radio Observatory Corp.)

PLATE XVIII (FIGURE 12.32).
The life-covered Earth. (Courtesy of NASA.)

PLATE XIX (FIGURE 13.12).
The Crab Nebula in color. (Courtesy of Hale Observatories.)

PLATE XX (FIGURE 14.5).
The galaxy M82, where evidence of a recent explosion can be seen.
(Courtesy of Hale Observatories.)

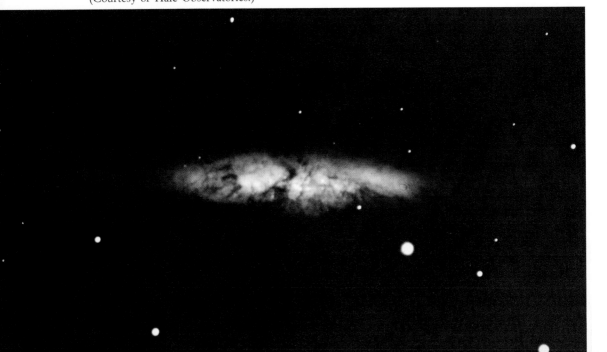

PLATE XXI (FIGURE 16.16).
The particle accelerator at Batavia, Illinois.
(Courtesy of Dr. R. R. Wilson, National Accelerator
Laboratory.)

A

B

PLATE XXII (FIGURE 16.17).
Scale of the tunnel of Figure 16.16 seen from
the air in (A), with the central building shown
in (B). (Courtesy of Dr. R. R. Wilson, National
Accelerator Laboratory.)

PLATE XXIII (FIGURE III.23).
The neutrino-detection experiment in the Homestake
Mine. (Courtesy of R. Davis, Jr., Brookhaven National
Laboratory.)

Chapter 16:
The Origin of the Universe

§16.1. Recapitulation

In the synopsis of this section at the beginning of Chapter 15, we set forth the equivalence,

Particle masses constant + Complex universal geometry $\equiv$

Particle masses variable + Simpler universal geometry.

This equivalence represents two ways to describe the universe, two ways that mathematically are strictly the same. In Chapter 15 we worked in terms of the first half of the equivalence, but in this chapter we will work in terms of the second half, since changing to this alternative way of describing the universe makes many aspects of cosmology easier to understand, especially those aspects concerned with the "origin" of the universe.

Recall also from Chapter 15 that, if $G_1, \ldots, G_n$ are points representing the positions of n galaxies at some moment of time, a polygon can be constructed by connecting G_1 to G_2, G_2 to G_3, . . . , G_{n-1} to G_n, G_n to G_1, all by straight lines. This can be done for the same set of galaxies at other moments of time. The resulting polygons all have the same shape. We chose one of them, at the moment t_0, as a standard of reference. The ratio of the polygon at the moment t to that at the moment t_0 was denoted by $Q(t)$. From Einstein's theory of gravitation, we were able to infer certain properties of $Q(t)$, especially for the case where the geometry of these galaxy polygons is the simple geometry of Euclid, which we referred to as case A, or as the Einstein-de Sitter model.

§16.2. The Meaning of the Expansion of the System of Galaxies

Given this brief recapitulation, let us consider just three galaxies. The polygon is then a triangle, and the observed situation is that the triangle increases its scale (Q increases) as time goes on. We have the situation in Figure 16.1. In the past the triangle was smaller than it is now. In the future the triangle will be larger than it is now. This prompts the question: If we go far enough back into the past, was the triangle ever shrunk down to nothing at all, as in Figure 16.2? The answer to this is yes, because Q was once zero. This followed in Chapter 15 from Einstein's theory of gravitation, as we saw from Figure 15.11 for the Einstein-de Sitter model, and from Figure 15.12 for all the three geometrical cases A, B, C.

Now exactly what do we mean by $Q(t)$ varying with time? How do we measure a change in the scale of our galaxy polygon? The answer to this question is that our basic standard, whether of an interval of time or of a spatial length, is set by the wavelength of the radiation from a specified quantum

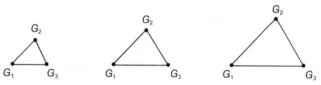

FIGURE 16.1.
In an expanding universe, the triangle formed by joining the positions of three galaxies was smaller in the past than it is now, and in the future will be larger than it is now.

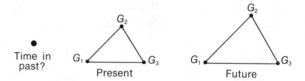

FIGURE 16.2.
Figure 16.1 raises the question: Was there a time in the past when the triangle was shrunk to a point?

659

§16.3. An Alternative
Description of the
Expansion of the
Universe

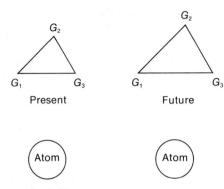

§16.3. An Alternative
Description of the
Expansion of the
Universe

FIGURE 16.3.
When we say that the triangle of Figure 16.1
increases with time, we mean that the *ratio* of
the triangle to a length scale determined by
atomic sizes (which are grossly exaggerated!) is
increasing, whereas the scale itself stays the
same.

transition of some chosen atom. In turn the wavelength of the radiation is
determined by the masses of the particles that constitute the atom, especially
by the mass of the electron. When one calculates the frequency, say, ν, of the
radiation emitted in the transitions of a certain atom, say, the Hα transition
of hydrogen, it turns out that ν is proportional to the electron mass. Thus,
writing m for the electron mass, we have ν proportional to m. Now the wavelength of the radiation is c/ν, and hence is proportional to the reciprocal of
the electron mass, $1/m$. It is $1/m$ which determines our physical scale, expressed
by the sizes of atoms, and by the unit of time given by radiation from atoms,
i.e., the unit of an atomic clock. When we say that distances between galaxies
increase with time, we mean distances measured with respect to $1/m$. We mean
that our galaxy polygons increase with respect to $1/m$ as unit, as in Figure
16.3.

§16.3. An Alternative Description of the Expansion
of the Universe

The significance of the equivalence given in §16.1 will now be clear. Its first
half requires $1/m$ to stay fixed, in which case the galaxy polygons increase
according to the scale factor $Q(t)$. But if we go by its second half, we can
take the galaxy polygons as staying fixed, so that $Q = 1$ at all times, in which
case $1/m$ must decrease, as in Figure 16.4. Can we devise physical experiments
that can distinguish between the situations depicted in Figures 16.3 and 16.4?
Local experiments could not do so, because they occupy intervals of time so
short that any change of $1/m$ during them would be quite negligible. But

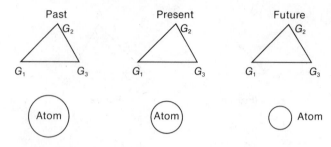

FIGURE 16.4.
An alternative to Figure 16.1 would be to keep the triangle fixed and
to suppose that the atomic dimensions (again grossly exaggerated) decrease with time.

suppose a terrestrial experimenter were able to live for a very long time, and
suppose he could take radiation from a certain atom—for example, the H and
K oscillations of calcium—and store the radiation somehow for future reference.
Then, after a long time had elapsed, the stored radiation could be compared
with new radiation from the same kind of atom. Would the oscillation frequency
of the old radiation be the same as that of the new radiation?

There is a sense in which this experiment *can* be carried out, and this is
illustrated in Figure 16.5. Light from a distant galaxy takes a very long time
to reach us, and we can think of passage through space as a form of storage.

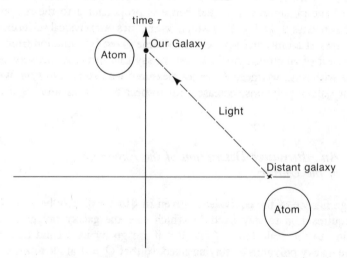

FIGURE 16.5.
We can say that light from a distant galaxy was generated at a time
when the masses of particles were less than they are at present, and
it is for this reason that radiation from a specified atomic transition is
redshifted with respect to present-day radiation from the same transition.

661

§16.3. An Alternative
Description of the
Expansion of the
Universe

When we receive light from such a galaxy, we can examine this "old" light, light that (for some galaxies) was generated billions of years ago. What do we find? We find that the old light does *not* have the same oscillation frequencies as the new light which we generate currently in the terrestrial laboratory. Why should we not regard this as a confirmation that the particle masses were different billions of years ago than they are today? Why bother with the Doppler shift and with the idea that the galaxies are rushing apart from each other? The redshift observation we have just described can be interpreted in this very different way, in terms of a change of $1/m$.

It is a characteristic of the physicist that he will become most uneasy at the thought that there are two different *but indistinguishable* interpretations of a phenomenon. He will react to such a situation by arguing that if there are two indistinguishable ways of describing an observation, then, however apparently different they may seem, the two ways must really be the same. So, instead of asking which of Figures 16.3 and 16.4 is the correct picture, we should really regard them as the same picture, *and we should adjust our physical ideas to make them the same.*

What would the foregoing imply? That the mass of a particle must be determined by its relationship to other particles, according to certain rules which are to be stated with full mathematical precision. These rules must be chosen so that our physical theories (for example, of gravitation) lead to exactly the same observable results for Figure 16.4 as for Figure 16.3. This demand for precise equivalence has necessitated considerable technical development, the details of which need not concern us here. For now it is sufficient to know that Figures 16.3 and 16.4 *can* be made rigorously equivalent to each other. Then we may proceed, if we wish to, using Figure 16.4 instead of 16.3 as our description of the universe.

A remarkable simplification emerges immediately. When we adopt the picture of Figure 16.4, the universal geometry becomes the same as local spacetime geometry, *and it remains the same no matter whether the spacelike behavior of our galaxy polygons in Figure 16.3 is of type A, B, or C.* Whatever the former situation, we now have the convenient situation that the geometry of the whole universe is the same as the local geometry that we studied in Appendix II.4.

The Einstein-de Sitter model has another important simplification. In the new picture the galaxies are uniformly spaced, as in Figure 16.6. Here it is useful to introduce the idea of a "smoothed-out" universe. The matter in the galaxies of Figure 16.6 can be imagined to be smoothed out so that everywhere the density of particles is the same. When this is done, we can draw a picture like Figure 16.6, but with the world lines now representing individual particles instead of galaxies, as in Figure 16.7. Clearly the individual particles of Figure 16.7 will be much closer together than the galaxies of Figure 16.6. (The decrease of separation would of course be much more marked than can be indicated in the figures.)

Notice that the average separation of particles in Figure 16.7 provides us with a new scale of length, say, L, which we could use for measuring spatial

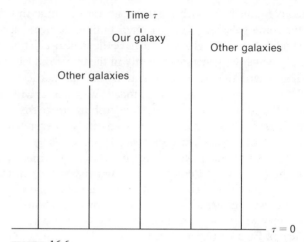

FIGURE 16.6.
In this alternative picture, and in the Einstein–de Sitter
model, the universe becomes static.

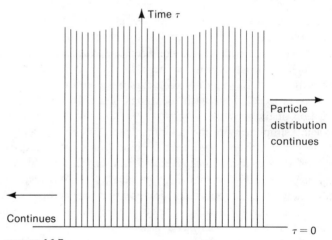

FIGURE 16.7.
We can think of the material of the galaxies as being smoothed out into
a uniform background of particles.

lengths, instead of using the scale determined by atoms. The scale L has the advantage of being the same at all times, whereas the scale set by radiation from atoms would now have the disadvantage of changing with time, because the masses of the particles making up the atoms are variable. Indeed, this variation is just the redshift effect appearing again in a new guise, as we already noted above. Notice too that the scale L is physical. We can also establish a

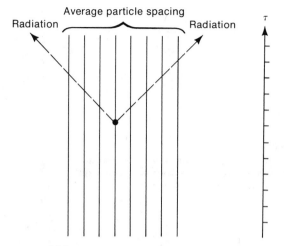

Average particle spacing

Radiation ⎯⎯⎯⎯⎯⎯ Radiation

τ

663

§16.3. An Alternative
Description of the
Expansion of the
Universe

FIGURE 16.8.
The average distance between particles in the smoothed-out
model of Figure 16.7 gives a spatial scale. The scale in which
time is measured is chosen so that radiation propagates at an
angle of 45° to the time axis, just as it did in our discussions
based on "local geometry."

physical scale for our time unit quite simply by requiring light and other forms
of radiation to propagate at 45° in Figure 16.7, as in Figure 16.8. When
measured in this way, we denote the time by τ. Another result which can be
proved exactly is that, for time defined in this way, the particle masses in the
Einstein–de Sitter model vary proportionately to τ^2. Let us see if we can gain
an insight into the way in which this result arises.*

The mass of particle a at the point A of Figure 16.9 is to be regarded as
being made up of contributions which travel at an inclination of 45° from other
particles, of which particle b in Figure 16.9 is an example. We also regard the
interactions, at any rate for the moment, as coming from the past, as light does.
First we ask: How will the contribution of such an interaction depend on the
distance r between particles a and b? Clearly we expect the contribution to
be less when r is large, and the two particles are widely separated, than it will
be when r is small. This suggests some kind of inverse relationship, but will
the contribution be proportional to the simple reciprocal, $1/r$, or to the inverse
square, $1/r^2$, or to some other form? At first sight one might suppose that the
inverse square law, $1/r^2$, would be correct, as it would be for the intensity
of radiation from a source at distance r. But the intensity of radiation itself
depends on the square of a more basic quantity, often referred to as the
amplitude of the radiation, which behaves as $1/r$. The interaction we are seeking

---

*Instead of worrying about the following details the reader may prefer to jump immediately
to §16.4.

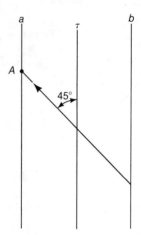

FIGURE 16.9.

The mass of particle *a* at the point *A* is to be regarded as made up of contributions from other particles *b*, traveling also at an angle of 45° to the time axis.

behaves like amplitude, not like intensity, and so varies like $1/r$. The situation therefore is that, to determine the mass of particle *a* at the point *A* of Figure 16.9, we have to add $1/r$ contributions for all other particles, of which particle *b* in Figure 16.9 is an example. Let us see how to go about making this addition, remembering that the particles are uniformly spaced with a separation L.

With center at the point *A* of Figure 16.9, draw a set of spherical surfaces with radii $L, 2L, 3L, \ldots$. Because in Figure 16.9 we are only showing one of the three dimensions of space, we cannot draw these spheres in the usual way. With two of the three spatial dimensions suppressed, the spheres appear as simple points, as in Figure 16.10. Eventually the spheres reach particles from which there is no contribution, because, for interactions propagated at 45° in our diagram, there is a limit to how far back we can go into the past, *since none of the particles exists before time zero.* Writing τ for the time at point *A*, we reach such particles for a spatial distance $r = \tau$. Evidently, then, the contributions to the mass of particle *a* at point *A* is limited to particles reached by the finite series of spheres $L, 2L, \ldots$, the last sphere in this series having a radius that is less than τ by no more than the small distance L. The situation is illustrated in Figure 16.10.

Next we consider the contribution to the mass from all those particles that are reached by the sphere of radius $(n + 1) L$, but which are not reached by the sphere of radius nL. Here n is to be thought of as a large integer, but not so large that $(n + 1) L$ exceeds τ. All the particles in question are essentially at a spatial distance nL from point *A*, and so each will make a contribution to the mass at *A* that is proportional to $1/nL$. The number of such particles will be proportional to the volume between the two spheres, which is $4\pi n^2 L^3$ to sufficient accuracy. Hence the total contribution from all the particles reached by the sphere $(n + 1) L$ but not by the sphere of radius nL must be proportional to the product $(1/nL) \times 4\pi n^2 L^3$, i.e., to $4\pi nL^2$.

665

§16.3. An Alternative
Description of the
Expansion of the
Universe

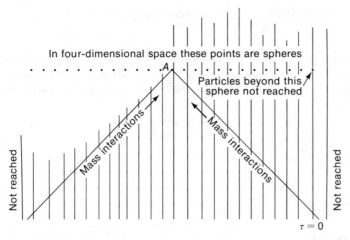

FIGURE 16.10.

With center at A, draw a set of uniformly spaced spheres. Because two of the three spatial dimensions are suppressed here, the spheres appear as simple points. There is a last sphere reaching particles which contribute to the mass of particle a at A.

The last step in the argument is to sum the contributions for all such pairs of spheres. The contributions vary from one pair to another like the integer n, and so the required summation is proportional to the series

$$S = 1 + 2 + \ldots + k.$$

PROBLEM:

Prove that $S = \dfrac{1}{2} k (k + 1)$.

[Hint: Assume the result to be true for a particular integer k. Add $k + 1$, and prove the result to be true for the integer $k + 1$. What further step is still needed?]

Using the result of the above problem, we can approximate S as $\frac{1}{2}k^2$, and we can write $k = \tau/L$, since L is small compared to τ. Thus substituting $k = \tau/L$ in $k^2/2$, we get $1/2 \, (\tau/L)^2$, which is proportional to τ^2. This is the answer we were seeking. Hence we understand how it comes about that the mass of a particle can vary with time, and we have understood in an elegant way why a redshift effect is observed in the radiation from distant galaxies.

§16.4. *The Redshift-Magnitude Relation of Hubble and Humason*

We now aim to prove a still more ambitious result, namely, the relation between the magnitude and the redshift that was plotted in Figure 15.14. Consider light received at time τ from a galaxy at distance r, as in Figure 16.11. Since the

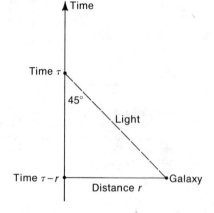

FIGURE 16.11.

Light received at time τ from a galaxy at spatial distance r must start its journey at time $\tau - r$.

scale for measuring τ was defined by the requirement that light should propagate at an angle of $45°$ in this figure, the light from the galaxy must start its journey at time $\tau - r$ in order to be received at time τ. Now from what we have just seen, particle masses at the time of emission of the radiation must be proportional to $(\tau - r)^2$, whereas the particle masses at the time of reception of the radiation are proportional to τ^2. We can express this result for electrons by writing $m(\tau - r)$ and $m(\tau)$ for the masses at times $\tau - r$ and τ, respectively, with $m(\tau - r)$, $m(\tau)$, satisfying the equation

$$\frac{m(\tau)}{m(\tau - r)} = \frac{\tau^2}{(\tau - r)^2}.$$

Using a similar notation, we write $\nu(\tau - r)$, $\nu(\tau)$, for the frequencies of the radiation emitted by a certain transition of a certain atom; for example, $\nu(\tau - r)$ could be the frequency of the Hα transition of hydrogen emitted at time $\tau - r$, and $\nu(\tau)$ would be the frequency of Hα emitted at time τ. From what was said above about the relation of the emitted frequency of the radiation to the electron mass, namely, ν proportional to m, we see that

$$\frac{\nu(\tau)}{\nu(\tau - r)} = \frac{m(\tau)}{m(\tau - r)}.$$

Now, the lefthand side of this equation is just the quantity $1 + z$, so

$$1 + z = \frac{m(\tau)}{m(\tau - r)} = \frac{\tau^2}{(\tau - r)^2}.$$

To obtain the different values of the redshift z for different galaxies, as determined by an observer living at time τ, we simply keep τ fixed in this formula and vary the distance r.

The next problem is to find how the magnitudes of such galaxies depend on r. The apparent brightness of a galaxy of intrinsic luminosity $\mathcal{L}$ at distance r is just $\mathcal{L}/4\pi r^2$, because our universal geometry is now the same as our familiar local geometry. However, in using this simple result, we must remember to choose $\mathcal{L}$ for time τ-r, the moment of emission of the light received at time τ. Now $\mathcal{L}$ can be shown to behave like the *square* of the particle masses. Thus $\mathcal{L}$ is proportional to $m^2(\tau - r)$, and so is proportional to $(\tau - r)^4$. [Luminosity means "energy emitted per unit time." Energy behaves like the particle masses, and so is proportional to $m(\tau - r)$. Unit time is proportional to $1/m(\tau - r)$. Hence energy divided by unit time is proportional to $m^2(\tau - r)$.]

It follows from the preceding paragraph that the energy flux f is proportional to $(\tau - r)^4/r^2$. In Appendix I.2, we saw that the magnitude, say, M, is defined by

$$M = -2.5 \log f + \text{a certain constant,}$$

which can be written in the form

$$M = -2.5 \log \left[\frac{(\tau - r)^4}{r^2 \tau^2} \right] - 5 \log \tau + \text{a certain constant.}$$

For τ constant, the term $-5 \log \tau$ can be taken with the last term on the righthand side; so we have

$$M = -2.5 \log \left[\frac{(\tau - r)^4}{r^2 \tau^2} \right] + \text{a constant,}$$

the constant being the same for every galaxy observed at the time τ.

The last step in this argument comes from combining this result for M with the above result for z. From $1 + z = \tau^2/(\tau - r)^2$, it is easy to see that

$$\frac{r}{\tau} = 1 - \frac{1}{\sqrt{1 + z}}.$$

PROBLEMS

1. Satisfy yourself that this expression for r/τ is correct. It is now easy to eliminate the distance r from the above equation for M.

2. Wherever r appears in the above equation for M, substitute $\tau(\sqrt{1+z}-1)/\sqrt{1+z}$ in its place, and hence show that M is related to the redshift z by the equation

$$M = 5 \log (\sqrt{1+z} - 1) + 2.5 \log (1 + z) + \text{constant},$$

the constant having no dependence on z.

The result of this second problem is the curve given in Figure 15.14, and reproduced here as Figure 16.12.

The result obtained for M in the above problem is a considerable achievement, since an extended technical course on relativity and cosmology is usually considered necessary in order to obtain it. Here we have used only very straightforward considerations.

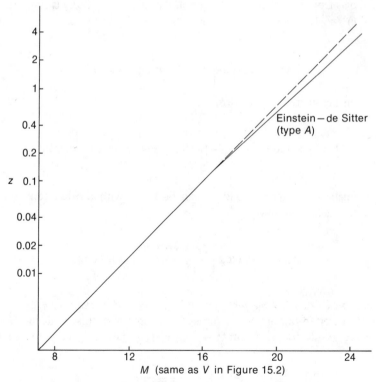

FIGURE 16.12.
The z-M relation for the Einstein–de Sitter model.

We have indeed now a clear picture of the structure of the universe in terms of the Einstein–de Sitter model. We have a simple understanding of why the galaxies exhibit the redshift phenomenon: because particle masses are increasing with time. Our galaxy polygons no longer shrink to zero at the initial moment $\tau = 0$. Geometry is simple at all times. In particular, there is no geometrical problem at $\tau = 0$; the geometry of Appendix II.4 applies at $\tau = 0$ just as it does at any other moment of time. Notice too the important point that all our particles are at rest. We have no need for the Doppler effect here. There is no conceptual difficulty of the kind we encountered at the end of Appendix II.8, where we considered the erroneous idea, which often arises out of the Doppler interpretation of the redshifts, that because all galaxies appear to recede from our own galaxy, we must therefore be at a "center" of the universe. No center appears in Figure 16.7. We also see what it is that makes the moment $\tau = 0$ so peculiar. Particle masses are zero at $\tau = 0$. They are zero because there were no interactions preceding $\tau = 0$.

§16.5. The Microwave Background

Consider the situation when τ is small, but not strictly zero. Because the masses of particles are then small, radiation emitted by atoms has low frequencies. Thus an atom of a kind which now emits visible light in the terrestrial laboratory would even emit radiowaves if τ were sufficiently small. Since the frequencies of oscillation of any radiation left over from the early history of the universe will not have changed with the passage of time, the geometry being now of simple Euclidean form, it is to be expected that such radiation, if it exists, will be observed at low frequencies only. We can calculate that the frequency distribution would have a form of the kind shown in Figure 16.13, but we cannot, as yet, calculate which of the various possibilities it would follow. These forms are known as *black-body curves*. Each one of them is characterized by a temperature on the usual °K system.

Low-frequency radiation does indeed arrive at the Earth from directions all over the sky. To within the errors of observation, it arrives isotropically, a result that is established to an accuracy better than about 0.1 per cent. The intensity of the radiation at various frequencies is measured by the quantity $I(\nu)$, which has been plotted on a relative scale in Figure 16.13 for the calculated possibilities. The actual observations are shown in Figure 16.14, and are seen to fit very well to a possibility corresponding to the temperature of 2.7°K. It is as if we were observing an ideal radiator with a temperature of 2.7°K distributed uniformly all over the sky. This indeed is just the interpretation one is inclined to make.

It is worth emphasizing the remarkable nature of this observation, first made by Arno Penzias and Robert Wilson in 1965. When we look at the sky, whether by means of visible light or radiowaves or any other form of radiation, we see all objects projected together on the celestial sphere. But we know that

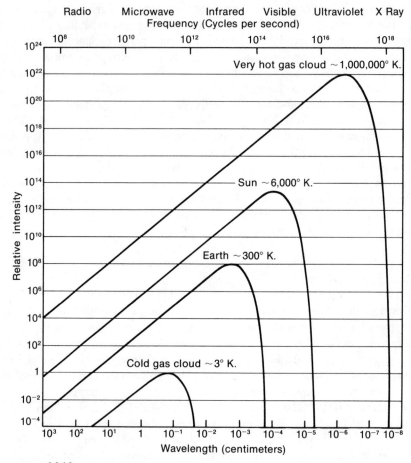

FIGURE 16.13.
Examples of "black-body" curves. (Except for being plotted on a logarithmic scale, these are basically the same curve as that of Figure II.1.)

one object may lie behind another. Faint stars are usually more distant than bright ones. Even the brightest and nearest galaxies still lie far behind the stars of our own galaxy. Faint galaxies usually lie behind bright galaxies. Behind all these objects, far behind, now comes this new low-frequency radiation. When we look out into space we are looking backward along the light cone shown in Figure 16.15. The microwave radiation comes from the most remote part of this cone, the part which comes close to the very beginning of the universe at the time $\tau = 0$.

So far it has not been possible to observe the microwave radiation at all frequencies. The portion of the black-body curve to the right (i.e., on the high-frequency side) of the maximum has not yet been well observed, for the

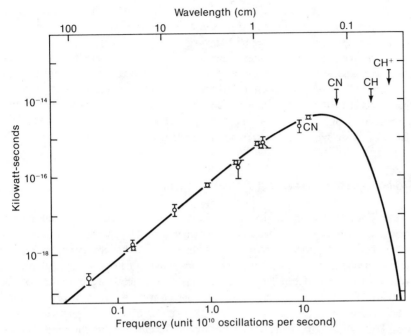

FIGURE 16.14.
The black-body curve for a temperature of 2.7 °K gives a close fit to the observed points. (With the lefthand scale interpreted in units of kilowatt-seconds, the curve represents the power in kilowatts falling on an area of $1 / \pi$ square kilometers.)

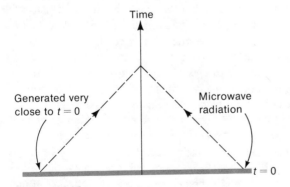

FIGURE 16.15.
The microwave radiation comes from the most remote part of the light cone, close to $\tau = 0$.

reason that radiation of the frequencies in question do not penetrate the Earth's atmosphere to ground level. Such observations would need to be done from a rocket or from a satellite. Eventually, no doubt, the required experiments will be performed. In the meantime some upper limits have been established, and are also shown in Figure 16.14.

The reader may well wonder, since tens of billions of dollars have so far been expended in space research, why such an exceedingly critical and exciting experiment has not yet been performed. The reason lies in part with the unsatisfactorily long lead times needed for satellite experiments, in part with the technical difficulty of the experiment itself, and in part with the lack of vision of those who have hitherto controlled the space program itself.

§16.6 Superparticles and the Hadronic Era

In many modern cosmological discussions, it is assumed that radiation was already present at $\tau = 0$. Although this may seem a somewhat peculiar idea, since nothing is supposed to have existed before $\tau = 0$ that could have generated the radiation, it is an idea with interesting consequences. Even though the radiation was only of low frequency, it must nevertheless have dominated the behavior of matter, just because particle masses were so close to zero at small τ. The radiation could have produced all manner of esoteric particles which nowadays can only be generated by powerful accelerators of the kind that has recently been built at Batavia, Illinois (Figures 16.16 and 16.17). The nature of the particles existing very near $\tau = 0$ would be exceedingly complex, and would belong to what has been called the "hadronic" era of the universe.

For FIGURES 16.16 and 16.17, see PLATES XXI and XXII.

Some physicists have wondered whether such complexities could have had an effect that is still observable in the present-day world. Could a bunching of particles near $\tau = 0$ have led to galaxies being eventually formed? This may seem an ambitious question, but we must bear in mind that other, apparently more straightforward attempts to explain the origin of galaxies have not had very much success. An interesting idea developed by Steven C. Frautschi at the California Institute of Technology is that perhaps the first particles near $\tau = 0$ were truly superparticles which eventually decayed after quite an appreciable time-interval into many ordinary particles, so many indeed that a single superparticle eventually gave rise to a whole galaxy.

There could be a connection here with the outbursts of radiogalaxies and QSO's, discussed in Chapter 14. Conventional attempts to explain these outbursts have not been particularly successful either. I have had for some years the lurking suspicion that the cascades of highly energetic particles responsible for our observations might be generated by the decays of some superparticle. It is essentially certain that our theories of the structure and content of the universe cannot be final. Remarkable discoveries must remain to be uncovered. It is possible that a road leading to a whole new territory may be found in the direction of these superparticles.

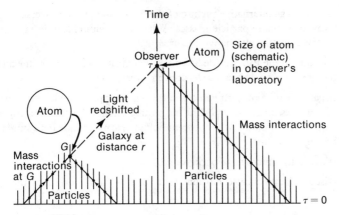

FIGURE 16.18.
Summary of results for the Einstein-de Sitter model. Mass interactions occuring back to $\tau = 0$ give a mass proportional to τ^2 for a particle in an observer's laboratory.

§16.7. Summary

The position we have reached is illustrated in Figure 16.18, again for the Einstein-de Sitter model. We have obtained an easily understood explanation of the redshift phenomenon. We have seen how early conditions in the universe provide an explanation of the background of longwave radiation (microwaves), observations of which are shown in Figure 16.14. We encountered the idea of a "hadronic" era near $\tau = 0$. All this is satisfactory. But can we believe that the situation of Figure 16.18 is correct? Can the universe in the large really be as simple and as crude as this? The tendency of astronomers is to answer affirmatively, but many would admit causes of disquiet. In the remainder of this book, we will examine suggestions which have been made for changing the picture of Figure 16.18, beginning in the next chapter with the steady-state model, and then in the last chapter using a new idea to extend Figure 16.18 in an interesting way.

General Problems and Questions

1. Within the framework of constant particle masses, discuss the "expansion of the universe."

2. To what conclusion concerning the origin of the universe does this expansion lead?

3. How may an alternative picture of the expansion of the universe be given in terms of variable particle masses? What major simplification of geometry is achieved in this alternative picture?

4. Working in the Einstein-de Sitter cosmological model, how in the alternative picture do particle masses depend on time?

5. A galaxy with a redshift value $z = 0.1$ has an apparent bolometric magnitude of $+16$. Assuming the redshift-magnitude relation for the Einstein-de Sitter model, what would be the apparent bolometric magnitude for a similar galaxy having a redshift value $z = 1.0$?

6. What is the microwave background?

7. Discuss the physical conditions at times close to the origin of the universe.

Chapter 17:
The Steady-State Model

§17.1. Broken Particle Paths

The situation in Figure 16.18 may be described by saying that the paths of particles are broken at the time $\tau = 0$. Why should all the "ends" of the paths happen to be lined up in this way? Why not contemplate a situation in which the ends occur at different values of τ, as in Figure 17.1?

This step raises the problem of the origin of matter more sharply than Figure 16.18, since in Figure 16.18 we can argue that there was no universe before $\tau = 0$, and hence that we are not obliged to explain the broken ends of Figure 16.18 in terms of the laws of physics. In Figure 17.1, the situation is different, however. The ends of the particle paths occur within the universe and must therefore be understood in physical terms. The kind of explanation that is needed is shown schematically in Figure 17.2, where interactions occur at the broken ends of the paths. These interactions cannot be electrical or gravitational, but must constitute a new kind of "field," as physicists describe it.

Given the general idea of Figure 17.2, it is possible to build a precise mathematical theory using methods that are standard in physics, methods based on an "action principle." The story of Newton and the brachistochrone—the problem invented by Leibniz and Bernoulli—may be recalled (Chapter 9). The

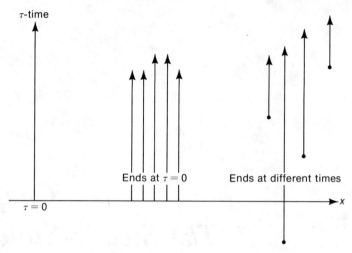

FIGURE 17.1.
Instead of the ends of the particle paths all falling at one moment of time, as in the big-bang model, why not contemplate that the ends may fall at different moments of time—even moments *before* $\tau = 0$?

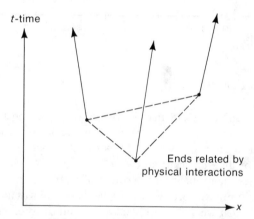

FIGURE 17.2.
In such a theory we consider that physical interactions relate the "end" of one particle to the ends of other particles.

branch of mathematics that Newton invented for solving this problem is the one used in formulating the action principle. There are standard ways of relating such an action principle to Einstein's gravitational theory, and of then working out the consequences for cosmology. When this is done, a quite remarkable result emerges, namely, that the universe settles itself into a steady-state

condition. The perfect cosmological principle of Bondi and Gold, enunciated
in Chapter 15, becomes applicable. We have the logical equivalence:

677

*§17.2. The
Steady-State
Expansion of the
Universe*

Ends of particle paths at different values of τ

$\Leftrightarrow$ Perfect cosmological principle.

§17.2. The Steady-State Expansion of the Universe

It is usual to discuss the steady-state model in terms of the expansion picture
of Figure 16.3, in which atoms are considered to have fixed sizes. The distances
between galaxies then increase with time, as in the usual expansion picture of
the universe, in which the time t is measured by an atomic clock. [It will be
recalled from Chapter 16 that the time τ was measured in terms of the average
spacing L of particles and the propagation of light.] In the picture of Figure
16.3, the galaxy polygons, obtained by joining a number of galaxies, maintain
the same shape at all times. However, the scale of the polygons, denoted by
$Q(t)$, changed with the time t. In the steady-state model there is a simple relation
between Q and t, namely, $Q = 10^{0.4343\,Ht}$. The Hubble constant H is more
important here than it was in the theory of Chapter 15. The relation $v = Hd$
discussed in Chapter 15 referred to a certain moment of time; so an observer
who calculated this relation at different times would, in the theory of Chapter
15, obtain different values of H. In the steady-state model, on the other hand,
H is the same at all times. H is a true constant, a property which follows from
the mathematical development referred to above.

The constancy of H can also be derived from the perfect cosmological
principle, which requires that no temporal moment, no value of t, shall be any
different from any other moment. If this is to be so, then H cannot vary with
t, otherwise we could distinguish the epoch by a measurement of H.

Using the estimate $1/H = 18 \times 10^9$ years, discussed in Chapter 15, the
above relation for Q turns out to require that every galaxy polygon doubles
its scale in a time of about 12.5×10^9 years.

PROBLEM:
Prove this result.

Yet in the steady-state model the average separation of all galaxies must always
be the same. If either this or any other observable property were dependent
on the epoch, the steady-state condition would be contradicted. How then, since
the distance between every pair of galaxies increases like $Q(t)$, are we to meet
this requirement? It can be met if new galaxies are steadily being formed, in

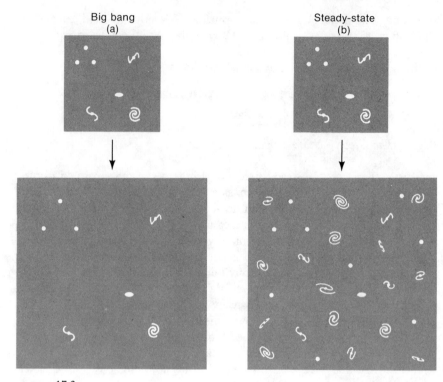

FIGURE 17.3.
In (a) the already-existing galaxies expand apart, this being the situation for the big-bang cosmologies, but in (b) new galaxies are born at the rate necessary to compensate for the steadily increasing distances between already existing galaxies. The density of galaxies is greatly exaggerated in this figure.

the manner of Figure 17.3. In this picture, new galaxies are born at the rate necessary to compensate for the increases of the distances between already-existing galaxies. As the already-existing galaxies spread out, new galaxies are born in the spaces between them.

§17.3. *The Exposure of the Steady-State Model to Disproof*

The steady-state model is more readily subject to test than any other of the cosmological models, the reason being illustrated in Figure 17.4. When we make observations, using any form of radiation, we look back into the past along what is called the light cone. Accordingly our observations sample the conditions that existed at earlier epochs, and if we go back along the light cone far enough, it is possible to examine the state of affairs at times significantly earlier than now. If any astronomical property is found to have been significantly

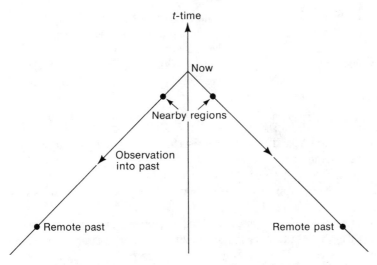

FIGURE 17.4.
By being compared with the state of affairs at a great distance—i.e., at a
significantly earlier epoch—the steady-state model can be tested for any
astronomical property.

different in the remote past, as shown in Figure 17.4, than it is in nearby regions,
we will have disproved the contention that the universe is steady in all respects.

Two difficulties stand in the way of using this logically straightforward
method as a test of the steady-state model. If the universe is changing with
time, if the steady-state model is wrong, then it will be better in principle to
disprove it by looking back a long way along the light cone rather than by
looking back only a small way. Changes during a comparatively short time
are likely to be much less than changes during a long time, and inherent sources
of inaccuracy in our observations will always obscure small changes. Inevitably,
then, we must expect to be concerned with observations at great distances.
Unfortunately, however, objects at great distances are usually very faint, and
therefore difficult to observe with precision. Spurious effects arising from
working at the limits of our instruments must not be misinterpreted as evolutionary changes.

A second difficulty comes from an uncertainty in the steady-state model itself.
On what scale is the universe supposed to be steady? Clearly not when taken
on the scale of the solar system, or on the scale of our galaxy, or even on
the scale of clusters of galaxies. The protagonists of the steady-state model
have never had an unequivocal way of answering this question. Bondi and Gold
supposed the universe to be steady on a scale not much greater than the distances
of nearby galaxies. Yet it would be possible, without sacrificing the most
important concepts of the theory, to consider that the "scale of steadiness,"

as we might call it, should be set much larger than this. Since in the theory there is no limit to either space or time, the scale on which we elect to work is rather arbitrary; steadiness, in the sense of maintained properties, might only set in when regions containing many millions of galaxies are considered. By extending the scale sufficiently, one could thus frustrate attempts to disprove the theory along the lines suggested above.

Let us consider these issues in terms of an explicit example, the counting of radio sources, discussed in Appendix V.1. The latest observational data were shown in Figure V.2. If the observations are taken to require an evolutionary model for the universe, then the steady-state theory in the restricted form of Bondi and Gold is disproved. There are ambiguities in the data, however, which were explained in Appendix V.1, so that the issue remains open, particularly if the "scale of steadiness" is taken to be large.

If it were not for the existence of the microwave background, discussed in Chapter 16, the steady-state model might reasonably be judged to stand well at the present time. The origin of the microwave background can be explained satisfactorily in terms of the "big-bang" cosmologies of Chapter 16. Conditions in the early universe near $\tau = 0$ were seen to be capable of generating the background radiation in these cosmological models (i.e., A, B, or C of Chapters 15 and 16). No such explanation can be given in the steady-state model, however, because *there is no origin of the universe* in the steady-state model. For the large-scale features of the universe, one epoch cannot be different from another epoch. This "steady" condition would be violated if the universe had an origin.

How, then, in the steady-state model can we explain the existence of the microwave background? The observed background consists of radiowaves with frequencies ranging from about 10^9 oscillations per second up to about 3×10^{11} oscillations per second. Many radio sources are known which generate such waves. So why should not the sources—i.e., discrete objects like radiogalaxies and QSO's—be responsible for the background? The difficulty with this apparently straightforward approach to the problem is that the known sources, counts of which are plotted in Figure V.2, fail to give an adequate intensity, particularly at the high-frequency end of the background. The only possible recourse would be to postulate the existence of a very large number of *undetected sources*, sources of very low intrinsic emission, say, with only a millionth of the intrinsic emission of the sources we do observe. There would need to be about 10^{14} such weak sources, which is about 100,000 times the total number of visible galaxies. Most astronomers find it unpalatable to assume the existence, not just of a new class of source, but of a class with a very great number of members. The criticism is a fair one. Even so, we must be on our guard against the ever-present tendency in astronomy to imagine that nothing exists in the world except the things which happen to be observable with present-day instruments. This point of view has been repeatedly wrong, and doubtless will turn out to be wrong again. But will it turn out to be wrong about the existence of a profusion of weak radio sources that would "save" the steady-state model? Beyond saying that an affirmative answer here seems doubtful, we can go no further with the argument.

The case against the steady-state model, resting largely on the problem of the microwave background, although not clear-cut, is strong enough to suggest that it may be more profitable to consider alternative ways of coming to grips with the unsatisfactory position that we reached at the end of Chapter 16. By our consideration of the steady-state model, we have now discussed one line of reasoning which avoids the conceptual difficulties associated with Figure 16.18. In the next, and final chapter, we consider another possible resolution of the problem of the origin of the universe.

General Problems and Questions

1. In what important respect, concerning the origin of particles, is the steady-state model similar to the usual cosmological models?

2. How do the predictions of the steady-state model differ from those of the usual models?

3. Describe the most critical observational objection to the steady-state model.

Chapter 18:
Final Considerations

At the end of a book like this, one would like to arrive at a stirring conclusion. Instead, we have arrived at an exceedingly mysterious situation. Let me first restate this situation, and then see if, by a few broad strokes, we can finally break through into a wider, albeit speculative, view of the whole subject of cosmology.

§18.1. The Present Dilemma.

Suppose that, instead of letting there be nothing before $\tau = 0$ in Figure 16.18, reproduced here as Figure 18.1, we seek to avoid the artificial breaks in the paths of the particles by extending them to earlier times, going backward indefinitely, as in Figure 18.2. We are immediately in trouble because our method of calculating the mass of a particle leads to an infinite result. Thus in Chapter 16 we found the mass of a particle to be proportional to a finite series, $S = 1 + 2 + 3 + \ldots + k$, in which the last member was defined in the manner shown in Figure 16.10: the product of the integer k and of the average interparticle spacing L is close to τ, so that, to sufficient accuracy, $k = \tau/L$.

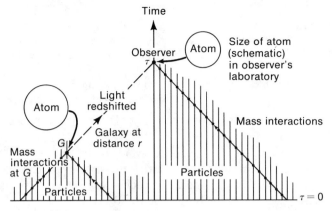

FIGURE 18.1
The summary of results for the Einstein–de Sitter model (from Figure 16.18).

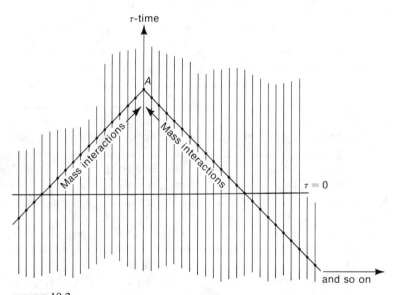

FIGURE 18.2
We can seek to avoid the breaks in the paths of particles by extending them indefinitely backward.

But in the situation of Figure 18.2 we have no such means of terminating the contributions to the mass from distant shells. That is to say, the series of spheres that we introduced in Chapter 16, with radii L, $2L$, $3L$, ..., can go on indefinitely, with the result that the calculated mass becomes proportional to a series $S = 1 + 2 + 3 + \ldots$, which also goes on indefinitely.

PROBLEM:

Satisfy yourself that, given any number x, however large, a sufficient number of terms of the series $1 + 2 + 3 + \ldots$ can be chosen so that their sum exceeds x.

[Hint: Start by considering the sum of a finite number of integers, $1 + 2 + 3 + \ldots + r$. Prove that this is $\frac{1}{2}r(r+1)$. Now increase r.]

We see from this abortive attempt to avoid the broken ends of the paths in Figure 18.1, that the result that the particle mass, m, is proportional to τ^2—a result necessary for the Einstein–de Sitter model—arose precisely because all the particles were taken to begin abruptly at $\tau = 0$.

PROBLEM:

Return to the discussion of the $M - z$ relation given in Chapter 16, namely,

$$M = 5 \log (\sqrt{1 + z} - 1) + 2.5 \log (1 + z) + \text{a constant},$$

and notice how the derivation of this relation, valid for the Einstein–de Sitter model, depended on the particle masses being proportional to τ^2.

Let us accept the lesson we have learned from Figure 18.2, return to Figure 18.1, and attempt to find a physical explanation of the abrupt beginning of the particle paths of Figure 18.1. This leads us now to the problem we considered in Chapter 17, and illustrated there in Figure 17.2. We saw that a mathematical law (determined by an "action principle" method) can be set up to describe the paths of particles with broken ends. Once we permit this mathematical law to operate physically, paths with broken ends at other values of τ also arise, as in Figure 17.1, and the general solution of the resulting physical problem leads completely away from the big-bang cosmologies. Once paths with broken ends are described by a sensible physical principle, we arrive inevitably at the steady-state model. Yet the existence of the microwave background appears to vitiate the steady-state model. The dilemma is, then, that normal physical and mathematical reasoning seem to have led us to a contradiction between theory and observation.

Many people are happy to accept this position. They accept Figure 18.1 without looking for any physical explanation of the abrupt beginning of the particles. The abrupt beginning is deliberately regarded as *meta*physical—i.e., *outside* physics. The physical laws are therefore considered to break down at $\tau = 0$, *and to do so inherently*. To many people this thought process seems highly satisfactory because a "something" outside physics can then be introduced at $\tau = 0$. By a semantic maneuver, the word "something" is then replaced by "god," except that the first letter becomes a capital, God, in order to warn us that we must not carry the enquiry any further.

Attempts to explain phenomena by means of metaphysical intrusions into the world have always failed in the past. At the beginning of the nineteenth century, it was thought impossible to synthesize organic molecules by normal chemical processes. Nowadays, a whole industry is based on doing so. The origin of life was another supposed breakdown of the physical laws, which point of view also appears to have collapsed. It is true that phenomena have been discovered in the past that have forced a widening of the physical laws, the discovery of radioactivity, for example, but widening the physical laws does not change their basic logic. Of course, one can argue that the origin of the universe is by its very nature a special case. Although to many this last contention appears respectable, I prefer personally to rely on past experience. I do not believe that an appeal to metaphysics is needed to solve *any problem of which we can conceive.*

Is the time ripe for the solution of the dilemma described above? This question is perhaps the most crucial of any in astronomy, perhaps of any in physics. I will explain why I believe that inevitably the time must be ripe.

It is a curious aspect of scientific research that no matter what stage may have been reached, whether the sophistication is that of 1800, 1900, 1950, or 1975, perfect understanding always seems to be just around the corner, although to the workers at any one time it is clear that the similar confidence of earlier generations was wholly misplaced. Why do we have this illusion of perfect truth always awaiting us around the next corner in what is obviously a long and tortuous road? Because we cannot conceive of a problem until we are close to its solution. The solution to the problems of which we can conceive do indeed lie around the next corner of the road. We take no account of the problems that will torment the distant future, for the good reason that we cannot yet conceive of them.

I take it, then, that since the problem associated with $\tau = 0$ in Figure 18.1 is one of which we can conceive, is a problem that can be formulated, the time must be ripe for its solution. What line should we take? Should we perhaps return to the steady-state model, and simply refuse to bow to the criticism based on the microwave background? This was the attitude I took ten years ago toward criticisms based on the counting of radio sources, and it has turned out now that those criticisms were not so strong as they first seemed. Perhaps the criticisms based on the microwave background will also turn out to have been overstated, and will weaken with the passage of time? Possibly so, but at this stage, after ten years or more of attempting to "tough it out," I prefer to try to break out of the restraining cycle of argument described above. Let us attempt to do so at what may at first sight seem an unprofitable point, the extension of Figure 18.1 to Figure 18.2, the extension which led to infinite mass values. Let us see if the infinities can be avoided in some way.

*Final
Considerations*

We know that there can be both positive and negative contributions to an electrical field, depending on the signs of the electric charges of the contributing particles. Aggregates of matter contain particles that have both positive and negative charge (protons and electrons) and hence make both positive and negative contributions to the field.

Let us try a similar idea for a field giving rise to particle masses, with positive and negative contributions coming from distinct, very large-scale aggregates, large even compared to the distances of remote galaxies. There is to be an important difference from the electrical case, however. An individual aggregate generates either a systematically positive contribution or a systematically negative contribution, not a mixture of the two. For example, suppose we consider positive and negative contributions to be separated on a cosmological scale at the time $\tau = 0$, as in Figure 18.3. Then, provided we make one important change of method, we arrive at calculated particle masses proportional to τ^2, just as we require for the Einstein–de Sitter model.

In Figure 18.4 we have a mass interaction of the form that was used in Chapter 16. The portions of the paths of particles a and b shown in the figure are considered to be on the same side of $\tau = 0$ (the positive side, say), and with τ at point A on the path of particle a at a later time than at point B on the path of particle b. In our previous method of calculation, the mass of particle a at the point A was considered to receive a contribution from point B, *but the mass of particle b at point B was not considered to receive a contribution*

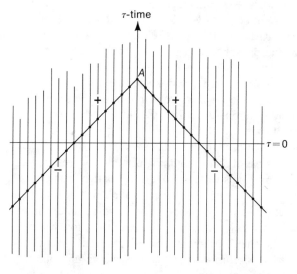

FIGURE 18.3.
Illustrating the hypothesis that positive and negative mass interactions are separated at time $\tau = 0$.

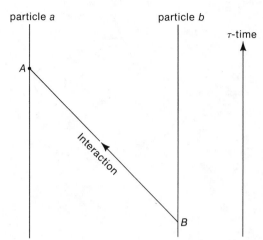

particle a   particle b

τ-time

A

Interaction

B

FIGURE 18.4
The situation of Chapter 16, where the mass of particle
a at the point A received a contribution from the
(earlier) point B of particle b, but the mass of b at B did
not receive a contribution from A.

from A; that is, the method of calculation had an undesirable asymmetry in it. Let us make the interaction of Figure 18.4 symmetric. In Newtonian terminology, action and reaction are made equal and opposite. This change is indeed a major improvement of method, and it leads to the required result that the particle masses are proportionel to τ^2.

PROBLEM:
Prove this last statement.

The dilemma set out above is therefore resolved by the system of Figure 18.4. The need for the particle paths to have broken ends has been avoided. But why is there a switch in the sign of the interaction at $\tau = 0$? Why is this particular moment so special? These are legitimate questions, and it is encouraging to find that they can be answered by a quite powerful generalization of the specialized arrangement of Figure 18.3.

§18.3. A General Formulation of the Mass Interaction

Consider the situation shown in Figure 18.5, where there are more than two + and − aggregates. (As always, we note the illustrative limitations imposed by the need to represent what is really a four-dimensional situation in terms of only two dimensions.) To determine the sign of the mass interaction between points A and B on the paths of particles a and b, shown in Figure 18.6, we multiply the signs appropriate to the regions in which A and B happen to fall. If A and B are both in + aggregates, then the sign of the interaction

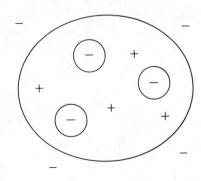

FIGURE 18.5.
Schematic representation of large
scale + and − aggregates.

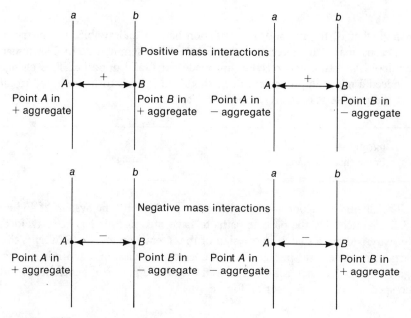

FIGURE 18.6.
The rule for determining the sign of the mass interaction between point *A* of *a* and
point *B* of *b* is that the sign is + if both the points *A* and *B* are in aggregates of the
same type; otherwise the sign is −.

is $(+1) \times (+1)$, and so is positive. If *A* and *B* are both in − aggregates, the
sign is determined by the product $(-1) \times (-1)$, and so is again positive But
if one point is in a + aggregate and the other in a − aggregate, the sign is
given by $(-1) \times (+1)$, which is negative. This is the meaning to be attached
to the ± signs of Figure 18.5.

Figure 18.5 is to be thought of as a schematic representation of the universe on a scale much larger than the portion of the universe accessible to practical observation. Indeed, we are to think of our observations of all the galaxies, even the most remote ones visible in the largest telescopes, as occupying only a comparatively *small element* of just one of the aggregates of Figure 18.5; for definiteness, let us say a + aggregate.

The interactions on a particle at an arbitrary point anywhere in the universe will in this picture be a complicated addition of contributions from all the various aggregates. If we make the sensible assumption that *on the average —* aggregates are as important as + aggregates, the combined effect of all inter-actions at an arbitrary place is as likely to be negative as positive. Regions where the contributions add to a positive total will be separated from regions with a negative total by surfaces on which the + and − contributions just cancel each other. The mass of a particle at the points on such surfaces will be zero.

In an extremely general way, we have now arrived at a crucially important result: the essential lesson we learn from the study of cosmology, especially from the study of the redshifts of the galaxies, is that *we happen in our element of the universe to lie near such a surface of zero mass.* Notice that the surface in question need not be a single unique surface, as we took it to be in Figure 18.3. To emphasize the point that our observations range over only a small element of the universe, let us show, in Figure 18.7, that Figure 18.3 is indeed

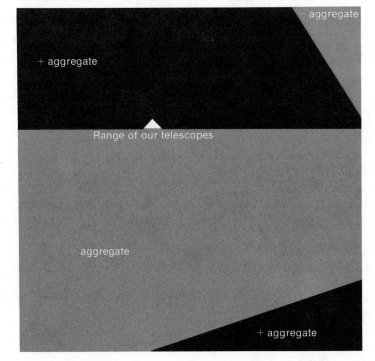

FIGURE 18.7
Our studies of cosmology may be concerned with only a small element of a much vaster universe.

only a small element of Figure 18.5, noticing that our studies of cosmology have been concerned only with this tiny region of a much vaster universe.

On the scale of this vaster universe, we can no longer use the second of the two "principles" of cosmology enunciated in Chapter 15. Instead, we must grapple with the full complexity of Riemannian geometry. It is only on the scale of the small triangle of Figure 18.7 that this so-called "principle" of cosmology holds good—indeed, it is just because we happen to be close to a zero-mass surface that it does so. Discovering the full-scale world geometry will therefore be a very difficult task; it will be far from easy to obtain the detailed forms of the zero-mass surfaces shown schematically in Figure 18.5.

Even so, it is surprising how much insight we can gain into the effects of the local zero-mass surface shown in Figure 18.3. Suppose light travels in the same sense on both sides of this surface, and suppose that galaxies and stars also exist on the "other side." Would we then expect to be able to observe these other galaxies? Unfortunately, not directly, because all the light from the stars of such galaxies must be strongly scattered, absorbed, and blurred at times close to $\tau = 0$ by particles lying between the stars. This strong blurring effect is caused by the smallness of the particle masses near $\tau = 0$. However, the blurred radiation would continue into "our half" and would indeed be observable. It would have just the black-body form we studied in §16.5. It would, moreover, have an energy content determined by the conversion of hydrogen into helium within the stars on the "other side." This energy can be calculated, and can be shown to lead to a temperature on the order of $3°K$ for the blurred radiation, just as the observed microwave background is found to have. Therefore, the existence of the microwave background (so apparently damaging to the steady-state model) favors the ideas presented here quite strongly. We are not required to postulate its existence *ad hoc,* as one must do in the usual big-bang cosmologies. With some justice, we might argue that the microwave background demonstrates the existence of the "other side" of the local zero-mass surface.

The concept of galaxies existing on the "other side" may also be related to problems of the clustering of galaxies on "our side." Astronomers have long suspected that galaxies are more clumped together than might reasonably be expected if only random factors are at work. Figures 18.8 to 18.12 relate to other problems that we have discussed already in preceding chapters and which may receive full solution only when conditions on the "other side" are incorporated into our astronomical theories.

In Figure 18.8 we have two strange galaxies, Markarian 474 with a redshift $z = 0.041$, a peculiar spiral NGC 5682 with $z = 0.0073$, and a QSO (quasar) with $z = 1.94$. According to orthodox astronomy, these objects are at very different distances away from us, and hence are unrelated physically to one another; their association together is only apparent, a chance projection on the sky. But why then is the spiral galaxy NGC 5682 so disturbed in appearance, as if it has been subject recently to strongly distorting forces? And Markarian 474 is also peculiar, as its designation implies (the Markarian catalogue is of unusual objects). It would be remarkable for a chance projection to associate not only

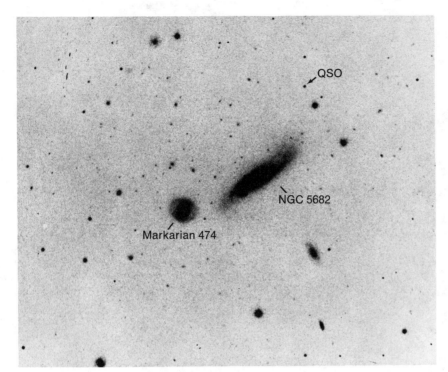

FIGURE 18.8.
The galaxy NGC 5682 has a redshift $z = 0.0073$, Markarian 474 has $z = 0.041$, and the QSO has $z = 1.94$. The association could be due to chance, but then it is strange that these objects should be so peculiar in their individual forms. (Courtesy of H. Arp, Hale Observatories.)

widely different values of the redshift, but also such peculiarities of internal form.

Much of what we have learned about astronomy in general is well brought out by Figures 18.9 to 18.11, which are all of NGC 1097, a galaxy best seen in the southern hemisphere. These photographs show how, by taking exposures in various ways, we discover the galaxy to have quite different properties. Figure 18.9, in the line Hα of hydrogen, shows a bright central nucleus and disc, surrounded by remarkably thin spiral arms. The bright condensations, strung like beads along the arms, are HII regions (Chapter 8), glowing clouds of ionized hydrogen made visible by associations of bright young stars not more than a few million years old. In the intermediate exposure of Figure 18.10, the arms broaden and strengthen in appearance, but still have myriads of HII regions. The pattern now suggests a dramatic swirling motion for the whole galaxy. In the long exposure of Figure 18.11, the central detail has been "burned out" of the photographic plate, but now a profusion of objects, star clusters and what are perhaps faint small galaxies, can be seen surrounding the whole system.

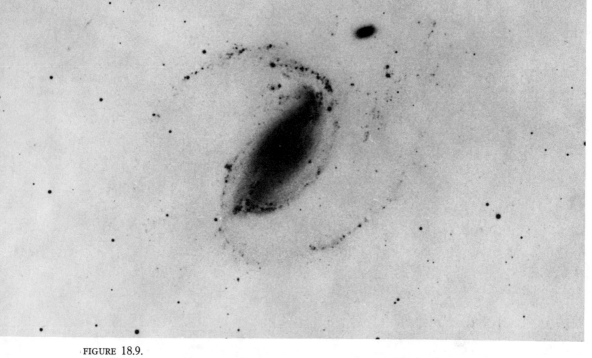

FIGURE 18.9.
The galaxy NGC 1097, in the line Hα of hydrogen. Notice the remarkably thin outer spiral arms with many Hɪɪ regions strung out along them. (Courtesy of H. Arp, Hale Observatories.)

FIGURE 18.10.
An intermediate exposure of NGC 1097, suggesting a swirling rotary motion. (Courtesy of H. Arp, Hale Observatories.)

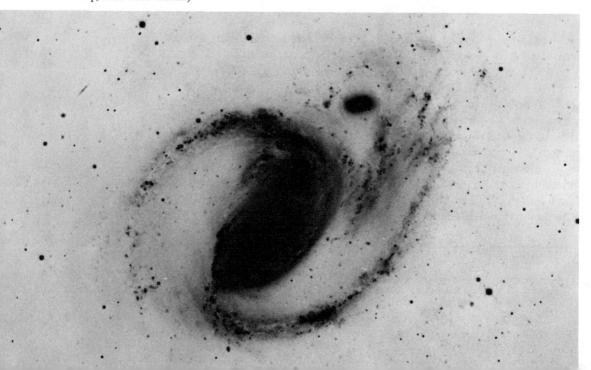

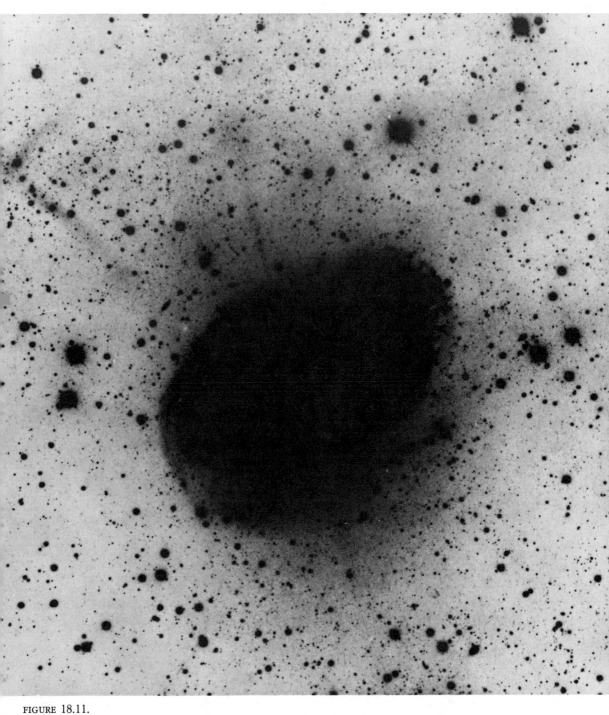

FIGURE 18.11.
A long exposure of NGC 1097. Note the remarkable streamers directed from the center. The streamer at upper left has a chain of five soft images, possibly small subsidiary galaxies, seemingly associated with it. (Courtesy of H. Arp, Hale Observatories.)

Two streamers from the center to the upper left are easily detected, and a counterstreamer in the opposite direction from one of these is probably also present. The leftmost upper streamer has a chain of five soft images, apparently small subsidiary galaxies, strung along its outer part, just before the streamer makes a sudden curve to the right. This chain appears to be associated with the streamer and hence with the main galaxy. What, one wonders, will the redshift values of the five images turn out to be?

The streamers suggest the presence of an explosive object at the center of the galaxy, an object perhaps similar in its physical properties to a QSO. The general halo of luminosity surrounding the galaxies could be an overlap of many streamers, older than the ones we see, which are almost certainly very young compared to the age of the galaxy itself—and probably young even compared to the rotation period of the galaxy. It is likely that we are observing the effect of explosions that have occurred during the last ten to a hundred million years.

General Problems and Questions

1. Rejecting a metaphysical origin of the universe, discuss the dilemma of the usual cosmological models.

2. How may this dilemma be avoided in a more broadly based theory that allows for the possible existence of plus and minus aggregates of mass?

3. What limitation does this broadly based theory imply for the range of astronomical observations?

4. What may be the origin of the microwave background?

Appendixes to Section VI

Appendix VI.1. The Form of the Mass Function near
a Zero Surface

The picture we arrived at in Figure 18.7 is one in which our observable universe, even taken to the most distant galaxies, is yet only a small element near a surface of zero mass. More precisely, what do we mean by "near"? Close enough to such a surface, the particle masses must vary more rapidly in a direction perpendicular to the surface than they do in the directions parallel to the surface. Provided we are close enough for variations parallel to the surface to be negligible, we say that we are "near" the surface. The mass m of a particle depends then only on the distance from the surface, which we denote by τ, measuring distance with the interparticle separation L as our unit, as in Chapter 16. Quite generally, we can write $m = \alpha\tau + \beta\tau^2 + \ldots$, where the unwritten terms are negligible provided τ is small enough. Here α and β are constants that do not depend on τ. These constants would in principle be calculable if we knew the precise distribution of the $+$ and $-$ aggregates of Figure 18.5. Since we do not, it might seem that nothing further can be said concerning α and β. Yet certain very general and important conclusions can be reached, according to whether α is zero or not.

It is possible to prove *rigorously* that if Einstein's theory of general relativity is to remain valid in our new situation, then the constant α must be zero. With

α zero, the mass m is proportional to τ^2, which is the condition required by the Einstein–de Sitter model. It follows that the apparent cosmological structure in any element of spacetime close enough to a surface of zero mass must be of the Einstein–de Sitter form *if* Einstein's theory of general relativity is correct.

As the distance τ from the zero surface increases, the mass m can develop appreciable variations in directions parallel to the surface. Should this happen, there is a departure from the Einstein–de Sitter model. Thus the latter model is only of *local* applicability, where by "local" we now imply the situation of Figure 18.7, in which the range of the model is considered to encompass the whole of the astronomer's "observable universe." Notice that we are now using the concept of "local" very differently from the way it was used in the Appendixes to Section II. The local geometry here is Einstein–de Sitter, and it extends to the most distant observed galaxies.

Remembering the classification of geometries given in Chapter 15, as τ increases, does the geometry remain Einstein–de Sitter, or does it become of class B or class C? The answer here is that we are no longer able to make the classification of Chapter 15, because one of the two principles on which the arguments of Chapter 15 were based need not be valid here. The "cosmological principle" of Chapter 15 will not be satisfied in the general situation of Figure 18.5. The large-scale geometry continues to be Riemannian, but it can be much more general than anything we considered in Chapter 15.

If we require Einstein's theory to be correct, this is as far as we need go. But since we have no certain knowledge that Einstein's theory is indeed correct,

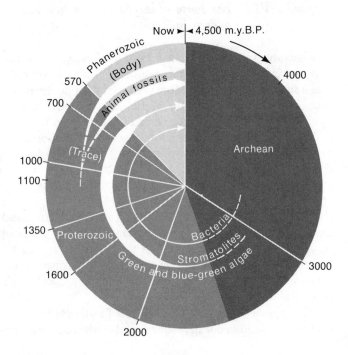

FIGURE VI.1.
The biological "time clock"
(from Figure 12.35).

it is reasonable to ask: what happens otherwise? The answer turns out to be that the constant α is not zero, and near enough to the zero surface the mass m is dominated by the term $\alpha\tau$. In this case the model is *not* that of Einstein and de Sitter. It turns out that the resulting local cosmological structure has the properties considered already in Chapter 12. Figure 12.35 has been reproduced here as Figure VI.1. An unorthodox theory, in which the gravitational constant in Newton's formula, Gm_1m_2/a^2, is considered to vary with time, was discussed in relation to Fig. 12.35. It is just this unorthodox theory which follows if the constant α is not zero. The cosmology near enough to the surface of zero mass is then of a type first proposed and discussed by Paul Dirac.

Appendix VI.2. What is Time?

There are many ways to set up systems of time and space measurements, which can be connected with each other by mathematical formulas, in such a way that, if we know from observation how to describe our everyday world in terms of one system, then we can work out how the world should be described in terms of any one of the other systems. Although for basic physics there is thus no unique way of measuring time and space, it is known that, both for our everyday experience and for astronomy generally, one specific time-space description turns out to be *simpler* than the others, and it has always seemed mysterious why it should be simpler.

We are now in a position to clear up the mystery. The simple description results from taking the time direction as being perpendicular to the zero-mass surface. The spatial directions are then parallel to the zero surface. In short, the zero surface decides the situation, since it is when we "travel" in the direction perpendicular to the zero surface that things change in a crucial way. The position is illustrated in Figure VI.2.

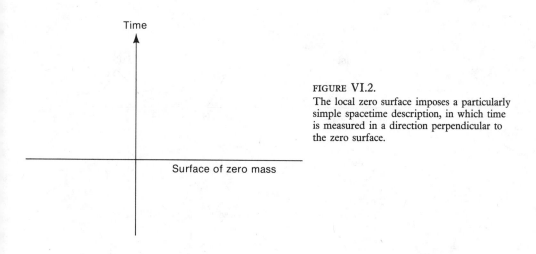

FIGURE VI.2.
The local zero surface imposes a particularly simple spacetime description, in which time is measured in a direction perpendicular to the zero surface.

Figures 16.3 and 16.4 are reproduced here as parts (a) and (b) of Figure VI.3. In (a), atoms are considered to remain of the same size and the galaxies to expand apart from each other. We found in Chapter 15 that the scale of a polygon formed by joining several galaxies, for example the three galaxies forming the triangle of Figure VI.3(a), could be described by a quantity Q which varied with the time t measured by means of radiation emitted by atomic transitions of a specified kind. We found that the geometries of types A, B, C, introduced in Chapter 15, behaved differently with respect to Q. The geometry of type B had the behavior reproduced here in Figure VI.4.

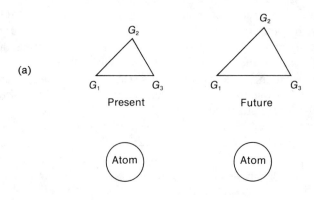

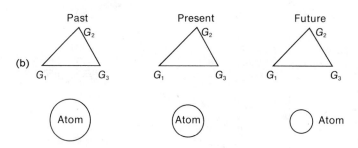

FIGURE VI.3.
Figures 16.3 and 16.4 are reproduced here as (a) and (b).

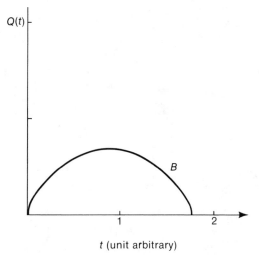

FIGURE VI.4.
The form of the $Q(t)$ curve for a world geometry of
type B.

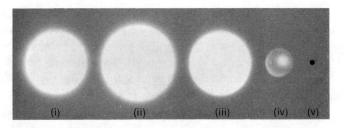

FIGURE VI.5.
The expansion and subsequent contraction of 'a local object without
internal pressure follows rules that are strikingly similar to those of a
world geometry of type B. The local object begins by expanding (i)
to maximum size (ii), then falls back (iii), and continues to fall (in)
until it eventually becomes a black hole (v).

These cosmological considerations have relevance to localized objects. Consider a spherical distribution of material of uniform density suddenly set in uniform outward expansion, as in Figure VI.5. If this impulsive outward motion is violent enough, the object will dissipate itself, as in an explosion. But for a modest outward impulse, the motion will eventually be checked by the gravitational attraction of the object on itself, and the expansion will be followed by contraction, just as in the cosmological example in Figure VI.4. Indeed, if we can neglect internal pressure within the object, then the motion can be described by a scale factor Q which has the same behavior as that of a cosmology

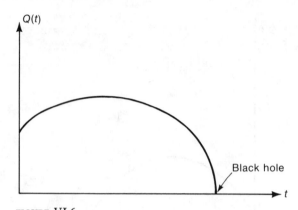

FIGURE VI.6.

The $Q(t)$ curve for such a local object reaches a maximum, and then turns over and decreases to zero. This is the phenomenon of the black hole.

of type B, except that Q does not start from zero, as in Figure VI.4. Instead we have the behavior of Figure VI.6. The critical aspect of Figure VI.6 is that Q declines to zero, which means that the object collapses into a point, where according to usual ideas it ceases to exist. It is said to have formed a *black hole*.

The question now arises whether we can proceed further, as we did in the cosmological case. Can we avoid the object's ceasing to exist in somewhat the same way that we resolved the dilemma of the origin of the universe? We resolved the latter dilemma by changing from the picture of Figure VI.3(a) to that of Figure VI.3(b), and by replacing Figure 18.1 by Figure 18.3—i.e., by adding a "second half" to the universe. The answer for a local object is that we can indeed proceed in a similar way. The "second half" which is thus introduced can be referred to as a *white hole*. The structure developed by a local collapsing object consists, then, not just of a black hole, but of a black hole together with a white hole.

Thinking back now to the problem of radio sources and of their outbursts, discussed in Chapter 14, it is natural to wonder if such phenomena might be connected with black holes and white holes.

The spatial scale of a black-hole/white-hole ranges from a few kilometers for a body with mass of stellar order to about $1/100$ of a light year for a body with mass of galactic order. The nature of the zero-mass surface requires the following sequence of events. The object collapses into a black hole as the zero surface is first approached. Immediately then, on crossing the surface, the body becomes a white hole. This first white hole does not persist, however, but changes for a second time to a black hole as a second condition of zero mass is reached by the particles of the body. Thereafter, a second white hole emerges, which would well be observable as an outburst of the kind that is found in QSO's and radio-galaxies.

Astronomers have recently come to believe that a black hole has been observationally detected. A source of x-rays known as Cygnus X-1 is coincident in the sky with the star HD 226868 (its number in the Henry Draper catalogue). Spectral analysis of the optical light from this star shows that it is a highly luminous star of class B (see Table II.1), with a technical subclass designation of 0Ib. Stars of this spectral type are known to be of large mass, not less than about $15\odot$. This fact is crucial to the following argument.

It will be recalled from Appendix II.8 that the wavelengths of the spectrum lines in the light emitted by a star will undergo a periodic oscillation if the star is a component of a binary system. The time taken for a complete oscillation of the lines is determined by (and so reveals to us) how long it takes the two components of the binary system to make a complete circuit of their orbits around each other, in the manner of Figure II.45. The spectrum lines of HD 226868 are found to have such an oscillation, with a cyclic period of 5.61 days. Only one set of comoving spectrum lines is actually observed; so the companion star to the B0Ib component must have too small a luminosity for its spectrum lines to be separately resolved. If the system were not emitting x-rays, the unseen companion could be taken simply to be a main-sequence star with mass appreciably less than $15\odot$. However, the x-ray emission from the system argues that the less-massive companion must be highly evolved, as in the binary systems discussed in Appendix III.7 for the x-ray sources Centaurus X-3 and Hercules X-1. It will be recalled that the x-ray emission from Centaurus X-3 pulsates in the very short time of 4.8 seconds, and that from Hercules X-1 pulsates in the even shorter time of 1.2 seconds. The x-ray emission from Cygnus X-1 has been found to undergo changes in times as short as 0.001 second; so the unseen companion must be still more evolved and compact.

The speed of motion of the B0Ib star in its orbit is unexpectedly large, too large for a system, such as Centaurus X-3 or Hercules X-1, in which the unseen companion is thought to be a white dwarf or a neutron star. If the line of sight from Cygnus X-1 to the Earth happens to lie in the plane of the binary orbit, then from the periodic shift of the spectrum lines we can calculate (by means of the Doppler shift) the speed of the B0Ib component; it is about 65 km per second, according to current observations. This orbital speed, and the period of 5.61 days, require the unseen component to have a mass at least $\frac{1}{4}$ of that of the B0Ib star, i.e., a mass of $3.75\odot$. No white dwarf or neutron star could have a mass as large as this, since the pressure from neither electrons nor neutrons could support such a mass against the strength of gravity. Instead of becoming a white dwarf or neutron star in the manner discussed in Chapter 7 (pp. 335-340), a mass of $3.75\odot$ would become a black hole. The black hole would have a radius of only about 12 kilometers. It would be surrounded by an x-ray-emitting region extending outward some 300 kilometers—large compared to the black hole itself, but still only some 5 per cent of the radius of the Earth.

One question remains. We have assumed that the line of sight to the Earth from this binary system lies in the plane of the binary orbit. Is this assumption critical for whether the object could be a black hole? The answer is no, because any other orientation of the orbital plane to the line of sight would merely *increase* the mass that the unseen companion must have, and we would thus be led to the same conclusion.

Appendix VI.4. The Problem of Primordial Helium and Deuterium.

We are now in a position to relate our development of cosmology to an important problem raised in Appendix III.4, namely, the origin of the element helium. Except for a certain class of star in which helium appears to be largely absent from the surface layers, helium is found in all objects in our galaxy. (The exceptional stars are ones in which helium is very likely segregated by gravity, and hence are probably not relevant to the problem of the ubiquitousness of helium in the universe.) Helium is also found in neighboring galaxies. Everywhere the abundance seems to be about the same: some 25 per cent of the mass of cosmic material consists of helium. If we were entirely sure that all the abundances were closely the same, or even if we could be certain that the abundance never falls below some particular value, 20 per cent, say, there would be a strong case for supposing that most of the observed helium was not generated in stars by the processes discussed in Section III. However, there is inevitably some ambiguity in the observations, because helium is an awkward element to deal with, essentially because its strongest quantum transitions emit radiation with frequencies in the far ultraviolet, which do not penetrate the Earth's atmosphere. Consequently, all measurements of the helium abundance are subject to some uncertainty. When we say that the mass fraction of helium is everywhere about 25 per cent, in some places the value could be 30 to 35 per cent, in other places it could be 15 to 20 per cent, without doing violence to the data. So we cannot be certain that there really is a standard cosmic value for the helium abundance, as is sometimes asserted in the astronomical literature. Nor do we know much about the helium abundance in galaxies other than quite nearby ones.

The main argument for supposing that helium might have originated very early in the history of the universe comes from the quantitative difficulty, already noted in Appendix III.4, of explaining an abundance of helium as high as 25 per cent in terms of processes within stars. If as much as 25 per cent of hydrogen had been converted to helium, and if the resulting energy production escaped from the galaxies as visible light, galaxies would be much brighter than they are observed to be. This argument had to be considered a strong one when it was thought that most of the radiation from galaxies is visible light, but it has been found recently that very strong radiation sometimes occurs at infrared frequencies. The argument is therefore not as strong as astronomers used to believe. However, the argument is certainly quite strong enough to make it attractive to seek a primaeval origin for most of the helium. Such a theory must still be regarded as the frontrunner. Yet at the same time we must note that a more conventional explanation of helium synthesis, in terms of straightforward nuclear processes within stars, has come recently to seem more plausible than it used to.

With this caveat, let us proceed to discuss the primaeval origin of helium, assuming that the particles present at early time τ are of the familiar kind, not the more esoteric superparticles discussed in Chapter 16. The relevant particles for the formation of helium are neutrons and protons. Neutrons are unstable through $n \longrightarrow p + e + \bar{\nu}_e$, decaying under laboratory conditions in a characteristic time of about 10 minutes. However, in the early stages of the universe, when the particle masses were much less than they are today, the time required for neutrons to decay was very much longer than this, so much so that the probability of neutrons and protons becoming associated together, $n + p \longrightarrow D$, was fairly high. The deuterons so formed become largely converted, by $D + n \longrightarrow T$ and by $D + p \longrightarrow {}^3He$, into tritium (T) and ${}^3He$. Then ${}^4He$ was formed by the addition of neutrons to ${}^3He$ and protons to T. In the Einstein–de Sitter model, the outcome can be shown to be a helium production amounting to 28 per cent of the mass of the primaeval material.

TABLE VI.1.
Some light nuclei that cannot be formed by synthesis within stars.

Nucleus	Approximate Mass Fraction
D	2×10^{-4}
${}^3He$	6×10^{-5}
${}^6Li$	10^{-9}
${}^7Li$	10^{-8}

Several other light nuclei are found, on the Earth, in the Sun, in the meteorites, with concentrations that are far too high to be explained by synthesis within stars. They are given, together with their mass fractions, in Table VI.1. Processes outside stars, or at the surfaces of stars, involving high-speed particles, can explain the origin in kind of these nuclei. Whether it is also possible by such processes to explain the mass fractions of these nuclei remains a matter of controversy, especially for deuterium. Because of its comparatively high mass fraction, D is harder to account for in this way than ${}^3He$, ${}^6Li$, ${}^7Li$. Hence D may also have originated from a primaeval synthesis.

We have been discussing the emergence of material from a zero-mass surface, the time-sense of the emergence being the same as the sense of the propagation of light. Material emerging on what we called "our side" of the surface is now seen to be essentially hydrogen and helium. This composition permits stars like the Sun to generate nuclear energy. Indeed, astrophysics as we understand it, and stellar evolution in particular, must be associated with regions of emergence from zero-mass surfaces.

As living creatures we are dependent on sunlight, and the Sun in turn is dependent on a plentiful supply of hydrogen. The need for hydrogen therefore forces us to live in the proximity of a zero-mass surface, in the manner of the small triangle of Figure 18.7. This is why our locality in the larger-scale universe must be close to just such a surface.

References for Section VI

Einstein's theory of gravitation, "Die Grundlage der allgemeinen Relativitäts-theorie," appeared in *Annalen der Physik,* 49 (1916), 769.

Riemann's great discoveries in geometry were published posthumously, in *Göttinger Abhandlungen,* 13, (1868), 133. The title of the work was *Über die Hypothesen welche der Geometrie zugrunde liegen.*

Other papers of relevance to this section are: K. Gödel, "Example of a New Type of Cosmological Solutions of Einstein's Field Equations of Gravitation," *Reviews of Modern Physics,* 21 (1949), 447; H. Bondi and T. Gold, "The Steady-State Theory of the Expanding Universe," *Mon. Notices Roy. Astr. Soc.,* 108 (1948), 252; F. Hoyle, "On the Origin of the Microwave Background," *Astrophysical Journal,* 196 (1975), 661.

The binary nature of Cygnus X-1 was discussed by B. L. Webster and P. Murdin, "Cygnus X-1: A Spectroscopic Binary with a Heavy Companion," *Nature,* 235 (1972), 37, and by C. T. Bolton, "Identification of Cygnus X-1 with HD 226868," *Nature,* 235 (1972), 271. For a recent discussion of the x-ray emission from a disc of material surrounding a rotating black hole, see K. S. Thorne and R. H. Price, "Cygnus X-1: An Interpretation of the Spectrum and Its Variability," *Astrophys. J. Letters,* 195 (1975), L101.

Glossary

Absolute magnitude: The absolute magnitude of a star is the magnitude the star would have if, instead of being at its actual distance, it were located at a distance from us of 10 parsecs, which equals 32.6 light years. A similar definition of absolute magnitude is used for galaxies and for other objects, such as globular clusters, although in fact another galaxy could not fit inside our galaxy at only 10 parsecs away from us.

Albedo: The sunlight that falls on a planet, satellite, or smaller body in the solar system is partly reflected and partly absorbed. The reflected fraction is the albedo. Since the light is not reflected uniformly in all directions, a practical calculation of the brightness of a planet or satellite must allow for the fact that the reflection depends on the angle between the direction from the Sun to the object and the direction from the Earth to the object. This dependence is known as a "phase effect."

Amino acid: A molecule with H, NH_2, and COOH, and a side chain R bonded to a carbon atom. The structure of R differs from one type of amino acid to another. There are many amino acids, of which 20 play a critical role in the biochemistry of life.

Ångstrom (Å): A unit of length equal to 10^{-8} centimeter.

Astronomical unit (AU): The average distance of the Earth from the Sun, close to 149.6 million kilometers.

Atomic number (mass number): The total number of neutrons and protons in an atomic nucleus is called the atomic number or the mass number.

Atoms: Bulk materials can be broken down into atoms, of which 103 different kinds are known at present. Of these, 20 undergo comparatively rapid radioactive decay, and so are not found in appreciable quantities in the materials of our everyday world. Atoms consist of a nucleus, which is very small in size (10^{-13} to 10^{-12} cm) and contributes most of the mass, surrounded by a cloud of electrons that extends farther out (to about 10^{-8} cm). The nucleus is made up of protons and neutrons. In a neutral atom, the number of protons is equal to the number of electrons. This number determines the chemical properties which define the type of the atom; for each value of this number, the atom is that of a distinct element.

Baryons: There are many forms of baryon, of which the proton and neutron are examples. Together with six others, the proton and neutron form an important family of eight baryons. The proton is the stable form for an isolated baryon, the neutron decaying into a proton in about 1,000 seconds, the other six decaying in a very much shorter time (10^{-10} sec). When several baryons are associated together, the stable form consists of a mixture of protons and neutrons. Such mixtures form the nuclei of atoms. These properties arise from the curious circumstance that the masses of the neutron and proton are nearly equal, whereas the masses of the other baryons are all much larger than that of the proton or the neutron.

Beta processes: The nucleus of an atom with any given total number of baryons adjusts the numbers of the protons and neutrons until the nucleus takes its most stable form. This happens by means of beta processes, which can either change a proton into a neutron or change a neutron into a proton.

Binary stars: A binary consists of two stars which move around a common center known as the barycenter. The line joining the stars always passes through the barycenter, and the orbit of each star is an ellipse with the barycenter as one of its two foci. Binary stars probably originate in the condensation process which gives rise to the stars themselves.

Black hole: Since gravitation tends to pull matter together, large-scale associations of material tend to evolve in such a way that the matter becomes more and more condensed, until in the ultimate limit a black hole is formed. Conventional ideas of geometry do not apply in the vicinity of a black hole, and it is this breakdown of conventional ideas that leads to the limit of the condensation process. The size of a black hole depends on the mass of material involved. A black hole with a mass equal to that of the Sun would be about 6 kilometers across, and one with a mass equal to that of our galaxy would be about 0.1 light year across.

Blueshift: Radiation from a moving object is said to be blueshifted when lines in the visible spectrum are shifted (*see* Doppler shift) toward the blue end of the spectrum. This happens when the object in question is approaching the observer.

Bolometric magnitude: When all the radiation from an object is included in the measurement of its magnitude, the result is called the bolometric magnitude. Thus the bolometric magnitude includes contributions from the radio band, the infrared, visual and ultraviolet light, x-rays, and γ-rays. Such measurements are difficult to make, because the atmosphere of the Earth is opaque, except in the visual and near infrared and in the radio band. Thus the bolometric magnitudes mentioned in astronomical literature are almost always estimates rather than direct measurements.

Carbon-nitrogen cycle: The carbon-nitrogen cycle is a sequence of nuclear reactions whose net effect is to convert hydrogen into helium. It is one of the two main processes responsible for the production of energy in most of the stars we see in the sky. The carbon and nitro-gen atoms promoting the cycle are only rarely changed (to oxygen), and even then there are further nuclear reactions that eventually regenerate nitrogen. Thus the carbon and nitrogen atoms behave catalytically.

Celestial sphere: The celestial sphere is an imaginary sphere with its center at the center of the Earth. It is usually considered to have a radius larger than the solar system but not as large as the distances of the stars. The precise radius of the celestial sphere is irrelevant to its purpose, however, which is to provide a surface on which all astronomical objects are projected. The celestial sphere moves with the Earth's motion around the Sun, causing the projected positions of stars and all other distant objects to show slight annual wobbles (*see* Parallax).

Cepheid variables: A star that undergoes regular radial pulsations, with its radius oscillating by about 10 per cent and its luminosity by about 100 per cent, the Cepheid variable is detected observationally from its characteristic variations of luminosity, the so-called light curve. The average intrinsic luminosity is related to the period of the variation in a known way. Hence an observational measurement of the period is equivalent to a measurement of the intrinsic luminosity. Knowledge of the intrinsic luminosity, taken with the observed apparent luminosity, allows us to calculate the distance of the star. Cepheid variables can thus be used as distance indicators. The typical surface temperature is not much different from that of the Sun. Periods range from about 2 days up to about 100 days, and intrinsic luminosities range from about 100 to about 10,000 times the luminosity of the Sun.

Chemical elements: *See* **Atoms**

Color temperature: The relative intensity distribution of the various colors (frequencies) in the light emitted by a star depends on the temperature of its surface. Observation of the relative intensity distribution can therefore be used to calculate the surface temperatures of stars, and such calculated temperatures are called color temperatures. Because the absorption of light by the atmosphere of the Earth prevents a ground-based observer from making measurements in the ultraviolet and far infrared, this method of calculating temperature becomes subject to error both for very hot stars and for very cool ones.

Coma: Coma is the failure of a telescope's objective, whether it is a lens system or a mirror, to bring objects that are slightly off the telescope's axis to a sharp focus. Correcting systems are used to minimize this defect in all large optical telescopes. The corrector in a conventional reflecting telescope consists of a series of glass lenses placed near the focal plane. In Schmidt-Kellner telescopes, the corrector is a sheet of glass placed across the telescope tube. Coma is not the only optical defect of the astronomical telescope, but it is the most serious.

Comet: A comet is thought to consist initially of one or more icy bodies with a total volume that is typically about 1,000 cubic kilometers. Ordinary water ice and dry ice are important components, but other normally volatile materials are also present. Embedded in these ices are myriads of small refractory particles, which remain as a swarm if the ices become evaporated when a comet approaches the Sun. This approach can happen because the highly elliptic orbits of comets are rather easily perturbed (changed in form), mainly by the gravitational effect of the planet Jupiter and perhaps also by the influence of nearby stars. The remarkable appearance of spectacular comets arises from the evaporation of the ices.

Constellations: From ancient times, the stars of the northern night sky have been divided into groups known as constellations. The groupings were usually decided by chance associations which happened to suggest the shapes of objects or people. From these chance associations, and from myths, the constellations received names, which are retained in modern astronomy because the names are convenient labels for specific bits of the sky. The constellations of the far south were named in modern times, and not so imaginatively as those in the north.

Cosmology, principles of: Literally, cosmology means the study of the cosmos, i.e., of anything outside the Earth. In recent years, however, cosmology has come more and more to mean the study of the largest-scale aspects of the universe. To facilitate this study, it is usual to make two simplifying postulates, which are known as the "principles" of cosmology. One principle holds that the same physical laws apply in every locality and at all times; the other principle states that the large-scale structure of the universe is the same in all directions (isotropy) and is also the same when seen from different localities (homogeneity).

Declination and right ascension: The declination and right ascension of an astronomical object are its latitude and longitude values measured in the following way. The object is considered to be projected on the celestial sphere. Projected also are the Earth's equator and the annual path along which the Sun moves in the sky. This gives two great circles on the celestial sphere, known as the celestial equator and the ecliptic, respectively. The declination and right ascension values for the projected astronomical object is then given by measuring both from and along the celestial equator. The measurement along the celestial equator is taken from the point where the Sun (and so the ecliptic) crosses the equator at the spring equinox. For a visual representation of this definition, see Figure 1.12.

Doppler shift: Given an atom of a specified element and a specified quantum transition of it, radiation generated by that transition in an assembly of such atoms has a well-defined frequency and wavelength. If an observer takes such an assembly that is at rest with respect to himself, he can use the frequency and wavelength of the radiation to establish units of time and of length; time measured by means of such a unit is called the observer's proper time. A second observer in motion with respect to the first can proceed in the same way, using similar atoms that are at rest with respect to himself, and that are therefore moving with respect to the first observer. The units of time and of length thus obtained by the second observer, and his proper time, are not the same as those of the first observer. The Doppler shift expresses the relation between the two systems. It does so in terms of the relative motion between the two observers.

Eclipse: The light from a distant object is said to be eclipsed when a second object crosses the line of sight from the Earth to the first object. A solar eclipse occurs when the Moon comes between the Earth and the Sun. A binary eclipse occurs when one star in a binary comes between the other star and the line of sight to the Earth. Such binaries are said to be eclipsing. The word eclipse is also used when the Earth blocks the light from the Sun to the Moon, giving a "lunar" eclipse—i.e., an eclipse for an observer situated on the Moon. Inconsistently, however, the word eclipse is not used when the Moon crosses the line of sight from the Earth to a star; in this situation, the star is said to be "occulted."

Ecliptic: *See* **Declination**

Electron: The electron is an electrically charged particle belonging to the family of leptons, a different family from the baryons. Electrons surround the nuclei of atoms, forming shell structures wherever possible. The electrons left over when all possible shells have been completed play a critical role in determining the chemical properties of the atom, Successive shells contain 2, 8, 8, 18, 18, and 32 electrons. Only in the heaviest atoms are all these shells completed.

Energy: The electrical energy consumed by a device with a power rating of 1 kilowatt, used for a time of 1 second, is the unit of energy known as the kilowatt-second. The kilowatt-second is related to the unit used most frequently in physics, the erg, by the equation

$$1 \text{ kilowatt-second} = 10^{10} \text{ ergs.}$$

Energy can exist in many forms. When energy is said to be "consumed," we mean that it has been changed from a useful form to a less useful form, not that energy has been destroyed. Some forms of energy, electrical energy, for example, are readily converted to other forms, but the conversion is usually not reversible. Gasoline and air are converted in a motor to water, carbon dioxide, and heat, but it is not easy to change the water, carbon dioxide, and heat so produced back again into gasoline and air. The concept of the "usefulness" of various forms of energy is dealt with in the branch of physics known as thermodynamics.

Equinoxes: There are two equinoxes each year, occurring when the position of the Sun on the celestial sphere is at one of the two points where the ecliptic intersects the celestial equator. The spring equinox occurs about March 21, the autumnal equinox about September 23. The lengths of day and night are usually considered to be equal at the equinoxes. This is not strictly true, because the Sun is not a mere point of light, and particularly because of the twilight caused by refraction of sunlight in the atmosphere of the Earth.

Evolution of stars: As nuclear reactions go on inside stars, their chemical composition changes—atoms change from one kind to another. The change affects both the structure of the star and the availability of nuclear fuel. The changing condition of the star is referred to as its evolution.

Fission: The electrical repulsion between the positive charges of the protons in the nuclei of atoms would cause them to fly immediately apart if it were not for the nuclear force which attracts protons and neutrons to each other. There is, however, a limit to the binding effect of the nuclear force. When the total number of neutrons and protons in the nucleus increases above about 210, the electrical repulsive effect becomes strong enough to cause the nucleus to disgorge bits of itself in the form of helium nuclei, a process known as alpha decay. As the total number of neutrons and protons is further increased, the electrical repulsion becomes still stronger, until, instead of merely losing helium nuclei, the nucleus divides into two pieces of comparable size. The latter process is called fission. Fission becomes more important than alpha decay when the total number of neutrons and protons exceeds 260. Fission can be induced by firing neutrons into a heavy nucleus. Such induced fission is more important than alpha decay for a total neutron-proton number exceeding about 230. Man-made reactors depend on induced fission.

Flare: Electrical energy is stored in the solar atmosphere at levels above those that give rise to normal sunlight. Conditions can arise in which the stored electrical energy, or a portion of it, is suddenly discharged. This process, known as a flare, is more complex than the discharge of a spark or of a lightning stroke, but the cases are analogous.

Flux: When a flat surface is exposed to a distant source of radiation, and when the direction to the source is perpendicular to the surface, the power that falls on a unit area is defined as the flux received from the source. Flux is like the power rating of a man-made device: it must be multiplied by a time interval to give energy. For a spherically symmetric (isotopic) source of radiation, the flux varies according to the inverse square of the distance from the center of the source.

Focal length: When a lens or mirror is used to bring a distant object to a focus, the object being on-axis, the distance of the focal point from the lens or mirror is the focal length of the system. Dividing the focal length by the diameter of the lens or mirror gives the focal ratio (*f* ratio).

Force: The lack of the correct concept of force held up the development of physics for two thousand years. A force acts on any body that does not move with constant speed in a straight line.

Frequency: The frequency of a continuing sequence of regular oscillations is the number of them that occurs in a unit duration of time. Frequency can be thus defined for many phenomena, such as the regular pulsations of a star, the bobbing of a float on a regular train of water waves, or the radiation emitted by a certain quantum transition of a certain kind of atom. Radiation of this last type can cause an electrified test particle, an electron, say, to undergo a sequence of oscillations. By observing the test particle, we can measure the frequency of the radiation.

Fusion: The term fusion is applied in astronomy to nuclear processes in which two or more nuclei come together in a reaction, or reactions, whose products contain a nucleus with more neutrons and protons than any one of the ingredients. Heavier nuclei are thus produced from lighter ones. Provided the nuclei so produced do not have an atomic number greater than about 60, fusion reactions yield energy. Fusion reactions take place within stars, where they supply energy. The most effective energy source comes from the fusion of hydrogen into helium.

Galactic coordinates: Like the system of right ascension and declination (*see* Declination), a system of galactic longitude and latitude can be set up on the celestial sphere. The galactic equator is the circle in which the central plane of our galaxy intersects the celestial sphere, and the galactic prime meridian is defined by the direction from the solar system to the center of the galaxy.

Galaxies: A galaxy is a large collection of stars occupying a region of space well-separated from other large collections of stars. A galaxy may also contain much gas and dust, or it may contain esoteric objects like black holes, but such nonstellar contents are not necessary to its definition. There is no clearly agreed-on lower limit to what constitutes a "large" collection of stars. Conventionally, a lower limit seems to be set at about 10 million; a collection of a million stars would usually be described as a cluster rather than as a galaxy. Whether or not a galaxy contains gas and dust is relevant to its type classification. Galaxies with more than a few per cent of their mass as gas and dust usually have flattened spiral forms, and are thence known as "spirals." Galaxies substantially without gas and dust are usually described as "ellipticals," although this gas-and-dust criterion for separating "spirals" and "ellipticals" is not strict. Galaxies tend to occur in groups or clusters, with numbers ranging from about 5 on the low side to about 1,000 on the upper side.

Gamma rays: Gamma rays consist of radiation of very high frequency, a hundred thousand or more times greater than the frequency of visible light.

Globular cluster: Globular clusters contain so many stars, 100,000 to a million in a typical cluster, that their inner parts usually appear on photographs as an amorphous distribution rather than as separated stars. The amorphous distribution usually has a more or less circular shape, and it is from this that the name is derived. In our galaxy there are on the order of 100 of these clusters. In giant galaxies there can be as many as 1,000 of them.

Gravitation: Gravitation causes matter to pull together, to condense into a tighter and tighter clump, unless this tendency is resisted by some other effect, for example, by pressure within the material. In Newton's theory, gravitation is described as a force, and an explicit mathematical formula is given whereby the gravitation between particles of matter can be calculated. In Einstein's theory, however, gravitation is not a force like pressure, but is instead described by means of a non-Euclidean geometry.

Half-life: Nuclei that undergo radioactive decay do so in a way that depends on the half-life (every radioactive species has its own characteristic half-life). After a time interval equal to the half-life, one-half of the nuclei have decayed. After a further time interval equal to the half-life, one-half of the remainder have decayed, leaving one-quarter in the original form. After yet another time interval equal to the half-life, one-half of the remainder have decayed, leaving one-eighth in the original form, and so on. (The number of nuclei of the radioactive species is assumed in these statements to be so large that statistical fluctuations can be ignored.)

Hertzsprung-Russell diagram: The Hertzsprung-Russell diagram is basically a plot in which the ordinate (vertical axis) of a point represents the intrinsic luminosity of a star and the abscissa (horizontal axis) represents the *reciprocal* of its surface temperature. Instead of the intrinsic luminosity, the absolute magnitude may be plotted, and instead of the reciprocal of the surface temperature, the spectrum type may be used (since the spectrum of a star is related in a known way to its surface temperature).

HI and HII regions: Hydrogen atoms are about 10 times more numerous in the gas clouds of our galaxy, and in those of most other galaxies, than all other kinds of atom combined. Regions where the hydrogen atoms are mostly neutral (electron and proton associated together) are known as HI regions. Regions where the hydrogen atoms are mostly ionized (electron and proton separated) are known as HII regions. Left to themselves, ionized atoms change into neutral atoms, emitting radiation when they do so. Consequently HII regions tend to disappear unless there is a rejuvenating agent. The rejuvenating agent is usually ultraviolet light from a hot star, which is absorbed by neutral atoms, causing a separation of electrons from protons. HII regions are systematically hotter and more diffuse than HI regions.

Ion: In a neutral atom, the number of electrons surrounding the nucleus is equal to the number of protons in the nucleus. Since the electron and proton have electric charges of equal magnitude but of opposite sign, the total charge of the neutral atom is zero. When the numbers of electrons and protons are not the same, the atom is said to be an ion. The more usual ionized condition is for the electron number to be less than the proton number, giving a "positive" ion. In rare cases the electron number is greater than the proton, and the ion is said to be of "negative" type.

These designations arise from the convention according to which the proton charge is taken to be positive and the electron charge taken to be negative.

Isotope: The number of protons in the nucleus of an atom determines the chemical element; the number of neutrons in the nucleus can be changed without the chemical nature of the atom being changed, except in certain fine details. Nuclei with the same number of protons, but different numbers of neutrons, are called isotopes of one another.

Laser: Normally atoms emit radiation independently of each other. It is possible, however, to set up a special condition in which radiation from an initial number of atoms causes other atoms to emit radiation, and to do so in the same direction as the initial radiation. The result is a directed beam of radiation which continues for as long as the atoms continue to maintain the special condition. When the radiation emitted is visible light, the system is called a laser. If radio waves are emitted, the system is referred to as a maser (*m* for microwave, replacing *l* for light).

Light cone: A pulse of light emitted at a particular moment from a point source travels in four-dimensional spacetime along a three-dimensional cone known as the light cone.

Light year: The distance traveled by light in a year of 3.156×10^7 seconds is the light year, about 9.46×10^{12} kilometers.

Local group: Galaxies tend to exist in groups or clusters, ranging from about 5 members on the low side to about 1,000 members on the upper side. Our galaxy is a member of a cluster with some 20 members known as the Local Group.

Logarithm: The equation $x = 10 \log x$ expresses the relation between any positive number x and its logarithm, $\log x$. It is by no means simple to use this equation to find $\log x$ when an explicit value of x is specified, or to find x when an explicit value of $\log x$ is specified. Such calculations can be avoided, however, by using an already prepared table of logarithms. Or nowadays one can simply have recourse to a pocket calculator, which solves the required equation in a fraction of a second.

Luminosity: The luminosity of a star, a galaxy, or some other astronomical object, is like the rating of a power station, and we could use the same unit of power in both cases, so many megawatts (MW), for example. To obtain the energy supplied in a specified time interval, the power rating must be multiplied by the time. Thus the energy emitted by a star of luminosity L in a time interval t is the product Lt.

Magnitude: When the word magnitude is used by itself, it usually means "apparent" magnitude. The apparent magnitude is a somewhat indirect way of stating what the flux (*see* Flux) from an astronomical object is measured to be. The observer first decides on a unit of power in terms of which the flux is then measured, say, F. The next step is to take the logarithm of F, log F (*see* Logarithm). The resulting value, log F, is next multiplied by -2.5, giving -2.5 log F, and a certain number, c, is finally added to give the apparent magnitude, -2.5 log $F + c$. The number c to be added is fixed by convention so that a specified standard object will turn out to have a prestated magnitude, say, $+1$. This conventional number to be added depends on the power unit in terms of which the flux was measured. With fluxes in kilowatts per square cm, the number is about -36.5. It is clear that an increase by a factor 10 in the measured flux changes the magnitude by -2.5, an increase by 100 in the measured flux changes the magnitude by -5, and so on.

Main sequence: Stars that generate the energy which they radiate by fusing hydrogen into helium in their central regions lie in a particular zone of the Hertzsprung-Russell diagram. When many such stars with variable masses are plotted in this diagram, the resulting points fall in a band that ranges from lower right to upper left. This band is the main sequence. If the plot is restricted to stars that have only recently formed, so that little of the initial hydrogen has been converted to helium, the points fall on a line instead of covering a band. This line is known as the zero-age main sequence.

Meridian: A meridian is an arc of a great circle extending from pole to pole on the celestial sphere.

Meteorite: The region of the solar system between Mars and Jupiter contains a number of small planetoids or asteroids, as well as a swarm of still smaller bodies. These bodies are constantly subject to gravitational perturbations from Jupiter and Mars which change their orbits, as do their occasional collisions with each other. From time to time, particularly because of Mars, these changes cause one of these bodies to take on an orbit which crosses the path of the Earth. Collision with the Earth is then possible. Such fragments hitting the Earth can penetrate through the atmosphere to ground level, where they can be retrieved. Retrieved fragments, known as meteorites, are divided into two classes, one class being of stony material, the other of an iron-nickel alloy. The stones can be subdivided into many varieties, the chemical analysis of which has provided much important information concerning the early history of the solar system.

Meteor: With the evaporation of the ices that form the nucleus of a comet, a swarm of tiny sub-pinhead solid particles is left, following and spreading out along the orbit of the comet. During the course of the year the Earth crosses several orbits of such evaporated comets, and small particles then enter the terrestrial atmosphere in large numbers. Because of their small size, these meteors, as they are called, do not penetrate to ground level; they are vaporized by friction with the atmosphere, and they show at night as momentary streaks of light in the sky.

Microwave background: The microwave background consists of radiation whose main wavelength distribution is from about 0.5 mm to about 20 cm, and which reaches the Earth smoothly from all over the sky. The radiation is believed by most astronomers to be a relic from a hot, dense condition of the universe, associated with what is often called the "big bang." It provides a new and important datum in the study of cosmology. The variation of flux with wavelength is consistent with a Planck distribution at a temperature of about $3°K$.

Milky Way: The Milky Way is a colloquial name for the part of our galaxy that can be seen in the night sky. The many distant stars along the plane of the galaxy produce a diffuse band of light crossing the sky in a great circle. Not all the galaxy can thus be seen, because of absorption of light by dust clouds.

Molecules: Although the charges of the electrons and protons in a neutral atom cancel each other, the electrical force which the atom exerts on an external charged particle is not exactly zero, because the electrons and protons of the atom have different spatial positions. Although the electrical force between well-separated neutral atoms (as in a diffuse gas) is weak, the force is comparatively very strong when the atoms are close beside each other (as in a liquid or solid). Indeed, the electrical force can bind neutral atoms together into composite structures called molecules. The simplest molecules, like CO and H_2, contain only two atoms, whereas complex biological molecules like proteins contain many thousands of atoms. Chemistry and biology are concerned with the permutations and combinations of these composite structures of atoms.

Neutrino: The neutrino is a particle with no electrical charge and with at most a very small mass; it belongs to the same family of particles as the electron, the leptons. There are several kinds of neutrino, two forms having been identified so far by experiment.

Neutron: A particle without electrical charge, and with a mass about a tenth of a per cent greater than the mass of the proton. The neutron is the partner of the proton in the nuclei of atoms.

Neutron stars: As their name implies, neutron stars are stars composed predominantly of neutrons. Typically, they have a radius of about 10 kilometers, although in mass they are comparable with the Sun. A chunk of their matter the size of a small sugar cube contains about 10^8 tons of neutrons. Under these extreme conditions, neutrons may take on the structure of a solid, so that starquakes analogous to earthquakes may occur. This possibility is supported by certain observations of pulsars, which are thought to be rotating neutron stars. Neutron stars are believed to be formed as a final stage in the evolution of supernovae. They are not much bigger than a black hole of the same mass. It is possible that supernovae also give rise to black holes.

Nuclear force: The nuclear force acts between baryons. It is the force that binds protons and neutrons together in the nuclei of atoms. Un-

like both gravitation and the electrical force, the nuclear force operates only at the short range of about 10^{-13} cm. Should a neutron or proton escape from a nucleus, the nuclear force ceases to act on it.

Nuclear reaction: When two nuclei overlap each other, as they may do for a short while in a collision, the nuclear force connects the protons and neutrons in one nucleus with those in the other nucleus. This may lead to a redistribution in which the protons and neutrons have a changed arrangement after the collision. The nuclei that emerge from the collision are then different from those which entered it, and a nuclear reaction is said to have taken place.

Nuclear shells: The neutrons and protons in a nucleus tend to form (separately) into shells, as do the electrons that surround the nucleus. But whereas the electron number in atoms with closed shells are 2, 10, 18, 36, 54, 86, the proton and neutron shell numbers are 2, 8, 20, 28, 50, 82, 126, 184. The latter are often referred to as "magic" numbers.

Open cluster: Open clusters usually contain a few hundred stars, which appear well-separated from each other in photographs. The stars of open clusters are believed to have had a common origin.

Parsec: A star at a distance of 1 parsec would appear to undergo an oscillation through an angle of 2 arc seconds in its position on the celestial sphere, due to the yearly motion of the Earth around the Sun. A distance of 1 parsec is equal to 3.26 light years.

Perihelion: The point in its orbit when a planet, asteroid, or comet is closest to the Sun is its perihelion. There is a corresponding definition for a body moving around the Earth, the word perigee then being used.

Perihelion rotation: According to Newton's theory of gravitation, a planet would follow a strictly elliptic orbit around the Sun, if the planet and Sun were alone. According to Einstein's theory of gravitation, the planet does nearly the same thing, except that the long axis of the planet's elliptic path turns very slowly, an effect known as perihelion rotation. The effect is largest for the planet Mercury. Observations of the orbit of Mercury have confirmed the Einstein theory. In practice, the problem is complicated by the fact that Mercury and the Sun are not alone.

The other planets also cause a perihelion rotation, and this further effect, predicted by both theories, has to be allowed for before the decision between the two theories can be made.

Periodic table: The periodic table is an arrangement of the elements (the different kinds of atom) according to the increasing number of protons in the nucleus. It is usual to arrange the table so that elements in each column have the same number of electrons left over after all possible electron shells have been completed. The advantage of this arrangement is that elements in the same column then have similar chemical properties.

Photosphere: The atoms emitting the radiation that escapes from a star are nearly all contained in a thin shell. It is usual to think of this shell as a surface. For a spherically symmetric star, the surface is a sphere, the photosphere, concentric with the star. The term is most often used in relation to the Sun, which is indeed a nearly spherically symmetric star.

Planck curve: Imagine a closed box with walls that are maintained at a fixed temperature. Because the walls emit and absorb radiation, a distribution of radiation with respect to frequency builds up within the box. This distribution, which is called a Planck distribution, depends only on the temperature. The distribution can be displayed on a graph, by plotting the energy content of small unit steps in the frequency. The resulting curve is known as a Planck curve.

Planet: There are nine planets moving around the Sun. A planet is not just any body that moves in a nearly elliptic orbit around the Sun. A minimum size for a planet is implied by the circumstance that the largest of the smaller bodies in the region between Mars and Jupiter, which also move around the Sun, are called minor planets or planetoids, or asteroids. The minimum size for a planet has never been precisely defined, however.

Precession: The rotation of the Earth causes it to bulge slightly at the equator. The gravitational forces which the Moon and Sun exert on the slightly flattened form of the Earth then produce a twist, technically called a "couple," which causes the axis of rotation to precess like that of a spinning top. The effect of the precession is to cause the rotation axis to move slowly around the surface of a cone,

completing a circuit in about 26,000 years. Thus the poles (the directions in which the rotation axis meets the celestial sphere) move slowly in circles on the celestial sphere. The direction toward the center of either of these circles is perpendicular to the plane of the Earth's orbital motion around the Sun.

Precession of the seasons: The celestial equator moves with the motion of the celestial poles. Therefore the two points where the celestial equator intersects the ecliptic also move. Since it is the positions of these two points which determine the spring and autumnal equinoxes, the seasons of the year change slowly with respect to the position of the Earth in its orbit around the Sun—the positions of the Earth at midsummer and midwinter interchange approximately every 13,000 years.

Prominence: The Sun has an atmosphere, the lower part of which is called the chromosphere, the upper part the corona. The corona extends far out, to several diameters of the Sun—indeed, the corona may not end at all (*see* Solar wind). The gases of the inner part of the corona, and still more of the outer part, are too diffuse and too hot to emit much visible light. Occasionally, however, owing to unusual effects of the gravitational and magnetic forces which control the solar atmosphere, a region of the gas becomes compressed, cooling and emitting visible light as its density increases. The compressed gas can then be seen and is known as a prominence, since it belongs to a region located high above the normal surface of the Sun.

Proper time: *See* **Doppler shift**

Proton: A particle with an electric charge equal in magnitude but opposite in sign to the charge of the electron. The proton is 1,836 times more massive than the electron, however, and it experiences the nuclear force, which the electron does not. The proton is the partner of the neutron in the nuclei of atoms.

Proton-proton chain: The proton-proton chain is a sequence of nuclear reactions whose net effect is to convert hydrogen into helium. It is one of the two main processes responsible for the production of energy in most of the stars we see in the sky, and is the more important of the two in stars of small mass. The proton-proton chain is largely responsible for the generation of energy in the Sun.

Protostar: A star that is to be, a star in the making.

Pulsar: Pulsars are believed to be rotating neutron stars (*see* Neutron star). They emit radiation, it is thought, from a relatively small spot on the star; this situation produces an effect rather like the beam of a lighthouse. The distant observer receives a pulse of radiation as the beam sweeps across his position on each rotation of the star. Pulsars emit radiation at all frequencies, ranging from the radio band to γ-rays. It is by means of their radio emission that pulsars are most conveniently discovered. About 150 of them are known.

Quantum transitions: Quantum transitions occur both for the electrons surrounding the nuclei of atoms and for the neutrons and protons of the nuclei, the physical principles governing the two cases being essentially the same. The electrons spend most of their time in one or another of a number of states. Occasionally, however, an electron undergoes a transition (jumps) from one state to another. When a transition occurs, a quantum of radiation is either emitted or absorbed. When many atoms undergo the same transition, the radiation involved has a well-defined wave structure. The protons and neutrons of the nuclei also have states, and radiation is similarly emitted or absorbed in transitions between them. Frequencies for nuclear transitions are usually on the order of a million times greater than the frequencies for the transitions of the electron states.

Quasar (QSO): A quasar is a compact object radiating with a power output comparable to, or greater than, that of a whole galaxy. Quasars are thought to have an inner structure that is about 1/100 of a light year across, and which may contain a black hole. The lines observed in the spectra (*see* Spectrum) of quasars come, however, from a comparatively diffuse gas, which probably spreads across several hundred light years. About 500 quasars are known.

Radar: Distances to some near-by objects can be measured by radar. The method consists in emitting a regular train of well-separated radio pulses toward the object in question. The object reflects a small fraction of the radio power back toward the observer, who measures the time which elapses between the emission of a pulse and the reception of its reflection. Distance is then given by multiplying the speed of radio waves by the elapsed time. The radio waves used are confined to a narrow band of frequencies. Large antennas are employed to direct the waves toward the object under investigation. Radar gives the most accurate method for measuring the size of the solar system, the planet Venus being the most suitable reflecting object for this purpose.

Radiative interaction: The influence of one electrically charged particle on another is known as a radiative interaction. If particle *a* is made to oscillate at frequency *v*, the speed of the motion being small compared to the speed of light, the radiative influence of *a* can make another electrically charged particle *b* also oscillate with the same frequency *v*. This property is often described by saying that the radiative interaction has a wave structure (*see* Waves). The radiative interaction goes from past to future, not from future to past, a property that is important in establishing the relation of cause and effect.

Radioactivity: An association of protons and neutrons, bound together by nuclear forces, reaches its most stable condition by radioactivity. This usually happens by the emission of helium nuclei (two protons and two neutrons in association), a process known as alpha decay, or by a process known as beta decay, in which a neutron changes to a proton with the emission of an electron, or by the inverse of the same process, with the emission of a positron. All these changes go by the name of "radioactivity."

Radiogalaxies: Radio waves of exceptionally large intensity are received from small patches on the sky. Many of these patches, but not all of them, are associated with galaxies, which are thus known as radiogalaxies. Radiogalaxies emit jets of highly energetic particles, often two jets in opposite directions, which come, it is believed, from quasar-like objects (*see* Quasar) situated at their centers. The radio waves are thought to be generated when the emitted particles impinge on external clouds of gas, with magnetic fields playing an important role in the emission process. The radio waves involved in the detection of radiogalaxies are usually of longer wavelength than the waves that carry the main energy of the microwave background (*see* Microwave background).

Redshift: Radiation from a moving object is said to be redshifted when lines in the visible spectrum are shifted (*see* Doppler shift) toward the red end of the spectrum. This happens when the object in question is receding from the observer.

Redshift-magnitude relation: The spectrum lines (*see* Spectrum) found in galaxies show a redshift, which may be described in the following way. Let λ be the wavelength of a particular line as measured in the terrestrial laboratory, and let $\lambda + \Delta\lambda$ be the wavelength of the same line (i.e., arising from the same transition of the same kind of atoms) measured for the galaxy in question. Then $z = \Delta\lambda/\lambda$ is the redshift of the galaxy. The value of z is the same whatever line of the spectrum is chosen. The redshift-magnitude diagram is a plot of the logarithm of z, $\log z$, against the magnitude of the galaxy. Usually the magnitude is represented as the abscissa (horizontal axis) and $\log z$ as the ordinate (vertical axis). When observations of many galaxies are thus plotted, an approximately straight-line relation between magnitude and $\log z$ is found. This straight line is the redshift-magnitude relation.

Reflection: In a vacuum, light can be made to propagate one-way, in any chosen direction, and the same is true for light propagating in a translucent material, provided the material is homogeneous—i.e., of the same composition at all places. However, when light propagating in a homogeneous medium encounters another, different homogeneous medium, the one-way condition is destroyed. Not all the light continues into the second medium. Light is turned back at the boundary of the media, and is said to be reflected. Light that continues into the second medium is said to be refracted. The directions of the reflected and of the refracted light are related to the direction of the incident light by the laws of reflection and refraction, which can be expressed by simple mathematical formulas. These formulas provide much of the information required in the construction of telescopes, microscopes, cameras, and other optical instruments. Similar considerations apply to forms of radiation other than visible light.

Refraction: *See* **Reflection**

Resonant nuclear reaction: A nuclear reaction occurs when the neutrons and protons in two colliding nuclei redistribute themselves during collision (*see* Nuclear reaction). The reaction is said to be resonant if the sum of the energies of the colliding nuclei plus the energy of their relative motion happens to coincide with the energy of a state of the compound nucleus, which is simply the nucleus that would be formed by taking all the protons and neutrons of both colliding nuclei and putting them together into one nucleus.

Right ascension: *See* **Declination**

Salpeter function: Stars differ one from another in mass—i.e., in the number of atoms which they contain—at the time of their birth, but not in a random way. The Salpeter function is a mathematical formula which describes the relation of the number of stars to their masses.

Satellite: Bodies which move around planets are known as satellites. Thus the Moon is a satellite of the Earth. A rocket payload expelled from the Earth that continues to move around the Earth is an artificial satellite, but a rocket payload escaping from the gravitational influence of the Earth is a space vehicle. The term satellite galaxy is sometimes used for small galaxies that are controlled by the gravitational influence of a larger galaxy.

Solar constant: The flux of the Sun, taken at the Earth's mean distance from the Sun, about 149.6 million kilometers, is the solar constant, about 1.39 kilowatt per square meter. The solar constant includes radiation of all frequencies. Since it has so far been measured from the Earth's surface, allowance must be made for ultraviolet light and infrared that have been absorbed by the terrestrial atmosphere. This difficulty is minimized by making measurements at a high-altitude desert station.

Solar cycle: A complex of phenomena that occur in the outer part of the Sun—sunspots, prominences, flares, the shape of the corona, magnetic fields—all show cyclic change with an average period of about 22 years. This solar cycle is believed to be related to the convection of material which takes place below the photosphere, but the precise cause of the cycle is not well-understood. Successive cycles are neither of equal duration nor of equal intensity. For a while in the eighteenth century the cycle largely died away.

Solar system: All the bodies that are moving around the Sun—planets, satellites, comets, asteroids—belong to the solar system. Occasionally "solar system" is used to refer to only these bodies, but more usually the Sun itself is considered to be also a member of the solar system.

Solar wind: The outer atmosphere of the Sun has no end. Its particles become less dense with increasing height above the photosphere, at first rapidly, and then more slowly, ultimately becoming a low-density wind of outward-moving particles. The density in the wind as it reaches the Earth varies markedly with time, sometimes being as low as 1 atom per cm^3 and sometimes as high as 1,000 atoms per cm^3. The high-density situations are associated with the occurrence of flares and other disturbances in the lower parts of the solar atmosphere.

Spectrum: When the light from an object—star, galaxy, planet, quasar, or whatever—is separated into its constituent frequencies, the resulting distribution of light is said to be the spectrum of the object. The frequencies are separated by a device known as a spectrograph, whose essential component is either a glass prism or, more usually in astronomy, a diffraction grating. Separation of frequencies is essentially a separation into colors, the colors of visible light ranging from violet (the highest frequency) through blue, green, yellow, to red (the lowest frequency). Light emitted by ionized atoms shows a continuous range of color, usually with an excess or deficit of light at certain frequencies, which are called "lines" of emission or absorption. Neutral atoms in a diffuse gas give only emission or absorption lines, although at high density, as in a heated solid substance, neutral atoms can also give a continuous range of color.

States of an atom: The paths which the electrons of an atom may follow can be arranged in bundles which have the property that if at a given moment the electrons happen to be in the paths of a certain bundle with certain specified probabilities, then the electrons will continue in the same bundle with the same probabilities for a comparatively long time. When arranged in accordance with this comparatively long-term constancy, the bundles are called stationary states, or more simply states. Transitions from one such state to another occur from time to time, however, and radiation is then emitted or absorbed (*see* Quantum transitions).

Solstice: Throughout the year the axis of rotation of the Earth maintains a nearly constant direction relative to the stars, but the direction of the line from the Earth to the Sun changes because of the Earth's motion around the Sun. The summer solstice occurs when the angle between the direction of the Sun and the axis of rotation is least. The winter solstice occurs when the angle is greatest.

Spectral classification: The most prominent spectrum lines of different atoms correspond to quantum transitions between states of different energy values; certain of the lines show up prominently only within certain temperature sequences. The appearance (or nonappearance) of these lines in the spectrum of a star therefore gives information about the surface temperature of the star. The classification of stars, essentially in a temperature sequence, according to which lines appear (and which do not) is known as spectral classification.

Spectroscopic parallax: Certain spectrum types among stars are associated with certain values of the intrinsic luminosity. This fact is established by working initially with stars whose distances are already known, usually from measurement of their trigonometric parallax (*which see*). Then, it is argued, other stars having similar spectrum types must have similar intrinsic luminosities. So a measurement of their apparent luminosities then allows us to calculate the distances of these other stars. Such distance calculations are said to be by the method of spectroscopic parallax.

Steady-state cosmology: Steady-state cosmology is a model for the large-scale features of the universe that is homogeneous for both time and space (*see* Cosmology, principles of). The model requires matter to be created continuously instead of all at once, as is required in the so-called big-bang cosmology. The strongest objection to the steady-state model comes from the existence of the microwave background radiation.

Supernova: Stars of large mass end their lives with catastrophic explosions, in which the power output for about a month is as great as that of a whole galaxy of stars. Thereafter the emission declines gradually, the luminosity decreasing by about 0.6 magnitude per month. The cause of supernovae appears to be complex. The explosion is believed to start with the collapse of the core of the star, which eventually becomes either a neutron star or a black hole. Then, from the highly collapsed core, a shock wave propagates into the outer regions, which have not yet experienced much collapse. The shock wave, which may arise from the absorption of a sudden flood of neutrinos from the core, heats the outer material. The latter material contains much oxygen, which is explosive on sudden heating to temperatures in the range 2 to $3 \times 10^9 \, ^\circ K$. It is possible that this sequence of events occurs only in a restricted class of massive star, not in all of them.

Synchrotron radiation: Synchrotron radiation is generated by charged particles moving with speeds close to that of light in a magnetic field. The paths of the particles are helices with axes parallel to the direction of the magnetic field. The essential characteristic of synchrotron radiation is that it contains a continuous distribution of frequencies, and that the main energy of the radiation is at frequencies that are much higher than the number of turns per unit time of the particles in their helices. The smaller mass of the electron causes electrons to be much more effective in the emission of radiation than protons, so that electrons are usually considered to be the source of cosmic synchrotron radiation. The name is derived from a man-made device, the synchrotron, in which the effect was first observed.

Tektite: Tektites are small dark pieces of glass, often with button-like shapes. Their chemical composition is essentially terrestrial, which makes it likely that they were formed by the collision of a large meteorite, or perhaps of a comet, with the Earth. The chemical composition of the lunar surface having proved to be very different from the Earth's, it is most unlikely that the tektites have come from the Moon.

Temperature: Any closed physical system, one which energy is neither added to nor taken away from, tends to share the available energy uniformly between all its components, between particle motions and radiation, for example. In a situation in which sharing has become complete, temperature is a measure of the average energy per component. Temperature is proportional to this average energy, however, not equal to it, the constant of proportionality being determined by the convention that there are to be 100 units of temperature between melting ice and boiling water (taken at a standard pressure of 76 cm of mercury).

Titius–Bode law: The numbers obtained from the Titius-Bode law $0.4 + (0.3 \times 2^n)$, with n taking the values 0, 1, 2, 3, 4, 5, 6, happen to match the radii (in astronomical units) of the planets from Venus ($n = 0$, $2^0 = 1$) to Uranus ($n = 6$), with $n = 3$ corresponding to the minor planets Ceres and Pallas. And the first term alone, 0.4 in the formula, matches the radius of the orbit of Mercury. The Titius-Bode law fails badly when $n = 7, 8$, are used for Neptune and Pluto.

Trigonometric parallax: The annual motion of the Earth causes the projections of the stars onto the celestial sphere to oscillate over small arcs which are parallel to the plane of the Earth's orbital motion around the Sun. For a star at a distance of 1 parsec (3.26 light years), the oscillation is through the angle of 2 arc seconds (*see* Parsec). For a star at a distance d parsecs (3.26d light years), the oscillation is through an angle of $2 \div d$ arc seconds. By measuring the angle of oscillation, the distance d can be inferred. This method of calculating distance, known as the method of trigonometric parallax, gives reliable results for distances up to about 100 parsecs, but tends to become inaccurate for larger distances, the angles of oscillation becoming then too small for good observational measurements to be made.

Universal time: An immediate simplifying effect of the assumption of homogeneity for the large-scale structure of the universe (*see* Cosmology) is that a system of universal time can then be defined. Observers in different galaxies use similar clocks, based on the same transition of the same kind of atom, and they all set their clocks to read the same time when the Hubble constant H, obtained from galaxies in their individual localities, has an assigned value, which is chosen by convention.

Visual magnitude: When the magnitude of an object (*see* Magnitude) is measured in terms of visual light only, the result is said to be the visual magnitude.

Waves: A wave has three basic properties, as follows. (1) At each spatial point there is an oscillation. The number of oscillations that occur in a unit time is the frequency ν. (2) There is a spatial ordering. If at one moment, at a given place, the wave is "up," at a nearby place the wave is "down." The spatial separation between adjacent places where the wave is "up" (or where it is "down") is the wavelength λ. (3) A train of waves can propagate. At one moment the waves have not yet reached a certain point. At a later moment they have passed the point. These three properties are possessed both by water waves and by the radiation emitted by charged particles. Because of the similarity of these basic properties, it is possible, by thinking about water waves, to obtain insight about waves of radiation. The speed of propagation c of the waves is related to ν and λ by the simple equation $c = \nu\lambda$.

X-rays: X-rays consist of radiation of high frequency, between about 100 and 100,000 times the frequency of visible light.

Index